“十三五”国家重点出版物出版规划项目
海 洋 生 态 科 学 与 资 源 管 理 译 丛

海洋生物酶

MARINE ENZYMES FOR BIOCATALYSIS

Antonio Trincone 著
刘伟治 刘 涛 杨 艳 律倩倩 译

海洋出版社

2018年·北京

图书在版编目（CIP）数据

海洋生物酶/刘伟志译. —北京：海洋出版社，2018. 6
ISBN 978-7-5210-0101-3

Ⅰ. ①海… Ⅱ. ①刘… Ⅲ. ①海洋生物-酶-研究 Ⅳ. ①Q55②Q178. 53

中国版本图书馆 CIP 数据核字（2018）第 099807 号

图字：01-2018-2727

责任编辑：方　菁
责任印制：赵麟苏
海洋出版社　**出版发行**
http：//www. oceanpress. com. cn
北京市海淀区大慧寺路 8 号　邮编：100081
北京文昌阁彩色印刷有限公司印刷　新华书店发行所经销
2018 年 6 月第 1 版　2018 年 6 月第 1 次印刷
开本：787mm×1092mm　1/16　印张：23. 5
字数：540 千字　定价：128. 00 元
发行部：62132549　邮购部：68038093　总编室：62114335
海洋版图书印、装错误可随时退换

目　次

第1章　深海中的宝藏

Claire W. Armstrong, Jannike Falk-Petersen and Inga Wigdahl Kaspersen

DOI：10.1533/9781908818355.1.3

摘要：本章介绍海洋生物勘探的简明发展史，阐述了当前在该领域，特别是与海洋酶类和药物相关的进展。同时，基于生态系统服务功能及其价值评估框架对海洋生物勘探展开讨论，某种程度上解释了在生物勘探活动上的公共投资。同时介绍了涉及环境保护与生物勘探活动冲突相关的问题，以及生物勘探促进自然保护的潜在可能性。

关键词：海洋生物勘探，海洋酶类，生态系统服务功能及价值

1.1　引言

至今，生物勘探主要与茂密的热带雨林有关，如亚马孙的热带雨林。人们希望通过对这些地区的生物勘探获得关乎人类福祉的重要物质，如药物，还有食品、化妆品、能源、工业产品等有价值的化合物。人们应该记得，生物勘探实际上是一个用来描述一个古老过程的新术语。最单纯的生物勘探模式并不是人类的创举。即使动物也知道通过选择性地摄食它们周边环境中的植物、土壤和昆虫来进行自我调节，实现疾病的防治。一些动物也能利用天然存在的化合物生产食物。

框1.1　动物自我医疗

切叶蚁利用微生物将纤维素分解成葡萄糖，这为它们的真菌群落提供了食物来源。相应地作为喂养蚂蚁的回报，真菌得到了稳定的食物来源，同时被蚂蚁共生细菌产生的抗生素所保护从而避免了发霉（Bot et al., 2002）。

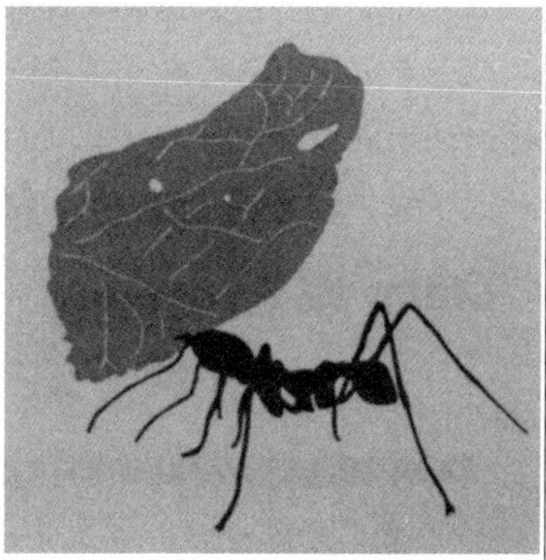

寻找可以生产具有经济价值化合物的动植物是全人类的努力方向。当前大部分的研究都是集中在陆地生物，因为海洋环境，特别是深海，在多年前对于人类来说是一个未知领域。我们对海洋生物的认知局限于潮汐或是渔业设备捕获的海洋生物。然而阅读《白鲸》、《海底两万里》等经典会发现，尽管探索手段匮乏，人类对巨浪之下的海洋着迷已久。当我们思考生物多样性时，我们倾向于将热带雨林看做物种丰富的标杆。但是论及生物多样性，海洋环境并不居于其后。海洋占地球覆盖面积达70%以上，海洋生态系统拥有现今发现的全部35个代表动物种群中的28种，并且有1/3的种群仅存于海洋环境中，这充分说明海洋环境拥有丰富的生物和化学多样性（Primack，2010）。此外，由于在海洋环境中具有比陆地更长期的寒冷环境，因此这里的生物往往可以产生一些有意义的物质，例如在相对低温时具有酶活的适冷酶。这种酶在工业上的应用可以降低工业生产的能耗。更为重要的是，这些酶还可以通过某些特定的温度控制其活性，相对于陆地环境来源的酶，海洋生物酶通常可以在较低的温度环境下控制其活性，这样同样也会减少能耗和避免化学物质的使用。

尽管陆地生物勘探历史更加悠久，但是仍然有很多药用或工业化合物长期以来都是从海洋生物中提取。在古代，人们收集地中海海螺以提取它们的分泌物用于印染。这种颜色被称为品蓝或帝王蓝，用它印染的任何衣物都极其昂贵，仅最高阶层的贵族能够穿戴。另一个例子是将鲸的龙涎香用作香水的固定剂，有些现代的香水依然含有龙涎香，但因它产自濒危物种抹香鲸，在20世纪90年代很多国家已禁用。

在20世纪中期，Giuseppe Brotzu、Werner Bergman 和 Robert J. Feeney 在海洋药物方面的工作进一步激发人们对大海潜在宝藏的兴趣（框1.2）。自从初创时期，开发了大批海洋药物，表1.1展示了获批和在临床前试验阶段的海洋药物。

截至2012年1月，共有7种药物来自海洋环境。一种药物在三期临床阶段，即最后的试验阶段；6种药物在二期临床试验阶段，意味着即将建立检测平台。除此之外有4种海洋化合物处于一期临床试验阶段，正在研究其安全使用方法及副作用，另有几种化合物处在临床前试验阶段（Gerwick and Moore，2012；Mayer et al.，2010）。

框1.2　生物勘探先锋

随着青霉素重要的属性被发现，Sardinia 的 Giuseppe Brotzu 教授在1945年提出了一个问题，那就是为什么伤寒病毒在他的城市比任何其他地方毒性都要低。他怀疑这与到城市居住者有游泳的习惯有关，而游泳的地点恰恰在城市下水道向海洋排放污水的地方，并且他测试了海水对 *Salmonella typhi* 的影响。最后他分离了一种真菌，*Cephalosporium acremonium*，它产生一种有效对抗所有革兰氏阴性菌的物质。但这却不能使任何一家意大利药品公司信服从而资助他下一步研究，Brotzu 缺乏更深入的研究。1948年他给牛津大学的研究小组寄去了关于 *C. acremonium* 培养液的同时附上了他发现的副本。在 Sir William Dunn 病理学院的研究者将有活性的化合物命名为头孢菌素。他们成功地从 Brotu 的培养基中分离头孢菌素母核。Eli Lilly 在1964年对母核进行修饰后获得有用的抗生素，同时产生了截至目前四代头孢菌素的第一个成员，从而建立了叫做头孢噻吩品牌。

因为头孢菌素在海洋环境中被发现，一些学者认为 Brotzu 应该被推选为海洋生物技术的先锋，但另外的学者认为 Werner Bergmann 是真正的海洋生物发现之父。在 Brotzu 研究发现后的几年，Werner Bergmann 和 Robert J. Feeney 报道了从 Caribbean 海绵 *Cryptotethia crypta* 分离出一种之前从未在自然环境中发现的核苷混合物。他们发现的证据表明这些核苷包含稀有阿拉伯糖而不是正常的核糖，同时成功获得其中一种核苷。Bergmann 和 Feeney 命名它为海绵毒素以纪念它是从海绵里分离出的，这也为其之后抗肿瘤活性的研究打下了基础（Bergmann and Feeney 1950；Gullo 1994；Campos Muñiz et al.，2007；Beg et al.，2011）。

表 1.1　由海洋微生物产生作为治疗剂的海洋天然产物及类似物和衍生物

临床状态	化合物名称	微生物来源	对应病症	备注
通过审批	阿糖胞苷（Ara-C）	海绵	癌症	海绵毒素的衍生物 在 1969 年被 FDA[①]批准
	阿糖腺苷（Ara-A）	海绵	抗病毒	海绵毒素的衍生物 在 1976 年被 FDA 批准 在 2001 年中断
	齐考诺肽	锥蜗牛	缓解疼痛	天然产物 在 2004 年被 FDA 批准
	艾日布林甲磺酸盐（E7389）	海绵	癌症	软海绵 B 的衍生物 在 2010 年被 FDA 批准
	ω-3-酸乙酯	鱼	高甘油三酯血症	ω-3-脂肪酸的衍生物 在 2004 年被 FDA 批准
	曲贝替定（ET-743）	海鞘	癌症	天然产物 在 2004 年被 EMEA[②]批准 在美国进行到临床Ⅲ期实验
	brentuximab vedotin（SGN-35）	软体动物	癌症	衍生物 在 2011 年被 EMEA 批准
临床实验Ⅲ期	缩酚酸肽	被囊动物	癌症	天然产物

续表

临床状态	化合物名称	微生物来源	对应病症	备注
临床实验Ⅱ期	DMXBA（GTS-21）	蠕虫	知觉 精神分裂症	衍生物
	Plinabulin（NPI2358）	菌类	癌症	衍生物
	Elisidepsin	软体动物	癌症	衍生物
	PM00104	裸鳃亚目动物	癌症	衍生物
	CDX-011	软体动物	癌症	衍生物
	PM01183	被囊动物	癌症	衍生物
临床实验Ⅰ期	Marizomib （盐孢菌酰胺 A； NPI-0052）	细菌	癌症	天然产物
	PM060184	海绵	癌症	衍生物
	SGN-75	软体动物	癌症	衍生物
	ASG-5ME	软体动物	癌症	衍生物

注：①美国食品药品管理局；②欧洲药品管理机构.

资料来源：http：//marinepharmacology. midwestern. edu/clinPipeline. htm，http：//www. ClinicalTrials. gov；Mayer et al.（2010）；Gerwick and Moore（2012）

虽然海洋在生物催化领域的贡献仍然有限且大多被忽视，但在生物催化领域已有大量的已知酶（Trincone，2010）。如图 1. 1 所示，与海洋酶类相关研究的实质性增长主要发生在近 20 年中。事实上，当前人类对深海仍知之甚少，预示未来海洋勘探及其在生物催化领域具有极大发展潜力（Armstrong，et al.，2013）。

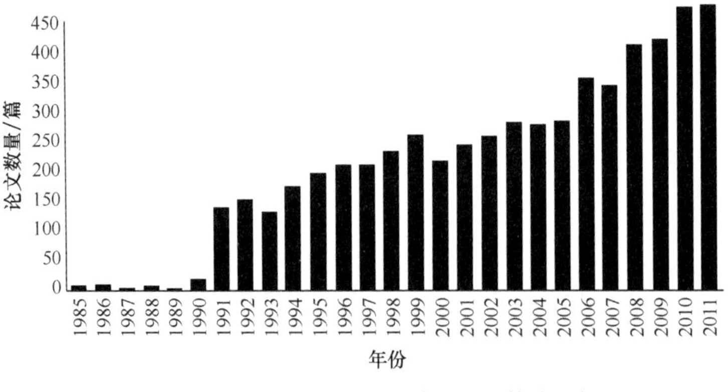

图 1. 1　1985—2011 年包含“海洋酶”的科学研究论文（知网）总数达到 5 528 篇

1.2 生态系统服务功能及海洋生物勘探

海洋生态系统通过多方面造福人类。除了供给鱼类和工业原材料，它们还是休养的基地，为气候调节做出了贡献，并且拥有丰富的可以用于生物勘探工业开发的基因资源。千禧年生态系统评估（2005）（MEA）是常用来判定造福人类的生态系统功能的框架。千禧年生态系统评估将这些服务功能划分为 4 个类别（图 1.2）。配置性服务是一些如食品、燃油、生化产品、天然药物、药剂等从生态系统中取得的产品。调节性服务受益于生态过程的调节，它包含了气候和水文调节及废弃物的生物修复。文化服务是得益于此的非物质方面的收益，比如愉悦的精神、社会关系、娱乐活动和认知发展。支持性服务对于其他生态系统服务发挥功能至关重要，它包含了营养物循环、初级生产、栖息地和氧气的产生（Millenium Ecosystem Assessment，2005；Defra，2007）。

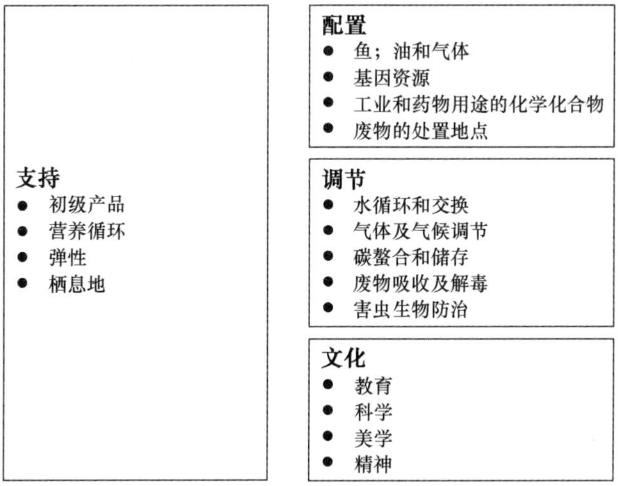

图 1.2　海洋生态系统的服务

发现及使用海洋酶用于生物催化被划分为配置服务。然而，针对海洋生物及化合物的研究也推动了认知的发展，比如推进文化服务，从中获取的知识能被转化成技术或用于制药产业（Beaumont et al.，2007）。比如在工业生态研究中模拟了自然生态系统后的工业体系，可以提高能源和原材料的效能和收益率，减少浪费及对环境的不利影响。在微观尺度上，着重于生物材料、生物催化或生物勘探能够潜在地改善其热力学效率、资源再生及经济的可持续性发展（Seager，2004）。由于酶有可反复利用及通过常规废弃物处理方式丢弃的优点，生物催化的应用减少了工业废弃物。公司因此得以避免高昂的废物处理步骤，减少使用对环境有害的物质，如重金属等（Woodyer et al.，2004）。此外，生物修复是指利用微生物自身新陈代谢加速自然降解的过程，是清除泄漏石油和其他海洋污染物的环境友好型方法（Anonymous，2002）。因此，生物勘探工业通过减少影响感官及食品安全的废物使用的方式，对生态系统的服务功能产生积极影响。这也有利于生态系统中生物体的健康，确保生物多样性，从而保护生态系统复原能力。

1.3 生物资源勘探经济学

在海洋生物勘探基础上开展的经济研究在很大程度上涉及知识产权问题，然而我们将更多专注于研究常规的海洋生物勘探，以及海洋生态系统造福人类的方方面面。

来自生态系统服务功能的经济价值经常使用所谓的总经济价值框架来展示（Pearce and Turner，1990）。如图 1.3 所示，该框架明确了与生态系统服务相关的不同类型的价值，如直接和间接使用价值构成的使用价值，还有期权价值和非使用价值。直接使用价值能通过研究与市场直接相互作用判断。来自海洋环境的典型例子是渔业、水产养殖和旅游业。间接使用价值是对直接价值的支持和补充，但并不明确地体现在市场上。例如来自海洋环境的栖息地和具有商业价值物种的食物。期权价值描述了自然环境在将来产生的潜在价值，然而期权价值组分—准期权价值，即预期在认知提升的将来社会将愿意为保留这样一个生态系统的价值。非使用价值包括现存的及遗产价值，其中现存的价值描述的是现在存在的，与使用价值无关的价值；而遗产价值指人类愿为子孙后代保留一个环境而付出的价值。

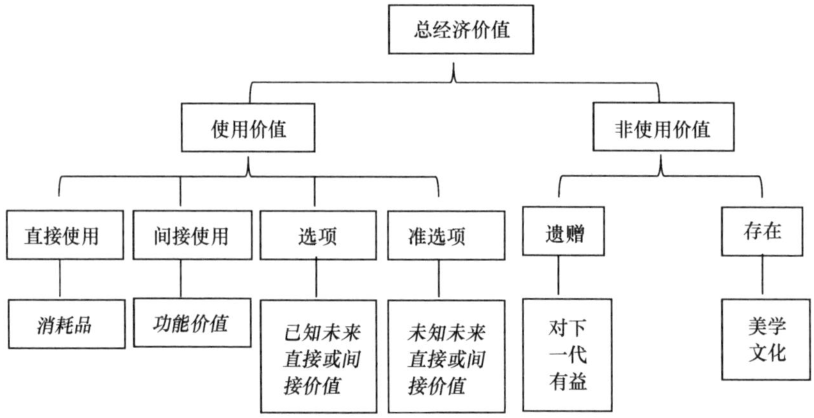

图 1.3 自然资源总经济价值

（来源于海洋生物勘探分支，衍生活动的价值用斜体表示）

来源于海洋生物勘探及相关活动的价值显而易见是直接使用价值，包含如药物、食品添加剂、工业原料。然而，相较于生产价值海洋生物技术公司有更高的股票价值（Armstrong et al.，2008），这表明在该领域投资有更高期望的回报，或者是期权价值。很明显这包含有期权价值，即人们愿意去保护海洋环境，以确保在将来的海洋生物勘探中获得回报。

图 1.4 中的逻辑能描述目前大多数生物勘探的调查和发展。通过采集以备筛查的样本，阐明了生物勘探自然而然地与生物多样性的直接联系。通常而言，采集，时而也有筛查和鉴定，都是由公共拨款资助，或是由大学、研究院和私企等公共机构合作开展研究的项目。对于海洋生物勘探和样本采集而言，需要昂贵的科考船和深海技术。进行调查对公共资金的迫切需求也暗示了公众看好海洋生物勘探。高昂的投资、发展的延滞、潜在的公共利益或是期权及准期权价值之间的衡量，解释了政府投资的合理性和公正性，例如，这反映在欧盟在 15 号立场文件对海洋生物技术方面的政策（Querellou, 2010）。

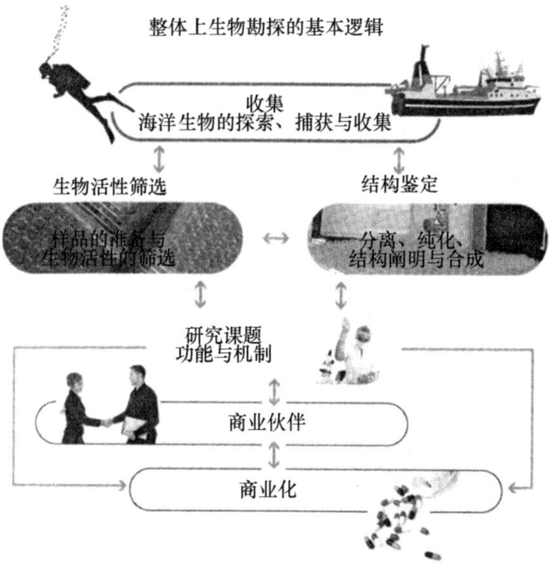

图 1.4　生物勘探的基本逻辑

1.4　生物勘探和可持续发展

尽管海洋生物勘探有很多积极作用，但仍不免有一些担忧，如新化合物研究对生态系统及其服务功能可能产生的消极影响（Bruckner, 2002; Synnes, 2007）。技术革新使人类能够触及先前绝无可能到达的独特的生态系统。伴随着调查研究与矿物开采，对新化合物的寻找给如深海热液喷口和珊瑚礁等生态系统增加了压力（Synnes, 2007）。因此，如不谨慎调查将会对这些系统的服务功能产生消极影响，因此会减少对人类的福祉。例如深海珊瑚生态系统生物多样性高，被认为是极有希望开发药物和生物技术相关化合物（Synnes, 2007）。此外，这些生态系统提供了支持服务例如鱼类栖息地，通过贮存 CO_2 进行气候调节（Foley et al., 2010）。同时它们也具有文化服务的功能，人们认为这些生态系统值得不被人类开发而独自存在，而且应该得到保护以利于子孙后代（Glenn et al., 2010）。既然蛋白质能够通过基因的复制而从基因工程生物体中获得，因此，只要样本采集小心合理，微生物勘探将是可持续的（Synnes，2007）。然而一些生物体需要被采集和培养。许多珊瑚生态系统中能够被用作治疗药剂的生物分布有限或者是生物量低。一些仅包含了少量的目标化合物，这就需要冒着减少自然群体的风险捕获大量生物样本（Bruckner，2002；Hunt and Vincent，2006）。

海洋水产养殖也给生态系统带来了消极的影响，比如引来疾病或入侵种，废物导致污染、养殖产品进入野生环境而给自然群体带来消极影响的风险（Bruckner，2002；Bert，2007）。合成化合物减少了采集对生态系统直接造成的压力，但是常常涉及对宿主生物比如细菌的基因修饰。基因修饰生物对人类健康及生态系统潜在消极影响的问题已得到关注（Anonymous，2003；WHO，2012）。

然而，保护生物多样性不仅对确保生态系统的功能极其重要，而且对生物技术产业本身也很重要（Synnes，2007）。渔业被认定为深海海洋生物多样性的主要威胁，包括对深海珊瑚的威胁。采矿工业对深海热液喷口铜锌矿床的调查也是一大威胁（Synnes，2007）。尽管因为对生态系统功能认知的局限，这些生态系统提供给人类的许多服务难以被量化（Armstrong and Falk-Petersen，2008），然而这些系统对生物勘探产业的价值可以通过在市场上明码标价的产品和大量在调查研究上投入的资源来体现。当人们认识到海洋生态系统是开展全新调查研究、获取新知和产品的资源，对深海中潜在的生物多样性价值认知可能是对保护多样性所付出努力的肯定与支持（Segura-Bonilla，2003）。

一些经济学家质疑是否可以利用通过鼓励一些生物公司针对这些多样性的生物栖息地进行可持续利用，他们可以支付一些费用，用于维持和保护当地的海洋环境，尤其是针对一些贫穷的地区（Barrett and Lybbert，2000）。对于海洋生物勘探，这可能与某些发展中国家的渔业发展没有监管及破坏自然栖息地相关。其他人更担心生物剽窃、小型不发达的团体被大型跨国公司滥用（Svarstad，2004）。在大洋中，特别是在有管辖权有争议的海域，担心多与法律不健全和缺少控制有关。例如采矿公司对在类似热液喷口等极端环境下的开采技术兴趣的提高，使保护这些具有独特物种，有利于生物多样性的环境的保护问题更加突出。

1.5 结论

本章介绍了与生物勘探相关的经济背景和开发海洋环境中有价值的化合物。本章溯源海洋活动历史，一直到当前呈现指数式增长的研究局面，例如海洋酶类。海洋化合物处在生态系统服务功能的框架内，它们的价值体现在直接使用价值和潜在期权价值两方面。尽管人们担心海洋生物勘探可能对易受破坏的海洋环境具有潜在威胁，用于产品合成、筛查与鉴定的有限量的需求不是一个主要的问题。然而，与此相关的基因修饰以及潜在污染问题却涉及更广的工业领域。

感谢

我们感谢 Rudi Caeyers 为生物勘探做的插图，也感谢 Tromso 大学的 Trond Jorgensen 和 Klara Stensvag 教授让我们分享他们在海洋生物勘探方面的知识。

参考文献

Anonymous 2002. Marine Biotechnology in the Twenty-first Century: Problems, Promise, and Products. National Academy Press, 117 pp.

Anonymous 2003. 'An open review of the science relevant to GM crops and food based on interests and concerns of the public.' Department of Trade and Industry. First report, 296.

Armstrong, C. W. and Falk-Petersen, J. 2008. Food for thought-habitat-fisheries interactions: a missing link? *Ices Journal of Marine Science* 65, 817-821.

Armstrong, C. W., Foley, N., Tinch, R. and van den Hove, S. (2013) Services from the deep: steps towards valuation of deep sea goods and services. *Ecosystems Services* 2, 2-13.

Armstrong, C. W., Kahui, V. and Aanesen, M. 2008. *Valuation of Marine Environment—The Lofoten-Vesterålen Coastal Area.* In Norwegian: *Økonomisk verdsetting av havmiljø-Anvendelse på havområdene i Lofoten-Vesterålen.* Norwegian College of Fishery Science, University of Tromsø, 26 pp.

Barrett, C. B. and Lybbert, T. J. 2000. Is bioprospecting a viable strategy for conserving tropical ecosystems? *Ecological Economics* 34, 293-300.

Beaumont, N. J., Austen, M. C., Atkins, J. P., Burdon, D., Degraer, S. et al., 2007. Identification, definition and quantification of goods and services provided by marine biodiversity: Implications for the ecosystem approach. *Marine Pollution Bulletin* 54, 253-265.

Beg, Q. Z., Al-hazimi, A. M., Ahmed, M. Q., Fazaludeen, M. F. and Shaheen, R. 2011. Resistant bacteria a threat to antibiotics. *Journal of Chemical and Pharmaceutical Research* 3, 715-724.

Bergmann, W. and Feeney, R. J. 1950. The isolation of a new thymine pentoside from sponges. *Journal of the American Chemical Society* 72, 2809-2810.

Bert, T. M. (ed.) 2007. *Ecological and Genetic Implications of Aquaculture Activities.* Springer, 548 pp.

Bot, A. N. M., Ortius-Lechner, D., Finster, K., Maile, R. and Boomsma, J. J. 2002. Variable sensitivity of fungi and bacteria to compounds produced by the metapleural glands of leaf-cutting ants. *Insectes Sociaux* 49, 363-370.

Bruckner, A. W. 2002. Biomedical compounds extracted from coral reef organisms: harvest pressure, conservation concerns, and sustainable management. In *Marine Biotechnology in the Twenty-first Century: Problems, Promise, and Products*, 117 pp.

Campos Muñiz, C., Cuadra Zelaya, T. E., Rodríguez Esquivel, G. R. and Fernάndez, F. J. 2007. Penicillin and cephalosporin production: A historical perspective. *Microbiologίa Alam* 49, 88-98.

Defra 2007. *An Introductory Guide to Valuing Ecosystem Services.* Department for Environment, Food and Rural Affairs, 65 pp.

Foley, N. S. , van Rensburg, T. M. and Armstrong, C. W. 2010. The ecological and economic value of cold-water coral ecosystems. *Ocean & Coastal Management* 53, 313-326.

Gerwick, William H. and Moore, Bradley S. 2012. Lessons from the past and charting the future of marine natural products drug discovery and chemical biology. *Chemistry & Biology* 19, 85-98.

Glenn, H. , Wattage, P. , Mardle, S. , Van Rensburg, T. , Grehan, A. and Foley, N 2010. Marine protected areas-substantiating their worth. *Marine Policy* 34, 421-430.

Gullo, V. P. (ed.) 1994. *The Discovery of Natural Products with Therapeutic Potential.* Butterworth-Heineman, 461 pp.

Hunt, B. and Vincent, A. C. J. 2006. Scale and sustainability of marine bioprospecting for pharmaceuticals. AMBIO: *A Journal of the Human Environment* 35, 57-64.

Mayer, A. M. S. , Glaser, K. B. , Cuevas, C. , Jacobs, R. S. , Kem, W. et al. , 2010. The odyssey of marine pharmaceuticals: a current pipeline perspective. *Trends in Pharmacological Sciences* 31, 255-265.

Millenium Ecosystem Assessment 2005. *Ecosystems and Human Well-being: Synthesis.* Island Press, 137 pp.

Pearce, D. and Turner, R. K. 1990. *Economics of Natural Resources and the Environment.* Pearson Education Ltd.

Primack, R. B. 2010. *Essentials of Conservation Biology.* Sinauer Associates, 601 pp.

Querellou, J. 2010. 'Marine Biotechnology: a new vision and strategy for Europe. Marine Board, Position paper 15, 91.

Seager, T. P. 2004. Understanding industrial ecology and the multiple dimensions of sustainability. In *Strategic Environmental Management for Engineers*, Bellandi, R. (ed.), pp 17-70.

Segura-Bonilla, 0. 2003. Competitiveness, systems of innovation and the learning economy: the forest sector in Costa Rica. *Forest Policy and Economics* 5, 373-384.

Svarstad, H. 2004. A global political ecology of bioprospecting. In *Political Ecology Across Spaces, Scales, and Social Groups*, pp 239-256.

Synnes, M. 2007. Bioprospecting of organisms from the deep sea: scientific and environmental aspects. *Clean Technologies and Environmental Policy* 9, 53-59.

Trincone, A. 2010. Potential biocatalysts originating from sea environments. *Journal of Molecular Catalysis B*: Enzymatic 66, 241-256.

WHO 2012. 20 questions on genetically modified foods, *http://www.who.int/ foodsafety/puhlications/hiotech HO questions/en/* (23. 08. 12).

Woodyer, R. , Chen, W. and Zhao, H. 2004. Outrunning nature: directed evolution of superior biocatalysts. *Journal of Chemical Education* 81, 126.

第2章　生物催化的基本原理和乐趣

Peter C. K. Lau and Stephan Grosse

DOI：10.1533/9781908818355.1.17

摘要：由于工业化世界的进展，我们正在努力开展对于可持续发展的探索。因此，有关为人类提供具可选择性、专一性、高产率以及环境友好型产品与服务的酶和微生物的问题，怎么强调也不为过。生物催化，被认为是一个包含了绿色化学12个原则的华丽辞藻。为了说明生物催化的基本原理和乐趣，对各种成功应用的范例进行了重点介绍，包括丙烯酰胺故事原型和一系列来自“美国总统绿色化学挑战奖”1996—2012年生物催化应用的例子。定义催化反应的基本参数也以使用手册的形式进行了介绍。最后，对无限的海洋环境如何能够最大限度地提供生物催化，天然和人工方法获取的生物催化剂之间的对比、以石油为基础的化工生产到以生物质为基础的可持续生产的转换，都进行了讨论。总而言之，生物催化注定会在满足社会和工业需求方面发挥日益重要的作用，同时实现环境的可持续性。

关键词：绿色化学，酶催化，海洋生物，微生物多样性，工业（白色）生物技术，生物质化工，可持续发展

2.1　引言

“并不存在一个可以命名应用科学的科学范畴。只有像树和果实一样，将科学和应用科学结合在一起才能支撑它。”（Louis Pasteur，1871）

据说在大约20年前发生了一个观念转换，它强调“需要重塑工业生产技术来阻止源头污染”同“从已经被破坏了的环境中去除污染物”相比（OECD，1998）。这是关于从污染源防治与污染控制的比较，在其他方面被称为管末处理技术。

污染防治是绿色化学（图2.1）12项原则的第一真髓，由波士顿的一个研究生团队中的Anastas和Warner（1998）提出的。剩余的原则覆盖原子经济的重要方面——产品中具有出现最大数量的原子（Trost，1991，1995）；最小化的能耗和促进可再生材料的使用等。

但更为值得注意的是12个原则为什么会在这里要重新介绍的原因。波利亚科夫和他在诺丁汉大学的同事，巧妙地（天才地）以一种更便于记忆的形式提供了12项原则的简化版本：PRODUCTIVELY（Tang et al.，2005）（图2.2）。

1. 防止污染物的产生优于污染物产生之后再处理
2. 合成方法应设计成能将所有的起始物质嵌入到最终产物中
3. 只要可能，反应中使用和生成的物质应对人类健康和环境无毒或毒性很小
4. 设计的化学产品应在保留原有功效的同时尽量使其无毒或毒性很小
5. 尽量不使用辅助性物质（如溶剂、分离试剂等），如果一定要用，也应使用无毒物质
6. 能源消耗越小越好，应从环境和经济方面考虑能源的消耗；合成的方法应该在常温常压的条件下完成
7. 只要技术上和经济上可行，使用的原材料应是能再生的
8. 应尽量避免不必要的衍生过程（如基团的保护，物理与化学过程的临时性修饰等）
9. 尽量使用选择性高的催化剂，而不是提高反应物的配料比
10. 设计化学产品时，应考虑当该物质完成其功能后，不再滞留于环境中，而可降解为无毒产物
11. 分析方法也需要进一步研究开发，使之能做到实时、现场监控，以防有害物质的形成
12. 化学过程中使用的物质或物质形态，应考虑尽量减少实验事故的潜在危险，如气体释放、爆炸和着火等

图2.1　绿色化学的12个基本原则

资料来源：Anastas and Warner，1998

绿色化学的原则
P 防止污染
R 可再生材料
O 减少衍生步骤
D 化学产品可降解
U 使用安全的合成方法
C 催化剂优先
T 常温常压环境
I 监控中间步骤
V 副产物少
E 因素，产物最大化
L 化学产物低毒性
Y 是的，是安全的
绿色工程的原则
I 本身安全无危险
M 减少物料多样
P 防治优于污染后处理
R 使用可再生材料和能源
O 输出设计
V 简单
E 有效利用物质、能源、空间和时间
M 能够满足需求
E 易分离
N 通过网络实现当地物质和能量的交换
T 检测设计的生命周期
S 整个产品生命周期中的可持续
绿色非洲的原则
G 不浪费已有财富
R 重视所有生命和人类健康
E 太阳能
E 确保可降解且无危害
N 创新思维
E 设计简单实用
R 必要的循环
A 合适的材料
F 副产物和溶剂少
R 使用催化剂反应
I 当地可再生原料
C 干净的空气和水
A 避免他人误解

图2.2　依据Tang等（2005）和Asfaw等（2011）的易记得“绿色”原则

E 在 PRODUCTIVELY 中指的是环境因素，由 Sheldon 引进，在这里，E = 千克废料/千克生产，原子效率的度量（Sheldon，2007）。简单说高的 E 代表了更多废弃物的产生，造成高 E 因素的罪魁祸首是精细化学品的制造商。

Anastas 和 Zimmerman（2003），一个夫妻团队，随后制定了绿色工程的 12 项原则，却被再次缩写为 IMPROVEMENTS（图 2. 2）。总之，“富有成效的改善”是代表 12 + 12 绿色化学与绿色工程原则的豪言壮语（Tang et al. ，2005）。最近，英国诺丁汉团队为了实现绿色非洲，将绿色化学与工程原则增加到 13 个（Asfaw et al. ，2011）。13 的意义与埃塞俄比亚日历上有一个月只有 5~6 d，其余月份均有 30 d 有关。为了避免他人错误地理解为有其他原则，本文的其他地方遵从最初的 12 原则。

顺着这 12 条的逻辑，我们提出了将“生物催化”作为一个词表达了关于绿色化学的所有。生物催化的已知属性表明下面是选择的理由。

- 高效率 – 高转数/率（10^8–10^{10}）。
- 选择性/特异性 – 立体/空间–选择性和对映选择性（手性）。
- 温和的条件 – 适宜的温度、pH 值及气压。
- 能耗低。
- 不被其自然属性所束缚（底物耐受性）。
- 在复杂混合物中的高选择性（无副反应）。
- 可生物降解的（天然生物制品）。
- 副产物少。
- 可大量生产。

最为重要的是，酶是天然酶。

在“酶合成简介”中，Hudlicky（2011）指出：“Anastas（与 Warner）的原则完全忽略生物方法；‘酶催化’或‘生物方法’这样的词离奇的‘缺席’了，虽然可以说酶类确实属于广义的催化”。然而，我们应该认识到这些，Tucker（2006）已经旗帜鲜明地指出，绿色化学“不受制于离子液体，微波化学，超临界流体，生物转化（生物催化），氟相化学反应，或任何其他新技术。绿色化学在使用的技术之外，但却在技术应用的目的和结果之内。”

绿色化学的核心是“一个新的对环境的优先考虑，一个由效率驱动与环境责任耦合的理论”，3 个 e 体现出了持续发展的根本支柱。本章重点讲述了大自然能够通过生物催化或酶为人类提供什么，和人类如何利用这些进行进一步获得什么。在此，广义上的生物催化被定义为使用酶来实现化学转换。这是个类似于生物转化的术语，它包含任何酶促反应，这个过程就像发酵或代谢工程的产物。首先，对定义一个生物催化反应的基本参数进行了讨论，接着论述生物催化的实例和面临的挑战。

2.2 定义生物催化反应的基本参数

本节讨论了各种对于描述生物催化极为重要的各种基本概念，同时结合由 Gardossi 等（2010）提出的指导方针，其中包括对标准酶学数据报告（STRENDA；http：//www.strenda.org）的建议。通过讨论它的特点，对生物催化反应进行更为全面的描述，优势与潜在的不足的了解将有望推动生物催化反应实现工业规模应用的可能，或是提供实现这一目标良好的工程指导。

2.2.1 酶的特性和来源

若是完整的细胞或纯酶，生物催化剂的特征必须被具体的反应类型所命名：EC 编号；如果适用的话。还有菌株保存编号，基因检索号等，在此提醒，酶催化特定的化学反应已被国际生物化学与分子生物学协会酶学委员会（IUBMB EC）分类，分类如下：

EC 1 氧化还原酶类：C–H，C–C，C=C 键的氧化；电子的转移。

EC 2 转移酶类：官能团的转移：醛，酮，酰基，磷酰基，或甲基。

EC 3 水解酶类：酯类，酰胺类，内脂类，内酰胺类，环氧化物，腈类，酸酐类，葡糖苷类的形成/断裂。

EC 4 裂解酶类：去除或增加 C=C，C=N，C=O 键。

EC 5 异构酶类：外消旋作用和差向异构化。

EC 6 连接酶类：C–O，C–S，C–N，C–C 键的形成或断裂，需要 ATP 裂解。

此外，亚类和亚亚类已经被用来进一步描述各种酶的作用（see Enzyme Nomenclature 1992，Academic Press，San Diego，California，ISBN 0－12－227164－5）。BRENDA（BRaunschweig ENzyme DAtabase；http：//www.brendaenzymes.org）是一种基于网络的宝贵工具，它利用酶结构和功能的数据等，提供广泛的酶信息。据报道，在 2011 年的更新中，来自 10 500 种不同生物的酶已被表征。而接近 75 000 酶名称存在同义词，每种酶可包含 15 个同义词（Scheer et al.，2011）。

通过分子生物学技术，例如 16S 核糖体 DNA 测序，来进行特定菌种鉴定是非常重要的，因为这会影响对微生物处理的适当性与安全性，以及基因操作方法的选择。菌株的形态学鉴定和脂肪酸分析往往是不够的。细菌菌株的错误分类在文献中很是普遍的。在组成病原菌的分枝杆菌属中，如结核杆菌在系统发生学上与革兰氏阳性属如红球菌属，诺卡氏菌属或节杆菌属有关。

2.2.2 酶活性及其测定

重要的参数是测定反应的初速度，同时尽可能地记录比活力。通常用 U（μmol·

$min^{-1} \cdot mg^{-1}$或 $katal \cdot kg^{-1}$）表示比活力单位。Katal 是每秒催化 1 摩尔底物转化的酶量（$mol \cdot s^{-1}$）。在分泌酶中，U/mL 培养物常见报道。在某些情况下，需要描述生物催化剂的干重或湿重时，需要特定的说明比较好。另一方面转换数是无量纲，指反应过程内，产物摩尔数与使用的催化剂的摩尔数的比例。

取决于所使用的方法（如分光光度法，用荧光光度计测荧光性），在最适 pH 值、温度、缓冲液浓度（离子强度），所需的金属或辅因子条件下进行酶活测定是至关重要的。适当的酶浓度也很重要（有时少实际上比多更好）因为酶比活力的精准测定需要过量的底物（S）和可能的双底物（S）的存在。读者可参考 Reymond（2005）对各种酶检测方法，尤其关于高通量筛选和其他综述及研究性论文（Goddard and Reymond，2004a，2004b；Mastrobattista et al.，2005；Schmidt and Bornscheuer，2005；Reymond et al.，2009；Lucena et al.，2011）。

有时酶的天然底物是未知或不易获得的，因此用人工底物来检测并描述酶活性。羧酸酯水解酶（EC 3.1.1.-）就是一个典型的例子。依据它们的属性和底物特异性将这些酶分成许多亚类，它们广泛存在于微生物、植物和动物中。实验室常用的底物是对硝基苯基（*p*-NP）酯的羧基和脂肪酸（例如对硝基苯基醋酸盐，对硝基苯基丁酸酯，和对硝基苯基棕榈酸酯等），在酶解后释放的易检测和定量的对硝基苯酚，可通过测量其在 410 nm 处的吸光度来定量。对于对硝基苯酯类底物的分析有时被当做一种酶分类的标准。例如角质酶（EC 3.1.1.74）表现出羧酸酯酶的亚型，专一性的针对高等植物角质层结构组成的角质，它是羟基和羟基环氧脂肪酸组成的聚酯。与脂肪酶（EC 3.1.1.3）偏好于长链脂肪酸（>C10）组成的酯类相反，角质酶对于短链（<C10）时活性更高。然而，这些研究结果大多是经验性的，可能还要跟特定底物的选择和测活方法的优化有关。虽然它们作为初筛可以为酶分类提供一些指导，但仍然需要深入考虑酶分子的性质和探索其可能的天然底物，这是至关重要的。

2.2.3 动力学参数（K_m，V_{max}，k_{cat}，k_{cat}/K_m）

如 Michaelis-Menten（米氏方程）（2.1）描述的，酶动力学核心参数是 K_m 和 V_{max} 值，与游离底物浓度（S）对酶促反应速度（v）有关。Briggs 和 Haldane（1925）对米氏方程进行推倒，引入了稳态近似方程（2.2），它假设初始反应结束后，整个反应在检测过程中，酶底物复合物（ES）达到一定数值并保持不变。注意，k_{on}是二阶速率常数，用单位时间内的浓度（$mol \cdot L^{-1} \cdot min^{-1}$）来描述酶和底物的结合，它取决于在反应中相互接触的底物和游离酶结合的难易程度。与此相反，k_{off}和 k_{cat}是一阶速率常数，用单位时间（s^{-1}，min^{-1}）内酶和底物复合物解离而产生酶和底物或酶和产物来分别描述。酶与底物复合物的解离常数 K_D 由 k_{off}和 k_{on}的商来定义，因此被视为一种浓度，就像 K_m 由个速率常数来定义（2.3）。

$$v = \frac{V_{max}[S]}{K_m + [S]} \tag{2.1}$$

$$\mathrm{E} + \mathrm{S} \underset{K_{\mathrm{off}}}{\overset{K_{\mathrm{on}}}{\rightleftharpoons}} \mathrm{ES} \xrightarrow{k_{\mathrm{cat}}} \mathrm{E} + \mathrm{P} \tag{2.2}$$

$$K_{\mathrm{D}} = \frac{k_{\mathrm{off}}}{k_{\mathrm{on}}} K_m = \frac{k_{\mathrm{off}} + k_{\mathrm{cat}}}{k_{\mathrm{on}}} \tag{2.3}$$

K_m和 $V_{\max}$可以通过 v/S（米氏方程图）的线性作图直接获得，更多的是通过一种线性变换来获得，例如双倒数作图（$1/v$ 对 $1/S$ 作图），Hanes 的（S/v 对 S 作图）或 Eadie-Hofstee（v/S 对 v 作图）。直接的线性作图方法，例如 Eisenthal Cornish-Bowden 图也被证明是有用的。如今，一些期刊（如生化杂志）转而采用基于计算机软件的非线性回归。然而，这里应该强调指出的是，为获得有用的动力学数据，可以使用一般的常识和适当的加权。

在一段时间内单一酶促反应中，一种底物不一定能完全转化产生一种产物，因为这一过程通常需要辅酶［如 NAD（P）H 作为氧化还原酶］、共底物（如水解酶对水的作用），或对于裂解酶（合成酶）技术上讲是可以产生两种产物或将两种产物转化为一种底物。当从动力学角度描述这些类型的反应时，定义一个限速因素很重要，这个限速因素可以是底物，共底物等，在一定时间内并且其他的条件都是过量。

在酶活测定中，底物浓度要远高于酶浓度而几乎保持不变，产物是线性积累，也不存在产物到底物的逆反应。特别是在使用不连续的试验条件时，确保在试验过程中产物累积是线性的这一点很重要，同时反应不能被限制或已经完成，否则会导致计算的动力学参数无法使用。底物浓度至关重要而且应该在 K_m值的 0.1～10 倍，对于未知酶常用试错法检测。转化的米氏方程（2.1）也表明，K_m值指反应过程中反应速度达到 $V_{\max}$一半时的底物浓度，在 $k_{\mathrm{off}} >> k_{\mathrm{cat}}$时，$K_m$也可等于酶-底物复合物的解离常数（$K_D$）。$k_{\mathrm{cat}}$通常被称为转换数，可直接由 $V_{\max}$和总酶浓度的商算出。转换数常被报道为单位时间内每分子酶产生的产物的分子数，如 mol/（mol s）。因此，如果产物和酶使用相同的单位，他们最终会相互抵消，k_{cat}将会成为时间的倒数（例如 s^{-1}）。k_{cat}/K_m的比值最初被称为“特异性”（Fersht，1984）或是“性能常数”（Dean et al，1996）并代表一个二阶速率常数（$M^{-1} \cdot s^{-1}$）被用于测定酶的催化效率。然而，Koshland（2002）指出这参数可能会产生误导，因为酶的特异性通常与一般的化学结构差异有关。通常情况下，k_{cat}/K_m的值越高，酶的性能就越好，因此这个比值可以用于比较酶或一种酶的不同底物。据说，碳酸酐酶的 k_{cat}/K_m的比率 $10^8 \sim 10^9$，已实现动力学的完美，反应不再由 k_{cat}限制，而是由酶和底物的扩散限制，因为它们一旦相遇就会形成酶-底物复合物。

2.2.4 抑制（K_i，K_d等）

在生物系统中酶的活性可以用多种方式来调节，如表达控制（例如诱导型和组成型），产生无活性前体酶（酶原），酶的磷酸化或通过代谢调控，诸如与酶抑制剂/激活剂分子的相互作用达到影响正常反应。这些相互作用可以是变构（多亚基酶如磷酸果糖激酶的调节），或是直接或间接地与酶活性位点相作用，并且它们通常是可逆的。最常见的抑制类型是：

竞争性抑制：结构与底物非常类似的抑制剂能直接与底物竞争酶分子上的结合位点，这种抑制剂仅与游离酶结合。因此，底物的表观 K_m 增大但 V_{max} 保持不变。增加底物浓度，最终可以降低甚至抵消抑制剂作用。

非竞争性抑制剂：抑制剂与游离酶和酶-底物复合物均能结合，而不是直接作用于酶活性位点。其表观 K_m 不受影响但 V_{max} 减小，这种作用不能通过提高底物浓度而消除。

反竞争性抑制：这种抑制剂只与酶-底物复合物结合因此 K_m 和 V_{max} 都会降低。

混合型抑制剂：这种抑制剂对酶和酶-底物复合物的亲和性不同，因而 K_m 和 V_{max} 均会受影响。

为了描述某种潜在抑制剂的结合能力，我们可以测量解离常数（K_d），K_d 也被称作与酶-抑制剂复合物的抑制常数（K_i）。根据可逆抑制剂的类型和由此抑制剂的结合对象（例如酶，酶-底物复合物等），有多种方式来检测这一常数。有时，抑制剂的结合会影响酶的光谱性质，例如，通过色氨酸残基的荧光猝灭可以测量相关的值（例如用咪唑苯酸盐测尿刊酸酶（urocanase）抑制剂作用，O'Donnell and Hug，1985）。实时结合常数的测量同样可以通过表面等离子体共振来完成，例如固定化酶与潜在抑制剂的相互作用。然而，在催化过程中最常见测定抑制剂浓度提高的影响是通过测定游离酶及酶-抑制剂复合物，从而算出 K_i。

应该提到的一点是，并非所有的抑制剂都是可逆的；一些抑制剂结合的非常紧密或不可逆，从而使酶失活。其中一个例子是苯甲基磺酰氟（PMSF）作为丝氨酸蛋白酶不可逆抑制剂（胰蛋白酶、胰凝乳蛋白酶、凝血酶）特异性的酯化活性位点的丝氨酸上的羟基。苯甲基磺酰氟常被用于细胞破碎或蛋白质纯化过程中抑制不需要的蛋白酶活性。

2.2.5 稳定性

在工业环境下，稳定性是催化剂的重要参数之一。即使是活性很高的酶，如果在可以想象的一些处理或生产过程后会导致其失活，该酶将没有实际应用价值。对于工艺研发过程中，基本的问题是需要设计酶来满足一些现有的生产方案，或是设计一些生产工艺过程适应酶的特性。我们可以区分几种形式的酶稳定性，例如化学稳定性（如 pH 值的影响、盐或溶剂浓度）、热稳定性（如升温引起的蛋白质可逆折叠）和动力学稳定性（描述在发生不可逆变性前的时间、剩余酶活性）——更广泛的综述可见 Illanes（1999）和 Polizzi 等（2007）。

然而，酶稳定性的研究通常是在实验室条件下（例如在简单的缓冲液中）进行。因此，这虽然可以提供一些指导，但他们掌握的有关评估酶是否适合工业生物过程的信息依然有限。在制备工艺条件下，酶的底物作用可能表现得很不一样，这与产物的累积，或者溶剂会以任何一种方式影响其活性：如提高活性（保护）或抑制活性（去稳定作用）（Illanes et al, 1996)。应当指出，酶的稳定性往往依赖于实验时酶的浓度，这又增加了另一个变量。

酶的热稳定性往往可以通过量热法或圆二色性（CD）光谱法进行测定。这经常被低估或忽视。对于后者，升温时，酶吸收圆偏振光的差别可由一特定波长测出（如 222 nm 处）。蛋白质因可逆的去折叠而失去其二级结构特征，因此其吸光度将会逐渐减少直至蛋白质完全去折叠。去折叠过程的平衡常数（K）描述的是去折叠（f_U）与折叠蛋白（f_F）组分的比率，ΔG 是吉布斯自由能的变化。在解链温度下，去折叠与折叠蛋白处于平衡状态（$K=1$）因此 $\Delta G=0$。

$$K=\frac{f_U}{f_F}=\frac{(T_m)f_U}{(1-f_U)} \quad \Delta G=-RT\ln K \tag{2.4}$$

因为未折叠蛋白相当于失活酶，因此不同酶 T_m 值的比较，提供了关于可能的最适合最大工作温度很好的参考。然而，T_m 也不是描述总体温度稳定性所必需的，因为总体温度稳定性可由酶的半衰期来表征（$t_{1/2}$）更合适。为了测定该参数，酶在一定温度下孵育，测定残余酶活力随着时间的变化。随着时间推移，剩余酶活力的半对数图可用于计算出温度依赖性变性的速率常数（k_d，线性回归斜率），基于此可以算出 $t_{1/2}$（$\ln 2/k_d$）。

然而，热稳定性只是评估酶的一个标准。例如盐，溶剂和 pH 值稳定性，都是相关的影响因素，而且可以遵循类似的方式来表征。由于所有这些参数都是相互依赖的，因此需要精确描述这些参数测定的反应条件，这样才可以进行催化相同反应酶的稳定性的比较。

2.2.6 特异性

酶催化的特性之一是对于底物和反应的特异性。酶通常具有对一种反应类型高度的特异性。然而，也有一些例外，例如复合酶展现出多种催化功能（例如丙酮酸脱羧酶）或水解酶也经常催化转移酶反应（Yonaha and Soda，1986）。除了消旋酶和异构酶，在作用于键连接的底物手性中心或从手性底物（例如 L-氨基酸氧化酶，D-氨基酸氧化酶）中形成手型产物的过程中，酶通常展现出绝对的立体专一性。与此相反，作用于远离底物手性中心基团的酶不一定显示出高的立体特异性，例如羧酸酯酶可转换 L-酪氨酸和 D-酪氨酸乙酯（Stoops et al.，1969）。

为了更好地表征酶的立体选择性，并从而量化评估其潜在的动力学拆分，Chen 等（1982）引入了无量纲的旋光异构体比率（E）表示为二级速率常数（k_{cat}/K_m)^R 和（k_{cat}/K_m)^{S}的商。这是测定一个酶拆分的“选择性”的量度（Faber，1995）。E 值可以用实验方法获得，通过测量剩余底物或相应产物的对映体过量百分数（ee）以具体的转化率算出，根据方程：

$$E = \frac{\ln[1 - c(1 + ee_p)]}{\ln[1 - c(1 - ee_p)]} \tag{2.5}$$

(Rakels et al. , 1993; Straathof and Jongejan, 1997)。

为了获得准确的 E 值，常用的规则是在反应达到50%转化率时停止反应。为简化计算方法，一个可用的程序在 ftp: //biocatalysis. uni-graz. at/ pub/enantio。

2.2.7 总结生物催化中的实用关键参数

低 K_m 导致底物抑制，而当解离常数（K_S 和 K_P）小于 0.1 mmol/L 时，低的 K_p 会引起产物抑制（图 2.3）。从 ES 到 EP 速率常数 k_{cat}，在理想情况下要尽可能的高。一种酶的比活力 1 U/mg，$k_{cat}=2\ s^{-1}$ 足以每天产生 1 mmol 产物（Koeller and Wong，2001）。

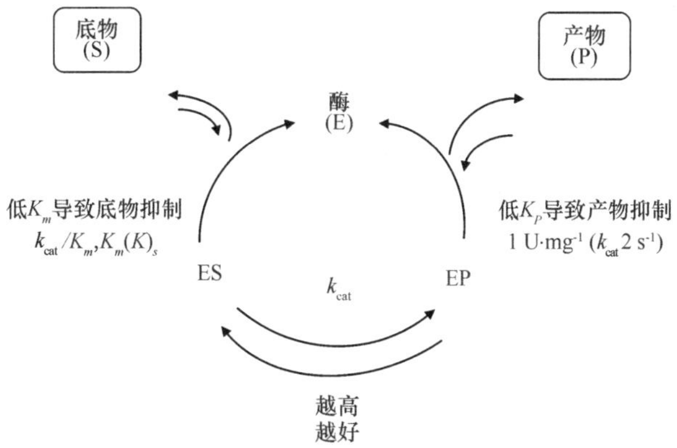

图 2.3 酶催化过程中关键的动力学参数

资料来源：根据 Koller 和 Wong（2001）改编

注：ES 和 EP 是酶（E）和底物（S）或产物（P）的复合物

关于对映体的比率 E，低于 25 的 E 值对于实际应用是不可接受；15~30 是中度至良好；超过 30 被认为是极好。E 值大于 200 很可能不能准确测量，归咎于无法准确地通过譬如 HPLC 或其他形式获取 ee 值，因为 ee_P 或 ee_S 的细微变化都能造成 E 值的显著变化（Faber，1995）。

2.2.8 生物催化剂的发展组件图

图 2.4 展示了生物催化剂发现基本要素的流程，生化特性，通过蛋白质工程可能的改进，和工业生物催化的一些加工路线。各个参数都在本文中讨论了。

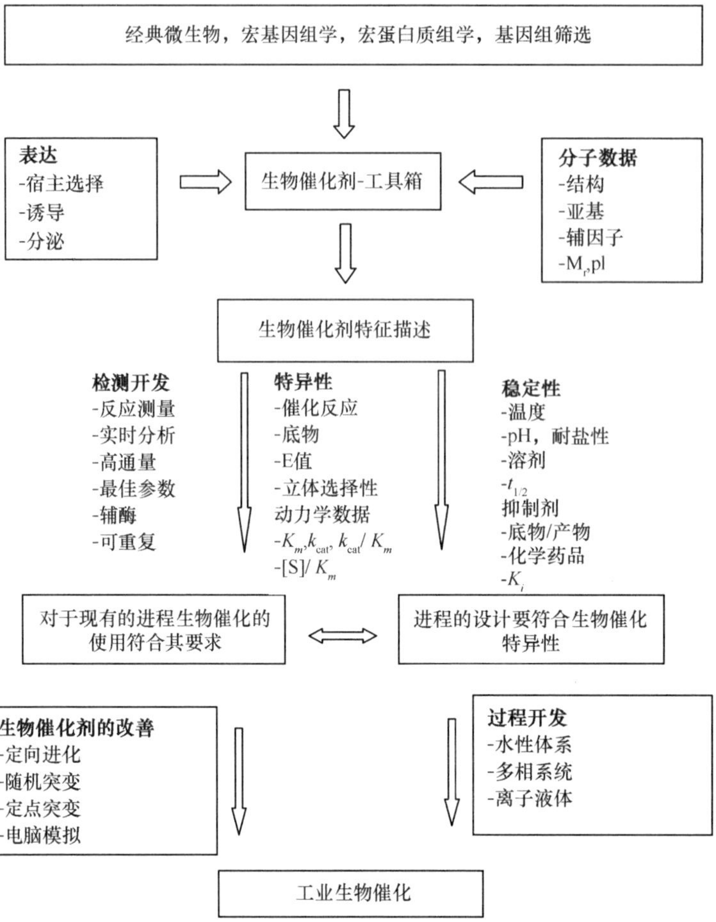

图 2.4　生物催化的基本原理——从发现到创新与应用

2.3　丙烯酰胺的故事重现

这不是关于 2002 年 4 月在各国政府食品机构中备受关注的饮食中的丙烯酰胺（例如，炸薯条和薯片中 $450\times10^{-12}\sim1\ 200\times10^{-12}$ 的丙烯酰胺），而是工业生产的丙烯酰胺。关于前者，一言以蔽之，高碳水化合物含量的食物经油炸，烘焙和烧烤等高温处理后，就会导致丙烯酰胺这种可能的人类致癌诱变剂的产生。Stadler 等（2002）表明，氨基酸，尤其是天冬氨酸，与羰基源如含还原糖的碳水化合物结合后经热处理（例如 180℃）产生美拉德反应，是释放丙烯酰胺的主要原因。反之称非酶促褐变（Zhang and Zhang，2007；Claus et al.，2008）。

在实验室条件下，纯的丙烯酰胺因聚丙烯酰胺凝胶在蛋白质或 DNA 电泳中的使用而被人熟知。为了降低成本，其中一位作者（PCKL）回想起不得不从加热的氯仿中重结晶微黄色丙烯酰胺——对今天“幼小”的这一代来说是非常不可思议的，他们会觉得实验室工作是多么可怕和愚蠢！工业上，丙烯酰胺是一种重要的常用化学品，是聚合物形式的典型代表，包括用于纸浆和纸类行业制造的聚合物、水处理系统所用的絮凝剂、灌浆和隧道工作中的封固剂或用在焦油砂油回收中。具有酰胺和乙烯基为功能团，双边缘丙烯酰胺分子被用作单体化学物质或水溶性聚合物生产的原料就不足为奇了（Kobayashi et al.，1992）。

全球化学催化法生产丙烯酰胺大约是 415 000 t/a，酶法至少生产 45 000～85 000 t/a（OECD，2001）。传统的丙烯酰胺化学合成涉及以铜盐为催化剂的水合丙烯腈法，替换了较旧的硫酸水解法，硫酸水解法还会产生大量的副产物硫酸铵。

$$H_2C=CHCN + H_2O \rightarrow H_2C=CHCONH_2 \tag{2.6}$$

与这个化学途径有关的问题和相关的工艺特点以及与酶法优点的比较见表 2.1。

表 2.1　化学与酶学丙烯酰胺生产进程的对比

项目	化学	酶学
反应	在强酸性或者碱溶液中高压下 200～300℃	在正常 pH 值或弱碱性水溶液介质及较温或稍低的温度下（<30℃）
底物特异性	对于化学-，立体-或者区域选择性有广阔的范围	对于化学-，立体-或者区域选择性有较窄的范围
丙烯酰胺浓度	～30%	48%～50%
催化剂的回收	不可行	通过酶固定可行
产品的提取及纯化	有机相萃取及产品多步骤纯化	蛋白去除即可
产率（单程）	70%～80%	～100%

资料来源：OECD，2001；Prasad and Bhalla，2010.

被称为腈水解酶途径和腈脱水酶（NHase）途径的两条微生物降解腈途径的阐明，为丙烯酰胺的酶合成法铺平了道路（见综述：Nagasawa and Yamada 1989；Kobayashi et al.，1992；Yamada and Kobayashi，1996；Yamada et al.，2001）。腈水解酶（EC 3.5.5.1 氨基水解酶）催化腈水解生产相应的酸和氨气如下：

$$RCN+2H_2O \longrightarrow RCOOH+NH_3$$

（这里 R 代表苯基式 α/β 烯基）

在氨基水解酶（EC 4.2.1.84）催化途径中，饱和脂肪腈代谢分为两步。首先酰胺转为酸加氨（如下：）

$$RCN+H_2O \longrightarrow RCONH_2$$

$$RCONH_2+H_2O \longrightarrow RCOOH+NH_3$$

（这里 R 代表烷基）

分离得到的具有分解丙烯腈的微生物假单胞菌 B23 是酶催化合成丙烯酰胺的重要里程

碑（Asano et al，1982）。这个微生物事实上是第二代氨基水解酶生产者，第一代被 Mitsubishi Rayon Co. Ltd 确认为革兰氏阳性 *Rhodococcus* sp. N774，并在 1985 年用于丙烯酰胺的生产（表 2. 2）。这两个铵盐水解酶都是含有低旋三价铁离子的非血红素铁酶。读者参阅了 Yamada 和 Kobayashi（1996）发表的相关文章，他们涉及了技术细节和菌种改良（无黏多糖产生的突变体一代）和生物过程技术（细胞固定），这都是把 B23 设计成工业通用的品种所必需的。前期由于使用相对用量无毒害的代用底物（如异丁腈）对 B23 进行了筛选，因为丙烯腈在吸收后对细菌有很大的毒害作用，从中我们付出了很多代价。正是由于部分筛选的标准，高氨基水解酶活性与少量或微量丙烯酸的生成即低酰胺酶活性菌种被筛选出来。在前期的实验中，当使用休眠的 B23 菌种在 10℃孵育 7. 5 h 后就有超过 400 g/L 的丙烯酰胺累积。之后与 Nitto 合作，在 1998 年生产了大约 6 000 t 丙烯酰胺作为第一年的生产规模（表 2. 2）。这是第一个将生物技术运用于石化工业制造化学产品的成功典范。

表 2. 2　丙烯酰胺生产的整体细胞生物催化方法的进步性优化

项目	第一代 红球菌 sp. N774	第二代 假单胞菌 B23	第三代 紫红色红球菌 J1	最新一代 诺卡氏菌 sp. 86-163
诱导剂	组成型	甲基丙烯酰胺	尿素	未知
丙烯酰胺的限值/%	27	40	50	50
丙烯酸的生成	非常少	基本检测不到	基本检测不到	<0. 3%
培养时间/h	48	45	72	60
原始活性/（$U \cdot mL^{-1}$）	900	1 400	2 100	>5 600
特异活性/（$U \cdot mg^{-1}$）（cells）	60	85	76	190
细胞产率/（$g \cdot L^{-1}$）	15	17	28	29
丙烯酰胺产量/（$g \cdot g^{-1}$）（cells）	500	850	>7 000	
总产量/（$t \cdot a^{-1}$）	4 000	6 000	>30 000	>150 000
丙烯酰胺最终浓度/%	20	27	40	35
产品上市首年	1985	1988	1991	1993

资料来源：Yamada et al.，2001；Zheng et al.，2010.

但是在 1992 年一篇综述文章（Kobayashi et al.，1992）中提到的，这个成功的故事并没有结束。为了提高丙烯酰胺的生产，一个来源于 J1 菌种的含钴氨基水解酶生物催化剂被分离出来，具体来说是具有高分子量（520 kD）来源于 *Rhodococcus rhodochrous* J1 的酶被分离出来，并被表征。J1 菌种的最初筛选是在苯甲腈作为唯一碳源和氮源的培养基中进行的。作为含钴的酶向培养基中添加钴离子对氨基水解酶的形成来说是不可少的。但在钴存在的情况下，加入不同的诱导物（如尿素或环己甲酰胺），可选择性地制备不同产量的氨基水解酶。

纯化的氨基水解酶与丙烯腈有很低的 K_m 值（1.89 mmol/L）和高的亲和力。这 3 个氨基水解酶含有的辅助因子钴和三价铁离子的不同与其结合活性位点少数的氨基酸有关（Yamada and Kobayashi，1996；Kobayashi and Shimizu，1998）。3 个酶的 α 亚基所有氨基序列的同源性在 60%~72%（45%~60%一致性），β 亚基同源性在 40%~67%（28%~56%一致性）。

在工厂生产规模，第三代性能优良的超级生物催化的效率是第二代菌种 B23 的 10 倍多。改进包括使用更经济的诱导物尿素，对 50%丙烯酰胺的耐受性，更高热稳定性的酶（达到 50℃）等。细胞外提取液总溶解蛋白可得到 50%以上的大量 H-HNase 和 100%的底物转化率。早在 1991 年就获得了超过 30 000 t 丙烯酰胺的年生产量。

2.3.1 前景展望

自首次开始使用酶工艺生产丙烯酰胺已经过去 27 年。技术已经成熟并得到广泛传播，每年生产量已超过 400 000 t。中国占其年产量的一半，至少有 9 个与此相关的工业基地（Zheng et al.，2010）。中国实验室使用的腈水合酶来源于土壤 *Nocardia* sp. 86-163，分离于山东省泰安地区的土壤样品（Zhang et al.，1998）。虽然 HNase 已经在原核和真核领域被发现（Prasad and Bhalla，2010；Marron et al.，2012），但是对它的研究探索依然在进行。藻酸盐固定的大肠杆菌系统表达的具有耐热性的来源于 *Comamonas testosteroni* 5-MGAM-4D 的腈水合酶，发现其在 5℃下有很好的丙烯酰胺单位生产效率，还可通过使用丙烯腈减少酶活的损失（Mersinger et al.，2005）。真核细胞 HNase 密码基因第一次在 *Monosiga brevicollis*（一种海洋单细胞领鞭虫类）中发现报道（Foerstner et al.，2008）。Brandon 和 Bull（2003）曾报道在深海沉积物中获得的放线菌中发现有腈水合反应活性。考虑到 HNase 在包括工业烟酰胺的生产中的多种工业用途，更多的关于丙烯酰胺类的成功事例将会被发现。

2.3.2 环境的影响

除了酶法生产丙烯酰胺的各种优势外，能量的消耗和二氧化碳的产生量已成为评估对环境影响的指标（OECD，2001）。表 2.3 显示相对化学方法，酶法总体是更为环境友好的工艺。虽然丙烯酰胺和丙烯腈都是有害的化合物，在封闭系统条件下谨慎化学控制和安全工艺可以避免对人类和环境不必要的伤害。

表 2.3　能量消耗和二氧化碳产生量的对比

<table>
<tr><th colspan="2" rowspan="2">项目</th><th rowspan="2">化学</th><th colspan="2">酶学</th><th rowspan="2">文献</th></tr>
<tr><th>第一代</th><th>第三代</th></tr>
<tr><td rowspan="4">CO_2产生量
（kg/kg 丙烯酰胺）

（t/t）</td><td>水蒸气</td><td>1.25</td><td>2.0</td><td>0.2</td><td rowspan="4">OECD，2001
Zheng et al.，2010</td></tr>
<tr><td>用电</td><td>0.25</td><td>0.25</td><td>0.1</td></tr>
<tr><td>原材料</td><td>2.3</td><td>2.3</td><td>2.3</td></tr>
<tr><td></td><td>1 500</td><td>300</td><td></td></tr>
<tr><td rowspan="4">能源
（MJ/kg 丙烯酰胺）

（MJ/t）</td><td>水蒸气</td><td>1.6</td><td>2.8</td><td>0.3</td><td rowspan="4">OECD，2001
Zheng et al.，2010</td></tr>
<tr><td>用电</td><td>0.3</td><td>0.5</td><td>0.1</td></tr>
<tr><td>原材料</td><td>3.1</td><td>3.1</td><td>3.1</td></tr>
<tr><td></td><td>1 900</td><td>400</td><td></td></tr>
</table>

2.4　总统绿色化学挑战奖

自 1996 年美国环境保护机构（EPA）实施总统绿色化学挑战奖项目（www.epa.gov/greenchemistry），以此通过绿色化学技术创新来推动污染预防，同时在工厂有广泛的适用性。包含更加清洁、经济、巧妙的化学技术，使用安全的原材料和获得安全更好的产物在内的挑战。每年有 5 种类别的荣耀奖项：学术，小型企业，绿色合成途径，绿色反应条件，绿色化学设计。总体上，奖励的技术工艺已经：

减少了超过 13 亿镑的有害化学物质和试剂的产生；

节约了超过 420 亿加仑的水；

减少了近 460 百万镑二氧化碳排到空气中。

（Source：EPA publication 744K10003，June 2010；EPA publication 744F12001，June 2012.）

表 2.4 列举了不同领域中的 21 位胜出者，他们将生物催化和微生物应用于绿色化学技术发展过程的全部或者其中的某个阶段。作者对编制列表的任何遗漏条目表示道歉，特别是考虑到使用可再生原料（如：biomass and microorganism）却没有清晰的酶使用意义的项目。Dow Agrosciences LLC 提出的一种由土壤微生物 *Sacharopolyspora spinosa* 制成的环境友好型杀虫剂，是多杀菌素发展成果的一个例子。

表 2.4　在生物催化领域的总统绿色化学技术挑战奖获得者（1996—2012 年）

类别	学术
标题	二氧化碳循环生物合成高级醇
技术	工程微生物（蓝藻或藻类）利用二氧化碳生产 3~8 个碳的醇作为燃料代替品
获奖者	生物技术 LLC 和加利福尼亚大学（J. Liao）
年份	2010
标题	利用脂酶进行温和可选择的聚合作用
技术	脂肪酶催化聚合反应产生一系列的多元醇聚酯
获奖者	理工大学（R. Gross）
年份	2003
标题	大规模有机合成中的酶
技术	使用简单酶的具有手性和立体选择的合成反应与使用葡糖基转移酶复杂多级酶反应合成低聚糖
获奖者	斯克里普斯研究所（C. H. Wong）
年份	2000
标题	使用微生物环境友好的合成催化剂
技术	基于工程大肠杆菌用糖生产顺-黏糠酸之后与氢化合生成己二酸，作为尼龙的构建模块单元
获奖者	密歇根大学（K. Draths and J. Frost）
年份	1998

类别	小型企业
标题	生物琥珀酸的下游应用和综合生产
技术	以下游纯化综合工艺为基础的细菌葡萄糖发酵产生琥珀酸和水
获奖者	BioAmber. Inc.
年份	2011
标题	微生物生产可再生石油燃料和化学物质
技术	工程微生物像精炼厂一样把糖发酵转换为烷烃类、烯烃类、脂肪醇类或脂类成为超净的柴油
获奖者	LS9. Inc.
年份	2010
标题	使用生物技术生产天然塑料
技术	工程微生物生产聚羟基烷酸作为可生物降解塑料
获奖者	Metabolix. Inc.
年份	2005
标题	李鼠糖脂生物表面活性剂：天然，低毒选择性合成表面活性剂
技术	铜绿假单胞菌被选择来生产李鼠糖脂生物表面活性剂
获奖者	Jenell 生物表面活性剂公司
年份	2004

类别	绿色反应条件
标题	使用转氨酶绿色生产西格列汀
技术	使用R-转氨酶合成治疗Ⅱ型糖尿病药物西格列汀的关键手性氨基
获奖者	Merck 和 Co. Inc. Codexis. Inc.
年份	2010
标题	3个生物催化剂定向进化生产立普妥中的活性成分手性阿托伐他汀
技术	通过酮还原酶和NADP依赖的葡萄糖脱氢酶为辅助因子的可再生系统拆分还原4-氯乙酰乙酸乙酯，然后通过卤代醇脱卤素酶催化［S］乙基-4-氯-3羟基丁酸的氰化作用，为阿托伐他汀的合成的中间产物
获奖者	Codexis. Inc.
年份	2006
标题	优化：改善纸循环的一种新酶技术
技术	酯酶水解多聚（乙烯基，乙酸）和来自纸的类似材料多聚（乙烯基，乙醇）使之可溶于水和无黏性
获奖者	巴克曼国际实验室
年份	2004
标题	微生物生产1，3-丙二醇
技术	工程微生物通过玉米淀粉糖代谢生产1，3-丙二醇，过程类似于Sorona的聚酯合成反应
获奖者	国际杜邦和杰能科公司
年份	2003
标题	自然工厂的聚乳酸工艺
技术	通过可再生材料谷物、乳酸和丙交酯进行聚乳酸的合成
获奖者	自然工厂公司（先前卡吉尔道公司）
年份	2002
标题	棉纺织品生物制剂：经济有效环保的制备工艺
技术	使用果胶裂解酶作为洗擦试剂降解果胶，从而从棉花中获得柔软的材料
获奖者	北美诺维信公司
年份	2001

类别	绿色化学设计
标题	酶减少了能量和木质纤维的使用，可满足制造高质量的纸和纸板的要求
技术	纤维素酶处理木质纤维增加原纤维的数量以增加其强度和质量
获奖者	巴克曼国际公司
年份	2012
类别	绿色合成途径
标题	制造辛伐他汀有效的生物催化工艺
技术	运用酰基转移酶和有效的酰基供体合成降胆固醇药物辛伐他汀
获奖者	克迪科思公司，Y. Tang（UCLA）
年份	2012
标题	经济可再生原料生产基础化学物质
技术	基因改造微生物糖类发酵生产 1，4-丁二醇
获奖者	Genomatica 公司
年份	2011
标题	生产化妆品和个人护理品的无溶剂生物催化工艺
技术	固定化酶（脂肪酶）处理不饱和脂肪酸生产酯
获奖者	伊士曼化学公司
年份	2009
标题	通过使用硫辛酸酰胺酶与蔬菜油的相互酯化生产低反式油脂
技术	通过固定化的脂肪酶使含有饱和脂肪酸的甘油三酸脂发生相互酯化作用
获奖者	Archer Daniels Midland 公司和 Novozymes
年份	2005
标题	基于植物细胞发酵提取法绿色合成制造紫杉酚
技术	通过植物细胞特别是来源于愈伤组织的紫衫细胞合成紫杉酚的活性成分紫杉醇
获奖者	Bristol-Myers Squibb 公司
年份	2004
标题	生物催化在制药上的实际运用
技术	使用酵母（*Zygosaccharomyces rouxii*）还原酶活性合成抗惊厥候选药物
获奖者	Lily 研究室
年份	1999

迄今为止记录的那些使用生物催化作用和微生物的获奖者代表占据了总体人数（17年88人）的25%。全细胞的使用，合成路径的改善，固定化的酶系统，生物催化剂的改进等都是绿色化学技术发展的一部分，它是化学、微生物学、化学工程、蛋白质工程和生物技术等的交叉点。

以下是特别挑选的生物催化作用的例子，其与纺织业、制药业和新兴生物经济息息相关。

2.4.1 生物洗练：一个舒适的故事

最舒适的棉织物几乎都是以纤维素为主，但它们的获得不得不以消耗环境为代价。染色前的预处理，传统方法是用氢氧化钠（达到5%）的洗涤液煮沸棉料。煮沸不仅消耗大量能量，碱处理棉料还要求大量的冲洗来中和，这样就消耗了大量的水。此外，经常接触腐蚀性的化学药品也导致纤维的破坏。生物洗练方法则是使用已开发的酶生物制剂（表2.4）来改善纺织品预处理工艺，使之更环保经济。

同属于果胶酶的果胶内切酶（EC 3.2.1.15）和果胶裂合酶（EC 4.2.2.2）可裂解来自于植物细胞壁的果胶杂多糖，已成为人们最喜爱的洗练液成分。这种洗练液不仅可以用于棉织物（Solbak et al.，2005），还用于清洗自然韧皮纤维（如麻和亚麻）（Ouajai and Shanks，2005；Akin et al.，2004，2007）的处理制剂。果胶裂合酶通过β消除反应催化断开果胶主链多聚半乳糖醛酸内部的α-1，4-糖苷键，并在非还原末端产生半乳糖醛酸酯。此酶促反应在pH值为8的室温下进行。

东加勒比国家研究所（生态学研究），一个欧洲调查咨询研究所，已经对Scourzyme的环境影响和益处进行了研究。比较了来源于芽孢杆菌中重组形成的果胶裂合酶与用化学试剂处理（与BioPrepration™是同义词）（表2.5）。可以看出酶催化工艺在各方面都有很明显的降低，例如使用酶工艺降低了30%的能量消耗。诺维信公司进一步报道出10 L Socurzyme（301 L）用于纺织厂带来的不同：这种方法确实减少了20 000 L用于漂洗的热水；每吨纱大约减少1 000 kg二氧化碳的排放。Nielsen等（2009）全面描述了对环境的影响。

表2.5 酶和化学工艺对环境的影响

影响类别	化学进程	酶学进程	变化
能源消耗/MJ（LHV）	5 450	3 810	1 640
全球变暖/kg（CO_2 eqv.）	382	281	101
酸化/g（SO_2 eqv.）	1 100	792	308
养分富集/g（PO_4 eqv.）	3 480	1 200	2 280
夏季烟雾/g（乙烯 eqv.）	331	238	93

对于以上研究工作，作者实验室已经表征了新的细菌果胶酶和通过定向进化获得的在大肠杆菌中表达具有耐热性特点的果胶裂合酶（A31G/R236F 突变体），其耐热性和活性都超过了野生酶，开始主要被用于麻和亚麻的自然纤维漂洗预处理（Xiao et al.，2008a，2008b；patent）。有趣的是，惊奇的发现仅一个氨基酸的替换（R236F，精氨酸变为苯丙氨酸）使蛋白溶解温度（T_m）明显增加 6℃。通常情况下这种变化不会超过 1～2℃（Kuchner and Arnold 1997）。这种双突变果胶裂合酶现在可以进行 1 000 L 发酵规模，每批次达到 1 kg 纯化后的酶，可处理超过 300 kg 的纤维素，如果被重复使用还可以处理更多。Solbakd 等（2005）优化了一种果胶裂合酶突变体（CO14），含有 8 个氨基酸突变（A118H，T190L，A197G，S208K，S263K，N275Y，Y309W 和 S312V），并且当在 50℃保持其同样的比活性，T_m 值比野生酶高 16℃。在这种情况下突变酶的最佳温度是 70℃，比普通酶高 20℃。生物洗练的效果比使用化学洗练明显要好。诺维信公司除了作为世界上主要的酶生产厂家，生产酶供制造厂和洗炼厂应用，在北京和马来西亚吉隆坡还有两个具有设备齐全的实验室，分别开展多样的应用研究并帮助消费者做到同样的应用（Nielsen et al.，2009）。

上述果胶水解酶已可以从植物相关细菌像黄单胞菌或者由含有腐烂植物材料的样品土壤中获得。Truong 等（2001）第一次从嗜冷水海洋细菌（假交替单胞菌属 ANT/505）得到了两个果胶裂合酶编码基因并进行了克隆。但是至今没有看到任何海洋来源的果胶水解酶具有可以成为现在陆地来源的代用品明显的优势。

2.4.2 1，3-丙二醇（PDO）：一个代谢工程故事

这种含有两个羟基，无色无味的液体事实上充满了满腹色彩，是利用代谢工程绿色生产化学物质平台的第一个例子。（图 2.5）这个 C3 上的二醇有许多可以开展的合成反应，包括聚酯、聚醚和聚氨酯等缩聚反应。最为人们所知的以丙二醇为基础单位的聚酯是聚对苯二甲酸丙二醇酯（PTT），商品名称为 Sorona 3GT，由二甲基对苯二酸酯或对苯二甲酸制成（Kurian，2005）。

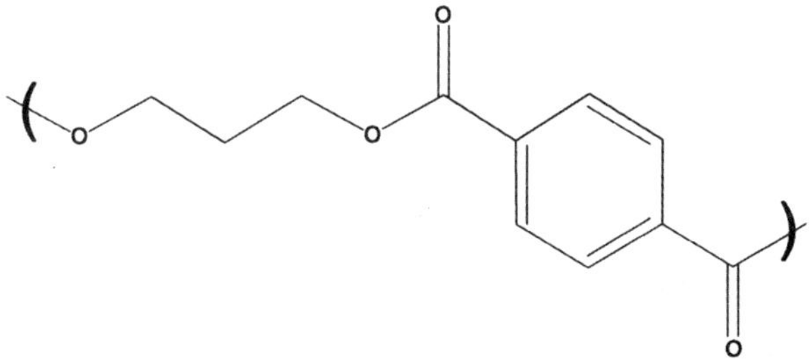

图 2.5　聚对苯二甲酸丙二醇酯或 3GT 聚体的结构

皇家荷兰壳牌公司市场化了 PTT，当做 Corterra 聚合物出售。很长时间丙二醇（PDO）

在胶黏剂、层压制品、洗涤剂和化妆品方面作为溶剂，提供一个长时间但无黏性的保湿效果（Zeng and Biebl，2002）。PDO/PTT 市场包括热塑性塑料、纺织品、地毯和室内装潢用品。

化学合成丙二醇

两个传统化学合成丙二醇的路线：使用乙烯氧化物或使用丙烯为原料进行的氧化催化产生丙烯醛而合成丙二醇（Zeng and Biebl，2002；Kurian，2005；Kraus，2008）。通过丙烯醛还可由甘油在压缩热水中加入大量硫酸催化合成。另外，甘油本身使用异构有机金属催化剂可以还原为丙二醇和乙醇混合物（Kraus，2008）。不必说，许多反应条件（高压，金属催化剂的使用等）与绿色化学的概念不相符。

葡萄糖生物合成丙二醇

在 1881 年就记录了有关微生物合成丙二醇的方法，负责的科学家是 August Freund，他用甘油发酵混合培养 *Clostridium pasteurianum*（Biebl et al.，1999）。其他的甘油发酵菌有：*Klebsiella*、*Clostridium*、*Citrobacter*、*Enterobacter* 和 *Lactobacillus*。在前面的这些微生物中，二醇的生产浓度在 81～87 g/L（reviewed in Zeng and Sabra，2011）。随后出现了一种重组大肠杆菌 W2042（Nakamura et al.，2000），与前面不同的是它使用更经济的原料 D-葡萄糖，而且有高达 135 g/L 的生产浓度（Nakamura and Whited，2003）。在 10 L 发酵罐葡萄糖流加培养的条件下，丙二醇的生产率为 3.5 g/(L·h)，且有 51%的产量。

以菌种 K12 为基础的工程化大肠杆菌是 Dupont/Genencor 公司的贡献完成，他们的工作获得了 2003 年总统绿色化学挑战奖（表 2.4）。并在 2007 年与 Tate、Lyle 一起获得美国化学会英雄化学奖。1996 年构想设立的美国化学社会英雄奖化学项目着重标出工业化学科学的重点，通过成功的商业创新和产品来改善对人类的福利。

感兴趣的读者可以参考查阅 Nakamura 和 Whited（2003）在代谢工程生产丙二醇项目上的权威综述，另请参阅 Celinska（2010）。简单点说，来自 *Saccharomyces cerevisiae* 的具有把葡萄糖转化为甘油的基因可以插入宿主细胞。来自 *Klebsiella pneumoniae* 具有把甘油转化为 3-羟基丙醛功能的基因也能被整合（图 2.6）。另外某些有助于氧化还原反应平衡的大肠杆菌宿主基因得以保留，最终优化了产品的生产。

最后的结果是一个有氧过程，它生产的丙二醇要多于无氧的甘油发酵（Zeng and Sabra，2011；Vickers et al.，2012）。从环境的角度来说，生物合成丙二醇减少了 40%的能量消耗，与石油合成丙二醇相比减少了超过 40%的温室气体的排放（http://www.duponttateandlyle.com/life_cycle.php）。DuPont Tate 和 Lily 公司每年的丙二醇生产量大约 67 500 t（135 000 磅，Erickson et al.，2012）。分别用于个人护理和流体运用新产品如：Zemea 和 Susterra 已被上市出售。除去始终如一的付出，一个可以从生物丙二醇的发展过程获得很有价值的经验就是 7 年努力地去改善效价（图 2.7）。

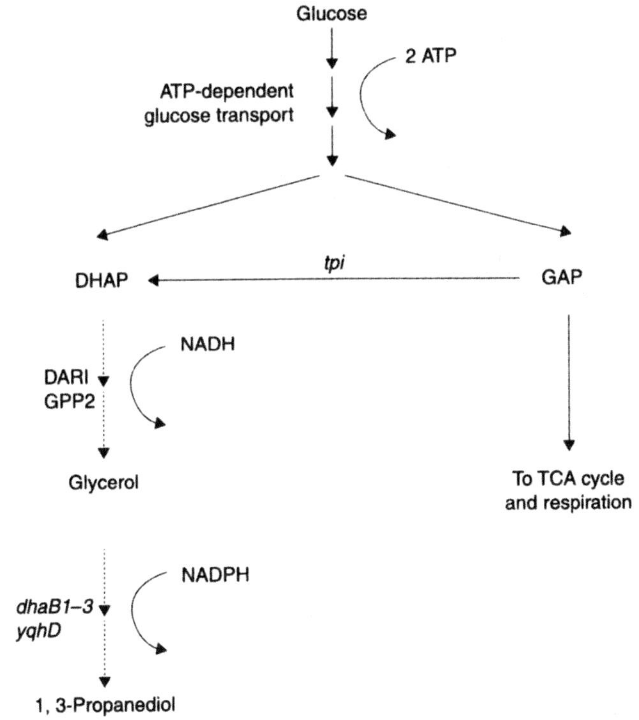

图 2.6　在大肠杆菌中基于 ATP 依赖性从葡萄糖合成 1，3-丙二醇的途径

基因名称：*tpi* 是磷酸丙糖异构酶基因；GAP 是 3-磷酸甘油醛；*dhaB*1-3，甘油脱水酶 B1、B2 和 B3 都来自 *Klebiella pneumoniae*；*yqhD* 是大肠杆菌内生的氧化还原酶；虚线代表生物基因编码的反应；DAR1 和 GPP2 分别是来自 *Saccharomyces cerevisiae* 的甘油 3-磷酸脱氢酶和甘油 3-磷酸磷酸酶 .

资料来源：Nakamura and Whited，2003.

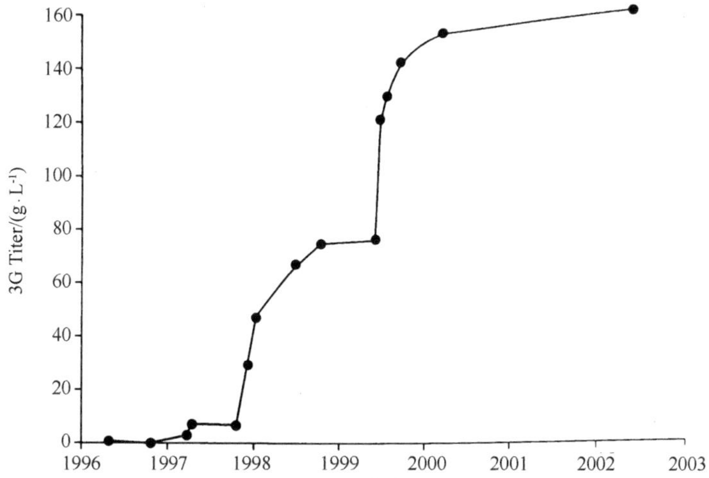

图 2.7　1，3-丙二醇的葡萄糖生产效价

资料来源：改编自 C. E. Nakamura and P. Soucaille. Engineering *E. coli* for the Production of 1，3-Propanediol，presented at Metabolic Engineering Ⅳ，Applied Systems Biology，2002，Castelvecchio，Pasoli，意大利 .

2.4.3 西格列汀：一个有关甜的故事

最近 Merck 公司成功制成的抗二型糖尿病药物西他列汀（商品名），是西格列汀片的活性成分，这是蛋白质工程合成手性胺的一个极端例子（Savile et al.，2010）。W-转氨酶（EC 2.6.1.x）又称为 ATA-117，之前仅知道该酶针对甲基酮和小的环形酮进行特异的 R-转氨作用。经过突变后可以容纳较大体积的底物，制备出西格列汀的前体（图 2.8）。除去 306 位缬氨酸替代了异亮氨酸，ATA-117（Savile et al.，2010）的 330 个氨基酸残基序列与来自 *Arthrobacter* sp. *KNK*168 的 R 对映选择转氨酶序列相同（Iwasaki et al.，2012）。根据这种酶在 340 nm 处的可见光吸收判断其是磷酸吡哆醛 PLP 依赖性的酶（Iwasaki et al.，2012）。

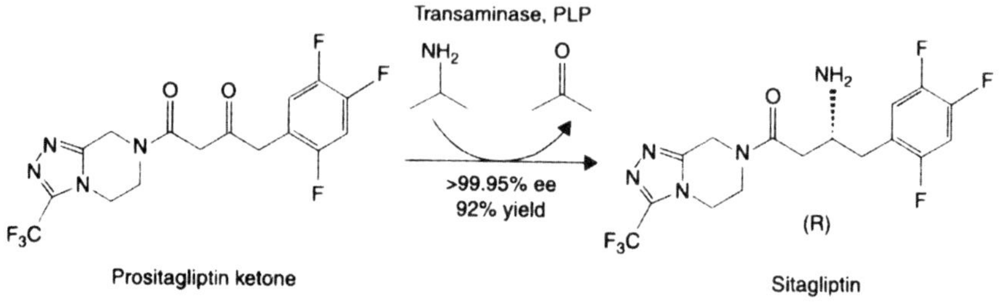

图 2.8 转氨酶催化酮底物合成抗糖尿病药物西格列汀

虽然单个氨基酸的突变体（ATA-117 很可能也来源于同样的 *Arthrobacter* sp. *KNK*168 菌种（Savile et al.，2010）），最终选取的 ATA-117 编码基因是经过大约 11 轮严格的进化筛选，包含有 27 个突变体的催化剂，它现在具有能够把 200 g/L 的西格列汀前体以超过 99.95%的转化率转化成为西格列汀的能力。最终确定的生物催化剂 ATA-117-11 不仅具有较大的底物范围，而且可以承受高浓度（1 mol/L）的异丙胺（为反应提供胺）和有机溶剂如：DMSO（50%）和丙酮。同时提高了对反应溶剂的温度（45℃）和 pH 值（7.5~8.5）的耐受性。简而言之，设计出的这种生物催化剂不仅能够催化所需要的反应，而且可以符合生产制造工艺的条件。

与当前使用的涉及基于以铑为基础的手性催化剂在高压下（250 psi）进行的烯胺不对称加氢反应进行的化学合成西格列汀的方法相比，基于生物催化的路线发现具有以下优势：整体收益率提高 10%~13%，生产效率（$kg \cdot L^{-1} \cdot d^{-1}$）提高 53%，整体废弃物降低 19%，避免了所有重金属和降低了整体工艺制备的花费。在生物催化工艺中虽然需要多功能的生物反应器，但是避免了使用专业的高压加氢反应装备（Savile et al.，2010）。

就像丙烯酰胺的故事一样，制造合成西格列汀同样是协作力量的一个例证。在之后的例子中有多个公司的合作，Codexis 公司提供的定向进化和 in silico 的计算设计，与 Merck 的化学合成处理工艺达到了一个双赢的境地。更多这样的例子将会继续出现。

最后 ATA-117-11 和（R）-AT 的存在的一个重要生化性质秘密是：后者被报道是以同源四聚体（Iwasaki et al.，2012）结构形式发挥活性，而前者尽管在凝胶过滤色谱中有一个小峰值显示是四聚物，但仍然推测其是一个二聚体结构（Savile et al.，2010）。此外，有趣的是所有蛋白质工程的工作和 in silico 计算分析等都是基于模拟的 ATA-117 的结构：该模拟的结构是基于 3 个模板，其对应的蛋白质数据库信息是：3DAA，来自 *Bacillus*. sp. YM-1；1 WRV：D-氨基酸转氨酶，来自 *T. thermophilus*；1AGE：支链氨基酸转氨酶，来自大肠杆菌的支链氨基酸转氨酶。它们的氨基酸序列与 ATA-117 的相似度分别为 28%、27%和 25%。假设这是我们已知其真实的三维结构。

2.4.4 辛伐他汀的生产：一个（LovD）的故事

他汀类药物是 3-羟基-3-甲基戊二酸单酰辅酶 A（HMG-CoA）还原酶的抑制剂，它是胆固醇合成中的调控和限制酶（Tobert 2003；Barrios-Gonzalez and Miranda，2010）。这些化合物能够减少"坏胆固醇"（低密度脂蛋白），现已经成为心血管疾病有效的预防药物。辛伐他汀是洛伐他汀（由 *Aspergillus terreus* 菌种自然产生的一种酮化合物）的一种半合成衍生物，前者能够更有效地治疗血胆固醇过高症（Tobert，2003）。从化学上，辛伐他汀在洛伐他汀侧链（图 2.9）2 号碳（C2）上含一个额外甲基。这两种药物由 Merck 生产商品名分别为 Zocor®和 Mevacor®。

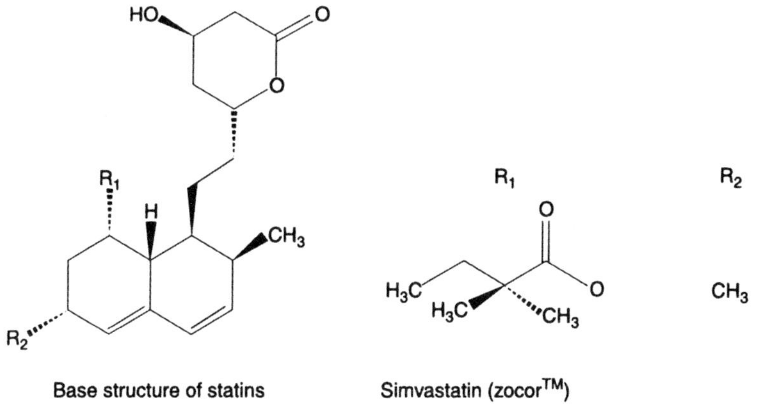

图 2.9　天然他汀类药物结构和它们的衍生物辛伐他汀

资料来源：Barrios-Gonzalez and Miranda，2010.

在 2007 年，Yi Tang 教授和他在 UCLA 的团队已经发展出了水解生产洛伐他汀，Monacolin J，进行全细胞生物催化方法合成辛伐他汀（图 2.10）（Xie and Tang，2007）。这种生物催化剂是来自洛伐他汀生物合成基因簇的一个含有 413 个氨基酸的酰基转移酶（LovD）（Xie et al.，2006），它是许多 α/β 折叠的水解酶超家族中的一员。这种蛋白有选择性地从各种酰基硫代酸酯上，将2-甲基丁酰侧链转移到 Monacolin J 的乙醇基 C8 上（钠盐或铵盐）。与其他硫酯相比，二甲基丁酰基-S-丙酸甲酯（DMB-S-MMP）是一个特别的动力学最优的酰基供体，而且对 Monacolin J 无明显的底物抑制作用（Xie and Tang，2007）。

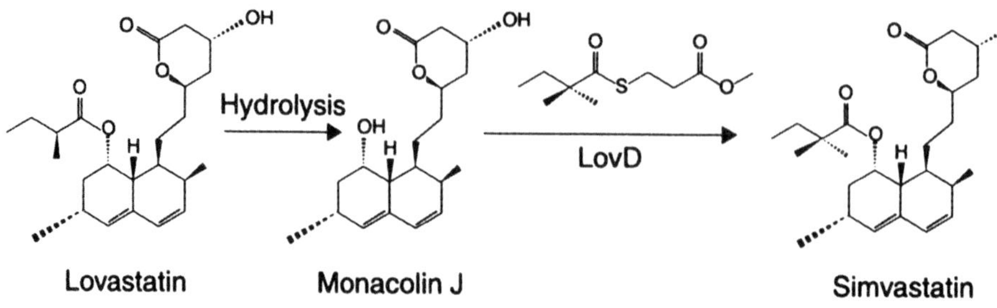

图 2.10　跨脂酶（LovD）催化 DMB-S-MMP 作为酰基供体合成辛伐他汀的最后步骤

就热稳定性而言，LovD 是一个未达到最优的生物催化剂。因此，开展了基于定向进化的蛋白质工程，获得了具有高催化效率和 T_m 值增加 9℃的热稳定性的突变体（Gao et al.，2009）。在最后的工作中，Codexis 公司从 UCLA 获得许可，开展了针对 LovD 的适应商业制造的优化工作。在体外进行 9 种交互影响的进化筛选后，在 216 种克隆文库中筛选了 61 779 个突变体，其中获得了一种有几千倍活力增加的 LovD 突变体（http：//www.epa.gov/opptintr/greenchemistry/pubs/pgcc/winners/gspa12.html）。同时随着热稳定性的提高和对产物抑制耐受度的改善，这种新工艺能在高浓度底物（75 g/L monacolin J）下达到超过 97%的辛伐他汀收率。酰基供体的使用量和为了提取与产物分离而使用的溶剂的量都缩减到最少。辛伐他汀合成中仅有的副产物 3-甲基巯基丙酸是可循环的。整体上这种新技术不仅改善了生产成本，而且减少了有害化学物质的使用，包括：叔丁基二甲基氯硅烷、碘甲烷、正丁基锂。已有超过 10 t 的辛伐他汀使用这种工艺生产出来。另外化学合成路线中，洛伐他汀中特定甲基的引入需要多步骤化学合成。其中涉及的保护步骤是利用选择性甲硅烷基化的，二甲基丁酰氯的酯化和去保护步骤等（Xie and Tang，2007）。通常，化学工艺合成由于保护和去保护作用而资源利用率不高，需要大量有毒化学物质并且整体收益率达不到 70%。

2.5　生物催化的挑战

俗话说：“这并不像初看到的那样简单。”利用生物催化剂，可以完成必要的生物转化是一件事——事实上不是一件小事。满足工艺要求和扩大反应是一场全新的比赛。也就是说，我们有必要的生物催化剂来做正确的化学过程吗？答案当然是看情况而定。

对于精细化学产品的生产，大部分的工业生物转化是由水解酶催化（44%），其次是氧化还原酶催化剂（30%），氧化还原生物催化剂包括用于不对称酮还原的 ketoreductases

(Straathof et al., 2002)。在我们分析的134种工业处理工艺中，裂解酶和转化酶一起的使用量与氧化还原酶是一样的（Straathof et al., 2002）。Kaul 和 Asano（2012）着重强调了硫酸化作用在化学制药行业的重要性，和需要利用酶的催化硫酸化作用来避免能源和化学物质的大量使用的过程，以及必要的保护和去保护过程。不用说，酶如磺酸基转移酶，尽管他们需要比如 *p*-对硝基苯硫酸盐或3′-磷酸腺苷 5′-磷酰硫酸（PAPS）作为供体，但是依然有望控制精确反应的选择性，而且能在温和条件下利用一个步骤完成那些需要复杂的化学反应完成的过程。近期，Paul 等（2012）的综述上了包括芳香基族磺酸基转移酶这个家族的酶。

甲基化反应代表了另一个挑战（Kaul and Asano, 2012）。通过化学手段，包括利用甲基卤化物和硫酸盐作为亲电甲基供体进行区域选择性往往是很困难的。尽管甲基转移酶是合适的催化剂，但是他们需要的辅因子 S-腺苷甲硫氨酸（俗称 SAM，它也是价格昂贵）已经阻碍了商业应用。甲基转移酶在 DNA 限制-修饰系统中的应用是最著名的，但是 O-甲基转移酶基因是在大环内酯类抗生素的生物合成途径被发现的，如红霉素和麻西那霉素（一种新的碱性大环内酯类抗生素）(Weber et al., 1989; Li et al., 2009)。

以上可以被视为新兴的用于化学合成的生物催化剂。Pollard 和 Woodley（2006）提出了几个包括“建立”、“扩大”和“新兴”的几个化学术语去指那些现成的生物催化剂，一个小的列表的扩展和简单的生物催化剂将对制药行业产生重要影响。已经建立的化学反应包括脂肪酶催化的水解，酯类的外消旋，酮还原。扩展的化学反应指氰基还原和氰醇合成。新兴的化学反应包括：转氨酶的转氨作用；烯酮还原酶还原烯醇；细胞色素 P450 (di) 羟基化；Baeyer-Villiger 的单加氧化作用，氢过氧化物单加氧化；环氧化物水解酶水解作用；加卤酶或细胞色素 P450 的环氧化作用；加卤酶催化卤代醇的形成（Pollard and Woodley 2006; and references therein）。

如前在2.4.3中对西格列汀故事的描述中，转氨作用的进展显然是很有说服力的（Savile et al., 2010）。Gao 等（2012）描述了一个最近的新发现，乳酸菌 NADH 依赖的烯酮还原酶有合成99%转换效率和98% 过量非对映体的（2R，5R）-二氢黄蒿萜酮的能力。利用真菌来源的羰基还原酶，完成将（R）-香芹酮完成转化达到大于99% 的转化效率，且是唯一产品，这是（1S，2R，5R）-二氢香芹醇合成的先决条件（Xi et al., 2012）。在细胞色素 P450 方面，已经在方法或手段上取得了稳步的进展，这允许天然或者合成化合物进行区部和立体选择氧化碳氢键的激活，以合成相应的醇类化合物。使用睾酮作为底物，Kille 等（2011）报道了通过迭代饱和突变获得的 P450（BM3）突变体（F87A）的进展，这个突变体具有立体化学选择性地将氧化睾酮转化成 2β- 和 15β-醇。用不同的方法，Larsen 等（2011）建立了化学辅助效用的概念验证，例如可可碱来控制 P450 细胞色素 CYP3A4 催化羟基化的选择性。末端烯烃的拆分环氧化作用也被证明，但转化率很低，收益不佳。尽管这50年来 P450 的研究是一个不断扩大的活跃研究领域，但在这里仅引用了几个最近的综述（O'Reilly et al., 2011; Urlacher and Girhard, 2012; Sakaki, 2012; Whitehouse et al. , 2012)。

氢过氧化物的单加氧酶（BVMO）系统，由于它的区域性、立体性和对映选择性（de Gonzalo et al. , 2010; Leisch et al. , 2011）比使用强氧化剂如内酯或酯的产物——高酸的氢过氧化物化学反应更闻名（ten Brink et al. , 2004）。最近在医药领域，它的应用为抗酸药物埃索美拉唑（抗酸剂）的合成开拓了一条道路，这种药物是一种质子泵抑制剂，包括其外消旋体奥美拉唑的磺化氧化作用，这最初由 *Acinetobacter* sp. NCIMB 9871 中进化而来的环己酮单加氧酶（CHMO）完成（Olbe et al. , 2003; Bong et al. , 2011）。在辅因子再生系统如葡萄糖 6-磷酸盐脱氢酶存在的情况下，工程化的 S-对映选择性 CHMO 加上其他改进的特性共同完成了这个反应（Bong et al. , 2011）。顺便，在辅因子再生方面已经取得了很大进展，一些已不再成为一个重要问题；然而，如上所述这些涉及 SAM 仍然是一个挑战（Zhao and van der Dork, 2003; Weckbecker et al. , 2010）。

由于将很快达到青少年状态，在室温及离子液态环境下的生物催化的应用预计将得到发展——争取更好。离子液态环境由于没有挥发性（接近零蒸汽压）、高热稳定性、耐燃性以及对于极性溶剂混溶的可调节性等而被视为绿色溶剂，已经将更高的选择性、更快的反应速率和提高酶稳定性赋予给了酶（representative reviews: Park and Kazlauskas, 2003; van Rantwijk and Sheldom, 2007）。近日，在离子液态环境中，首次单加氧酶催化反应（Ammoeng™ 102，在一个四价盐与一个 *N*，*N′*-alkylimidazolium methylsulfate，[bmim] $MeSO_4$）被报道在由耐热性的苯基丙酮单氧酶催化的外消旋苄基酮的氧化反应中得到提高的 *E* 值（Rodriguez et al. , 2010）。重要的是，通过离子液体的使用，底物的上样量至少翻了 1 倍（在水介质中是 50 mmol/L。而在 10% Ammoeng™ 102 中是 120 mmol/L）。尽管如此，在工厂中，除了其他挑战，离子液体相对较高的成本（超出在实验室规模中传统溶剂的 5~20 倍；Tadesse and Luque, 2011）是一个问题，将产品从高黏度的溶剂中分离就是其中一个例子。因此至今，在生物催化中离子液体的应用相当有限（Quijano et al. , 2010）。除了一些水解和氧化还原酶反应，在其他领域这个相对较新的系统的广泛应用将能显示出更多的新特性。有趣的是，在生物质处理中，例如通过木质纤维材料的预处理来提供生物燃料生产可利用发酵糖的过程中，除了选择性提取木质素作为可再生资源外，离子液体也已经取得了显著进展（Mora-Pale et al. , 2011; Tadesse and Luque, 2011; Vancov et al. , 2012）。此外，工业的规模化应用要求具有良好的经济效益和流程的优化，这需要进一步投入。

具有高活性、保质期长、在有机溶剂中特别稳定属性的酶，天然的或者非天然的，其在任意一个生物催化反应中应用还有很长的路要走。固定化酶能克服操作和稳定储存的问题，同时解决高价格商品的回收和再利用问题。交联酶聚合（CLEAs）技术，是最初交联酶晶体（CLECs）概念的变体，已成为固定化酶的一个有趣的选择（Sheldon, 2011a, b）。除了涉及结晶，该过程首先是酶在水缓冲液中的简单沉淀（硫酸铵或叔丁醇），其次是与双官能团的试剂如戊二醛或作为硅前体的硅酸有机酯共聚体交联。已经具有良好的可回收性和长期储存稳定性（直到4个月）腈水合酶的报道。成功的“cleations”包括许多水解酶、氧化还原酶、裂解酶、转氨酶（Sheldon，2011a，b）。包含两个或两个以上的酶的交联酶聚合体已被成功地用于级联反应过程和磁稳定流化床中的新的磁化交联酶聚合（后者是一个建立的过程），两者都在发展。

显然，还有很多其他的挑战。除了单步反应，多组分的反应和过程都出现了上升的趋势（Wohlgemuth，2010）。Woodley 和同事回顾了针对这些多酶工程，在体内生物反应器的模式和工程设计指导原则的策略和发展（Santacoloma et al.，2011；Xue and Woodley，2012）。虽然不是一个新概念，但纳米生物催化是另一个发展（Kim et al.，2008）。对于新纳米材料（纳米纤维、纳米管、纳米粒子）的需求会推动这种技术（Illanes et al.，2012；Sathishkumar et al.，2012）。

因为生物催化还不是一个主流技术，其商业化应用的经济评估是不可避免的。对评价生物催化过程中催化剂的生产成本，Tufvesson 等（2011）最近提供了宝贵的指导方针。此外，对生产力的范围提出建议：例如，对大容量、低价格商品的化学物质，2 000~10 000 kg（产品）/kg（固定化酶），对低容量、高价值的医药产品，50~100 kg（产品）/kg（固定化酶）。

一般来说，酶的成本一直是一个棘手的问题，好消息是现在成本正在下降。与20世纪80年代的高价相比，在分子生物学实验室常用的限制性内切核酸酶 *Eco*RI 目前成本约每100 units 60美分。在生物质加工/生物燃料贸易方面，“明显而现实”的问题是处理过程中酶成本的降低。图2.11显示了在过去10年中，测算的利用预处理玉米秸秆生产每加仑乙醇的成本发生了根本性的改善（Dean et al.，2006）。在酶公司，这种改善并非没有投资（Genencor International and Novozymes）。改善酶的混合物以及通过基因改造或补充关键酶来提高酶耐受的工作温度是降低成本的主要因素。

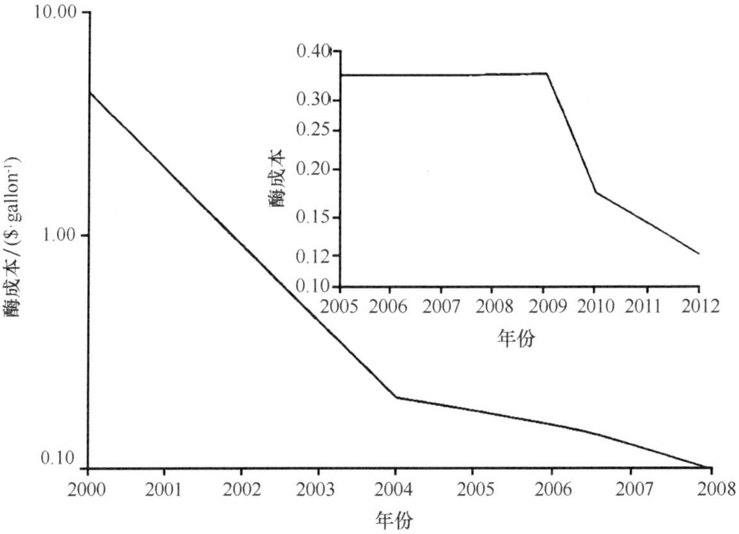

图2.11　纤维素酶预算成本的减少

资料来源：Dean et al.，2006；McMillan，2011.

2004 年初，据报道通过超过 20~30 倍成本的减少，使乙醇生产低于每加仑 0.50 美元的一个有效的成本（US DOE，2006）。2010 年的目标和实际酶的成本为每加仑乙醇 0.17 美元，基于酶水解产糖量 90%，2012 年的目标是每加仑降 5 美分到 0.12 美元（McMillan，2011）。继续努力有望进一步降低酶的成本到每加仑乙醇 0.10 美元或更少。另一方面，最近，一个学术性的技术-经济分析提出一个更高的成本——根据最高的糖转化量，以前有文献报道，基于真菌酶（多种纤维素酶）的生产成本是每加仑 0.68 美元，考虑到之前文献报道的糖的转化率和发酵产量，成本是每加仑 1.47 美元（Klein-Marcuschamer et al.，2012）。这意味着需要更多额外的努力来降低酶对生物燃料生产成本的贡献。但我们不要忘记，对于生态环境和工业持续发展，成本不是最根本的。环境储蓄和社会效益是至关重要的。

2.6 先天资质与后天培养

在这个讨论中，后天培养是指通过蛋白质工程的各个方面努力来改进酶的特性和性能。先天资质指的是，虽然有克隆或基因组整合在同源或异源系统中表达，但未改造过的酶的状态或编码基因。

近几年来，许多酶工程策略已经实施了，包括各种形式的定向进化技术（Stemmer 1994；Arnold，1998，2009；Reetz，2009），半理性的设计技术（Fox et al.，2007；Lutz，2010），最近开始的基于计算机的从头开始或创新蛋白质设计（Zanghellini et al.，2006）。这些都对酶工程领域有所贡献，包括对生物催化的机制的理解或者解析如何满足工业需求。读者可以阅读最近该领域权威性的综述（Romero and Arnold，2009；Lutz，2010；Strohmeir et al.，2011；Quin and Schmidt-Dannert，2011；Jochens et al.，2011；Dalby，2011；Reetz，2011；Lewis et al.，2011；Bornscheuer et al.，2012；Wang et al.，2012；Liszka et al.，2012）。

你得到你所选择的是模仿达尔文进化论的定向进化方法的关键或核心。最初这需要一个相对较大的不同变量选择的组合文库，但是这已经被半理性设计的方法大大改进了，这些方法包括 CASTing（组合活性部位饱和度测试）技术，迭代饱和突变的方法（Reetz and Carballeira，2007；Reetz et al.，2010a），这是一个所有的在整个酶结合口袋内相关氨基酸的饱和突变，该策略不同于在远离蛋白质的结合口袋的饱和突变。做像这样一个集中库有效地减少了用于筛查库的大小（例如 100~1 000 的 cast）。基于可从 X 射线晶体学数据中获得的 B-因子的蛋白质耐热改造，B-FIT 方法似乎是一个有效的策略（Reetz et al.，2010 b）。

顾名思义，半理性设计需要蛋白质结构的知识和使用计算预测算法；蛋白质序列-活性相关性（ProSAR）就是一个典型的例子，这种基于蛋白质序列活性关系分析的策略被有效地用于进化细菌的卤代醇脱卤素酶（Fox et al.，2007）。此外，一个高度挑战性的方法是酶的从头设计，这种方法需要一个反应过渡态的计算机建模 、量子力学模拟并构建蛋白质支架等。近期，Richter 等（2011）以磷酸丙酮异构酶（TIM）作为例子提供了辅助使用 Rosetta3 算法从头设计酶的教程。另外，到目前为止已经有 3 个主要的从头设计酶的例子：retro-aldolase（Jiang et al.，2008）、Kemp 果胶酶（Rothlisberger et al.，2008）和 Diels-Alderase（Siegel et al.，2010）。Retro-aldolase 在非天然底物中（4-羟基-4-（6-甲氧基-2-萘基）-2-丁酮）催化碳碳键的断裂；Kemp 果胶酶参与一种来自碳中质子转移的模式反应，通过切割氮氧键导致氰基苯酚产物形成；Diels-Alderase 形成两个碳碳键，例如二烯或亲二烯体（取代烯烃）中环己烯的形成。据推测，这些里面没有已知的天然酶。然而，有一个模板，例如新的 Kemp 果胶酶 KE59 已经通过定向进化获得，表现出了提高约 2 000 倍的催化效率（Khersonsky et al.，2012）。

除了蛋白质工程，生物勘探的宏基因组学或简单地从自然环境中分离和表征新菌株对新生物催化剂的发现也是有利的策略。毫不例外，海洋环境已经被几个团队很好地综述过（Leary et al.，2009；Trincone 2010，2011，2012；Imhoff et al.，2011；Kennedy et al.，2011；Dionisi et al.，2012；Villarreal-Chiu et al.，2012；Zhang and Kim，2012）。南极洲东部的 Nella 峡湾的海洋沉积物被很好地分析了，发现含有相对较高生物多样性的可培养细菌，包括 α-变形菌门、γ-变形菌门、拟杆菌和可产生低温活性水解酶放线菌：脂酶、几丁质酶等（Yu et al.，2011）。不仅细菌是酶的来源，单细胞浮游生物如可产生碳酸钙垢的（颗石藻）海洋球石藻，被发现也是一个水解酶的丰富来源，如酸/碱性磷酸二酯酶和磷酸单酯酶（Reid et al.，2011）。从海洋宏基因组 DNA 中筛选了 PhnX 和 PhnY，前者编码 α-酮戊二酸 / Fe（Ⅱ）依赖的双加氧酶，后者编码 Fe（Ⅱ）依赖的水解酶，这两种酶被发现按顺序作用的机制去断裂碳-磷键，这跟之前报道的机制是完全不同的（McSorley et al.，2012）。2-氨基甲基磷酸被裂解成无机磷酸盐和甘氨酸，断裂后 C-P 键重新生成了 O-P 键。这充分说明酶系统获得无机磷酸盐的独特能力，一种来源于广泛分布在海洋环境中化学稳定的磷酸酯，也是一种基本的细胞组成成分。有趣的是，最近在海洋细菌磷酸酯代谢相关基因的计算机分析中提示它们既是重要的 C-P 生产者，也是磷酸酯的消费者（Villarreal Chiu et al.，2012）。

它们是否合成萜类化合物（Guella et al.，2010；Cane and Ikeda，2011）参与或其他天然产品（Imhoff et al.，2011），海洋环境肯定是含有许多未开发的制备有用化合物的酶。第一个来源海洋的“BVMO”和一个偏好 NADH 作为辅助因子的酶被报道是从 *Stenotophomonas maltophila* 菌的 PML168 菌株中获得的，该菌株来自英格兰的泥盆纪潮间带岩石的表面（Willetts et al.，2012）。用 3-乙酰吲哚到吲哚乙酸乙酯的显色转换检测酶活性。按照顺序，这个假定的独特的 BVMO 与其说是典型的 1 型和 2 型 BVMO 倒不如说是一种黄素单加氧酶（FMO）（Leisch et al.，2011）。有趣的是，Jensen 等（2012）独立研究报道了相同的蛋白质。它是由具有单一组成的 NADH 依赖的具有硫醚氧化能力的黄素单加氧酶。在氢过氧化物氧化反应中，该酶不具有转化标准底物的活性，如环己酮、环戊酮和苯乙酮，但能够氧化紧密的/融合类的环丁酮（双环醇（3.2.0）hep-2-en-6-one）。总之，这种新酶几乎不具备 BVMO 的性质，但是更像是日益增长的 FMO 家族的成员（van Berkel et al.，2006）。顺便说一下，什么是海洋细菌和与之相对的分离也需要解决，像之前讨论的关于陆地或海洋鞘氨醇单胞菌（Cavicchioli et al.，1999）或已知寡养单胞菌属（Ryan et al.，2009）。

除了我们已经找到的，大自然是否为我们提供了其他生物催化剂？答案可能是非常肯定的。通过严谨的对微生物基因组序列数据库的分析，已经从高温 *Thermanaerobacter tengcongensis* 中获得了一个有吸引力的耐热性的拆分酯酶，它优于在实验中逐渐形成的来源于常温微生物的酶，T_m 相对于嗜常温的酶升高 10~12℃（Grosse et al.，2010）。在从高温 *Acidothermus cellulolyticus* 11 B 中发现了一个高度耐热性的醛醇氧化酶（名为 HotAldO）基因组后，Winter 等（2012）提出了“热否？”的问题。它的 T_m 是 84℃，其半衰期在 75℃ 是接近 2 h。逆向进化设计耐热性的氧化酶去获得一个常温的醛醇氧化酶的工作是失败的。不正确的蛋白质工程的问题被提出。不管怎样，在这两种情况下，当然也不是非常特殊的例子，嗜热微生物不仅具有超越它们常温微生物酶的生物催化的特点，并且清楚地展示出了它可以丰富蛋白质工程的作用。

采取“纯天然”的路线，具有盐耐受性、嗜压性或冷适应性的这些海洋酶可能比通过改造获得更容易。通过自然进化和对极端环境的适应，大自然已经提供了解决相关的工业应用问题的方法。了解热稳定性的机制和任何其他蛋白质性能是很重要的事，找到解决工业难题最有利的解决方案是另一回事。选择就像是一个岔路口，解决方案是获得它。

曾经的一篇社论标题：自然教育我们，但可以是更好（Eltis and Withers，2008）。这或许是正确的，从一个不知道的狭隘的角度来看，在普通实验室条件下，我们仅可培养 1% 的细菌（Schloss and Handelsman，2005；Whitman et al.，1998）。对于所有意图和目的，我们的底线是了解自然规则。我们认为我们知道会更好。

2.7 结论

从发现到表征，从分子机制到蛋白质设计，从有机合成到工业应用，生物催化是一个科学领域“the tree that binds and bears（it）”（指开放引用）。在生物化学、微生物学、分子生物学、有机化学的十字路口上开展工作，即使不是最佳的，生物催化至少是一个可行的、跨学科和可持续的技术。工业和学术合作伙伴之间日益增加的相互交叉再怎么强调都不为过，最近辛伐他汀绿色化学奖的故事说明了这一点。在蛋白质工程的努力和从头设计变得越来越激烈的时候，生物催化据说将掀起第三次浪潮（Bornscheuer et al.，2012）。另一方面，大自然正在耐心地等待，如同针对大多数沉默微生物开始开展的基因组分析和相关的包括宏基因组在内的相关方法，如宏蛋白组学。（图 2.4；Simon and Daniel，2011）。选择性的对基因组开发可能是一个好机会，如多种新的和没有表征过的基因序列借助以千计数的基因组以指数形式增加。最后，我们重申需要分离和表征环境中新的微生物，因为我们不仅想要丰富我们知道知之甚少的微生物多样性和群落结构，而且也因为新的微生物可能具有新的有趣的代谢或生物催化属性。最近两个新分离的海洋生物，*Maricurvus nonylphenolicus* 和 *Tropicibacter phthalicus* 降解环境污染物的能力，比原来表征过的更有用（Iwaki et al.，2012a，b）。

在结论中，Louis Pasteur 也说：无限小的角色可能作用是无限大的。不用说，估计 3.67×10^{30}微生物（Whitman et al.，1998）的海洋环境提供了一个巨大的基因库和探索与创新的无限可能性。优良生物催化种类的浪潮即将到来。

感谢

我们感谢 Hélène Bergeron 在准备某些图上的帮助。P. C. K. L. 感谢 FQRNT 在中心绿色化学和催化作用部分的支持。

参考文献

Akin, D. E., Condon, B., Sohn, M., Foulk, J. A., Dodd, R. B. and Rigsby, L. L. (2007) Optimization for enzyme-retting of flax with pectate lyase. *Indust. Crops Prod.* 25, 136-146.

Akin, D. E., Henriksson, G., Evans, J. D., Asamsen, A. P. S., Foulk, J. A. and Dodd, R. B. (2004) Progress in enzyme-retting of flax. *J. Natural Fibers.* 1, 21-47.

Anastas, P. T. and Warner, J. C. eds. (1998) Green Chemisty, Theory and Practice. Oxford University Press.

Anastas, P. T. and Zimmerman, J. A. (2003) Design through the 12 principles of green engineering. *Environ. Sci. Tech.* 37, 95A-101A.

Arnold, F. H. (1998) Design by directed evolution. *Acc. Chem. Res.* 31, 125-131.

Arnold, F. H. (2009) How protein adapt. Lessons from directed evolution. Cold Spring Harbor Lab Symposium on Quantitative Biology, *Vol. LXXIV*, 41-46.

Asfaw, N., Chebide, Y., Ejigu, A., Hurisso, B. B., Licence, et al., (2011) The 13 principles of green chemistry and engineering for a greener Africa. *Green Chem.* 13, 1059-1060.

Asano, Y., Yasuda, T., Tani, Y. and Yamada, H. (1982) A new enzymatic method of acrylamide production. *Agric. Biol. Chem.* 46, 1183-1189.

Barrios-Gonzalez, J. and Miranda R. U. (2010) Biotechnological production and applications of statins. Appl. *Microbiol. Biotechnol.* 85, 869-883.

Biebl, H., Menzel, K., Zeng, A.-P. and Deckwer, W.-D. (1999) Microbial production of 1, 3-propanediol. *Appl. Microbiol. Biotechnol.* 52, 289-297.

Bong, Y. K., Clay, M. D., Collier, S. J., Mijts, B., Vogel, M. et al., (2011) Synthesis of prazole compounds. *US patent WO* 2011071982.

Bornscheuer, U. T., Huisman, G. W., Kazlauskas, R. J., Lutz, S., Moore, J. C. and Robins, K. (2012) Engineering the third wave of biocatalysis. *Nature* 485, 185-194.

Brandao, P. F. B. and Bull, A. T. (2003) Nitrile hydrolysing activities of deep-sea and terrestrial mycolate actinomycetes. *Antonie van Leeuv.* 84, 89-98.

Briggs, G. E. and Haldane, J. B. S. (1925) A note on the kinetics of enzyme action. *Biochem. J.* 19, 338-339.

Cane, D. E. and Ikeda, H. (2012) Exploration and mining of the bacterial terpenome. *Acc. Chem. Res.* 45, 463-472.

Cavicchioli, R., Fegatella, F., Ostrowski, M., Eguchi, M. and Gottschal, J. (1999) Sphingomonads from marine environments. *J. lnd. Microbiol. Biotechnol.* 23, 268-272.

Celinska, E. (2010) Debottlenecking the 1, 3-propanediol pathway by metabolic engineering. *Biotech. Adv.* 28, 519-530.

Chen, C.-S., Fujimoto, Y., Girdaukas, G. and Sih, C. J. (1982) Quantitative analyses of biochemical kinetic resolutions of enantiomers. *J. Am. Chem. Soc.* 104, 7294-7299.

Claus, A., Carle, R. and Schieber A. (2008) Acrylamide in cereal products : A review. *J. Cereal Sci.* 47, 118-133.

Dalby, P. A. (2011) Strategy and success for the directed evolution of enzymes. *Cwrr. Opin. Struct. Biol.* 21, 473-480.

de Gonzalo, G., Mihovilovic, M. D. and Fraaije, W. M. (2010) Recent developments in the applications of Baeyer Villiger monooxygenases as biocatalysts. *Chembiochem.* 11, 2208-2231.

Dean, A. M., Shiau, A. K. and Koshland Jr., D. E. (1996) Determinants of performance in the isocitrate dehydrogenase of Escherichia coli. *Prot. Sci.* 5, 341-347.

Dean, B., Doege, T., Valle, F. and Chotani, G. (2006) Development of biorefineries - technical and economic considerations. In Biorefineries-Industrial Processes and Products, Status Quo and Future Directions, Vol. 1. Kamm, B., Gruber, P. R. and Kamm, M. (eds). *Wiley-VCH*, 67-81.

Dionisi, H. M., Lozada, M. and Olivera, N. L. (2012) Bioprospection of marine microorganisms: biotechnological applications and methods. *Rev. Arg. Microbiol.* 44, 49-60.

DOE/SC-0095 (2006) Breaking the biological barriers to cellulosic ethanol——A joint research agenda.

Eltis, L. D. and Withers, S. G. (2008) Nature teaches but can be bettered. *Curr. Opin. Chem. Biol.* 12, 115-117.

Erickson, B., Nelson, J. E. and Winters, P. (2012) Perspective on opportunities in industrial biotechnology. *Biotech. J.* 7, 176-185.

Faber, K. (1995) Biotransformations in Organic Chemistry——A Textbook. Springer-Verlag, Berlin.

Fersht, A. R. (1984) Basis of biological specificity. *Trends Biochem. Sci.* 9, 145-147.

Foerstner, K. U., Doerks, T., Muller, J., Raes, J. and Bork, P. (2008) A nitrile hydratase in the eukaryote Monosiga brevicollis. *PLos One* 3, e3976.

Fox, R. J., Davis, S. C., Mundorff, E. C., Newman, L. M., Gavrilovic, V. et al., (2007) Improving catalytic function by ProSAR-driven enzyme evolution. *Nat. Biotechnol.* 15, 2785-2794.

Gardossi, L., Poulsen, P. B., Ballesteros, A., Hult, K., Svedas, V. K. et al., (2010) Guidelines for reporting of biocatalytic reactions. *Trends Biotechnol.* 28, 171-180.

Gao, X., Ren, J., Wu, Q. and Zhu, D. (2012) Biochemical characterization and substrate profiling of a new NADH-dependent enoate reductase from Lactobacillus casei. *Enzyme Microbial Technol. SI*, 26-34.

Gao, X., Xie, X., Pashkov, I., Sawaya, M. R., Laidman, J. et al., (2009) Directed evolution and structural characterization of a simvastatin synthase. *Chem. Biol.* 16, 1064-1074.

Goddard, J. -P. and Reymond, J. -L. (2004a) Enzyme assays for high-throughput screening. *Curr. Opin. Biotechnol.* 15, 314-322.

Goddard, J. -P. and Reymond, J. -L. (2004b) Recent advances in enzyme assays. *Trends Biotechnol.* 22, 363-370.

Grosse, S., Imura, A., Boyd, J., Wang, S., Kubota, K. et al., (2010) Nature versus nurture in two highly enantioselective esterases from *Bacillus* cereus and Thermoanaerobacter tengcongensis. *Mol. Biotechnol.* 3, 65-73.

Guella, G., Skropeta, D., Di Giuseppe, G. and Dini, F. (2010) Structures, biological activities and phylogenetic relationships of terpenoids from marine ciliates of the genus Euplotes. *Mar. Drugs.* 8, 2080-2116.

Hudlicky, T. (2011) Introduction to enymes in synthesis. *Chem. Rev.* 111, 3995-3997.

Illanes, A. (1999) Stability of biocatalysts. *Elect. J. Biotech.* 2, 15-30.

Illanes, A., Altamirano, C. and Zuniga, M. E. (1996) Thermal inactivation of immobilized penicillin acylase in the presence of substrate and products. *Biotechnol. Bioeng.* 50, 609-616.

Illanes, A., Cauerhff, A., Wilson, L. and Castro, G. R. (2012) Recent trends in biocatalysis engineering. *Biores. Technol.* 115, 48-57.

Imhoff, J. F., lbes, A. and Wiese, J. (2011) Bio-mining the microbial treasures of the ocean: new natural products. *Biotechnol. Adv.* 29, 408-482.

Iwaki, H., Nishimura, A. and Hasegawa, Y. (2012a) *Tropicibacter phthalicus* sp. nov., a phthalate-degrading bacterium from seawater. *Curr. Microbiol.* 64, 392-396.

Iwaki, H., Takada, K. and Hasegawa, Y. (2012b) Maricurvus nonylphenolicus gen. nov., sp. nov., a nonyl-degrading bacterium from seawater. *FEMS Microbiol. Lett.* 327, 142-147.

Iwasaki, A., Matsumoto K., Hasegawa, J. and Yasohara, Y. (2012) A novel transminase, (R) -amine: pyruvate aminotransferase, from *Arthrobacter* sp. KNK168 (FERM BP-5228): purification, characterization, and gene cloning. *Appl. Microbiol. Biotechnol.* 93, 1563-1573.

Jensen, C. N., Cartwright, J., Ward, J., Hart, S., Turkenburg, J. P. et al., (2012) A flavoprotein monooxygenase that catalyses a Baeyer-Villiger reaction and thioether oxidation using NADH as the nicotinamide cofactor. *Chembiochem.* 13, 872-878.

Jiang, L., Althoff, E. A., Clemente, F. R., Doyle, L., Rothlisberger, D. et al., (2008) De novo computational design of retro-aldol enzymes. Science 319, 1387-1391. Jochens, H., Hesseler, M., Stiba, K., Padhi, S. K., Kazlauska, R. J. and Bornscheuer, U. T. (2011) Protein engineering of c^fi-hydrolase fold enzymes. *Chembiochem.* 12, 1508-1517.

Kaul, P. and Asano, Y. (2012) Strategies for discovery and improvement of enzyme function: state of the art and opportunities. *Microbiol. Biotechnol.* 5, 18-33.

Kennedy, J., O'Leary, N. D., Kiran, G. S., Morrissey, J. P., O'Gara, F. et al., (2011) Functional metagenomic strategies for the discovery of novel enzymes and biosurfactants with biotechnological applications from marine ecosystems. *J. Appl. Microbiol.* 111, 787-799.

Kille, S., Zilly, F. E., Acevedo, J. P. and Reetz, M. T. (2011) Regio-and stereoselectivity of P450-catalysed hydroxylation of steroids controlled by laboratory evolution. *Nat. Chem.* 3, 738-743.

Kim, J., Grate, J. W. and Wang, P. (2008) Nanobiocatalysis and its potential applications. *Trends Biotechnol.*

26, 639-649.

Khersonsky, O., Kiss, G., Rothlisberger, D., Dym, O., Albeck, S. et al., (2012) Bridging the gaps in design methodologies by evolutionary optimization of the stability and proficiency of designed Kemp eliminase KE59. *Proc. Natl. Acad. Sci.* 109, 10358-10363.

Klein-Marcuschamer, D., Oleskowicz-Popiel, P., Simmons, B. A. and Blanch, H. W. (2012) The challenge of enzyme cost in the production of lignocellulosic biofuels. *Biotechnol. Bioeng.* 109, 1083-1087.

Kobayashi, M., Nagasawa, T. and Yamada, H. (1992) Enzymatic synthesis of acrylamide: a success story not yet over. *Trends Biotech.* 10, 402-408.

Kobayashi, M. and Shimizu S. (1998) Metalloenzyme nitrile hydratase: structure, regulation and application to biotechnology. *Nat. Biotechnol.* 16, 733-736.

Koeller, K. M. and Wong, C. -H. (2001) Enzymes for chemical synthesis. *Nature* 409, 232-240.

Koshland Jr., D. E. (2002) The application and usefulness of the ratio kcat/KM. *Bioorg. Chem.* 30, 211-213.

Kraus, G. A. (2008) Synthetic methods for the preparation of 1, 3-propanediol. *Clean* 36, 648-651.

Kuchner, O. and Arnold, F. H. (1997) Directed evolution of enzyme catalysts. *Trends Biotechnol.* 15, 523-530.

Kurian, J. V. (2005) A new polymer platform for the future-Sorona from corn derived 1, 3-propanediol. *J. Polym. and the Environ.* 13, 159-167.

Larsen, A. T., May, E. M. and Auclair, K. (2011) Predictable stereoselective and chemoselective hydroxylations and epoxidations with P450 3A4. *JACS* 133, 7853-7858.

Leary, D., Vierros, M., Hamon, G., Arico, S. and Monagle, C. (2009) Marine genetic resources: a review of scientific and commercial interest. *Mar. Policy* 33, 183-194.

Leisch, H., Morley, K. and Lau, P. C. K. (2011) Baeyer-Villiger monooxygenases: more than just green chemistry. *Chem. Rev. Ill*, 4165-4222.

Lewis, J. C., Coelho, P. S. and Arnold, F. H. (2011) Enzymatic functionalization of carbon-hydrogen bonds. *Chem. Soc. Rev.* 40, 2003-2021.

Li, S., Anzai, Y., Kinoshita, K., Kato, F. and Sherman, D. H. (2009) Functional analysis of MycE and MycF, two O-methyltransferases involved in the biosynthesis of mycinamicin macrolide antibiotics. *Chembiochem.* 10, 1297-1301.

Liszka, M. J., Clark, M. E., Schneider, E. and Clark, D. S. (2012) Nature versus nurture: developing enzymes that function under extreme conditions. *Annu. Rev. Chem. Biomol. Eng.* 3, 77-102.

Lucena, S. A., Lima, L. S., Cordeiro, L. S. A., Sant'Anna, C., Constantino, R. et al., (2011) High throughput screening of hydrolytic enzymes from termites using a natural substrate derived from sugarcane bagasse. *Biotechnol. Biofuels* 4, 51.

Lutz, S. (2010) Beyond directed evolution-semi-rational protein engineering and design. *Curr. Opin. Biotechnol.* 21, 734-743.

Marron, A. O., Akam, M. and Walker, G. (2012) Nitrile hydratase genes are present in multiple eukaryotic supergroups. *PLos One* 7, e32867.

Mastrobattista, E., Taly, V., Chanudet, E., Treacy, P., Kelly, B. T. and Griffiths, A. D. (2005) High-throughput screening of enzyme libraries: in vitro evolution of a ft-galactosidase by fluorescence-activated sorting of double emulsions. *Chem. Biol.* 12, 1291-1300.

McMillan, J. (2011) Enzyme solicitation support and validation task. Biochemical platform review. National Renewable Energy Laboratory.

McSorley, F. R., Wyatt, P. B., Martinez, A., DeLong, E. F., Hove-Jensen, B. and Zechel, D. L. (2012) PhnY and PhnZ comprise a new oxidative pathway for enzymatic cleavage of a carbon-phosphorus bond. *JACS* 134, 8364-8368.

Mersinger, L. J., Hann, E. C., Cooling, F. B., Gavagan, J. E., Ben-Bassat, A. et al., (2005) Production of acrylamide using alginate-immobilized E. coli expressing Comamonas testosteroni 5-MGAM-4D nitrile hydratase. *Adv. Synth. Catal.* 347, 1125-1131.

Mora-Pale, M., Meli, L., Doherty, T. V., Linhardt, R. J. and Dordick, J. S. (2011).

Room temperature ionic liquids as emerging solvents for the pretreatment of lignocellulosic biomass. *Biotechnol. Bioeng.* 108, 1229-1245.

Nagasawa, T. and Yamada, H. (1989) Microbial transformations of nitriles. *Trends Biotech.* 7, 153-158.

Nakamura C. E. et al., (2000) US patent 6, 013, 494. Method for the product, of 1, 3-propanediol by microorganisms.

Nakamura, C. E. and Whited, G. M. (2003) Metabolic engineering for the microbial production of 1, 3-propanediol. *Curr. Opiti. Biotech.* 14, 454-459.

Nielsen, P. H., Kuilderd, W., Zhou, W. and Lu, X. (2009) Enzyme biotechnology for sustainable textiles. In Sustainable Textiles. *Life Cycle and Environmental Impact*, Blackburn, R. S. (ed.), 113-138.

O'Donnell, P. S. and Hug, D. H. (1985) Tryptophanyl fluorescence quenching of urocanase from Pseudomonas putida by acrylamide, cesium, iodide, and imidazolepropionate. *Photochem. Photobiol.* 42, 107-112.

OECD (1998) Biotechnology for clean industrial products and processes. Towards industrial sustainability.

OECD (2001) The application of biotechnology to industrial sustainability.

Olbe, L., Carlsson E. and Lindberg, P. (2003) A proton-pump inhibitor expedition: the case histories of omeprazole and esomeprazole. *Nature Rev. Drug Discovery* 2, 132-139.

O'Reilly, E., Kohler, V., Flitsch, S. L. and Turner, N. J. (2011) Cytochrome P450 as useful biocatalysts: addressing the limitations. *Chem. Commun.* 47, 2490-2501.

Ouajai, S. and Shanks, R. A. (2005) Morphology and structure of hemp fibre after bioscouring. *Macromol. Biosci.* 5, 124-134.

Park, S. and Kazlauskas, R. J. (2003) Biocatalysis in ionic liquids-advantages beyond green technology. *Curr. Opin. Biotechnol.* 14, 432-437.

Paul, P., Suwan, J., Liu, J., Dordick, J. and Linhardt, R. J. (2012) Recent advances in sulfotransferase enzyme activity assays. *Anal. Bioanal. Chem.* 403, 1491-1500.

Polizzi, K. M., Bommarius, A. S., Broering, J. M. and Chaparro-Riggers, J. F. (2007) Stability of biocatalysts. *Curr. Opin. Chem. Biol.* 11, 220-225.

Pollard, D. J. and Woodley, J. M. (2006) Biocatalysis of pharmaceutical intermediates: the future is now. *Trends Biotechnol.* 25, 66-73.

Prasad, S. and Bhalla, T. C. (2010) Nitrile hydratases (NHases): at the interface of academia and industry. *Biotechnol. Adv.* 28, 725-741.

Quijano, G., Couvert, A. and Amrane, A. (2010) Ionic liquids: applications and future trends in bioreactor technology. *Biores. Technol.* 101, 8923-8930.

Quin, M. B. and Schmidt-Dannert, C. (2011) Engineering of biocatalysts: from evolution to creation. *ACS Catal.* 1, 1017-1021.

Rakels, J. L. L., Straathof, A. J. J. and Heijnen, J. J. (1993) A simple method to determine the enantiomeric ratio in enantioselective biocatalysis. *Enz. Microb. Technol.* 15, 1051-1056.

Reetz, M. T. (2009) Directed evolution of enantioselective enzymes: an unconventional approach to asymmetric catalysis in organic chemistry. *J. Org. Chem.* 74, 5757-5778.

Reetz, M. T. (2011) Laboratory evolution of stereoselective enzymes: a prolific source of catalysts for asymmetric reactions. *Angew. Chem. Int. Ed. Engl.* 50, 138-174.

Reetz, M. T. and Carballeira, J. D. (2007) Iterative saturation mutagenesis for rapid directed evolution of functional enzymes. *Nat. Protocols.* 2, 891-903.

Reetz, M. T., Prasad, S., Carballeira, J. D., Gumulya, Y. and Bocola, M. (2010a) Iterative saturation mutagenesis accelerates laboratory evolution of enzyme stereoselectivity: rigorous comparison with traditional methods. *JACS* 132, 9144-9152.

Reetz, M. T., Soni, P., Fernandez, L., Gumulya, Y. and Carnalleria, J. D. (2010b) Increasing the stability of an enzyme toward hostile organic solvents by directed evolution based on iterative saturation mutagenesis using the B-FIT method. *Chem. Comm.* 46, 8657-8658.

Reid, E. L., Worthy, A. C., Probert, L, Ali, S. T., Love, J. et al., (2011) Coccolithophores: functional biodiversity, enzymes and bioprospecting. *Mar. Drugs.* 9, 586-602.

Reymond, J. L. (2005) Enzyme Assays: High-throughput Screening, Genetic Selection and Fingerprinting, Wiley-VCH, Weiheim.

Reymond, J. L., Fluxa, V. S. and Maillard, N. (2009) Enzyme assays. *Chem. Commun.* 34-46.

Richter, F., Leaver-Fay, A., Khare, S. G., Bjelic, S. and Baker, D. (2011) De novo enzyme design using Rosseta3. *PloS ONE* 6, el9230.

Rodriguez, C., de Gonzalo, G., Fraaije, M. W. and Gotor, V. (2010) Ionic liquids for enhancing the enantioselectivity of isolated BVMO-catalysed oxidations. *Green Chem.* 12, 2255-2260.

Romero, P. A. and Arnold, F. H. (2009) Exploring protein fitness landscapes by directed evolution. *Nat. Rev. Mol. Cell Biol.* 10, 866-876.

Rothlisberger, D., Khersonsky, O., Wollacott, A. M., Jiang, L., DeChancie, J. et al., (2008) Kemp elimination catalysts by computational enzyme design. *Nature* 453, 190-195.

Ryan, R. P., Monchy, S., Cardinale, M., Taghavi, S., Crossman, L. et al., (2009) The versatility and adaptation of bacteria from the genus Stenotrophomonas. *Nat. Rev. Microbiol.* 7, 514-525.

Sakaki, T. (2012) Practical applications of cytochrome P450. *Biol. Pharm. Bull.* 35, 844-849.

Santacoloma, P. A., Sin, G., Gernaey, K. V. and Woodley, J. M. (2011) Multi-catalyzed processes: next generation biocatalysis. *Org. Process Res & Develop.* 15, 203-212.

Sathishkumar, P., Chae, J., Unnithan, A. R., Palvannan, T., Kim, H. Y. et al., (2012) Laccase-poly (lactic-co-glycolic acid) (PLGA) nanofiber: highly stable, reusable, and efficacious for the transformation of diclofenac. *Enzyme Microbial Technol.* 51, 113-118.

Savile, C. L., Janey, J. M., Mundorff, E. C., Moore, J. C., Tam, S. et al., (2010) Biocatalytic asymmetric synthesis of chiral amines from ketones applied to sitagliptin manufacture. *Science* 329, 305–309.

Scheer, M., Grote, A., Chang, A., Schomburg, I., Munaretto, C. et al., (2011) BRENDA, the enzyme information system in 2011. *Nucl. Acids Res.* 39, D670–D676.

Schloss, P. D. and Handelsman, J. (2005) Metagenomics for studying unculturable microorganisms: cutting the Gordian knot. *Genome Biol.* 6, 229.

Schmidt, M. and Bornscheuer, U. T. (2005) High-throughput assays for lipases and esterases. *Biomol. Eng.* 22, 51–56.

Sheldon R. A. (2007) The E factor: fifteen years on. *Green Chem.* 9, 1261–1384.

Sheldon, R. A. (2011a) Characteristic features and biotechnological applications of cross – linked enzyme aggregates (CLEAs). *Appl. Microbiol. Biotechnol.* 92, 467–477.

Sheldon, R. A. (2011b) Cross-linked enzyme aggregates as industrial biocatalysts. *Org. Process Res. Devel.* 15, 213–223.

Siegel, J. B., Zanghellini, A., Lovick, H. M., Kiss, G., Lambert, A. R. et al., (2010) Computational design of an enzyme catalyst for a stereoselective bimolecular Diels–Alder reaction. *Science* 329, 309–313.

Simon, C. and Daniel, R. (2011) Metagenomic analyses: past and future trends. *Appl. Environ. Microbiol.* 77, 1153–1161.

Solbak, A. I., Richardson, T. H., McCann, R. T., Kline, K. A., Bartnek, E et al., (2005) Discovery of pectin–degrading enzymes and directed evolution of a novel pectate lyase for processing cotton fabric. *J. Biol. Chem.* 280, 9431–9438.

Stadler, R. H., Blank, I., Varga, N., Robert, E, Hau, J. et al., (2002) Acrylamide from Maillard reaction products. *Nature* 419, 449.

Stemmer, RM. (1994) Rapid evolution of a protein in vitro by DNA shuffling. *Nature* 370, 389–391.

Stoops, J. K., Horgan, D. J., Runnegar, M. T. C., De Jersey, J., Webb, E. C. and Zerner, B. (1969) Carboxylesterases (EC 3. 1. 1). Kinetic studies on carboxylesterases. *Biochemistry* 8, 2026–2033.

Straathof, A. J. J. and Jongejan, J. A. (1997) The enantiomeric ratio: origin, determination and prediction. *Enz. Microb. Technol.* 21, 559–571.

Straathof, A. J. J., Panke, S. and Schmid, A. (2002) The production of fine chemicals to transformation. *Curr. Opin. Biotechnol.* 13, 548–556.

Strohmeier, G. A., Pichler, H., May, O. and Gruber-Khadjawi, M. (2010) Application of designed enzymes in organic synthesis. *Chem. Rev.* 111, 4141–4164.

Tadesse, H. and Luque, R. (2011) Advances on biomass pretreatment uisng ionic liquids. An overview. *Energy & Environ. Sci.* 4, 3913–3929.

Tang, S. L. Y., Smith, R. L. and Poliakoff, M. P. (2005) Principles of green chemistry: Productively. *Green Chem.* 7, 761–762.

ten Brink, G. J., Arends, W. and Sheldon, R. A. (2004) The Baeyer–Villiger reaction; new developments toward greener procedures. *Chem. Rev.* 104, 4105–4124.

Tobert, J. A. (2003) Lovastatin and beyond: the history of the HMG–CoA reductase inhibitors. *Nat. Rev. Drug Discovery* 2, 517–526.

Trincone, A. (2010) Potential biocatalysts originating from sea environments. *J. Mol. Catal. B–Enzym.* 66,

241–256.

Trincone, A. (2011) Marine biocatalysts: enzymatic features and applications. *Mar. Drugs.* 9, 478–499.

Trincone, A. (2012) Some enzymes in marine environment: prospective applications found in patent literature. *Recent Pat. Biotechnol.* 6, 134–148.

Trost, B. M. (1991) The atom economy—a search for synthetic efficiency. *Science* 254, 1471–1447.

Trost, B. M. (1995) Atom economy—a challenge for organic synthesis: heterogeneous catalysis leads the way. Angew. *Chem. Int. Ed. Engl.* 34, 259–281.

Truong. L. V., Tuyen, H., Helmke, E., Binh L. T. and Schweder T. (2001) Cloning of two pectate lyase genes from the marine Antarctic bacterium *Pseudoalteromonas haloplanktis* strain ANT/505 and characterization of the enzymes. *Extremophiles* 5, 35–44.

Tucker J. L. (2006) Green Chemistry: a pharmaceutical perspective. *Org. Process Res. Devel.* 10, 2001–2005.

Tufvesson, P., Lima–Ramos, J., Nordblad, M. and Woodley, J. M. (2011) Guidelines and cost analysis for catalyst production and biocatalytic processes. *Org. Process Res. Develop.* 15, 266–274.

Urlacher, V. B. and Girhard, M. (2012) Cytochrome P450 monooxygenases: an update on perspectives for synthetic applications. *Trends Biotechnol.* 30, 26–36.

US DOE (2006) Breaking the biological barriers to cellulosic ethanol. A joint research agenda. DOE/SC–0095. P130.

van Berkel, W. J. H., Kamerbeek, N. M. and Fraaije, M. W. (2006) Flavoprotein monooxygenases, a diverse family of oxidative biocatalysts. *J. Biotechnol.* 124, 670–689.

van Rantwijk, F. and Sheldon, R. A. (2007) Biocatalysis in ionic liquids. *Chem. Rev.* 107, 2757–2785.

Vancov, T., Alston, A. -S., Brown, T. and McIntosh, S. (2012) Use of ionic liquids in converting lignocellulosic materials to biofuels. *Renwable Energy* 45, 1–6.

Vickers, C. E., Klein–Marcuschamer, D. and Kromer, J. O. (2012) Examining the feasibility of bulk commodity production in *Escherichia coli. Biotech. Lett.* 34, 585–596.

Villarreal–Chiu, J. F., Quinn, J. R and McGrath, J. W. (2012) The genes and enzymes of phosphonate metabolism by bacteria, and their distribution in the marine environment. *Front. Microbiol.* 3, 1–13.

Wang, M., Si, T. and Zhao, H. (2012) Biocatalyst development by directed evolution. *Biores. Technol.* 115, 117–125.

Weber, J. M., Schoner, B. and Losick, R. (1989) Identification of a gene required for the terminal step of erythromycin A biosynthesis in Saccharopolyspora erythraea (Streptomyces erythreus). *Gene* 75, 235–241.

Weckbecker, A., Groger, H. and Hummel, W. (2010) Regeneration of nicotinamide coenzymes: principles and applications for the synthesis of chiral compounds. *Adv. Biochem. Eng/Biotechnol.* 120, 175–242.

Whitehouse, C. J. C., Bell, S. G. and Wong, L. L. (2012) P450BM3 (CYP102A1): connecting the dots. *Chem. Soc. Rev.* 41, 1218–1260.

Whitman, W. B., Coleman, D. C. and Wiebe, W. J. (1998) Prokaryotes: the unseen majority. *Proc Natl. Acad. Sci. USA* 95, 6578–6583.

Willetts, A., Joint, I., Gilbert, J. A., Trimble, W. and Wuhling, M. (2012) Isolation and initial characterization of a novel type of Baeyer-Villiger monooxygenase activity from a marine organism. *Microbiol. Biotechnol.* 5, 549–559.

Winter, R. T., Heuts, D. P. H. M., Rijpkema, E. M. A., van Bloois, E., Wijma, H. J. and Fraaije, M. W. (2012) Hot or not? Discovery and characterization of a thermostable alditol oxidase from Acidothermus cellulolyticus 11B. *Appl. Microbiol. Biotechnol.* 95, 389–403.

Wohlgemuth, R. (2010) Biocatalysis-key to sustainable industrial chemistry. *Curr. Opin. Biotechnol.* 21, 713–724.

Xi, C., Gao, X., Wu, Q. and Zhu, D. (2012) Synthesis of optically active dihydrocarveol via a stepwise or one-pot enzyme reduction of (R) – and (S) –arvone. *Tetrahedron: Asymmetry.* 23, 734–738.

Xiao, Z., Bergeron, H., Grosse, S., Beauchemin, M., Garron, M. –L. et al., (2008a) Improvement of the thermostability and activity of a pectate lyase by single amino acid substitutions, using a strategy based on melting–temperature–guided sequence alignment. *Appl. Environ. Microbiol.* 74, 1183–1189.

Xiao, Z., Boyd, J., Grosse, S., Beauchemin, M., Coupe, E. and Lau, P. C. K. (2008b) Mining Xanthomonas and Streptomyces genomes for new pectinase–encoding sequences and their heterologous expression in *Escherichia coli*. *Appl. Microbiol. Biotechnol.* 78, 973–981.

Xie, X. and Tang, Y. (2007) Efficient synthesis of simvastatin by use of whole–cell biocatalysis. *Appl. Environ. Microbiol.* 73, 2054–2060.

Xie, X., Watanabe, K., Wojcicki, W. A., Wang, C. C. and Tang, Y. (2006) Biosynthesis of lovastatin analogs with a broadly specific acyltransferase. *Chem. Biol.* 13, 1161–1169.

Xue, R. and Woodley, J. M. (2012) Process technology for multi–enzymatic reaction systems. *Biores. Technol.* 115, 183–195.

Yamada, H. and Kobayashi, M. (1996) Nitrile hydratase and its application to industrial production of acrylamide. *Biosci. Biotech. Biochem.* 60, 1391–1400.

Yamada, H., Shimizu, S. and Kobayashi M. (2001) Hydratases involved in nitrile conversion: Screening, characterization and application. *Chem Record.* 1, 152–161.

Yonaha, K. and Soda, K. (1986) Applications of stereoselectivity of enzymes: synthesis of optically active amino acids and alpha–hydroxy acids, and stereospecific isotope–labeling of amino acids, amines and coenzymes. *Adv. Biochem. Eng. /Biotech.* 33, 95–130.

Yu, Y., Li, H., Zeng, Y. and Chen, B. (2011) Bacetiral diversity and bioprospecting for cold – active hydrolytic enzymes form culturable bacteria associated with sediment from Nella Fjord, eastern Antarctica. *Mar. Drugs.* 9, 184–195.

Zanghellini, A., Jiang, L., Wollacott, A. M., Cheng, G., Meiler, J. et al., (2006) New algorithms and an in silico benchmark for computational enzyme design. *Prot. Sci.* 15, 2785–2794.

Zeng, A. P. and Biebl, H. (2002) Bulk chemicals from biotechnology: The case of 1, 3-propanediol production and the new trends. *Adv. Biochem. Eng. /Biotechnol.* 74, 239–259.

Zeng, A. –P. and Sabra, W. (2011) Microbial production of diols as platform chemicals: Recent progresses. *Curr. Opin. Biotechnol.* 22, 749–757.

Zhang, Y., Fang, R. and Shen, Y. (1998) Studies on strain producing nitrile hydratase. *Gongye Weishengwu.* 28, 1–5.

Zhang, C. and Kim S. K. (2012) Application of marine microbial enzymes in the food and pharmaceutical industries. *Adv. Food Nutr. Res.* 65, 423–435.

Zhang, Y. and Zhang, Y. (2007) Formation and reduction of acrylamide in Maillard reaction: A review based on the current state of knowledge. *Crit. Revs. Food Science and Nutri.* 47, 521–542.

Zhao, H. and van der Dork, W. (2003) Regeneration for cofactors for use in biocatalysis. *Curr. Opin. Biotechnol.* 14, 583–589.

Zheng, R.-C, Zheng, Y.-G. and Shen, Y.-C. (2010) Acrylamide, microbial production by nitrile hydratase. In Encyclopedia of Industri Biotechnology: Bioprocess, Bioseparation, and Cell Technology, Flickinger, M. C. (ed.), 1–12.

第3章 海洋酶催化剂与其稳定性研究：分子方法

K. K. Pulicherla VIT University, India and K. R. S. Sambasiva Rao Acharya Nagarjuna University, India

DOI：10. 1533/9781908818355. 1. 71

摘要：酶催化生物体内多种生物化学反应使其能够配合完成生命过程，并且使生物体拥有独特的体系，这是其适应特定环境的必要条件。由于在冷环境适应中发挥的主导作用，海洋嗜冷微生物产生的嗜冷酶在最近几年引起了高度关注。嗜冷酶的热力学稳定性，与之相关的灵活性或者说可塑性，以及在分子水平上的催化效率的研究是当今微生物研究的重要内容。对分子内氢键和离子键、氨基酸组成、表面疏水性、螺旋结构稳定性以及核心组装这些结构特点的研究，揭示了通过对不同小区域结构的调整形成复杂结构的基础，这些结构调整是为了通过降低酶的热稳定性，提高分子灵活性，从而提高酶在低温环境中的催化效率。嗜冷酶在极端微生物新陈代谢中是非常重要的，同时也是蛋白折叠、催化作用和生物技术应用这些基础研究的重要模型。本章中主要讨论分子水平海洋生物催化的进展、低温适应以及它们的工业应用。

关键词：嗜冷酶，嗜冷微生物，极端微生物，冷适应蛋白，果胶酶

3.1 引言

地球上大部分生态区域是低温环境，细菌、酵母菌、单细胞藻类和真菌等多种微生物拥有在低温环境成功繁殖的生存能力。嗜冷微生物来源的酶革新了低温海洋生物技术。地球表面大约71%被海洋覆盖，90%的海洋环境年平均温度在5~10℃。据估计，地球生态区包括一系列（大约3. 67×10^{30}个）已确定的在陆地环境没发现的微生物。尽管低温环境对于生化反应存在明显的抑制作用，这些嗜冷生长旺盛微生物的繁殖、生长和运动与常温环境生存的微生物的速率类似。海洋嗜冷微生物通过在结构上精细的调整，如在膜结构、结构蛋白和酶，以适应低温环境。通过这些方面结构调整以补充低温造成的严重影响。由

于具有独特的代谢途径和有研究价值的生命活动，如今海洋微生物被认为是治疗和其他工业用酶的重要来源。基本的代谢驱动力和生物催化剂是所有微生物适应周围环境的重要研究目标，来源于嗜冷微生物的海洋生物催化剂也不例外。

1840 年，参与南极洲海军探险的植物学家 Joseph Dalton Hooker 首次记录了嗜冷微生物。1884 年 Certes 报告了低温环境下细菌的生长，1887 年 Forster 测定了来自鱼的嗜冷细菌在低温环境下的生长和繁殖。Ingraham 根据 0℃ 的快速生长定义嗜冷细菌，而不是根据它们的最大或最适宜生长温度。后来，Ingraham 和 Stokes 再次把专性嗜冷微生物定义为这样的细菌，这些细菌不仅在 0℃ 可以快速生长，而且在 20℃ 以下生长最快。而 1975 R. Morita 提出的嗜冷的微生物的概念被广泛接受。他用 psychrophile 这个词来描述这些微生物，这些微生物的最重要生长温度参数，也就是说最低，最适宜，最高生长温度分别是低于 0℃、15℃、20℃，然而有较高的最适宜，最大量的微生物被定义为 psychrophic。psychrotroph 这个术语意思是吃冷的或结霜的食物并且通常在食品工业上描述能够破坏或污染冷冻食品的微生物。1902 年 Schmidt-Nielsen 在 0℃ 培养出了微生物，并命名为嗜冷微生物，即微生物在低温环境中旺盛的繁殖，如微生物能够在接近冰点温度持续的旺盛生长，它们没有温度调节，不能在常温环境生长。在希腊，*psychro* 意思是冷，*philic* 意思是喜欢。这些微生物在低温活性酶的帮助下产生一种适应极端环境的机制，这些微生物被命名为极端微生物，并且在 1974 年被 Macelroy 第一次提到，他认为这些微生物可以在对常温微生物来说是致命的环境中生长和增殖。很多微生物都属于嗜冷微生物，比如革兰氏阴性和阳性细菌、古生菌、酵母、很多真核细胞生物（像藻类、植物、昆虫、海洋和陆生无脊椎动物）和鱼类。第一个冷适应酶的 X-射线结构在 1994 年被解析。与其他来源的酶相比，来自海洋微生物的冷适应酶由于它们具有潜在的工业应用（Khawar Sohail Ricardo，2006）而成为现在的研究热点，即使在低温环境中也具有最高的生物催化活性。

3.2 嗜冷酶潜在的工业应用

酶的潜在工业应用可以追溯到很多年前，酶法在现代生物技术工业领域发挥着举足轻重的作用。大部分工业酶是从常温微生物分离来的，但是具有热稳定或有嗜热特性的酶被认为是更理想的生物催化剂。但在有些工业应用中，酶促反应不得不在低温环境下进行。在这种情况下，冷适应酶比常温或嗜热酶更实用（表 3.1）。在工业进程中，嗜冷酶具有像节约能源、减少不稳定或挥发性物质的使用、避免酶的污染和在后续处理过程中容易失活的优势，因此在工业领域，大家更倾向于采用嗜冷酶生产相关产品。嗜冷酶的工业应用涉及很多方面，接下来我们将讨论一些有重要潜力的工业应用酶。

表 3.1　嗜冷酶的来源和优点

酶	冷活性酶来源	应用	优点
α-淀粉酶	*Pseudoalteromonas haloplanktis*	麦芽糖的生产，提高糖浆果糖含量，脱浆，洗衣，纸涂层	保证产品质量
β-乳糖酶	*Arthrobacter psychrolactophi*, *Arthrobacter* sp. C2-2, *Pseudoalteromonas* sp. TAE 79b, *Pseudoalteromonas haloplanktis* TAE79, *Pseudoalteromonas* sp. 22b, *Planococcus* sp., *Carnobacterium piscicola* BA, *Arthrobacter* sp.	结构改变	保证产品质量，保持新鲜
β-葡聚糖苷酶	SB *Paenibacillus* sp. Strain C7	结构改变	保持新鲜
过氧化氢酶	*Vibrio salmonicida*, *Pseudoalteromonas haloplanktis*, *Vibrio rumoiensis* S-1, *Pseudoalteromonas* sp. DY3	废水处理，消除工业进程中产生的过氧化物的影响	非常有价值的应用于南极和北极的废水处理
纤维素酶	*Pseudoalteromonas haloplanktis* TAC125	去污剂配方，果汁净化	保持香味，应用在自来水处理
DNA 连接酶	*Pseudoalteromonas haloplanktis*	生物分子研究	应用于低温表达
DNA 聚合酶	*Cenarchaeum symbiosum*	分子生物学研究	应用于低温表达
核酸内切酶	*Vibrio salmonicida*	分子生物学研究	应用于低温表达
酯酶	*Psychrobacter* sp. ANT300, *Pseudomonas* sp. strain B11-1, *Acinetobacter* sp. strain No. 6	废水处理，去污剂的组成	易挥发和热敏感材料的应用
果糖-1，6-二磷酸-醛缩酶	*Vibrio marinus*	结构修饰	保持新鲜
脂肪酶	*Pseudomonas* sp. strain KB700A, *Pseudomonas* sp. strain B11-1	废水处理，去污剂配方	应用于易挥发和热敏感材料，应用于自来水处理
氧化酶	*Pseudoalteromonas* sp. strain AS-11	防腐	易挥发和热敏感材料，防腐
果胶裂解酶	*Pseudoalteromonas haloplanktis* strain ANT/505	果汁净化	保持新鲜

续表

酶	冷活性酶来源	应用	优点
蛋白酶	*Yersinia ruckeri*, *Xanthomonas maltophilia*, *Pseudomonas fluorescens* strain 164/03, *Pseudoalteromonas issachenkonii* UST041101-043, *Pseudomonas fluorescens* 114, *Colwellia psychrerythraea* strain 34H, *Escherichia freundii* B. 3038, *Pseudomonas* sp. strain TAC Ⅱ 18, *Pseudoalteromonas* sp. strain AS-11	去污剂配方，鱼皮肤去除，皮制品工业	保证产品质量，应用于自来水处理
RNA 聚合酶	*Shewanella violacea* DSS12, *Pseudomonas syringae* Lz4W	分子生物学研究	应用于低温表达
超氧化物歧化酶	*Pseudoalteromonas haloplanktis* TAC125	药学研究，活性氧消除	热敏感的药物学应用

3.2.1 去污添加剂

众所周知，像蛋白酶、脂酶、α-淀粉酶和纤维素酶，这些重要的工业用酶是在去污剂中作为添加剂使用，和生物磨料及石洗一起应用于纺织物清洗，也应用在去污剂配方中。在洗涤剂工业，冷水洗涤的一个优点是低温环境能减少能耗和对衣物的破坏。纤维素酶和淀粉酶这类酶在洗衣去污剂中的应用可以提高洗涤剂的洗涤效率和确保环境安全(Pulicherla et al.，2011)。

3.2.2 纺织工业

冷适应纤维素酶在纺织工业用于生物抛光和石洗进程。在纺织产品中，经常有棉花纤维线头从主纤维伸出，降低了布料平滑度并改变了衣服表面的外观。在不断洗涤后，这些问题会增多并进一步破坏衣服的质量。因此，在合适的条件下用纤维素酶预处理，去除伸出的线头，可以减少起球并提高织物的耐受力和柔软度。以上问题可以通过使用冷适应蛋白酶进行改进，这些酶在低温下有较好的催化效率。在褪色牛仔裤的生产中，即使微量的纤维素酶应用可以替代大量的磨光石。此外，纤维素酶在该领域的使用可以降低处理的成本。

3.2.3 食品工业

冷适应酶在食品工业领域的应用优势涉及多个方面。果胶酶能自然降解果胶状物质，在需要果胶清除的工业领域有着巨大的应用。主要应用于果汁、咖啡和茶的加工，浸软植物和蔬菜组织，植物纤维脱胶，蔬菜油提取，酿酒工业上的果胶层消除，也有被应用于生产糖尿病患者检测一下低甲氧基的说法。果胶酶的添加能够降低果汁的黏度，提高果肉的挤压能力，使胶状物结构瓦解，以容易获得较高产量的果汁。来自嗜冷微生物的果胶酶很容易储藏。β-乳糖酶是被广泛应用于乳糖不耐症的另一个重要的酶。小肠中少量的乳糖就能导致消化不适，被称作乳糖不耐症，占世界人口 2/3 的人都具有这个问题。因此，如今市场上的无乳糖牛奶需要通过多种化学和物理方法加工而成。降解牛奶中乳糖的首选方法是采用来自嗜冷菌的β-乳糖酶处理。而且，牛奶中乳糖的消除提高了甜度，增加了溶解性，抑制了浓缩奶和冰淇淋中乳糖结晶。纤维素酶在商业上应用于咖啡加工，因为它能在咖啡豆干燥时水解纤维素。淀粉酶在工业加工中有广泛的应用，如食品加工，发酵和医药工业。淀粉酶是淀粉转化成寡糖的关键酶，可以广泛应用于把淀粉转化成果糖和葡萄糖。嗜冷糖苷酶广泛应用于烘焙工业，在食品加工后能保持剩余活性的物质，同时像正常的糖苷酶一样，在储存时不改变产品的结构（Pulicherla et al.，2011）。

3.2.4 生物修复

用微生物减少如溢油和废水处理这样环境污染的方法是很传统的，但是一种相对其他物理方法来说是更可行的选择。在温带地区，温度季节性明显的变化降低了微生物降解的有效性，如对有机污染物油和脂质的降解。然而，通过对污染环境中接种特殊的嗜冷微生物并进行放大培养，有助于提高顽固化学物质的生物降解。由于它们酶的高催化效率以及在低温与温和条件下催化的专一性，嗜冷微生物对生物治理修复是一个理想的选择。嗜冷酶也被应用于废水处理消除有害化合物，像硝酸盐、碳水化合物、芳香化合物、重金属以及生物大分子，如纤维素、几丁质、木质素、蛋白质和甘油三酸酯（Gerday，2000）。

3.3 低温生活环境对生命的影响

在探讨酶对低温环境的适应前，有必要了解低温对环境、细胞和分子的影响。通过导致的物理化学参数的改变，温度和温度波动对生命有重要的影响，这些参数包括：pH 值、盐度、气体溶解度、压力、黏度、水活性和氧化还原电位。任何细胞膜的基本结构都是脂质双分子层，双分子层有合适的流动性以允许重要膜蛋白的移动和渗透。这个双分子层的功能基础是液晶相，但是低温会降低膜的流动性，对膜的物理特性和功能产生相反的影响。通常在低温下，有机体会产生较高分子量的不饱和、多不饱和、甲基支链脂肪酸，或者产生一个较短的酰基链，带有高比例的顺式不饱和双键和碳末端 2 号碳支链脂肪酸。膜的物理特性取决于脂质化合物的组成。因此，它的组成与微生物的生活环境息息相关。组成成分的改变被认为是增强膜流动性的关键，它通过引进空间位阻改善膜的组合模式或者减少膜之间的相互作用来增加其流动性。增加脂肪头组分，蛋白质和非极性胡萝卜素的含量也可以增加膜流动性。

研究者也从基因组和蛋白质组水平来揭示嗜冷微生物的适应机制。嗜冷微生物的基因组包含冷休克结构域，推测其在 RNA 稳定方面起作用。Liu 等（2012），在对 *Listeria monocytogenes* 研究时发现，在低温生长时 *lhkA* mRNA 的水平提高了。

在蛋白水平长期研究已经揭示了嗜冷微生物适应低温环境的秘密。这个蛋白复合物被称为降解体，包含了多聚核苷酸磷酸化酶和 RNA 解旋酶，保证了细胞内 RNA 的稳定性。北极细菌的降解体包含了另一个核糖核酸外切酶，称为核糖核酸酶 R（RNAse R），它替代了多聚核苷酸磷酸化酶。至今关于核糖核酸酶 R 的复合体还没被完全了解，但是可以确信的是核糖核酸酶 R 可以降解 RNA 成大量次级结构。Purusharth 等（2005）推测不需要 ATP 的解旋酶有助于细胞在低温环境下能量的节约，而这对于大多数解旋酶是需要的。

嗜冷微生物另一个重要的普遍特征是对嗜冷蛋白一级序列的修饰。这是为了抵抗低温的主要分子修饰，它不仅局限于对嗜冷微生物本身，同时它的表达有助于其他微生物适应骤降的温度变化。冷激活蛋白有两类，冷适应蛋白和冷休克蛋白。将近 20 种冷适应蛋白（CAPs）已经被确认，并且它们都是在稳定生长期被合成的，用于在低温维持细胞周期。冷休克蛋白（Csps）也被称为冷诱导蛋白，在转录和翻译水平调控着持家蛋白，也作为分子伴侣参与 mRNA 二级结构折叠。从转录和翻译水平上，冷休克蛋白的表达与很多因素有关。大多数表达是为了克服冷休克造成的影响（Georlette，2004）。相反地，有一些其他的特殊分子，它们与冷休克蛋白有相同的功能，被称为抗冻蛋白。抗冻蛋白（AFPs）或冰结构蛋白（ISPs）是一种多肽，它们的独特结构允许某些无脊椎动物、植物、藻类和细菌在 0℃以下生长。抗冻蛋白（AFPs）分子大小差异较大，但都能够通过一个大的互补表面与冰晶结合，然后产生热滞后效应，保证在较低的温度情况下微生物可以生长。在一些耐冻微生物中，AFPs 可以抑制细胞外冰晶的重结晶。一些细菌抗冻蛋白的特点是它们依赖于 Ca^{2+}。高浓度的胞外多糖已经在南极洲海洋细菌中被发现。胞外多糖可以调整细菌细胞的物理化学结构，参与细胞表面黏着，并帮助固定水分。它们还有助于营养的保存和富集，保护胞外酶避免冷变性，也可以作为冷冻保护剂。

在细胞水平，温度降低引起细胞黏度升高，盐溶解度降低，气溶性升高。当温度降低时，生物缓冲液的 pH 值也随之改变。同时，这也降低了微生物的生长率，膜流动性和稳定性。然后依次改变被动和主动渗透（电子和质子运输、营养摄入和离子通道调节），识别过程和环境感知。

在分子水平，温度的降低影响蛋白溶解度和氨基酸的带电性，接着改变蛋白-蛋白相互作用。低温处理会引起酶结构的改变和活性丧失。低温最重要的影响是抑制许多酶催化的化学反应，这在 1889 年 Svante Arrhenius 就提出了。

由于活跃能不足，而为了克服在 0~4℃的反应的能垒，嗜冷微生物在进化过程中产生一系列的应对策略，包括提高能量浓度，进化出不依赖于温度的酶，或者控制酶的扩散。

大多数嗜冷酶伴随着稳定性降低时，最适温度值（T_{opt}）发生改变，和其他热稳定酶相比，由基态到过渡态之间的活化自由能降低（$\Delta G^{\#}$），嗜冷酶会拥有高的反应速率（k_{cat}）。等式可以写成这样：$\Delta G^{\#}=\Delta H^{\#}-T\Delta S^{\#}$

这里 $\Delta H^{\#}$是活化焓的改变，$\Delta S^{\#}$是活化熵的改变，T 是绝对温度。目前报道的大多数冷适应酶的焓较低，导致反应率对温度的依赖性降低。同时在低温时，维持较高的反应速率（Khawar Sohail Ricardo，2006）。

当温度从 37℃降到 0℃，酶的活性维持恒定的。在温度降低时，离子相互作用依然很强且保持着其相应的柔性，即使在温度降低的情况下依然保持。转化率或者特定酶活的提高抵消了低温对酶催化效率的影响。尤其当温度改变的时候，底物结合情况也应该被考虑在内。高的柔性，热稳定性和特定酶活是嗜冷酶的特性。这样嗜冷微生物已经演变了几种策略来弥补低新陈代谢效率，使其在低温下可以生存。

3.4 嗜冷酶催化位点关键氨基酸残基的保守性

相对于常温酶，嗜冷酶催化位点有关的氨基酸是高度保守的。与常温酶相比，嗜冷酶的催化口袋更大，更容易接近配体。这可能与活性位点边缘的无规则卷曲上氨基酸的缺失有关，通过改变无规则卷曲上环的结构的变化，或在活性位点附近用较小的侧链结构取代大侧链的氨基酸残基。活性位点外的结构改变被认为是调整催化残基的动力学特性，引起低温活性的结构适应性（Feller and Gerday，1997）。每个蛋白家族采用他们自己的策略，通过用一个或者很多这些结构改变来降低它们的稳定性。但是这些氨基酸取代更多的发生在活性位点之外，因为他们在引起整体 3D 结构改变的同时，也要维持功能。深入对这些酶结构改变和它们在冷适应酶催化活性决定因素方面的研究进展缓慢，一个原因就是因为迄今已经解析的嗜冷酶的 3D 结构数据较少。因此，三维模键研究对前期研究是非常有用的，有助于我们获取研究特定的修饰导致的蛋白结构的改变，而这有助于维持低温下蛋白的稳定性。

3.5 嗜冷酶的柔性

柔性被定义为蛋白构象之间相互转化的波动的总和。和常温微生物相比，嗜冷微生物的酶有较多的柔性区域，因此有助于提高底物的亲和性。但是，有时酶的高灵活性会导致酶发生热失活。酶的高柔性赋予了酶在低温下可以完成精细和快速的蛋白质构象转换。蛋白结构的柔性有两种，全局和局部柔性。全局柔性与蛋白的整体结构有关，局部柔性是指蛋白的某些特定区域的柔性。整体折叠导致了蛋白质的错误折叠的增加，这是嗜冷酶高活性低稳定性的原因（Khawar Sohail and Ricardo，2006；Feller and Gerday，1997）。酶的柔性也受到活性口袋附近无规则结构的影响。这些无规则结构使蛋白质与溶剂有较好的相互作用，并降低构象的紧密性。有时催化位点附近无规则卷曲上添加或删除氨基酸残基有助于提高它与底物的结合（Marx，2004）。同时，在活性位点插入一些氨基酸也有利于二聚体的解聚。柔性也受高自由度、缺少分支和电荷相互作用的影响（Smalas et al.，2000）。

3.6 嗜冷酶稳定性的影响因素

如上所述，在低温下快速繁殖的生物通常要经历以上修饰以适应低温环境，这包括在酶动力学、超分子和亚细胞结构和生理学方面一系列的生物化学调整，以及在超分子和亚细胞方面的调整，如膜流动性、离子通道、聚合和解聚的作用。

静电作用在蛋白质的分子内作用是非常重要的，因为它们具有作用力强和长距离的特性。离子键、氢键和范德华力这样的作用在低温下是稳定的，并且它们的形成是放热形式的。所有的这些类型中，蛋白质结构中最强最稳定的相互作用是在两个带相反电荷的侧链间的离子键（Feller and Gerday，1997）。金属离子键，通常是钙离子键，能提供一些其他弱的相互作用无法超越的额外的稳定性，甚至高于二硫键。钙离子键可以在几个二级结构或蛋白质的结构域间起桥接作用（Georlette，2004）。

精氨酸在温度适应上起着主要作用，这个稳定作用归因于精氨酸的胍基提供的电荷共振。和其他常温酶相比，嗜冷酶的精氨酸-赖氨酸比率较低，并且通常和水形成相互作用，因此影响整个酶结构的柔性。和赖氨酸相比，精氨酸被发现主要是通过提供较多的静电作用，可以提高酶热稳定性。这也与其在解释电子云密度时高 B-因子有关。

氢键是最多的非共价键相互作用。单独的氢键作用对蛋白稳定是弱的，但是大量氢键的相互作用对蛋白质的折叠结构和稳定性起着主导作用。蛋白质分子内的相互作用和它与环境的相互作用间的平衡被打破时，就会导致蛋白质去折叠。至今嗜冷酶中氢键相关的信息研究报道是很少的（Feller et al.，1997）。

Makhatadze 和 Privalov 在 1996 年把疏水作用定义为非极性基团对水分子的移除作用和相同基团间相互作用的联合作用。嗜冷酶的表面包含许多疏水氨基酸残基，这一特性在以下这些冷适应酶中存在，如甘油醛-3-磷酸脱氢酶，柠檬酸合成酶，虾碱性磷酸酶，AHA，胰蛋白酶和核苷酸激酶。由于水分子的低熵值和极性基团周围的笼状结构，导致蛋白质结构被破坏。在低温时，负熵增加进而降低水分子的流动性，因此增加酶的耐受力和柔性。疏水作用从以下两方面改善嗜冷蛋白质构象的柔性和热稳定性，也就是疏水残基之间的相互作用（核心疏水性）以及疏水残基和水分子之间的相互作用（表面疏水性）。埋藏在嗜冷酶中的氨基酸比来自常温和嗜热的同源酶的侧链要小，疏水性也较低（Marx，2004）。有助于稳定性的焓变是由蛋白质内部疏水残基之间的距离决定的。范德华力相互作用的减小和内部基团运动的增加会破坏嗜冷酶的结构。

芳香环之间的相互作用有助于提高热稳定性。酪氨酸、苯丙氨酸和色氨酸的芳香环有助于内部能量，并且因为苯环上的部分负电荷和 C-H 端的部分正电荷有偶极。这有利于芳香环之间的相互作用或芳香环和赖氨酸、精氨酸之间的相互作用。常温枯草杆菌蛋白酶或 α 内酰胺酶中高度保守的芳香环的相互作用在它们的嗜冷同源物中是缺失的，说明这些改变的相互作用跟酶的低稳定性相关。

如果蛋白质末端没有埋藏在结构内部或被弱相互作用抑制，它通常是蛋白质去折叠首选的位点。*A. haloplanctis* α-淀粉酶的 C-末端比常温微生物同源蛋白的长，并且在水溶液中自由存在。*P. immobilis* β-乳糖酶的 N-末端 α-螺旋结构和三文鱼胰蛋白酶的 C-末端螺旋结构在分子表面不是稳固存在的，表明这个松散的末端与整体弱的热稳定性有关（Smalas et al.，2000）。

盐桥给冷适应蛋白质提供了额外稳定性，盐桥是指非氢原子距离在 2.5~4.0 Å 之间的离子对。研究表明，冷适应酶中盐桥的数量是降低的。盐桥对稳定冷适应酶提供的能量是 12~21 kJ/mol。极端微生物中盐桥数量由大到小依次是嗜冷微生物、常温微生物、嗜热微生物，增加的数量与热稳定性是相关的。Met 的高含量是影响稳定性的另一个因素，这也在冷适应酶中有发现。

3.7 嗜冷酶的稳定性、活性和柔性之间的关系

嗜冷酶表明高的催化活性通常与低的热稳定性相关。分子结构的柔性或可塑性允许其在低能量消耗的情况下有更好的互补，这解释了嗜冷酶的高特殊活性。蛋白质结构中某些部位控制着蛋白质的稳定性，而其他区域则对在环境温度下蛋白质的柔性至关重要，这确保酶在环境温度下有一个最佳的催化效率。因此，这可能是嗜冷蛋白获得高催化活性与良好稳定性的关键。柔性的提高被认为是嗜冷酶的主要结构特点（Salvino，2006）。

3.8 嗜冷酶氨基酸组成分析

嗜冷酶的结构分析数据主要是通过从已知氨基酸序列的同源蛋白比较研究中获得。大量的针对嗜冷酶、常温酶和嗜热酶的同源蛋白的氨基酸组成分析研究有助于理解结构改变和蛋白质稳定性之间的联系。Arg 代替 Lys 更多地出现在常温和嗜热微生物中，而 Asn 更多地发现在嗜冷微生物中。低的 Arg/Arg+Lys 比例被认为是嗜冷蛋白质的决定因素。此外，Val 比较倾向于出现在嗜热微生物和常温微生物的 β-折叠片，而在嗜冷微生物中，Val 出现在埋藏区域和 β-折叠的频率比较低（Feller and Gerday，2003）。

将嗜冷酶与常温和嗜热同源蛋白的 X 射线结构进行比较，并详细分析了与低温活性相关的结构参数。高分辨率的嗜冷 α-淀粉酶的 X-射线结构与它最接近的常温结构同源蛋白作比较。研究整体 3D 结构时发现嗜冷微生物的 α-淀粉酶和常温微生物的 α-淀粉酶的显著不同在于无规则卷曲的数量和特点。在 A 结构域，loop 结构 1（50-56），4（216-227）和 5（268-272）在嗜冷微生物的结构中是不存在的。*A. haloplanctis* 的 α-淀粉酶的 273 号

残基是甘氨酸，而在人类 α-淀粉酶中是丙氨酸。哺乳动物 α-淀粉酶的 loop 7 结构在 *A. haloplanctis* 的 α-淀粉酶中不再是一个 loop 结构。Loop 8 结构在 *A. haloplanctis* 的 α-淀粉酶中是变短的，它仅对应着 316-318 残基，因此在序列比对中产生了一个位移。

在 B 结构域中两个无规则的卷曲结构域，发现了区别。与 *A. haloplanctis* 的 α-淀粉酶相比，在哺乳动物中，Loop 2 的 105~112 残基位于额外的二硫键附近。这也许是嗜冷微生物 *A. haloplanctis* 的 α-淀粉酶比哺乳动物的酶更具柔性的原因。另一个不同出现在 loop 3 区域，loop 3 包括位于 B 结构域表面的 138~146 残基。这个区域在 *A. haloplanctis* 的 α-淀粉酶中部分的缺失了，与之对应的是 118~120 残基。在结构域 C 中，loop 10 比相对应的哺乳动物 α-淀粉酶的 loop 结构的一级序列多出两个残基。最后一个不同是在 loop11 中发现的，似乎与 *A. haloplanctis* 的 α-淀粉酶一级序列氨基酸残基的缺失有关。表 3.2 总结了嗜冷 α-淀粉酶的结构特点（Nushin et al.，1998）。

表 3.2　α-淀粉酶的结构特点

α-淀粉酶	嗜温菌	嗜冷菌
盐桥	26	18
二硫键	5	4
疏水性	59.7	61.6
酸性氨基酸/%	21.3	18.2
碱性氨基酸/%	10.9	9.9
α 螺旋耦极电荷 Na 负电	4	2
Ca 正电	6	3
脯氨酸含量	21	13
精氨酸含量	28	13
精氨酸相关的 H 键	42	20
核心簇疏水性	103	47
0-芳香环相互作用	19	24
C 端非约束残基	19	6
静电荷	-17.9	-19
总氨基酸	~530	~477

参考文献

Georlette, D.（2004）Some like it cold: biocatalysis at low temperatures. *FEMS Microbiology Reviews* 28, 25-42.

Feller, G. and Gerday, C.（1997）Psychrophilic enzymes: molecular basis of cold adaptation. *Cell. Mol. Life Sci.* 53, 830-841.

Feller, S. E. , Yin, D. , Pastor, R. W. and MacKerell A. D, Jr. (1997) Molecular dynamics simulation of unsaturated lipid bilayers at low hydration: parameterization and comparison with diffraction studies. *Biophys. J.* 73 (5), 2269-2279.

Feller, G. and Gerday, C. (2003) Psychrophilic enzymes: hot topics in cold adaptation. *Nature Reviews, Microbiology 1*, 200-208.

Gerday, C. (2000) *Cold-adapted enzymes: From Futtdamentals to Biotechnology.* Elsevier Science Ltd. , Vol. 18, 103-107.

Khawar Sohail, S. and Ricardo, C. (2006) Cold-adapted enzymes. *Annual Review of Biochemistry 75*, 403-433.

Liu, S. , Graham, J. E. , Bigelow, L. , Morse II, P. D. and Wilkinson, B. J. (2002) Identification of Listeria *monocytogenes* genes expressed in response to growth at low temperature. *Appl. Environ. Microbiol. 68,* 1697-1705.

Makhatadze, G. I. and Privalov, P. L. (1996) On the entropy of protein folding. *Protein Sci.* 5, 507-510.

Pulicherla, K. K. , Mrinmoy, G. , Suresh, P. K. arid Sambasiva Rao, K. R. S. (2011) Psychrozymes - the next generation industrial enzymes. *J. Marine Sci. Res.* 1 (1), 1-7.

Purusharth, R. I. , Klein, F. , Sulthana, S. , Jager, S. , Jagannadham, M. V. et al. (2005) Exoribonuclease R interacts with endoribonuclease E and an RNA-helicase in the psychrotrophic bacterium Pseudomonas *syringae* LZ 4W. *J. Biol. Chem.* 280, 14572-14578.

Marx, J. C. (2004) A perspective on cold enzymes: current knowledge and frequently asked questions. *Cellular and Molecular Biology* 50 (5), 643-655.

Nushin, A. , Feller, G. , Gerday, C. and Haser R. (1998) Structures of the psychrophilic *Alteromonas haloplanctis* α-amylase give insights into cold adaptation at a molecular level. *Structure* 12, 1503-1516.

Schmidt Nielsen, S. (1902) Ueber einige psychorphile Mikroorganismen und ihr Vorkommen. *Centr. Bakteriol. Parasitenk.* , *Abt.* Ⅱ. 9, 145-147.

Salvino, D'Amico. (2006) Psychrophilic microorganisms: challenges for life. *EMBO Reports* 4, 385-389.

Smalas, A. O. , Leiros, H. K. , Os, V. and Willassen, N. P. (2000) Cold adapted enzymes. *Biotechnology Annual Review* 6, 1-57.

第 4 章　海洋生物酶在精细化学物质生物合成中的应用

Halina Novak and Jennifer Littlechild, Biocatalysic Centre University of Exeter, Exeter

DOI：10.1533/9781908818355.1.89

摘要：这一章将讨论酶在精细化工新药结构单元分子生物合成中的应用。我们会重点关注在海洋环境中发现的新颖的酶在可持续化学这一重要领域中的应用。被克隆和通过分离纯化得到的来源于海洋细菌、古细菌及大型海藻及病毒的海洋生物酶会被作为具体的例子进行阐述。这些酶的活性包括：加卤酶、脱卤素酶、乙醇脱氢酶、氨基酰化酶、蛋白酶、酯酶、脂酶。这些酶的生物化学和结构方面研究将会在本章中讨论，同时涉及相关酶类催化机制以及进化的多样性。

关键词：海洋生物酶，生物催化，精细化学合成

4.1　引言

4.1.1　生物催化作用

长时间可以反复使用的生物催化剂的主要优点是在温和的水环境下能够保持很高的活性。在工业上，许多溶剂都是有毒和有害的，而且 80%的溶剂消耗都来源于工业生产，因此，迫切需要减少溶剂的消耗（Woodley，2008）。在化学合成中利用酶催化不仅可以减少溶剂的消耗，并且避免了废弃物的处理及有机溶剂提取的步骤。相对于传统的有机合成，生物酶催化通过绿色的途径降低了精细化工合成过程中的成本（Ran et al.，2007；Tao and Xu，2009）。

生物体系内的反应条件是严格控制的，这样才能够保证酶发挥最大的催化效率。这就意味着酶在严格的条件下，如高温，pH 值，某些溶剂的存在，会使酶变得不稳定（Polizzi et al.，2007）。生物催化的这些缺点可以通过基因工程、分子生物学、重组方法来解决，

这可以用低成本获得大量的酶。目前已经发展了多种实验技术去尝试提高酶的活性或者改变酶的底物特异性，以获得满足工业需要的酶。这包括了靶基因的随机突变及筛选与定向蛋白质工程（Cherry and Fidantsef, 2003；Morley and Kazlauskas, 2005；Eijsink et al., 2005）。酶的固定化技术也被用于酶的生物催化中，可以解决其稳定性差的问题，从而提高酶的使用效率（Mateo et al., 2007）。

采用“绿色化学”的方法，制备大量结构多样性的分子在精细化工合成中具有大量的需求。因此，新颖的酶或者蛋白质的发掘变得越来越重要。具有特定活性的商业化酶的发掘已成为生物催化方面最大的限制。

4.1.2 海洋环境

近年来，海洋环境被认为是新型酶和代谢物未开发的宝库（Trincone, 2011）。海洋大约占地球表面积的71%，广袤的海洋环境中生活着适应不同环境的植物，无脊椎动物，脊椎动物以及海洋微生物（Kennedy et al., 2008）。海洋植物或者动物细胞内外的海洋微生物拥有大量的酶和复合物来维持其共生生活方式。这使得海洋微生物成为新型工业酶发掘的重要资源。

虽然只有少量的海洋生物酶被表征，但其中部分酶蛋白对于现代科研有很大贡献。比如从对虾中发现的对虾碱性磷酸化酶（SAP），来源于 Thermococcales 的 DNA 聚合酶和从 *Aequorea victoria* 中发现的绿色荧光蛋白 GFP 等，它们已成为细胞生物学必备的工具被广泛应用（Niehaus et al., 1997；Prasher et al., 1992；Cubitt et al., 1995；De Backer et al., 2002）。

在海洋环境中，最极端的环境已经被深入地研究过（Synnes 2007）。比如酸性湖泊，深海热泉，冰山盐水带和死海，这里生活着被称为极端的生物和独特的生物群体。只能生活极端环境微生物的极端海洋环境，其中包括能在60~120℃间可以生长的嗜热菌，能在-15~10℃之间生存的嗜冷菌，能承受38 MPa的适压菌，能在 pH≤3 的环境中生存的嗜酸菌，能在 pH≥9 的环境中生存的嗜碱菌，能在 NaCl≥0.2 mol/L 的环境中生存的嗜盐菌。适应在相对应的极端环境下生存的这些极端微生物已经具有了某些特定的特征，因此，它们能够在相对于温和条件的极端环境下生存下来（Champdoré et al., 2007）。它们通常具有更复杂的底物特异性，并且会参与宿主的一些新的代谢途径。

冰海是世界上最冷的栖息地之一，温度范围为-1.8~-30℃。冰海的界面被认为是分离嗜盐菌和嗜冷菌的关键环境，而这两种菌是可以产生耐盐酶和适冷酶。海洋环境并非都是营养丰富的，这表明生活在冰海的海洋微生物种群已适应了冰海环境，它们采用冬眠或者其他的方式来克服恶劣环境（Hall-Stoodley et al., 2004）。嗜冷酶在工业中的应用可以避免反应物与产物不稳定的问题，同时降低了反应的能耗成本。嗜冷菌通常会暴露在高盐、高压、高氧的环境中，这些条件能够使嗜冷菌的蛋白变得更加稳定（Gomes and Steiner 2004）。一种嗜冷酯酶（在0℃时依然保持了50%的活性）已经从假交替单孢菌

(*Pseudoalteromonas arctica*) 的嗜冷菌中被克隆，表达和纯化（Al Khudary et al.，2010）。这种酶拥有了广泛的底物活性，比如脂肪酸（包括 C2-C8 的脂肪酸），在 pH 7.5，25℃的环境中表现出最高活力。

对来源于嗜冷菌的蛋白质结构研究中发现，这些蛋白质在低温的环境中能够表现出更高的蛋白质柔性（Feller，2003）。Georlette 等（2004）曾经综述过关于酶的活力，稳定性以及柔性之间的精细平衡能够控制酶促反应动力学。嗜冷菌同其在常温下的同源蛋白的比较显示它们拥有更少的离子相互作用，氢键及疏水相互作用。嗜冷酶在低温下表现出极强的特异性活性，提示其在工业上具有广阔的应用前景（Cavicchioli et al.，2002.，Siddiqui and Cavichiolli，2006）。

耐盐和酶在有机溶剂中活力之间的直接联系已经被研究过。这使得来源于嗜盐菌（可以在浓度不小于 0.2 mol/L NaCl 的环境下生长）的酶非常有潜力在工业中广泛使用，并且可以替代传统的有机合成方式（Marhuenda-Egea and Bonete，2002；Trincone，2011）。海洋中高盐的环境，如盐湖和死海、盐海，都对发掘新型海洋生物酶提供了理想的环境。相对于陆地生物体内提取的氧化还原酶，从海洋微生物中提取的酶具有更丰富的立体选择性（Trincone，2011）。这些酶在高盐浓度中能够有更高的活性和热稳定性。这个例子说明来源于海洋生物群体的酶具有新型的特征。

海洋生物工程是这样被描述的：从整体、细胞和分子水平探索海洋生物的生存能力，运用现代生物技术的手段去推动对海洋生物系统的认识和了解，以此来解决当前面临的问题（Lee and Burrill，1994）。这也包括从海洋生物中寻找新型的可以用于精细化学合成的海洋生物酶（Querellou et al.，2010）。

4.2 卤代酶与脱卤酶

4.2.1 富含卤素的海洋环境

海洋藻类能合成多种多样的生物活性化合物，并且是有机卤素的最大生产制造者（Field et al.，1995；Valverde et al.，2004）。这包括用作生物合成构建模块以及制药行业感兴趣的卤化萜烯，它们通常显示抗菌和拒食素活性（Kurata et al.，1998；Cabrita et al.，2010）。例如藻类物种能合成卤代萜烯的有珊瑚藻属、凹顶藻属、海头红属，这些物种都含有钒溴代过氧化物酶（Butler，2005）。有机卤素是有毒性的，因此推测其可以抑制藻类表面微生物的聚集和生长，阻止了食草动物对其的摄入（Nightingale et al.，1995）。一些多毛纲的管状蠕虫也被认为能产生一系列结构不同的卤代化合物（Fielman et al.，2001）。海洋中丰富的有机卤素为藻类和海洋微生物产生卤代酶和脱卤酶提供了一个独特的潜在环境。

4.2.2 卤代酶和卤代过氧化物酶

卤化药物可以通过多种方式改变其活性，包括改变他们的专一性和影响药物分子穿过细胞膜的扩散能力。例如，从海洋放线菌提取的天然氯化抗癌剂比非氯化形式具有更强的活性（Feling et al.，2003）。在制药行业大约20%的药品是卤化的（Yarnell，2006），这使得作为生物催化剂——卤化在制药工业中是一个令人关注的目标（Eustáquio et al.，2008）。

催化化合物卤化的酶分为两类：卤代过氧化物酶和卤代酶。卤代过氧化物酶进一步分成两类：含血红素的卤代过氧化物酶和含钒的卤代过氧化物酶。这些酶能够在过氧化氢（H_2O_2）和卤化物离子（F^-、Cl^-、Br^-）存在的情况下催化有机分子的卤化（F −、Cl − Br）。第一次是从海洋生物体中提取出来的卤代过氧化物酶属于氧化还原酶超家族（Colonna et al.，1999）。

首个富含血红素的卤代过氧化物酶是在海洋真菌（*Caldariomyces fumago*）中被发现的。该酶的生化特征已经被表征，其高分辨率的（1.9 Å）晶体结构已经被解析（Sundaramoorthy et al.，1998）。该结构显示其由8个螺旋组成的一种新的折叠模式。血红素卤代过氧化物酶的催化机制已经建立，Glu^{183}氨基酸是用作酸碱催化剂，切断O–O键。这与使用His作为酸碱催化剂的其他氧化物酶不同。过氧化氢和卤化物离子都是完成氯化物氧化反应所必需的（Wagenknect and Woggon，1997）。

钒卤代过氧化物酶

钒是保持此类卤代过氧化物酶活性的一个必要元素。钒依赖的溴代过氧化物酶在各种各样的海洋藻类中被发现，包括跑叶藻、齿缘墨角藻、海带、珊瑚藻等。最近相关酶在细菌中发现，例如在通过生物合成萘吡酮霉素的链霉菌基因簇中，CNQ-525菌株可合成氯化杂萜（Winter et al.，2007）。

钒依赖的卤代过氧化物酶在其生活环境中的作用仍是有争议的。然而，通过体外化学转化(E)-(+)-橙花叔醇产生海洋天然产物 α-synderol，β-synderol，γ-synderol，证明珊瑚藻溴代过氧化物酶参与生物合成溴化环状倍半萜烯（Carter-Franklin et al.，2003）。

这些酶还将溴化一系列其他非天然的化合物 Carter-Franklin and Butler，2004）。利用龙须菜来源的溴代过氧化酶催化（E）-4-phenyl-buten-2-ol 生成溴代醇的立体化学特异性与通过化学反应过程形成的有着显著不同（Coughlin et al.，1993）。其他反应是没有化学立体特异性的，但酶促反应随反应物和反应条件的变化而变化。

从结构的层面，对几种钒依赖的氯代过氧化物酶和溴代过氧化物酶的催化机制及不同底物结合状态进行了研究。这些酶的主要结构特点是它们的 α-螺旋桶状的结构。然而这些酶的聚合状态是多种多样的，如跑叶藻溴代过氧化氢酶是二聚体，珊瑚藻溴代过氧化物酶是十二聚体。跑叶藻和珊瑚藻酶通过二聚体来形成两个活性中心，每个亚基帮助建立活性中心的通道和钒结合区域通道。

跑叶藻二聚溴代过氧化氢酶通过3个二硫键的连接来维持二聚体的稳定性（Weyand et al.，1999），而在珊瑚藻来源的同源酶中，这些半胱氨酸不保守，它的结构由6个二聚体有序的整合成十二聚体（Brindley et al.，1998；Isupov et al.，2000）。每个单体的N-末端区域组成一个26 Å 的腔形结构，这个腔没有特定的电荷或疏水性质，这使得它不可能结合金属离子，而只发挥结构的作用。在12个单体中，每个都由19个 α-螺旋组成，螺旋长度为6~26的氨基酸残基，8个 3_{10}螺旋和14个 β 链主要参与 β-hairpins 的形成。单体的一个表面是平面，二聚后表面之上形成了一个中心区域，导致在每个二聚体中心有两个由4个 α-螺旋组成的束。活性中心由来源于两个单体的氨基酸残基组成。来源于一个单体的氨基酸残基主要参与结合底部活性部位，它们主要是结合钒酸或磷酸，而来源于另一个单体的氨基酸残基构成活性中心的顶部区域。然后二聚体相互作用形成十二聚合体，该十二聚合体直径约150 Å，有23立方对称点群。

不同酶之间参与钒的结合的氨基酸残基是保守的。这些保守的残基在藻类溴代过氧化物酶和真菌氯化过氧化物酶中的是 Arg^{545}、His^{551}、Arg^{406}、Ser^{483}、Lys^{398}、Glu^{484}（这些顺序来源于珊瑚的溴代过氧化物酶）。在瑚藻溴代过氧化物酶中，曾在氯化过氧化物酶中结合氯的 Phe 被 His^{478}取代，Tyr 被 Arg^{395}替代，Gly 被 His^{485}替代。这些氨基酸残基在酸性磷酸酶中也是高度保守的（Hemrika et al.，1997；Littlechild et al.，2000）。对珊瑚溴代过氧化物酶结合磷酸和矾酸的活性中心进行了结构分析。最初获得结构解析的酶的活性部位结合的矾酸被磷酸代替。这是由于在结晶时使用了高浓度的磷酸盐。珊瑚钒依赖的溴代过氧化物酶结合钒酸盐的结构也已经获得解析。除了直接与钒酸盐结合的 His^{553}侧链 N 有0.63 Å 的偏移，其余活性部位的基本结构保持不变。

在结合矾酸盐的复合物结构中，矾酸是三方晶系的双锥体，这个结构中钒酸氧和 His 的咪唑处于轴向位置，3个钒酸氧处于顶点位置。钒酸的构象是不稳定的，在结合和非结合的状态之间变化。钒酸原子和氧离子之间的距离比在其他钒脱卤酶结构中的距离要长。

晶体结构的研究为理解钒依赖的脱卤酶的作用机制提供了一些信息。作为一个自由的卤化剂如 HOBr、Br_2、Br_3^-，或者是一个复杂的酶-Br 复合物或 V_{enz}-OBr 类，这些酶作为溴化剂的本质可以被描述为一个"Br^+-like 中间态"（Butler，1997）。用扩展 X 射线吸收精细结构（EXAFS）研究泡叶藻钒溴代过氧化物酶（Christmann et al.，2004），结果表明：结合溴的活性部位接近钒酸的中心。在酶促反应中，这使得溴原子在卤氧化反应的第一步中就可以直接进攻过钒酸盐中心。对于钒依赖的卤素过氧化物酶是直接结合底物，还是只是催化溴化剂或氯化剂的形成，接着在酶的活性部位与底物发生作用的机制一直在争论中。这常常用来解释许多底物缺乏立体构象特异性的原因。现已证实来源于 *A. Nodosum* 的溴代过氧化物酶，存在吲哚和溴化苯酚红之间的竞争（Tschirret-Guth and Butler，1994）。此外，同一研究小组通过荧光猝灭实验证明了吲哚也是结合活性部位的。此外，来源于 *A. nodosum* 的溴代过氧化物酶可以催化局部的特异性的吲哚溴化（Martinez at al.，

2001)。这些结果提供了反应发生在酶的活性部位最直接的证据。最近，珊瑚溴代过氧化氢酶在溴存在的环境中结晶，在同步辐射光源收集到衍射数据。这直接显示溴跟酶的结合，它与钒酸和 Arg^{397} 之间形成距离为 2.8~3.1 Å 的氢键。溴结合可以引起活性部位氨基酸残基的变化。Leu^{337}（其 C-末端有 4.7 Å 的偏移）和 Phe^{373} 的偏移被发现。结构分析发现结合磷酸、钒酸和溴导致了酶催化活性中心口袋的略微变小，这创造了一个更加疏水环境来促使催化反应的完成（Littlechild et al.，2009）。

所有上述描述的卤化酶在商业生物催化方面都有潜在的应用。其中有些酶被表征的比较多，这只会增加人们对其在生物催化商业应用领域的兴趣。人们关注的不仅仅是不同的酶在不同的卤代化合物的天然合成途径中的作用，而且更关注不同的酶在改变其合成化合物的生物活性方面的变化性的修饰及应用。如果分离纯化酶应用于生物转化，酶的稳定性往往是一个主要问题。钒依赖的卤代过氧化物酶在结构上更加牢固，而且酶对过氧化氢具有更强耐受性。但它们常常缺乏立体化学选择性，这个结论已经由通过快速监测到的珊瑚溴代过氧化物酶对多溴化酚红到溴酚蓝的活性得到验证。不同的钒卤代酶之间底物结合特异性有所不同，这与组成活性部位的氨基酸残基的不同有关。珊瑚藻和泡叶藻钒溴代过氧化物酶活性部位的 5 个高度保守的氨基酸残基并不参与钒的结合。在钒酸位点 7.5 Å 的范围内，珊瑚酶有 3 个带电氨基酸，而泡叶藻酶有 3 个亲水氨基酸，并且在这个区域的没有带电氨基酸。在真菌 *Curvularia* 的钒依赖的卤代过氧化物酶中，未见有结构上保守的氨基酸。这些酶同时还可以催化具有商业应用前景的磺化氧化反应，而不是环氧化作用反应（Littlechild，1999）。

4.2.3 脱卤酶

多种微生物通过进化获得了降解作为唯一碳源的卤代脂肪族化合物，各种各样的脱卤酶已被鉴定和表征，包括卤代醇脱卤酶、卤代烷脱卤酶和卤酸脱卤酶。

2-卤酸脱卤酶是特别引起关注的，因为其可产生纯的 2-羟基烃酸对映异构体，这是精细化工行业合成的重要构建模块（Schoemaker et al.，2003，Haki and Rakshit，2003）。2-卤酸脱卤酶可分成两个亚类。Ⅰ部分包括 D/L 卤酸脱卤酶（D/L-HAD）和 D-卤酸脱卤酶（D-HAD），Ⅱ部分包含 L 卤酸脱卤酶（L-HAD）。D-HADs 与 L-HADs 是一对特定的对映异构体，并且通常可以制备在 α 碳原子位置具有反向不对称性的产物，而 D/L-HADs 使用两个对映异构体作为底物，并且在 α 碳原子位置以反向不对称构型进行催化。

L-卤酸脱卤酶

L-HADs（EC 3.8.1.2）催化短链 L-2-卤代烷酸转变为 D-2 卤代烷酸。虽然许多从陆地微生物来源的 L-HADs 的生化和结构特征得到表征（Novak，2011；Novak，et al.，2013a），但最近研究报道了从海洋细菌中发现两个 L-HADs。这些 L-HADs 有一些令人关注的独一无二的生化和结构特点。

其中一个 L-HADs（Pin L-HAD）是从 *Psychromonas ingrahamii*（Pin）中分离出来的，这种细菌是 1991 年，从阿拉斯加的埃尔森咸水湖中分离出来的。原始水样本是从接近 -10℃ 的海冰表层提取。对于 *Psychromonas ingrahamii*（Pin），最低生长温度记录是-12℃下，240 h 内生长一代（Breezee et al.，2004）。尽管该酶是从嗜冷微生物中分离得到的，但酶活的最佳温度为 45℃，T_m 值为 85℃（Novak，et al.，2013b）。结构模型显示 Pin L-HAD 并没有太多嗜冷酶的特点。相对于常温的同源酶，一个更加疏水的表面被发现，这是符合其他嗜冷酶特点的性质。同时，也从模型中发现了一些嗜热酶的特点，如大量盐桥的存在。在海洋嗜冷微生物中，人们对这种相对耐热酶的存在产生了极大兴趣。

第二种 L-HAD 从海洋中的 *Rhodobacterareae*（Rhb）家族细菌中提取，这种多毛类蠕虫来自英国阿盖尔郡的特拉利海滩。这种酶的生物化学和结构特征都已经被表征了。尽管 L-HAD 的底物特异性类似于之前的 L-HAD 的特征，但其活性中心有明显的不同。共价键结合氯丙酸和氯醋酸的酶与底物的复合物结构证实了 Asp^{18} 是主要的催化残基。以前针对 L-HADs 底物结合特征研究提出的是“Lock down”机制。不同的是，DehRhb 似乎并不符合这个机制，被 His^{183} 活化的关键水分子和 Glu^{21} 在其他 L-HADs 中不存在。DehRhb 酶代表新型的脱卤酶，其是还未被研究的 L-HAD 家族的酶。DehRhb 结构的研究将为了解 L-HAD 酶的机制提供新的认识，因为它是同已知的陆地微生物来源的 L-HADs 不同的新酶。

通过实施溶剂孵育和热转变实验来评估 DehRhb 和 Pin L-HAD 的溶剂稳定性。后一种实验测量不同条件下蛋白质的熔化温度。虽然相较于之前研究的 L-HADs，两种酶都有更好的溶剂稳定性，DehRhb 是报道的首个在有机溶剂中孵育 1 h 后，活性增加了的 L-HAD 酶。热转移实验也表明 DehRhb 对溶剂具有较好的耐受性。研究发现其他耐盐微生物产生的蛋白质对有机溶剂也具有高稳定性（Trincone，2011；Marhuenda-Egea and Bonete 2002）。由于工业过程往往是在有大量有机溶剂存在的情况下进行的，耐受有机溶剂的酶是非常有优势的。有机溶剂可以改善底物的特异性和溶解度（Schmid et al.，2001）。

这些海洋来源的 L-HADs 酶在工业上具有重要的应用前景，尤其是在制药行业生产光学纯的药物中间体，包括羟基烃和卤代烷酸。同之前 L-HADs 酶相比，这些海洋 L-HADs 酶的独特特征说明海洋环境是挖掘具有独特性的新酶的重要资源。

4.3　乙醇脱氢酶

乙醇脱氢酶不仅能够用来分解乙醇，还可以用来高产率的合成新的手性乙醇。乙醇脱氢酶能够应用于制药工业中手性乙醇的生产。一个含锌的耐热乙醇脱氢酶已从好氧的古细菌 *Aeropyrum pernix* 中分离出来。它直接从基因组中克隆出来，并在 *E. coli* 中高效的异源表达（Guy et al.，2003）。这是从日本 Kodakara-Jima 岛的沿海硫质喷气孔中分离到的一种微生物。它是异养微生物，最适生长温度在 90~95℃，pH 值 7.0。这种酶的生化特性已经被表征，并且获得了高分辨的晶体结构（Littlechild et al.，2003）。这种酶是四聚体聚合物，其中含有锌，Ⅰ型的 ADH 单体分子量是 39.5 kDa。*A. pernix* ADH 每个单体与邻近的单体形成 4 个离子键，加上相对称的等价物，在四聚物中总共有 16 个离子键。在 *A. pernix* ADH 每个亚基的作用界面都有大量的疏水相互作用。同时每个单体之间都会形成一个二硫键，这对于维持四聚体的稳定性也是很重要的，而且在过量锌离子缺少的情况下也会形成。这确保 *A. pernix* ADH 在超过 90℃的高温下 1 h 依然保持 50%的活性。该酶也具有很好的溶剂稳定性，即使在 60%的乙腈或者二氧六环孵育的情况下，依然可以保持超过 50%以上的活性。该酶对一级和二级乙醇有活性，最佳的链长度是 C4-C5，对大环状醇类如环己醇和环辛醇具有很高的催化活性。针对修饰的环己酮的研究表明，该酶对 3-甲基-环己酮有最大的活性，而对 2-甲基-环己酮无催化活性（Guy，2004）。

因为 ADH 酶在常温下就已经有活性，因此没有必要在高温下使用该酶。如果有足够的酶，那么在次佳温度下实施催化过程也是可行的。ADHs 是辅因子依赖的酶，在细胞中，烟酰胺辅因子在很多酶中是持续循环利用的，提供稳定的氧化还原态的供应。当一个纯化的 ADH 被用来生产二级醇时，还原态的辅因子必需持续供应。此反应最常用的酶系统是甲酸脱氢酶。在此过程中，甲酸盐被氧化成二氧化碳可以重新产生烟酰胺辅因子（NAD^+）。Ngamsom 等（2010b）在微反应器中利用电化学的方法重新生成辅因子。

4.4 L-氨基酰化酶

很多具有药学活性结构都是含氮化合物，这类化合物能够从 L 或 D-氨基酸中分离出来。非天然氨基酸的需求也有很大增长。例如，针对此过程，从厌氧的古细菌 *Thermococcus litoralis* 中发现的嗜热的海洋酶氨基酰化酶。在混合物中，这种酶只能够利用混合物中的一种对映体，从而产生自然界中的 L-氨基酸。L-氨基酰化酶已经被克隆，并且在 *E. coli* 中高效表达。这使它的生化特征和晶体结构能够被很好地研究（Toogood et al.，2002a；Hollingsworth et al.，2002）。该氨基酰化酶是同源四聚体，单体分子量是 43 kDa。同来源于 *Pyrococcus horikoshii* 中的氨基酰化酶有 82%的序列同源性，同来源于 *Sulfolobus solfataricus* 的羧肽酶有 45%的序列同源性。Thermoccocus L-氨基酰化酶与细菌的氨基酰化酶是不同的，因为像丁基丙二酸盐这样的抑制剂对它的活性只有轻微影响。

重组酶对热比较稳定，70℃下半衰期是 25 h，并且能够被固定化和装进柱状反应器中，且能够在 60℃下，将底物持续转换成产物，在长达 10 d 的反应时间内酶的活性不丢失（Toogood et al.，2002b）。这种酶主要是针对含 *N*-苯甲酰氨基酸和 *N*-氯乙酰氨基酸的底物，偏爱含疏水基团，不带电或者带弱电的氨基酸，像 Phe、Met 和 Cys。目前，Dow Pharma/Chirotech 已经用 L-氨基酰化酶的方法来大批量的生产光谱纯的 L-氨基酸和氨基酸类似物。这种酶已经被固定到塑料制备的微反应器中。能够进行快速的底物筛选，避免底物和产物抑制的问题。这种稳定的 L- 氨基酰化酶在 80℃的条件下活性可保持 10 h（Ngamsom et al.，2010a）。

4.5 蛋白酶、酯酶和脂酶

一种新型的半胱氨酸蛋白酶已经被鉴定并克隆到 *E. coli*。这种酶被鉴定为焦谷氨酰羧肽酶，并应用在蛋白质组学分析中去封闭多肽和蛋白质（Singleton et al.，1999；Singleton and Littlechild，2001）。

4.5.1 酯酶和脂酶

酯酶和脂酶能够用在立体专一的水解，酯基转移，酯合成和其他有机生物合成反应中。这种酶已经广泛地被化学家所应用，因为它们在有机溶剂中比其他酶类更稳定。酯酶催化酯键的形成和断裂。脂酶能够通过界面活化这一现象与酯酶区分出来。因为界面活化只在脂酶中观察到。脂酶倾向于更多不溶于水的底物，典型的包括由长的脂肪酸链组成的甘油三酯。然而，酯酶更倾向于水解简单的酯类，通常只能够水解脂肪酸少于 6 个的甘油三酯。

一种极度嗜热的深海古细菌 *Pyrococcus furiosus* 中有一个溶血磷酸酯酶，它的特异性对中等链长度的对硝基苯酯有最佳活性，对对硝基苯己酸盐（C6）有最大活性。这种酶在 80℃时活性最高，85℃时的半衰期为 96 min。因此，它在新型药物中间体的生产领域具有潜在的应用（Mallett，2004）。

海洋病毒也是潜在的在工业催化领域具有应用前景的新型酶开发的重要来源。颗粒大小在 160～200 nm 的 EhV-86 裂解病毒能够感染 *Emiliania huxleyi*。它是 *Coccolithoviridae*，属于 *Phycodnaviridae* 家族，有一个大的圆形基因组，含有大约 407 339 个碱基对，其中包含 472 个预测基因或蛋白质编码序列。基于序列相似性和蛋白质功能匹配性，只有 66% 和 14% 的基因被注释。一些编码酯酶和脂酶的基因已经被鉴定，病毒独有的编码鞘脂类生物合成路径的基因也已经被鉴定。

4.6 结论

本章列举了一些来源于海洋生物的酶被成功应用的例子。它们的成功应用说明生物工程是一个欣欣向荣的领域。

参考文献

AL Khudary，R.，Venkatachalam，R.，Katzer，M.，Elleuche，S. and Antranikian，G.（2010）A cold-adapted esterase of a novel marine isolate，*Pseudoalterotmonas arctica*：gene cloning，enzyme purification and characterization. *Extremopbiles* 14，273-285.

Breezee, J. , Cady, N. and Staley J. T. (2004) Subfreezing growth of the sea ice bacterium '*Psycbromonas ingrabamii*', *Microb. Ecol.* 47, 300-304.

Brindley, A. A. , Dalby, A. R. , Isupov, M. N. and Littlechild, J. A. (1998) Preliminary X-ray analysis of a new crystal form of the vanadium-dependent bromoperoxidase from Corallina officinalis. *Acta Crystallogr. D* 54, 454-457.

Butler, A. (1997) Vanadium-Dependant Redox Enzymes: Vanadium Haloperoxidases, for Comprehensive Biological Catalysis, British Academic Press, Ed. M. Sinnott, p427-437.

Butler, A. (2005) Molecular Approaches in Marine Pharmacology. Department of Chemistry and Biochemistry, Research Final Reports, California Sea Grant program UC San Diego, Santa Barbara.

Cabrita, M. T. , Vale, C. and Rauter, A. P. (2010) Halogenated compounds from marine algae. *Mar. Drug.* 8, 2301-2317.

Carter-Franklin, J. N. , Parrish, J. D. , Tschirret-Guth, R. A. , Little, R. D. and Butler, A. (2003) AJ Vanadium haloperoxidase-catalyzed bromination and cyclization of terpenes. *J. Am. Chem. Soc.* 125, 3688-3689.

Carter-Franklin, J. N. and Butler, A. (2004) Vanadium bromoperoxidase-catalyzed biosynthesis of halogenated marine natural products. *J. Am. Chem. Soc.* 126, 15060-15066.

Cavicchioli, R. , Siddiqui, K. S. , Andrew, D. and Sowers, K. R. (2002) Low-temperature extremophiles and their applications. *Curr. Opin. Biotechnol.* 13, 253-261.

Champdoré, M. , Staiano, M. , Rossi, M. and D'Auria, S. (2007) Proteins from extremophiles as stable tools for advanced biotechnological applications of high social interest. *J. Royal Soc. Interface.* 4, 183-191.

Cherry, J. R. and Fidantsef, A. L. (2003) Directed evolution of industrial enzymes: an update. Curr. Opin. Biotechnol. 14, 438-443.

Christmann, U. , Dau, H. , Haumann, M. , Kiss, E. , Liebisch, P. et al. , (2004) Substrate binding to vanadate-dependent bromoperoxidase from Ascophyllum nodosum: a vanadium K-edge XAS approach. *Dalton Trans.* 21, 2534-2540.

Colonna, S. , Gaggero, N. , Richelmi, R. and Pasta, P. (1999) Recent biotechnological developments in the use of peroxidases. *Trends Biotechnol.* 4, 163-168.

Coughlin, P. , Roberts, S. , Rush, C. and Willetts, A. (1993) Biotransformation of alkenes by haloperoxidases-regiospecific bromohydrin formation from cinnamyl substrates. *Biotechnol. Lett.* 15, 907-912.

Cubitt, A. B. , Heim, R. , Adams, S. R. , Boyd, A. E. , Gross, L. A. and Tsien, R. Y. (1995) Understanding, improving and using green fluorescent proteins. *Trends Biochem. Sci.* 20, 448-455.

De Backer, M. , McSweeney, S. , Rasmussen, H. B. , Riise, B. W. , Lindley, R and Hough, E. (2002) The 1.9 Å crystal structure of heat-labile shrimp alkaline phosphatise. *J. Mol. Biol.* 318, 1265-1274.

Eijsink, V. G. H. , Gaseidnes, S. , Borchert, T. V. and Van den Burg, B. (2005) Directed evolution of enzyme stability. *Biomol. Eng.* 22, 21-30.

Eustáquio, A. S. , Pojer, F. , Noel, J. P. and Moore, B. S. (2008) Discovery and characterization of a marine bacterial SAM-dependent chlorinase. *Nat. Chem. Biol.* 4, 69-74.

Feling, R. H. , Buchanan, G. O. , Mincer, T. J. , Kauffman, C. A. , Jensen P. R. and Fenical, W. (2003) Salinosporamide A: a highly cytotoxic proteasome inhibitor from a novel microbial source, a marine bacterium of the new genus *Salinospora*. *Angewandte Chemie International Edition* 42, 355-357.

Feller, G. (2003) Molecular adaptations to cold in psychrophilic enzymes. *Cell Mol. Life Sci.* 60, 648-662.

Field, J. A., Verhagen, F. J A. and Jong, E. D. (1995) Natural organohalogen production by *Basidiomycetes*. *Trends in Biotecb.* 13, 451-456.

Fielman, K. T., Woodin, S. A. and Lincoln, D. E. (2001) Polychaete indicator species as a source of natural halogenated organic compounds in marine sediments. *Environ. Toxicol. Chem.* 20, 738-747.

Georlette, D., Blaise, V., Collins, T., D'Amico, S., Hoyoux, A., et al., (2004) Some like it cold: biocatalysis at low temperatures. *FEMS Microbio. Rev.* 28, 25-42.

Gomes, J. and Steiner, W. (2004) The biocatalytic potential of extremophiles and extremozymes. *Food Technol. Biotechnol.* 42, 223-235.

Guy, J. E. (2004) PhD thesis, University of Exeter.

Guy, J. E., Isupov, M. N. and Littlechild, J. A. (2003) Crystallization and preliminary X-ray diffraction studies of a novel alcohol dehydrogenase from the hyperthermophilic archaeon *Aeropyrum pernix*. *Acta Crystallogr. D* 59, 174-176.

Haki, G. and Rakshit, S. (2003) Developments in industrially important thermostable enzymes. *Bioresour. Technol.* 89, 17-34.

Hall-Stoodley, L., Costerton, J. W. and Stoodley, P. (2004) Bacterial biofilms: from the natural environment to infectious diseases. *Nature Rev. Microbiol.* 2, 95-108.

Hemrika, W., Renirie, R., Dekker, H., Barnett, P. and Wever, R. (1997) From phosphatases to vanadium peroxidases: a similar architecture of the active site. *Proc. Acad. Sci.* 94, 2145-2149.

Hollingsworth E. J., Isupov, M. N. and Littlechild, J. A. (2002) Crystallization and preliminary X-ray diffraction analysis of L-aminoacylase from the hyperthermophilic archaeon *Thermococcus litoralis*. *Acta Crystallogr. D* 58, 507-510.

Isupov, M., Dalby, A., Brindley, A. A., Izumi, Y., Tanabe, T., et al., (2000) Crystal structure of dodecameric vanadium-dependent bromoperoxidase from the red algae Corallina officinalis. *J. Mol. Biol.* 299, 1035-1049.

Kennedy, J., Marchesi, J. R. and Dobson, A, D. W. (2008) Marine metagenomics strategies for the discovery of novel enzymes with biotechnological applications from marine environments. *Microb. Cell Fac.* 7, 27.

Kurata, K., Taniguchi, K., Agatsuma, K. and Suzuki, M. (1998) Diterpenoid feeding-deterrents from *Laurencia saitoi*. *Phytochemistry* 47, 363-369.

Lee, K. B., Burrill, G. S. (1994) Biotech 95: reform, restructure and renewal. The industrial annual report. Palo Alto, Ernst and Young, p24-78.

Littlechild, J. (1999) Haloperoxidases and their role in biotransformation reactions. *Curr. Opin. Chem. Biol.* 3, 28-34.

Littlechild, J., Garcia-Rodriguez, E., Dalby, A. and Isupov, M. (2000) Structural and functional comparisons between vanadium haloperoxidase and acid phosphatase enzymes. *J. Mol. Recognit.* 15, 291-296.

Littlechild, J. A., Guy, J. E. and Isupov, M. N. (2003) The structure of an alcohol dehydrogenase from the hyperthermophilic archaeon *Aeropyrum pernix*. *J. Mol. Biol.* 331, 1041-1051.

Littlechild, J., Garcia Rodriguez, E. and Isupov, M. (2009) Vanadium containing bromoperoxidase – insights into the enzymatic mechanism using X-ray crystallography *J. Inorg. Biochem.* 103, 617-621.

Mallet, 2004. PhD thesis, University of Exeter.

Marhuenda-Egea, F. C. and Bonete, M. J. (2002) Extreme halophilic enzymes in organic solvents. *Curr. Opin. Biotech.* 13, 385-389.

Martinez, J. S., Carroll, G. L., Tschirret-Guth, R. A., Altenhoff, G., Little, R. D. and Butler, A. (2001) On the regiospecificity of vanadium bromoperoxidase. *J. Amer. Chem. Soc.* 123, 3289-3294.

Mateo, C., Palomo, J. M., Fernandez-Lorente, G. and Guisan, J. M. (2007) Improvement of enzyme activity, stability and selectivity via immobilization techniques. *Enzyme Microb. Technol.* 40, 1451-1463.

Morley, K. L. and Kazlauskas, R. J. (2005) Improving enzyme properties: when are closer mutations better? *Trends Biotech.* 23, 231-237.

Ngamsom, B., Hickey, A. M., Greenway, G. M., Littlechild, J. A., Watts, P. and Wiles, C. (2010a) Development of a high throughput screening tool for biotransformations utilising a thermophilic 1-aminoacylase enzyme. *J. Mol. Catalysis B: Enzyme.* 63, 81-86.

Ngamsom, B., Hickey, A. M., Greenway, G. M., Littlechild, J. A., McCreedy, T., et al., (2010b) The development and evaluation of a conducting matrix for the electrochemical regeneration of the immobilised co-factor NAD (H) under continuous flow. *Org. Biomol. Chem.* 8, 2419-2424.

Niehaus, F., Frey, B. and Antranikian, G. (1997) Cloning and characterisation of a thermostable alpha-DNA polymerase from the hyperthermophillic archaeon *Thermococcus* sp. TY. *Gene* 204, 153-158.

Nightingale, P. D., Malin, G. and Liss, P. S. (1995) Production of chloroform and other low-molecular-weight halocarbons by some species of macroalgae. *Limnol. Oceanogr.* 50, 680-689.

Novak, H. R. Biochemical and Structural Characterisation of Dehalogenases from Marine bacteria. (2011) PhD thesis University of Exeter.

Novak, H. R., Sayer, C., Isupov, M. N., Paszkiewicz, K., Gotz, D., Mearns-Spragg, A., Littlechild, J. A. (2013a) Marine Rhodobacteraceae L-haloacid dehalogenase contains a novel His/Glu dyad that could activate the catalytic water FEBS J. 280, 1664-1680.

Novak, H. R., Sayer, C., Panning, J., Littlechild, J. A. (2013b) Characterisation of an L-haloacid dehalogenase from the marine psychrophile *Psychromonas ingrahamii* with potential industrial applications. *Marine Biotech.* (in press)

Polizzi, K. M., Bommarius, A. S., Broering, J. M. and Chaparro-Rigger, J. F. (2007) Stability of biocatalysts. *Curr. Opin. Chem. Biotechnol.* 11, 220-225.

Prasher, D. C., Eckenrode V. K., Ward W. W., Prendergast, F. G. and Cormier, M. J. (1992) Primary structure of the *Aequorea victoria green-fluorescent protein. Gene* 111, 229-233.

Querellou, J., Borresen, T., Boyen, C., Dobson, A., Hofle, M., et at., (2010) "Marine biotechnology: a new vision and strategy for Europe". *In European Science Foundation*, Marine Board position paper no. 15 (ed. Niall McDonough), pp. 1-91. Beernem, Belgium.

Ran, N., Zhao, L., Chen, Z. and Tao, J. (2007) Recent applications of biocatalysis in developing green chemistry for chemical synthesis at the industrial scale. *Green Chem.* 10, 361-372.

Schmid, A., Dordick, J. S., Hauer, B., Kiener, A., Wubbolts, M. and Witholt, B. (2001) Industrial biocatalysis today and tomorrow. *Nature* 409, 258-268.

Schoemaker, H. E. , Mink, D. and Wubbolts, M. G. (2003) Dispelling the myths-biocatalysis in industrial synthesis. *Science* 299, 1694-1697.

Siddiqui, K. S. and Cavicchioli, R. (2006) Cold-adapted enzymes. *Anttu. Rev. Biochem.* 75, 403-433.

Singleton, M. , Isupov, M. and Littlechild, J. (1999) X-ray structure of pyrrolidone carboxyl peptidase from the hyperthermophilic archaeon *Thermococcus litoralis*. *Structure* 7, 237-244.

Singleton, M. R. and Littlechild, J. A. (2001) Pyrrolidone carboxylpeptidase from *Thermococcus litoralis*. *Methods Enzymol.* 330, 394-403.

Sundaramoorthy, M. , Terner, J. and Poulos, T. L. (1998) Stereochemistry of the chloroperoxidase active site: crystallographic and molecular-modeling studies. *Chem. Biol.* 5, 461-473.

Synnes, M. (2007) Bioprospecting of organisms from the deep sea: scientific and environmental aspects. *Clean Technol. Environ. Policy* 9, 53-59.

Tao, J. and Xu, J. H. (2009) Biocatalysis in development of green pharmaceutical processes. *Curr. Opin. Chem. Biotechnol.* 13, 43-50.

Toogood, H. S. , Hollingsworth, E. J. , Brown, R. C. , Taylor, I. N. , Taylor, S. et al. , (2002a) A thermostable L-aminoacylase from Thermococcuslitoralis: cloning, overexpression, characterization, and applications in biotransformations. Extremophiles 6, 111-122.

Toogood, H. S. , Brown, R. C. , McCague, R. , Brown, R. C. , Taylor, I. N. , et al. , (2002b) Immobilization of the thermostable L-aminoacylase from *Thermococcus litoralis to generate* a reusable industrial biocatalyst. *Biocat. and Biotrans.* 20, 241-249.

Trincone, A. (2010) Potential biocatalysts originating from sea environments. *J. Mol. Cat. B: Enzymatic.* 66, 241-256.

Trincone, A. (2011) Marine biocatalysts: enzymatic features and applications. *Mar. Drug.* 9, 478-499.

Tschirret-Guth, R. A. and Butler, A. (1994) Evidence for organic substrate binding to vanadium bromoperoxidases. *J. Am. Chem. Soc.* 116, 411-412.

Valverde, C. , Orozco, A. , Becerra, A. , Jeziorski, M. C. , Villalobos, B. and Solis, J. (2004) Halometabolites and cellular dehalogenase systems: an evolutionary perspective. *Int. Rev. of Cytol.* 234, 143-199.

Wagenknect, H. and Woggon, W. (1997) Identification of intermediates in the catalytic cycle of chloroperoxidase. *Chem. Biol.* 4, 367-372.

Weyand, M. , Hecht, H. J. , Keiss, M. , Liaud, M. F. , Vilter. H. and Schomburg, D. (1999) X-ray structure determination of a vanadium-dependant haloperoxidase from *Ascophyllum nodosum* at 2. 0 Å resolution. *J. Mol. Biol.* 293, 595-611.

Winter J. M. , Moffitt, M. C. , Zazopoulos, E. , McAlpine, J. B. , Dorrestein, P. C. and Moore, B. S. (2007) Molecular basis for chloronium-mediated meroterpene cyclization. *J. Biol. Chem.* 282, 16362-16368.

Woodley, J. M. (2008) New opportunities for biocatalysis: making pharmaceutical processes greener. *Trends Biotechnol.* 26, 321-327.

Yarnell, A. (2006) Nature's X-factors. *Chemical and Engineering News* 84, 12-14.

第5章　利用宏基因组策略从海洋生态系统中发现具有生物技术应用的新酶

Jonathan Kennedy, Lekha Menon Margassery, John P. Morrissey, Fergal O'Gara and Alan D. W. Dobson, University College Cork, Cork, Ireland

DOI：10.1533/9781908818355.2.109

摘要：由于地球化学和物理学的多样性，主要由微生物和病毒占据的海洋环境也拥有地球上最多样化的生命形式。海洋微生物的生存环境可以从严酷的海底环境到与其他海洋生物共生。生态多样性反过来导致了新陈代谢的多样性。现在人们已经普遍认识到海洋微生物组成了一个庞大的对人类有益的酶和代谢产物库。人类如何获取、探索并最终开发这个潜在的库是一个挑战。当前有限的培养微生物的能力是一个主要的障碍。但基因组和宏基因组学技术为海洋生物资源的开发提供了策略。本章描述和评价了基于不同的基因序列和以功能为基础的方法。在鉴定新酶方面，此方法已经取得了成功，但是仍存在一些问题。此领域正处于摇篮期，但是发展迅速。在现在和未来，筛选及海洋微生物基因表达新技术的发展都可能会推动具有独特性质的新酶的产生。

关键词：海洋酶，异源表达，生物发现，筛选

5.1　引言

海洋约占地球总面积的70%，体积近15亿 km^3，是化学和生物多样性的丰富来源。海洋是世界上已知的生物圈中最大的生态系统，每毫升海水中含有大约 5×10^6 个细菌（Kennedy et al., 2011）。据估计，整个海洋环境中约含有 3.6×10^{30} 个微生物，这个肉眼看不见的群体约占总生物量的90%（Whitman et al., 1998）。海洋环境构成了多样化的海洋生态系统，既有高静水压力的深海环境，如半深海（200~4 000 m）、深海（4 000~6 000 m）和远洋（>6 000 m）环境，也有沿着洋脊分布的温度高达400℃的热液喷口。实际上，据估计世界海洋的平均深度为3 800 m，大约60%的地球表面被深海覆盖（Brunnegarda et al., 2004），世界上超过50%的原核生物被认为是生活在次级海底沉积物的环境中（Karl,

2007)。因此，海洋生态系统注定是一个低温的生态系统，平均温度约为3℃，在北冰洋（-1.7~5℃）和南极海洋（-2.2~0.55℃）温度更低。此外，我们必须牢记这些极端条件往往是同时存在的，例如，深海海洋系统中的微生物经受着低温和约38 MPa的压力。因此适应寒冷和高压的能力是其生长所必需的（El-Hajj et al.，2010）。这些多样化生态系统中的微生物明显有一些在生活和生存方面强加的细胞过程压力，如营养吸收、整体代谢活动、能量产生和与之相关的增长率（Siddiqui and Cavicchioli，2006）。除了面临这些挑战，微生物还需要一些机制，如在有限营养条件下的竞争机制，以及食草动物捕食的潜在威胁或病毒感染情况下的生存机制（Lauro et al.，2009）。细菌要在这些条件下生存，就必须适应并发展独特的细胞生物化学和新陈代谢方式，确保其具有非常规的生理生化和立体化学特性的酶体系，如增强耐盐、耐高温、耐低温、耐高压等特性，这也许可以满足工业生产高温、高压环境酶的需求（Trincone，2011）。

5.2 宏基因组学

Jo Handelsman首次描述了宏基因组学——应用于从环境中提取的未经培养的微生物基因组分析，他直接提取并克隆环境中的DNA（Handelsman，2004）。因此对于海洋生态系统，它涉及来自海洋宏基因组DNA的宏基因组文库基因，它包含了不同的异源表达体系，和后续基于功能为基础或以序列为基础的筛选方法（Kennedy et al.，2008）（图5.1）。对于以功能为基础的筛选，宏基因组文库的筛选是直接基于特定的表型。能够表现出所需活性的酶或其他生物活性物质（如环状结构，失去或出现颜色变化）的克隆被分离筛选出来，这可以允许我们可能去发现那些具有对已知催化具有特殊活性的新酶的全基因信息（Uchiyama and Miyazaki，2009）。或者这些文库可以基于序列为基础的分析方法，通过以PCR为基础的方法靶向这些编码特定酶的特定基因。PCR引物的设计根据特定已知靶基因的保守序列，PCR扩增以文库或者单克隆为模板进行。

片段在PCR扩增后进行测序和分析。例如，分析漆酶保守的铜离子结合位点，通过PCR成功地从海洋微生物宏基因组中克隆到一个新型的漆酶基因（Fang et al.，2011），该方法相关的例子在后面会进行介绍。另一种方法涉及大规模宏基因组DNA测序，这种方法最经典的例子是Venter团队对马尾藻进行的宏基因组测序项目。据此识别了大量的新基因（Venter et al.，2004）。在这两种情况下，通过DNA测序识别推测基因后，将基因导入到一个适当的宿主中异源表达，对重组蛋白进行生化表征。这种方法是欧盟工作的重点部分，并受到MicroB3项目http：//www，microb3. eu/home的资助。大规模测序的数据将用于从海洋基因组中识别新的具有生物技术潜力的酶。

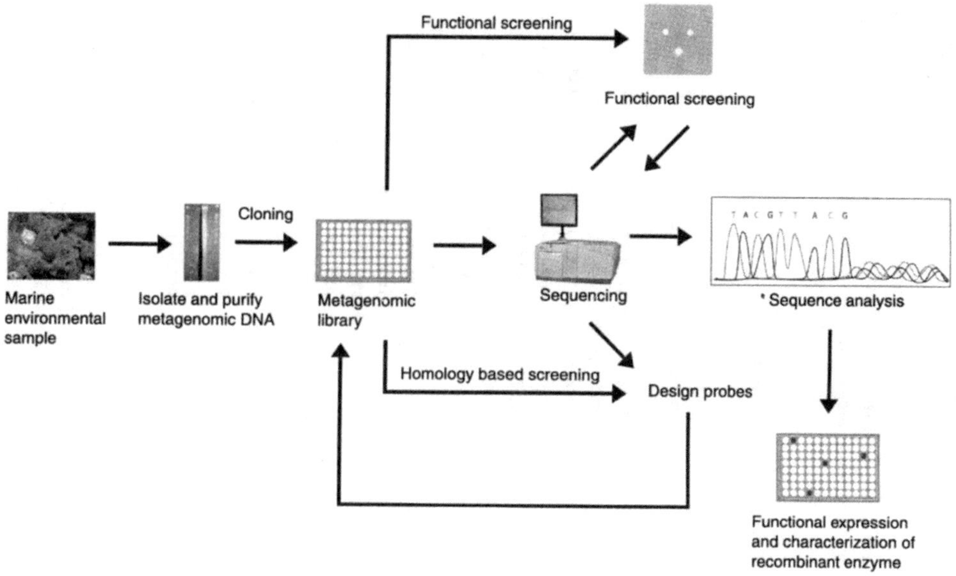

图 5.1　新型生物催化的宏基因组分析方法

5.3　以序列为基础的筛选——“全基因组扫描”

在当前后基因组时代，基因序列的信息可以从大量基因组测序项目中获得，因此，发现并最终开发海洋微生物代谢产物的机会是巨大的。在写本文时，已经有 11 578 个基因组项目上传到了基因组线上数据库（GOLD）（http：//www.genomesonline.org），目前已经登记的全基因组超过 2 967 个。在这些已经登记的基因组中，迄今由 Gordon and Betty Moore Foundation（http：//www.camera.calit2.net/mircogenome）基金项目支持的大约 180 个海洋细菌基因组已经发布。这些海洋微生物基因组序列为我们提供了前所未有的机会去开发这些细菌的生物工程应用潜力。可以预测在不久的未来，更多新酶的编码基因将被发现。近期的研究中，全基因组序列项目明显促进了海洋细菌来源的可用于生物技术的酶的鉴定。例如，在一种新的 *Alteromonas* 中发现了 6 种新的琼脂水解酶（Oh et al.，2011），在 *Glaciecola* sp. 4H-3-7 菌株中鉴定到多种糖苷水解酶基因，他们具有在生物燃料开发领域利用的潜能（Klippel et al.，2011）。

筛选细菌基因组数据库中的基因，获得具有新型生化特征有潜力的酶的编码基因，在新酶发现中是非常有吸引力的一个方向。从基因组中鉴定酶最常用的方法是根据已知功能的同家族同源蛋白的序列相似性。常规的方法是序列-序列比对，如 FASTA（Pearson and Lipman，1988）或 BLAST（Altschul et al.，1990）以及生物信息学数据开发工具如 MetaGene（Noguchi et al.，2006）、GeneMark（Lukaskin and Borodovsky，1998）或者 Manatee (www. manatee. sourceforge. net/jcvi)；或者通过批量—序列比对的方法，例如 PSI-BLAST (Altschul et al.，1998) 或 HMMER (Eddy，1998) 以及运用 SDISCOVER (http：//www. njit. edu/~ wangj/sdicovery. html) 和 MEME (Multiple EM for Motif Elicitation) (http：//meme. nbcv. net) 等工具。通过这些方法，可以选出候选基因，后续通过克隆与异源表达来评价生化水平的功能。例如 Janssen 团队利用这样的方法鉴定了一种推测的新型环氧化物水解酶，这组酶有对映选择性生物催化的能力，具有生物降解有机化合物的潜力 (van Loo et al.，2006)。在这项研究中，通过序列比对确定 10 个假定的微生物基因可能具有环氧化物水解酶功能的基因，并用克隆的方法在大肠杆菌中进行表达。结果表明，4 种新的环氧化物水解酶具有非常规的对映体偏好，底物为内消旋环氧化物和/或末端芳香族环氧化物，这些酶特别适合生成对映 (*S*，*S*) -二醇和 (*R*) -环氧化物。

现在公认的是，许多罕见微生物中的新酶是它们自身复杂的生态系统中的一部分。它们实际上很少出现在宏基因组文库中。基于 phi29 聚合酶多重扩增的全基因组扩增 (WGA) 可以帮助解决这个问题。多重取代扩增反应适合鸟枪法扩增单个细胞基因组 DNA，因此可以检测未经培养的微生物群落中的单个微生物的生化潜力 (Stepanauskas and Sieracki，2007)。在这种方法中，微生物细胞首先用特异的流式细胞荧光分选技术 (FACS) 分离，分离得到的单个细胞进行 WGA。对扩增的基因组进行测序，并分析具有催化和代谢潜能的基因。最近 Hentschel 团队用此方法鉴定了 *Poribacteria* (地中海水绵联合体 *Aplysina aerophoba* 的组成部分) 的单细胞基因组 (Siegl et al.，2011)。用这种方法测得近 1.6 Mb 的 DNA 序列，从基因组中可以鉴定多种酶，包括多种硫酸酯酶和肽酶。因此这种方法具有现实意义，可以获得不经培养而发现海洋微生物中全部生物酶的潜力。

5.4 基于序列的鉴定——宏基因组 DNA

如前文提到的，宏基因组由一个给定的生态系统中所有微生物基因组组成。然而，并不是典型的完整基因组序列集合 (如通过培养获得)，而是以部分基因组碎片的形式存在。Sherman 团队的报道证实了来自海洋环境共生体群落的宏基因组 DNA 序列数据的重要性。此微生物群落是 *Ecteinascidia turbinata* 共生的微生物，通过该宏基因组共鉴定出 25 个基因组成的非核糖体肽合成酶 (NRPS) 核心合成途径簇，能够合成抗癌药 ET-743 (Yondelis^{R})。随后该团队证明了一种与被囊动物共生的 γ-变形菌 *Candidatis Endoecteinascidia frumentensis* 可以产生四氢异喹啉核心 ET-743 分子 (Ratha et al.，2010)。

在酶的发现方面，根据海洋环境中分离得到的宏基因组 DNA 序列，采用与基于基因组的全基因组扫描类似的策略来鉴定新酶 (Imhoff et al.，2011)。以序列为基础分析海洋宏基因组文库鉴定到多种酶，包括肽酶 (Cottrell et al.，2005)、烷烃水解酶 (Xu et al.，2008) 以及一种延胡索酸酶 (Jiang et al.，2010)。以肽酶为例，来自马尾藻全基因组序列 (WGS) 数据库，包含了 10.45 亿对碱基的宏基因组 DNA 序列，超过 120 万个开放阅读框

（Venter et al.，2004）的肽酶是通过对海洋噬胞菌属样细菌 BLAST 分析水解酶，确定其降解可溶性高分子有机物的潜力，后续通过大肠杆菌异源表达检测到其具有肽酶活性（Cottrell et al.，2005）。延胡索酸酶催化延胡索酸生成 L-苹果酸的可逆反应，用于工业生产 L-苹果酸。从海水基因组文库中筛选并鉴定到一种新的延胡索酸酶，根据序列文库推断其基因，在大肠杆菌表达并检测重组酶。结果显示，酶在 pH 值 8.5，温度为 55℃时具有最高活性，这使其可以成为在较高温度下潜在生产 L-苹果酸的酶（Jiang et al.，2010）。

另外一个例子是最近从 Juan de Fuca Ridge 的 Mothra 热泉喷口的宏基因组文库中分离的一种新的淀粉酶（Wang et al.，2011）。将酶克隆并在大肠杆菌中重组表达，检测到重组蛋白具有淀粉酶活性，在 90℃ pH 值 7.5 时具有最高活性与良好的热稳定性，在 90℃保持 4 h 后仍具有 50%的活性。

基于序列同源性的筛选，包括以 PCR 为基础靶向与已知序列有相似序列的方法，可以从海洋环境中成功鉴定出编码新酶的基因。运用这种方法的一个例子是最近报道的从南海微生物宏基因组文库中分离出的一种新的漆酶（Fang et al.，2011）。研究者根据保守的铜离子结合位点设计 PCR 引物，鉴定出可能存在于宏基因组 DNA 的漆酶序列，然后在大肠杆菌中异源表达，重组蛋白显示出碱依赖的氯离子耐受的漆酶活性（Fang et al.，2011）。采用类似的方法，从海洋宏基因组中克隆得到了其他的酶包括甲壳质酶（LeCleir et al.，2004）、枯草杆菌蛋白酶、木聚糖酶及脂肪酶（Acevedo et al.，2008）。

有一种有趣的方法，在未来或许可以证明是有效的，叫作“合成宏基因组学”的方法。这种方法包括用生物信息学的方法检索 NCBI 数据库中的序列，然后检验假定的基因。用这种方法识别的基因通常是没有被表征的，标注为推断或者假设，或者基因来自未知的生物或来自环境宏基因组资源。这些基因可以针对特定异源表达的宿主进行优化，在化学合成 DNA 后连接到表达载体后在异源表达系统（如大肠杆菌或酿酒酵母）中进行功能鉴定。这种方法成功用于卤代甲基转移酶（MHT）的鉴定，此酶具有从多种生物资源中生成产甲基卤化物的潜力，从而可以将可再生的碳源转移至化学品和液体燃料中（Bayer et al.，2009）。Voigt 团队推断并化学合成了 89 个从多种资源中鉴定出的 *MHT* 基因，这些基因除了来自细菌和真菌外，还有来自植物和其他未在 NCBI 序列数据库中识别的物种。这些基因经过优化后在大肠杆菌和酿酒酵母中表达。这些推断的基因在两种表达系统中进行了筛选，结果显示，94%的重组蛋白表现出多种卤素转移酶活性（Bayer et al.，2009）。尽管 770 万个全球海洋抽样（GOS）序列分析预测已知的所有细菌包含约 612 万个蛋白，覆盖几乎所有已知的细菌蛋白家族，但仍有 1 700 个蛋白簇与已知的蛋白家族没有任何同源性（Yooseph et al.，2007）。因此有可能出现“合成宏基因组学”方法，从这 1 700 个蛋白簇中鉴定新的酶。

最近描述的另一种有趣的方法是根据用磁珠捕获杂交底物序列。这包括使用简并引物扩增部分内部基因，然后将此部分基因固定在表面有链霉亲和素的磁珠上，用这些磁珠作为探针，去靶向宏基因组 DNA 中的全基因（Meiring et al.，2010）。这种方法成功地用于克隆来自土壤的一种新的细菌多铜氧化酶，但是很明显该方法也可以在海洋宏基因组文库中应用。

5.5 以功能为基础的筛选

尽管有一些成功的例子，以序列为基础的鉴定方法被普遍认为具有一定限制性，因为只有同源序列存在的基因可以被靶向，因此对于一些新的基因很难被检测到。在此背景下，多数人认为以功能为基础的鉴定方法在全新的生物催化活性的新酶发现方面更有潜力（Simon and Rolf, 2011）。功能鉴定是基于宏基因组文库中基因克隆表达并鉴定其活性，这主要基于我们可以获得期待的表型（Kennedy et al.，2011）。尽管比以序列比对为基础筛选方法需要更多的时间和付出，但这种方法可以鉴定到全新的酶。许多成功的例子表明以功能为基础的鉴定已经从海洋宏基因组文库中克隆了多种新酶。

研究者已经从南海宏基因组文库中分离出两种新的酯酶。其中一种是具有新的催化作用，可以催化 p-硝基苯，并在多种溶剂（如乙醇、甲醇和二甲基甲酰胺）中具有高稳定性（Chu et al.，2008）。此外，研究者分离了一种在食品添加剂中具有应用潜力的新型酯酶。此基因来自水深超过 100 m 的海底沉积物的宏基因组文库。最初根据其具有水解三丁酸甘油酯活性将其分离，随后在大肠杆菌中表达后发现酶在温度 40℃，pH 值 8.5 时活性最高，最适底物为 *p*-硝基苯丁酸酯（Peng et al.，2011）。

利用添加 1%丁三酸甘油酯的 LB 培养基，从海绵属动物 *Haliclona simulans* 建立的宏基因组文库中已经鉴定到 20 种显示出脂肪水解酶活性的克隆（Lejon et al.，2011）。从南海基因组文库中也鉴定出两种新的脂肪水解酶，其最适温度为 40~50℃，最适底物为 *p*-硝基苯丁酸酯（Hu et al.，2010）。根据功能推测的方法，用添加 0.1%七叶苷水合物和 0.25%铁铵盐的 LB 培养基从南海海水宏基因组文库中克隆了一种新的 β-葡萄糖苷酶。结果显示，大肠杆菌异源表达的 β-葡萄糖苷酶在高浓度 NaCl 中仍有良好的稳定性，而且在葡萄糖浓度高达 1 mol/L 的条件下仍有 50%的活性（Zemin et al.，2010）。在另一种海绵动物 *Hyrtios erecta* 宏基因组中也发现一种新的具有中等热稳定性的酯酶，这种酶属于 SGNH 水解酶超家族。通过功能鉴定后进行克隆，在添加 1%吐温的 LB 培养基中观察到周围有粉末状的圆环，证明有酯酶活性（Okamura et al.，2010）。最后，最近从韩国江华岛的滩涂沉积物宏基因组文库中分离出一个耐高盐环境的酯酶亚家族（Jeon et al.，2012），此酶重组表达后显示出高活性而且耐受高盐，在 3 mol/L NaCl 或 KCl 中仍有 50%的活性。

考虑到蛋白酶占大约 40%的市场，在不同的工业领域，包括废水净化、洗涤、食品和制药都有应用（Gupta et al.，2002）。令人惊讶的是并没有太多用宏基因组学分离蛋白酶的报道。其中一个原因可能是功能鉴定需要用到含有脱脂牛奶的琼脂板。有一些报道称这种方法可以成功完成主要来自土壤环境微生物的分离（Waschkowitz et al.，2009），但也有关于该筛选方法的质疑，因为已经有关于糖苷水解酶的活性造成干扰的报道（Jones et al.，2007）。已经有一些利用功能为基础的方法从宏基因组文库中分离得到海洋蛋白酶的报道。其中之一是从深海沉积物宏基因组文库中分离并表征了一种蛋白酶（Lee et al.，2007），从基因序列推断出的氨基酸序列包含 His-Gly-X-X-His 序列，这是锌离子依赖的金属蛋白酶活性位点，这表明此酶可能是一种锌离子依赖的金属蛋白酶。在大肠杆菌重组表达纯化后，此酶在 pH 值 7.0，温度为 50℃时有最高活性，其活性受到金属离子螯合剂（如 EDTA、EGTA、1，10-邻菲咯啉）的抑制。此酶同时具有水解偶氮酪蛋白和纤维蛋白活性，说明其可能成为治疗血栓的药物。

最近又有从南极海洋沉积物宏基因组文库中分离到一种新的蛋白酶（Zhang et al.，2011）的报道，这种枯草蛋白酶样的丝氨酸蛋白酶在 60℃ pH 值 9 时活性最高。钙可以增加其热稳定性，在 50℃条件下 2 h 仍有 73%的活性。最近还报道了利用添加了 1%脱脂奶粉的 LB 固体培养基，从两种来自海绵 *Haliclona simulans* 宏基因组文库中分离得到两种蛋白酶（Lejon et al.，2011）。因此，尽管用脱脂牛奶的方法会遇到困难，但是根据功能鉴定来自海洋生态系统的宏基因组文库中的基因，研究者分离、鉴定和表征了多种新的其他蛋白酶。

在染料生产中成功用于鉴定来自土壤宏基因组文库的新的生化活性。例如上述文库中克隆表达出一种深棕色-黑色样的颜色，随后显示所包含的基因编码两种对革兰氏阴性菌和革兰氏阳性菌都具有抑制作用的金霉素 A 和 B（Gillespie et al.，2002）。来自土壤文库的紫色和深蓝色染料的克隆证实是分别可以产生抗菌的紫色杆菌素（Brady et al.，2001）和靛玉红/靛蓝（MacNeil et al.，2001）。采用同样的方法从深海沉积物文库中得到一个生产红棕色素的克隆（Huang et al.，2009），随后证明是一种推测的 4-羟基苯丙酮酸双加氧酶（HPPD）。此基因与先前描述的从土壤中得到的 HPPD 基因完全不同，但是与从深海细菌 *Idiomarina loihiensis* 推测的氨基酸序列有显著的相似性。克隆的基因所编码的新的推测的 HPPD 在大肠杆菌中表达后可以产生黑色素，其原理是通过 *p*-羟基丙酮酸在宿主体内转换为尿黑酸，随后积累并聚合形成黑色素。

另一个以颜色为基础的被证明是成功的功能筛选方法是从陆地宏基因组文库中筛选纤维素酶活性，此方法对海洋宏基因组可能同样有效。这种方法通常是添加 0.1% Ostazin brilliant red 羟乙基纤维素，如果表现出纤维素酶活性，则可见黄色的圈。从多种不同的陆地生态系统的宏基因组文库中已经鉴定到了多种纤维素酶（Ilmberger and Streit，2010），但是从海洋宏基因组文库中确定的纤维素酶的报道则很少。其原因还不清楚，可能仅仅是因为研究工作还没有集中到此领域。这或许有些令人惊讶，纤维素酶在生产生物乙醇这一再生能源的过程中有重要作用，现有的酶的不足之处在于不能高效地降解纤维素。新的纤维素水解酶已经从海洋环境中分离，例如来自印度半岛海岸海绵 *Dendrilla nigra* 细菌 *Marinobacter* sp.（MS1302）。另外一种纤维素水解酶是从中国厦门海域分离的海洋细菌 *Paenibacillus* sp. BME-14。最近一种新的纤维素水解酶成功地从鲍肠内微生物基因组文库中克隆（Kim et al.，2011）。此酶显示出与 *Vibrio alginolyticus* 膜外蛋白 100%类似，但与已知纤维素酶序列同源性低。重组酶在37℃、pH值7时有最高活性。因此，这种酶的出现预示着海洋环境的宏基因组可能包含大量新的纤维素水解酶基因。

5.5.1 以功能为基础鉴定的局限性

用宏基因组发现新酶最大的局限性之一是酶没有生物技术应用的潜能，例如管家基因大量出现在细菌基因组中，因此被更多地分离出来（Vieites et al.，2009）。尽管事实如此，估计基因文库中少于0.000 001%的酶可以通过功能为基础的鉴定方法检测到（Ferrer et al.，2008）。例如，如果我们认为细菌在海水中的浓度可以达到10^6/mL，同时假设每个基因组中平均含有3 000个基因，如果40%基因编码有催化活性的蛋白（Dinsdale et al.，2008)，那么在这个样本中可能共有3×10^9个基因，推测1.2×10^9个反应。在当前传统的鉴定技术下，估计实际只能确定几百种活性（Fernandez-Arrojo et al.，2010）。功能鉴定的方法也一定程度上受到鉴定到的酶数量的限制，活性易检测的酶可以轻松地筛选，这就解释了为什么从宏基因组文库中分离到如此多的脂肪酶的克隆。至今已经有超过80个宏基因组酯酶和脂肪酶被报道。

有趣的是，有时基因在最初特定功能鉴定中检测到有活性，但是在异源宿主中表达时表现出其他的新活性。最近报道的一个例子是一种新的羧酸酯酶具有水解β-内酰胺抗生素活性，此基因来自土壤宏基因组文库，编码此蛋白的基因在含有丁三酸甘油酯的琼脂板培养基中最初鉴定具有分解脂肪活性，但是重组蛋白在大肠杆菌中表达时发现其同时具有靶向p-硝基苯酯和水解β-内酰胺抗生素（如头孢菌素、头孢菌素Ⅱ、头孢噻吩和头孢唑啉）活性（Jeon et al.，2011）。

异源基因表达

以功能为基础的筛选方法的另外一个限制因素是要依靠宏基因组中克隆基因、转录并翻译编码多肽的能力，由于翻译的多肽是有活性的，因此给予宿主一个新的表型。为了促进这种方法，研究者开发了多种不同的异源宿主表达宏基因组文库，包括常用的大肠杆菌表达系统和其他替代的表达系统包括*Pseudomonas putida*、*Ralstonia metallidurans*、*Agrobacterium tumefaciens*、*Burkholderia graminis*与*Caulobacter vibrioides*（Craig et al.，2010，Wexler and Johnston，2010，Craig and Brady，2011）。考虑到会出现不同水平的基因表达，因此需要上述不同的表达系统来满足基因表达的需求。根据不同的分类、来自不同文库的DNA样品选择不同的表达系统。例如，导入的外源DNA在大肠杆菌宿主转录系统可能干扰宿主基因的转录，容易导致毒性基因的随机表达，并伴随着毒性作用（Warren et al.，2008）。此外，非常清楚的是大肠杆菌只有7种RNA聚合酶σ因子，与含有超过15种σ因子的其他细菌（如链霉菌、单假胞菌和根瘤菌）相比，需要其他特殊σ因子的基因不能在大肠杆菌内表达。事实已经证明，当大肠杆菌作为宿主时，来自32种不同细菌的基因只有40%能表达（Gabor et al.，2004）。大肠杆菌翻译起始时偏爱密码子AUG，使用率是 91%。而其他物种更常用 UUG 或 GUG，因此非 AUG 起始密码的基因可能不能被大肠杆菌有效地识别（Uchiyama and Miyazaki，2009）。出于这种考虑，科学家采用了多种方法以帮助提高外源基因在大肠杆菌中的表达水平。这些方法包括应用不同的核糖体蛋白识别高 GC 含量的基因，例如非硫化紫色光合细菌 *Rhodopseudomonas palustris* 的核糖体蛋白效率更高（Bernstein et al.，2007），以及导入具有额外 tRNA 的质粒解决稀有密码子的问题（Makrides，1996）。因此，很显然当使用以功能为基础的鉴定方法时，选用多种宿主可能增加活性的检测范围。

有些团队使用穿梭质粒解决表达问题，使用 *E. coli* 构建文库，然后转入第二宿主中进行表达和功能筛选（Troeschel et al.，2011）。例如穿梭质粒 pEBP18 在 *E. coli*、*Pseudomonas putida* 和 *Bacillus subtilis* 宿主中复制（Troeschel et al.，2011），pLAFR3 可以在 *E. coli* 和 α-proteobacerium *Rhizobium leguminosarum* 宿主中复制（Li et al.，2005），pRK7813 可以在 *E. coli* 和 *Sinorhizobium meliloti* 中复制（Wang et al.，2006），pCCERI 可以在大肠杆菌中复制，在 *P. putida* 和 *Streptomyces lividans* 中生长（Lejon et al.，2011）。在利用这些表达系统时，由于不同的宿主对宏基因组 DNA 内的基因有偏好，因此这些基因优先被识别。例如，在大的插入文库中，距离质粒启动子较远的基因只能靠自身启动子表达，而这些基因在宿主中是有功能的。有意思而且有潜力的解决此问题的方法是运用含有双向启动子的质粒以允许其双向转录，例如 Rupp 课题组用此方法从土壤和堆肥中分离脂肪酶、淀粉酶、磷脂酶和双加氧酶的表达系统，这成功地增加了获得阳性克隆的数量（Lammle et al.，2007）。

筛选系统的改进

虽然如前文所述，以功能为基础的鉴定方法的优势在于不依赖已知蛋白家族的同源蛋白，因此增加了发现全新酶的可能性，其缺点是受到现在高通量鉴定数量的限制（Simon and Rolf，2011），以及如前文所述，现在很多鉴定方法是鉴定可以易于检测活性的酶。琼脂固体培养基进行功能检测的一个优势在于可大规模检测，如果整合机器人技术，实验室可在一天内完成数千个宏基因组文库克隆的筛选。然而，受到阳性克隆反应活性检测方法灵敏度的影响，以及根据水解范围/圈还是出现颜色通常是不确定的。检测这些克隆的活性时常出现问题。这些反应通常需要酶到细胞外或底物转运至细胞内才能鉴定活性。当然，一些方法已经在研究，这可以帮助解决琼脂板为基础的鉴定方法出现的问题，包括使用细胞溶解物，通过将细胞在 96 孔板中培养，然后通过玻璃珠等物理方法或通过蛋白提取试剂等化学方法制备溶菌产物，然后将底物加入。这些方法可以显著增加反应的灵敏度（Suenaga et al.，2007）。

5.6 结论

考虑到我们正在研究的其他环境中新的具有生物技术应用的酶，海洋环境仍有待开发。这需要逐步改变，在本章中我们简述了多种生物方法确定一些新的酯酶、脂肪酶、蛋白酶等。通过表征这些酶揭示其具有优势的特点，例如关于某些产生抑制的条件的耐受性，从而说明了探索此环境的合理性。毫无疑问，酶的需求推动了一些新的灵敏的有效的高通量鉴定策略的发展，包括以序列和功能为基础的方法。

以 DNA 为基础的技术获取海洋多样性的潜能反映在快速增长的海洋微生物基因组序列的数量和海洋宏基因组序列数据的增加水平。不同的策略均具有优势和局限性，已经在（宏）基因组资源中得到了应用。从某种程度上说，通过同家族为基础的基因芯片或者 PCR/探针为基础的方法搜索序列数据，识别候选基因是可行的。这些基因可以克隆到适当的载体上，在宿主中表达并表征其功能。这种策略最主要的局限是不能完全识别具有全新特征/模式的酶，因为现有的方法是根据识别相似的 DNA 序列预测功能。尽管如此，发现更多比现有的酶更强的突变酶，这一领域仍需要付出更多努力。替代以序列为基础的筛选方法主要是以功能为基础的筛选方法识别（宏）基因组 DNA 片段编码的新酶。此方法的最主要的优越性是研究者不需要拘泥于对已知活性酶的了解，这样可以识别新基因，编码以前未发现的酶。然而，成功地筛选到新酶依靠设计合适的筛选方法，这个方法是依赖有强大且可以容易于高通量机器人检测表达系统兼容，同时可以成功表达此外源基因的表达系统。这些都是必须克服的困难，因为大多数筛选方法都是基于已知的酶优化的，但是对于新酶不一定可行，而且通常的表达宿主不一定能高效表达外源基因。在两个方面都取得了进展，可以预见这一领域（筛选方法和宿主）在未来几年内将有大的发展。实际上，除了基因识别，识别编码有效生物活性和功能基因最高效和直接的方法是表达然后用生物化学的方法表征。为了达到这一目的，一个具有更广范围的表达宿主系统和更有效的高通量筛选系统是需要的。整合新方法，例如合成生物学与新的生物发现策略结合，也会推动更多新酶的发现。

致谢

感谢海洋改变战略和科技创新策略下属博福特海洋研究奖（2006—2013），感谢海洋学院，国家发展计划 2007—2013 海洋研究子项目的基金支持。

参考文献

Acevedo, J. P., Reyes, F., Parra, L. P., Salazar, O., Andrews, B. A. and Asenjo, J. A. (2008) Cloning of complete genes for novel hydrolytic enzymes from Antarctic seawater by use of an improved genome walking technique. *J Biotechnol* 133, 277-286.

Altschul, S. F., Gish, W., Miller, W., Myers, E. W. and Lipman, D. J. (1990) Basic local alignment tool. *J Mol Biol* 5, 403-410.

Altschul, S. F., Madden, T. L., Schäffer, A. A., Zhang, J., Zhang, Z., et al., (1998) Gapped BLAST and PSI-BLAST: a new generation of protein database search programs. *Nucleic Acids Res* 25, 3389-3402.

Bayer, T. S., Widmaier, D. M., Temme, K., Mirsky, E. A., Santi, D. V, and Voigt C. A. (2009) Synthesis of methyl halides from biomass using engineered microbes. *J Am Chem Soc.* 131, 6508–6515.

Bernstein, J. R., Butler, T., Shen, C. R. and Liao, J. C. (2007) Directed evolution of ribosomal protein S1 for enhanced translational efficiency of high GC *Rhodopseudomonas palustris* DNA in *Escherichia coli. J Biol Chem* 282, 18929–18936.

Brady, S. F., Chao, C. J., Handelsman, J. and Clardy, J. (2001) Cloning and heterologous expression of a natural product biosynthetic gene cluster from eDNA. *Org Lett* 3, 1981–1984.

Brunnegarda, J., Grandel, S., Stahl, H., Tengberg, A. and Hall, P. O. J. (2004) Nitrogen cycling in deep-sea sediments of the Porcupine Abyssal Plain, NE Atlantic. *Progress in Oceanography* 63, 159–181.

Chu, X., Haoze, H., Guo, C. and Sun, B. (2008) Identification of two novel esterases from a marine metagenomic library derived from South China Sea. *Appl Microl Biotechnol* 80, 615–625.

Cottrell, M. T., Yu, L. and Kirchman, D. L. (2005) Sequence and expression analyses of cytophaga-like hydrolases in a western artic metagenomic library and the saragasso sea. *Appl Environ Microbiol* 71, 8506–8513.

Craig, J. W. and Brady, S. F. (2011) Discovery of a metagenome-derived enzyme that produces branched-chain acyl carrier proteins from branch-chain α-keto acids. *Chembiochem* 12, 1849–1853.

Craig, J. W., Chang, F. Y., Kim, J. H., Obiajulu, S. C. and Brady, S. F. (2010) Expanding small-molecule functional metagenomics through parallel screening of broad-host-range cosmid environmental DNA libraries in diverse proteobacteria. *Appl Environ Microbiol* 76, 1633–1641.

Dinsdale, E. A., Edwards, R. A., Hall, D., Angly, F., Breitbart, M., et al., (2008) Functional metagenomic profiling of nine biomes. *Nature* 452, 629–632.

Eddy, S. R. (1998) HMMER: biological sequence analysis using profile hidden Markov models, *http://hmmer.org/*

El-Hajj, Z. W., Allcock, D., Tryfona, T., Lauro, F. M., Sawyer, L., et al., (2010) Insights into piezophily from genetic studies on the deep-sea bacterium, *Photobacterium profundum* SS9. *Ann N Y Acad Sci* 1189, 143–148.

Fang, Z., Li, T., Wang, Q., Zhang, X., Peng, H., et al., (2011) A bacterial laccase from marine microbial metagenome exhibiting chloride tolerance and dye decolorization ability. *Appl Microbiol Biotechnol* 89, 1103–1110.

Fernandez-Arrojo, L., Guazzaroni, M-E., Lopez-Cortes, N., Beloqui, A. and Ferrer, M. (2010) Metagenomic era for biocatalyst identification. *Curr Opinion Biotechnol* 21, 725–733.

Ferrer, M., Beloqui, A., Vieites, J. M., Guazzaroni, M-E., Berger, I. and Aaron, A. (2008) Interplay of metagenomics and in vitro compartmentalization. *Microb Biotechnol* 2, 31–39.

Gabor, E. M., Alkema, W. B. and Janssen, D. B. (2004) Quantifying the accessibility of the metagenome by random expression cloning techniques. *Appl Environ Microbiol* 6, 879–886.

Gillespie, D. E., Brady, S. F., Bettermann, A. D., Cianciotto, N. P., Liles, M. R., et al., (2002) Isolation of antibiotics turbomycin A and B from a metagenomic library of soil microbial DNA. *Appl Environ Microbiol* 68, 4301–4306.

Gupta, R., Beg, O. K. and Lorenz, P. (2002) Bacterial alkaline protease: molecular approaches and industrial applications. *Appl Microbiol Biotechnol* 59, 15–32.

Handelsman, J. (2004) Metagenomics: application of genomics to uncultured microorganisms. *Microbial Mol Rev* 68, 669-685.

Hu, Y., Fu, C., Huang, Y., Yin, Y., Cheng, G., et al., (2010) Novel lipolytic genes from the microbial metagenomic library of the South China sea marine sediment. *FEMS Microbiol Ecol* 72, 228-237.

Huang, Y., Lai, X., He, X., Cao, L., Zeng, Z., et al., (2009) Characterization of a deep-sea sediment metagenomic clone that produces water-soluble melanin in *Escherichia coli*. *Mar Biotechnol* 11, 124-131.

Ilmberger, N. and Streit, W. R. (2010) Screening for cellulase-encoding clones in metagenomic libraries. *Methods Mol Biol* 668, 177-188.

Imhoff, J. F., Labes, A. and Wiese, J. (2011) Bio-mining the microbial treasures of the ocean: new natural products. *Biotechnol Adv* 29, 468-482.

Jeon, H. J., Kim, S-J., Lee, H. S., Cha, S-S., Lee, J. H., et al., (2011) Novel metagenome-derived carboxylesterase that hydrolyzes β-actam antibiotics. *Appl Environ Microbiol* 77, 7930-7836.

Jeon, H. J., Lee, H. S., Kim, J. T., Kim, S-J., Choi, S. H., et al., (2012) Identification of a new subfamily of salt-tolerant esterases from a metagenomic library of tidal flat sediment. *Appl Microbiol Biotechnol* 93, 623-631.

Jiang, C., Wu, L-L., Zhao, G-C., Shen, P-H., Jin, K., et al., (2010) Identification and characterization of a novel fumarase gene by metagenome expression cloning from marine microorganisms. *Microbial Cell Factories* 9, 91.

Jones, B. V., Sun, F. and Marchesi, J. R. (2007) Using skimmed milk agar to functionally screen a gut metagenomic library for proteases may lead to false positives. *Lett Appl Microbiol* 45, 418-420.

Karl, D. M. (2007) Microbial oceanography: paradigms, processes and promise. *Nat Rev Microbiol* 5, 759-769.

Kennedy, J., Marchesi, J. R. and Dobson, A. D. W. (2008) Marine metagenomics: strategies for the discovery of novel enzymes with biotechnological applications from marine ecosystems. *Microb Cell Fact* 7, 27.

Kennedy, J., O'Leary, N., Kiran, S., Morrissey, J., O'Gara, F. (2011) Functional metagenomic strategies for the discovery of novel enzymes and biosurfactants with biotechnological applications. *Appl Microbiol* 111, 787-799.

Kim, D., Kim, S-N., Baik, K. S., Park, S. C., Lim, C. H., et al., (2011) Screening and characterization of a cellulase gene from the gut microflora of abalone using metagenomic library. *J Microbiol* 49, 141-145.

Klippel, B., Lochner, A., Bruce, D. C., Davenport, K. W., Detter, C., et al., (2011) Complete genome sequence of the marine cellulose- and xylan-degrading bacterium *Glaciecola* sp. strian 4H-3-7 + YE-5. *J Bacteriol* 193, 4547-4548.

Lammle, K., Zipper, H., Breuer, M., Hauer, B., Buta, C., et al., (2007) Identification of novel enzymes with different hydrolytic activities by metagenome expression cloning. *J Biotechnol* 127, 575-592.

Lauro, F. M., McDougald, D., Williams, T. J., Egan, S., Rice, S., et al., (2009) A tale of two lifestyles: the genomic basis of trophic strategy in bacteria. *Proc Natl Acad Sci USA* 106, 15527-15533.

LeCleir, G. R., Buchan, A. and Hollibaugh, J. T. (2004) Chitinase gene sequences retrived from diverse aquatic habitats reveal environment specific distributions. *Appl Environ Microbiol* 70, 6977-6983.

Lee, D. G., Jeon, J. H., Jang, M. K., Kim, N. Y., Lee, J. H., et al, (2007) Screening and characterization of a novel fibrinolytic metalloprotease from a metagenomic library. *Biotechnol Lett* 29, 465-472.

Lee, H. S. , Kwon, K. K. , Kang, S. S. , Cha, S-S. , Kim, S-J. and Lee, J-H. (2010) Approaches for novel enzyme discovery from marine environments. *Curr Opinion Biotechnol* 21, 353-357.

Lejon, D. P. H. , Kennedy, J. and Dobson, A. D. W. (2011) 'Identification of novel bioactive compounds from the metagenome of the marine sponge *Haliclona simulans*' . In *Handbook of Molecular Microbial Ecology II: Metagenomics in Different Habitats*, de Bruijn, F. J. (ed.), 553-562. Wiley-Blackwell.

Li, Y. , Wexler, M. , Richardson, D. J. , Bond, P. L. and Johnston, A. W. (2005) Screening a wide host-range metagenomic library in tryptophan auxotrophs of *Rhizobium leguminosarum* and of *Escherichia coli* reveals different classes of cloned *trp* genes. *Environ Microbiol* 7, 1927-1936.

van Loo, B. , Kingma, J. , Arand, M. , Wubbolts, M. G. and Janssen, D. B. (2006) Diversity and biocatalytic potential of epoxide hydrolases identified by genome analysis. *Appl Environ Microbiol* 72, 2905-2917

Lukaskin A. and Borodovsky, M. (1998) GeneMark. hmm: new solutions for gene finding. *Nucleic Acids Res* 26, 1107-1115.

MacNeil, I. A. , Tiong, C. L. , Minor, C. L. , August, P. R. , Grossman, T. H. , et al, (2001) Expression and isolation of antimicrobial small molecules from soil DNA libraries. *J Mol Microbiol Biotechnol* 3, 301-308.

Makrides, S. G (1996) Strategies for achieving high-level expression of genes in *Escherichia coli. Microbiol Rev* 60, 512-538.

Meiring, T. , Mulako, I. , Tuffin, M. I. , Meyer, Q. and Cowan, D. A. (2010) Retrival of full-length functional genes using subtractive hybridization magnetic bead capture. *Methods Mol Biol* 668, 287-297.

Noguchi, H. , Park, J. and Takagi, T. (2006) MetaGene: prokaryotic gene finding from environmental shotgun sequences. *Nucleic Adds Res* 34, 5623-5630.

Oh, G, De Zoysa, M. D. , Kwon, Y-K. , Heo, S-J. , Affan, A. , et al, (2011) Complete genome sequence of the agarase-producing marine bacterium strain S89, representing a novel species of the genus *Alteromonas. J Bacterial* 193, 5538.

Okamura, Y. , Kimura, T. , Yokouchi, H. , Meneses-Osorio, M. , Katoh, M. , et al, (2010) Isolation and characterization of a GDSL esterase from the metagenome of a marine sponge-associated bacteria. *Mar Biotechnol* 12, 395-402.

Pearson, W. R. and Lipman, D. J. (1998) Improved tools for biological sequence comparison. *Proc Natl Acad Sci USA* 85, 2444-2448.

Peng, Q. , Zhang, X. , Shang, M. , Wang, X. , Wang, G. , et ah, (2011) A novel esterase gene cloned from a metagenomic library from neritic sediments of the South China sea. *Microb Cell Fact* 10, 95.

Ratha, C. M. , Jantob, B. , Earlb, J. , Ahmedb, A. , Hub, F. Z. , et ah, (2010) Meta-omic characterization of the marine invertebrate microbial consortium that produces the chemotherapeutic natural product ET-743. *ACS Chem Biol*, PMID: 21875091.

Siddiqui, K. S. and Cavicchioli, R. (2006) Cold-adapted enzymes. *Annu Rev Biochem* 75, 403-433.

Siegl, A. , Kamke, J. , Hochmuth, T. , Piel, J. , Richter, M. , et al. , (2011) Single-cell genomics reveals the lifestyle of Poriobacteria, a candidate phylum symbiotically associated with marine sponges. *ISME J* 5, 61-70.

Simon, C. and Rolf, R. (2011) Metagenomic analyses: past and future trends. *Appl Environ Microbiol* 77, 1153-1161.

Stepanauskas, R. and Sieracki, M. E. (2007) Matching phylogeny and metabolism in the uncultured marine bacteria, one cell at a time. *Proc Natl Acad Set USA* 104, 9052-9057.

Suenaga, H., Ohnuki, T. and Miyazaki, K. (2007) Functional screening of a metagenomic library for genes involved in microbial degradation of aromatic compounds. *Environ Microbiol* 9, 2289-2297.

Trincone, A. (2011) Marine biocatalysts: enzymatic features and applications. *Mar Drugs* 9, 478-499.

Troeschel, S. C., Drepper, T., Leggewie, C., Streit, W. R. and Jaeger, K-E. (2011) 'Novel tools for the functional expression of metagenomic DNA'. In *Metagenomics, Methods and Protocols, Methods in Molecular Biology* 668, Streit, W. R. and Daniel, R. (eds), 117-139. Humana Press.

Uchiyama, T. and Miyazaki, K. (2009) Functional metagenomics for enzyme discovery: challenges to efficient screening. *Curr Opirt Biotechnol* 20, 616-622.

Vietes, J. M., Guazzaroni, M. E., Beloqui, A., Golyshin, P. N. and Ferrer, M. (2009) Metagenomic approaches in systems microbiology. *FEMS Microbiol Rev* 33, 236-255.

Venter, J. C., Remington, K., Heidelberg, J. F., Walpern, A. L., Rusch, D., et al., (2004) Environmental genome shotgun sequencing of the Sargasso Sea. *Science* 304, 66-74.

Wang, C., Meek, D. J., Panchal, P., Boruvka, N., Archibald, F. S., et al., (2006) Isolation of poly-3-hydroxybutyrate metabolism genes from complex microbial communities by phenotypic complementation of bacterial mutants. *Appl Environ Microbiol* 72, 384-391.

Wang, H., Gong, Y., Xie, W., Xiao, W., Wang, J., et al., (2011) Identification and characterization of a novel thermostable gh-57 gene from metagenomic fosmid library of the Juan De Fuca Ridge hydrothermal vent. *Appl Biochem Biotechnol* 164, 1323-1338.

Warren, R. L., Freeman, J. D., Levesque, R. C., Smailus, D. E., Flibotte, S. and Holt, R. A. (2008) Transcription of foreign DNA in *Escherichia coli*. *Genome Res* 18, 1798-1805.

Waschkowitz, T., Rockstroh, S. and Daniel, R. (2009) Isolation and characterization of metalloproteases with a novel domain structure by construction and screening of metagenomic libraries. *Appl Environ Microbiol* 75, 2506-2516.

Wexler, M. and Johnston, A. W. B. (2010) 'Wide host-range cloning for functional metagenomics'. In *Metagenomics: Methods in Molecular Biology*, vol 668 Streit, W. R. and Daniel, R. (eds), 77-96. Humana Press.

Whitman, W. B., Coleman, D. C. and Wiebe, W. J. (1998) Prokaryotes, the unseen majority. *Proc Natl Acad Sci USA* 95, 6578-6583.

Xu, M., Xiao, X. and Wang, F. (2008) Isolation and characterization of alkane hydroxylases from a metagenomic library of Pacific deep-sea sediment. *Extremophiles* 12, 255-262.

Yooseph, S., Sutton, G., Rusch, D. B., Halpern, A. L., Williamson, S. J., et al., (2007) The Sorcerer II Global Ocean Sampling expedition: expanding the universe of protein families. *PLos Biol* 5, e16.

Zemin, F., Fang, W., Liu, J., Hong, Y., Peng, H., et al., (2010) Cloning and characterization of a b-glucosidase from marine microbial metagenome with excellent glucose tolerance. *J Microbiol Biotechnol* 20, 1351-1358.

Zhang, Y., Zhao, J. and Zeng, R. (2011) Expression and characterization of a novel mesophilic protease from metagenomic library derived from Antarctic coastal sediment. *Extremophiles* 15, 23-29.

第6章　海洋生物酶生物加工技术

Sreyashi Sarkar, *Sayani Mitra*, *Arnab Pramanik*, *Jayanta Debabrata Choudhury*, *Anirban Bhattacharyya*, *Malancha Roy*, *Kaushik Biswas*, *Anindita Mitra*, *Debashis Roy and Joydeep Mukherjee*, *Jadavpur University*, *Kolkata*, *India*

DOI：10.1533/9781908818355.2.131

摘要：本章重点介绍了将新型海洋活性酶向商业化生物加工技术转化的努力。在生产过程中，培养物可以是以悬浮培养或固定化方式进行培养。圆柱形罐是最常见的反应器，可以替代搅拌型反应器和没有机械搅拌装置的反应器。生物培养器主要有3种运行模式：批次培养，补料分批培养和连续培养。固态发酵是指微生物培养在坚实、潮湿基材上。包含新颖设计元素的生物培养器已应用于研究和在生物培养器中富集海洋微生物以取得对环境因素的良好控制。目前主要有3种不同的策略（1）模仿自然环境；（2）刺激未培养的微生物或生产有价值的代谢物；（3）控制沉淀物/水界面的氧化还原条件。实验室生物反应器规模化制备海洋酶的一些实例包括：由 Antarctic *bacillus* 菌生产的蛋白酶，*Teredinobacter turnirae* 的固定化培养和潮间带γ-变形菌门的生物膜培养。还比如有嗜热菌 *Pyrodictium abyssi* 生产的木聚糖酶和在填充床培养器中培养的海洋真菌 *Beauveria bassiana* 生产的L-谷氨酰胺酶。激烈热球菌 *Pyracoccus furiosus* 在连续培养中产生的糖化酶，以及从加压环境中培养的嗜压 *Shewanella* sp. 中得到的对苯二酚氧化酶。从嗜热 *Methanococcus jannaschii* 中获得的丙酮酸羧化酶。

关键词：生物加工，生物培养器，海洋微生物，固态发酵，填充床培养器，连续培养，生物膜

6.1　引言

微生物对于海洋环境中空间和营养的竞争是一种强大的选择压力，从而导致进化。进化促使海洋微生物产生多种酶系统以适应复杂的海洋环境。因此，海洋微生物酶能够产生具有独特性能的新型生物催化剂（Ghosh et al.，2005；Zhang and Kim，2010）。

具有潜在商业化前景的海洋酶类的例子，如包括基于酶的防污涂料（Kristensen et al.，2010），利用海水环境中的海洋细菌产生的淀粉酶对海洋微藻的糖化作用（Matsumoto et al.，2003）和酶对革兰氏阳性和革兰氏阴性细菌生物膜的降解作用，相关酶已获取美国专利（Manyak et al.，2010；Nijland et al.，2010）。

本章将重点介绍新型海洋酶活性转化为商业化生物加工的一些尝试。因此，对于在生物培养器中进行的实验室阶段的研究进行了极为详细的介绍，以便于感兴趣的研究者能够从所引用的参考文献中获取更多信息。本章首先介绍海洋生物培养的生物培养器技术的研究现状，随后介绍几种生物加工技术成功应用的酶。

6.2 传统的培养，生物培养器的配置和操作模式

常规培养物可以是悬浮的（自由的）或是被固定在培养基中。将细胞的活动限制在一定的范围内被称为细胞固定化。固定化的细胞培养物相比悬浮培养具有一些优点，如可获得较高细胞浓度，细胞的再利用，并能够实现高细胞浓度和高流速的组合，从而实现较高的生产效率（Shuler and Kargi，2003）。

在各种不同生物培养器的配置中，圆柱形罐是在生物加工中最常见的培养器。对于培养器设计的挑战主要在于能够提供足够的气体混合，为需要氧气的生物加工过程提供氧气。用于厌氧培养的培养器通常结构很简单，没有喷射或搅动的装置，而搅拌培养器的替代装置含有不带机械搅拌功能的反应容器。在空气泡塔培养器中，通风和气体混合由气体鼓泡搅拌实现，这比机械搅拌需要更少的能量。气升式反应器不需要机械搅拌就可以实现，所以剪切水平比搅拌装置显著降低。填充床培养器（PBR）含有填充有催化剂的管道，通常用于固定化或颗粒状生物催化剂。培养基可以在柱子的顶部或底部补充，并在颗粒之间形成连续的液相。与搅拌培养器相比，由于颗粒摩擦而造成的对填充床的损坏明显降低。当填充床在上行模式运行，并且催化剂珠粒具有适当的尺寸和密度时，填充床由于颗粒的向上运动，从而可以以很高的流速扩展。这是流化床培养器运行的基础（Doran，1995）。

生物培养器的操作方式是影响培养器性能的重要因素，目前生物培养器的操作模式主要包括 3 种：分批、补料分批和连续模式。操作模式的选择策略对底物转化、产物浓度、易污染程度和工艺的可靠性具有重要影响。

分批培养在封闭的系统中进行，底物只在培养开始时加入，产物只在培养结束后取出。经典的混合培养器是一个搅拌的培养罐；然而，混合培养堆也可以是鼓泡塔，气升式或者是其他设置。分批培养是用于工业发酵的常见模式，但是除了发酵时间外，生产周期和周转时间（所需要的灭菌、接种等），都导致整体生产能力的降低，并增加了生产成本。“取出和填充”培养是发酵的潜在替代方式，在该培养方式中，可以取出全部或部分培养基，同时周期性地补充新鲜培养基，因此避免了周转时间。

在分批补料操作中，通过营养物质的间歇或连续进料，补充培养器中的成分，并可控

制底物浓度范围。开始时，底物溶液相对较稀，随着转化的进行，添加更多的营养物质，避免了高增长率。这对于在快速生长期时需氧量较高的培养中是非常重要的，有利于培养器的质量转化能力。

在连续培养中，如果容器混合均匀，产品中的成分和组成与培养器内液体的成分是相同的。因此，当连续培养器用于培养可自由悬浮的细胞时，催化剂可连续地从产物流的容器中取出。同时生成更多的细胞来代替那些取出的细胞（Doran，1995）。

固态发酵（SSF）意味着将微生物培养在固体和不存在游[illegible]的湿润基质上，也就是，平均水活动要明显低于1.0。在一种更广泛的定义中，S[illegible]可以被看做是包括将微生物培养在底物浓度达到最大的液相中，或是培养在惰性载体上等方式。SSF，一种环境友好型生物工艺，与完全水相的工艺相比，在批量生产化学产品和酶方面，存在诸多优势，如简易的下游操作，能量需求降低，产生废水少，产品产率高，体积生产率增加，产品回收率提高；反应器设计简单等优点。影响SSF投入应用的主要原因是工程问题，该过程目前与标准化之间仍有较大差距，同时结果的重复性较差。

6.3 专业生物加工技术

生活在海洋沉积物中的微生物通常仅有非常有限的生存空间，具有特定的物理化学条件。只有在这些特定条件下，他们从代谢反应中得到获取的热力学能量才能够满足他们的需求。要在实验室中保存或者培养这些微生物，必须高度重视温度，压力和液体/气体通量等。Lang 等（2005）的综述描述了海洋原核生物和真菌生长行为以及产生新代谢产物或酶的潜力对温度/压力的依赖性。继分批式培养，补料分批式和/或连续培养类型之后，最新出现的摇瓶培养和可控生物培养器培养也已被描述。对最大生物量，特定生长率和（次）最佳产率等数值也有报道。利用嗜温微生物来加强生物加工技术的研究是生物加工的目标。从嗜冷和超嗜热微生物中分离出低温活性酶和热稳定酶是许多研究的目标。对于生物工程研究者来说，从海洋深处，如近1 100 m深处和从热液喷口处发现的嗜压菌是一项特殊的挑战（Lang et al.，2005）。除了所描述的传统方法外，具有新颖设计元素的生物培养器已被用于研究和富集海洋微生物，实现对环境因素的良好控制（Zhang et al.，2011）。继Zhang等在2011年的报道之后，表6.1列出的是已在培养器系统中进行了体外模拟的一些工艺。一种最合适的系统的选择依赖于实际条件。

表6.1 海洋微生物进程及封闭环境的体外模拟

进程及环境	反应器	文献
潮间带河口	螺旋盘生物反应器	Sarkar et al.，2008
从水体表面下沉到深海	加压微观	Grossart and Gust，2009

续表

进程及环境	反应器	文献
冷水渗透	持续流速反应器，持续高压反应器，膜生物反应器	Girguis et al. , 2003 Meulepas et al. , 2009 Zhang et al. , 2010
热泉风口	持续高压反应器，气体生物反应器，发酵罐	Imai et al. , 1999；Mukhopadhyay et al. , 1999；Houghton et al. , 2007；Postec et al. , 2007
高压环境	金刚石钻细胞	Aertsen et al. , 2009

注：每一个特别因子类型的环境模拟由文献表明 .

根据不同的研究目标，3 种不同的策略区分如下：①模拟天然环境；②刺激未培养的微生物或生产感兴趣的代谢物；③控制在沉淀物/水界面的氧化还原条件。Zhang 等（2011）列举了六大类适合于海洋生态系统研究的传统培养器（表 6.2）。

表 6.2　用于海洋微生物培养的常规生物反应器的组成

生物反应器组成	特性	文献
搅拌釜反应器	批量和持续培养	Muffler and Ulber，2008； Sarkar et al. , 2008
旋转盘/鼓生物反应器	在盘/鼓上，生物膜形成，在盘上收集/移除快速生成的膜，模拟有氧或者无氧环境	Konopka，2000
空气上升反应器	气体基质通过瓶底促使其混合	Meulepas et al. , 2009
膜反应器	有选择性的膜被应用去保留生物量	Meulepas et al. , 2009
持续高压反应器	基于高压泵去产生高压力，高压力气体或者连接压缩机去制备高气体压力	Wright et al. , 2003 Pradillon et al. , 2004 Parkes et al，2009 Zeng et al. , 2009 Deusner et al. , 2010 Zhang et al. , 2010
生物电子化学系统	生物电子化学反应在时间空间上有区分	Logan et al. , 2006；Zuo et al. , 2008；Clauwaert and Verstraete，2009

注：这些特别反应器组成的应用由相应的文献报道 .

设计一种有效的生物培养器用于海洋微生物的培养，要综合考虑温度、压力和构件材料等物理因素。

6.3.1 温度

超嗜热微生物培养方案的发展遇上了一些有趣的问题，这些问题在培养传统嗜中温微生物时是没有遇到过的。可能最主要的问题是目前对超嗜热微生物的生长和代谢所得到的信息相对较少。生物化学和酶学的研究通常因为低生物量产率而受到限制，而很多超嗜热微生物都存在这个问题。在这一方面，大量生物质的生产对一些容积效率差的发酵来说呈现出一些不利因素，包括危害物的处理，生物发酵产生的大量硫化氢所带来的腐蚀性的处理等附加问题。因此，超嗜热菌的培养难题成为主要的技术障碍（Andrade et al.，2001）。

尽管全球深海中的温度都很低，对于温度可达到 1 000℃的热液喷口的研究成为热点。为了保证在实验室内培养从热液喷口发现的超嗜热菌并保持正确的培养温度，可以将培养器置于可以控温的培养箱中，或是放置在可以控温的水保护层中。这些方法通常可以控制温度在 1~80℃。如果培养温度必须更高，热水不可能实现的，可以使用热空气培养箱（Zhang and Kim，2010）。

6.3.2 压力

海洋微生物能够在高达 110 MPa 流体静压的条件下生存，这比大气压要高出 3 个数量级。对于高压培养器的设计，需要同时考虑高流水静压和高气压。嗜压细菌的营养底物在大气条件下很容易溶解，流水静压能够改变其基因的表达，以实现最大的细胞生长。根据不同的研究目的，已建成不同类型的高流水静压的培养器：加压恒化，模式加压和连续模式超高压培养器。对于某些嗜压微生物，在大多数情况下它们的主要营养物（如甲烷或氢气）是气态形式，导致在大气压下是难溶的。因此，体外培养需要高气压，以增强其代谢活动和生长速率（Zhang and Kim，2010）。三菱重工提供一种岸上设施，能够培养生活在深海泥中的微生物。这个设备可以保证深海微生物被培养在与深海相同的环境中（压力条件和温度条件），而不必暴露在陆地环境中

（http：//www.mhi.co.jp/en/products/expand/hightemperature_ and_ high-pressure_tank_ supply_ result_ 03.html#anchor Pagetop）。

6.3.3 构造材料

一些海洋微生物在他们的生长过程中特别需要高浓度钠离子。在空气中高盐水平可能会导致奥氏体钢的腐蚀等问题，因此需要额外加入钼以达到抗腐蚀的目的。铁和不锈钢的腐蚀受到脱氮和硫酸盐还原过程的影响，当铁浸入缺氧液体中时，其阴极可以产生氢气。对于高压培养，一些材料如 PEEK（聚醚醚酮）塑料和不锈钢材料已被应用于容器和管道的构建。表面形成的附着和生物膜会影响很多海洋微生物的代谢和酶制剂的生产。某些微生物只有附着在表面时才能达到最高产出目标产物的速率，这种微生物的培养需要特殊的

培养器。在反应器的内部，载体材料（如很多不同的高分子材料），已被检测出能够提供足够的表面以形成生物膜层（Sarkar et al.，2011；Zhang et al.，2011）。

6.4 海洋酶制剂生物加工技术

本节介绍生产微生物来源的海洋酶制剂所采用的特定生物加工方法的现有数据。通过参照美国和欧洲专利数据库（可免费获得部分）以及研究期刊的有关资料，本节侧重于多种不同酶的介绍，包括生产特异性催化剂的嗜极微生物一些有趣的案例。这些独特的催化剂与不同加工行业中普遍流行的催化剂相比，更能够在极端条件下发挥功能。

6.4.1 蛋白酶

哈氏弧菌（*Vibrio harveyi*）被用来在海水/ Zobell 基础培养基（补充脱脂奶粉）中生产胞外蛋白酶。对 3 种不同搅拌速度（300 r/min，500 r/min 和 700 r/min）和 3 种不同空气溢流速率（0.2 L/(L·min)，0.5 L/(L·min) 和 0.8 L/(L·min) ——每分钟每升培养基的空气体积）进行了研究，以确定在 1.5 L 发酵罐（Applikon，Holland）中的最佳工艺条件。通过加入几滴灭菌葵花油控制泡沫。生长曲线定义为生物量的对数［$Y=\ln(X/X_0)$］对时间（t）的函数，μ 由 Gompertz 模型来确定［式（6.1）］

$$y = A\exp\left\{-\exp\left(\frac{\mu e}{A}(\lambda - l) - 1\right)\right\} \tag{6.1}$$

其中 μ 为比生长速率；λ 为滞后时间，渐近 A。加入脱脂牛奶的 Zobell 培养基可以将胞外酶制剂生产效率提高 5 倍。以 300 r/min 未检测到明显活性，而当气体流速为 0.8 L/(L·min)，转速为 500 r/min 或 700 r/min 时，活性明显增加（Estrada-Badillo and Marquez-Rocha，2003）。

Kumar 等（2004）对由 *Bacillus clausii* 分泌的一种碱性蛋白酶的分批培养工艺进行了研究，发现在 5 L 生物培养器中（Model KF5L，Kobiotech，Inchon，Korea），氧化剂和 SDS 在一定范围生产条件下对其活性没有影响。培养过程中控制搅拌速率在 400 r/min 而不控制 pH 值，同时空气流速保持在 1.5 $v/(v\cdot\text{min})$。消泡剂 A（Sigma）用于减少泡沫的形成。酶活性随着发酵时间和搅拌速率的增加而增加。蛋白酶产量的增加取决于有效的质量转化系数，这有赖于最佳的通气速率和搅拌速率。

Narinx 等（1997）在 LH2000 发酵罐（10 L）中，8℃的条件下 6 d 培养了一株 Antarctic *Bacillus* TA39。发现在静止期，细胞的数量在 25℃下仅为在 4℃下的一半，而蛋白酶的分泌量还达不到 4℃条件下的 1/3。大约 20 h 的滞后期之后，在 4℃条件下的对数生长期时的倍增时间为 9 h，而在 25℃条件下为 2 h（图 6.1）。

产金属蛋白酶菌株 *Hyphomonas jannaschiana* 的深层培养是在有氧条件下并且机械搅拌或空气喷布器的条件下完成的（Weiner et al.，1996）。通常需要培养 20~50 h（图 6.2）。

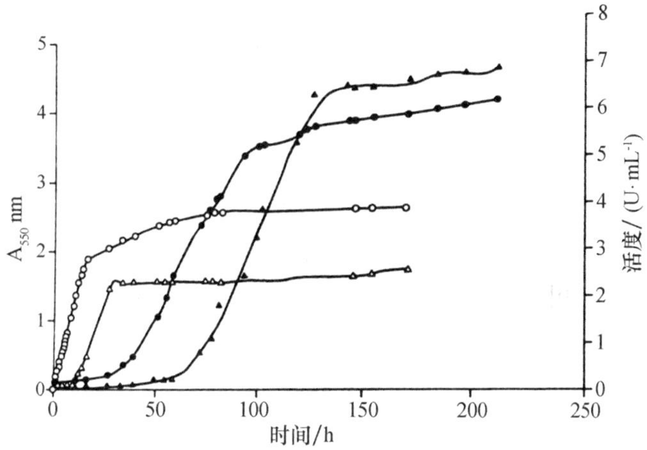

图 6.1　南极菌株 *Bacillus* TA39 在 4℃（实心圈）和 25℃（白圈）的生长及枯草杆菌蛋白酶 S39 在 4℃（实心三角）和 25℃（空三角）的释放

（经过牛津大学出版社允许后重印）

在最佳情况时，培养约 17 h 后可在培养物上清液检测到胞外酶的活性。细胞也可以在静止生长期的早期被收获，并重新悬浮于小体积的培养基中，以在培养基中诱导产生外切蛋白酶。

Elibol 和 Moreira（2003）利用褐藻酸钙珠，将海洋船蛆细菌 *Teredinobacter turnirae* 的整个细胞进行固定化，以生产碱性蛋白酶。将细胞悬浮液在无菌条件下加入到无菌的海藻酸钠溶液中，以达到所需的细胞/藻酸盐的比率。然后将得到的混合物滴加到 $CaCl_2$ 溶液并在该溶液中硬化。对多次使用钙藻酸盐基质中固定化 *T. turnirae* 细胞的可行性进行了研究，将藻酸盐粒珠连续用于 8 批次的培养，每次持续 72 h。随着循环培养次数的增加，蛋白酶产量增加，并在 3 个循环后达到最大值，然后随加工时间的增加持续降低，单位体积总产量增加了 3.5 倍。

在 Beshay 和 Moreira（2003）进行的下一个研究中，作者们选择在不同的无机材料上固定化 *T. turnirae* 细胞的最佳条件，评价了在重复分批培养固定化生物催化剂中碱性蛋白酶的产率。在相同的培养条件下，固定细胞生产的碱性蛋白酶的活性是自由悬浮培养的 2.3 倍。重复使用 *T. turnirae* 细胞的可能性研究是通过重复性分批培养 16 d（7 个周期）来进行的。固定化细胞产生的碱性蛋白酶的活性逐渐增加，在 4 个周期后达到稳定，直到第 7 周期结束一直保持恒定（图 6.3）。

El-Gendy 等研究了影响青霉菌 Morsy1 在 SSF 培养基中生产角蛋白酶的理化参数（2010）。测定了 *Penicillium* spp. Morsy1 在 SSF 中和不同的农业和家禽废弃物中的角蛋白酶活性，发现 Morsy1 在 pH 值 6.0 时角蛋白酶产量最大，随后开始下降。真菌产生的胞外酶达到高效价的最适 pH 值是其生长 pH 值，在 26℃培养条件下蛋白酶的产量最高。

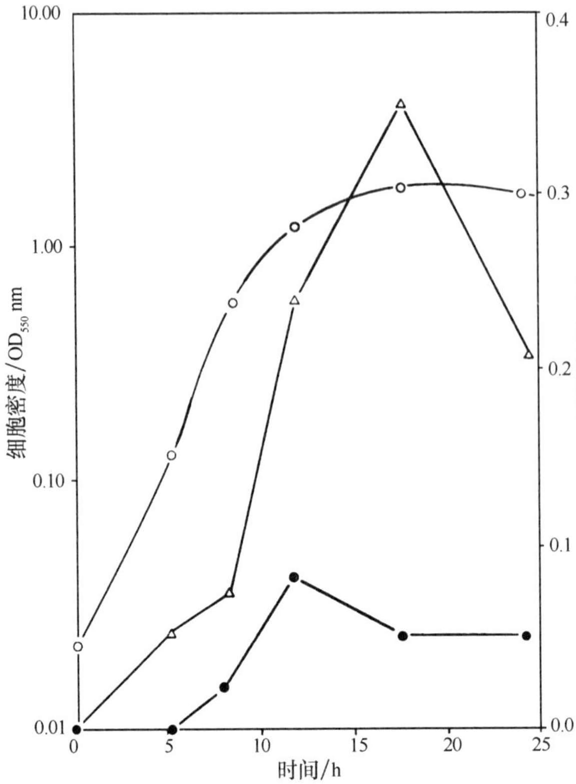

图 6.2　由 *Hyphomonas jannaschiana* 产生的热稳定碱性金属蛋白酶

注：圆形为在 550 nm 测定细胞生长的吸光度值，空心三角形为胞外蛋白酶活性，实心圆形为胞质蛋白酶活性，蛋白酶活性由偶氮酪蛋白方法决定，样品活性单位（U/mL）（经过美国专利和商标办公室允许后重印）.

聚甲基丙烯酸甲酯（PMMA）锥柱形烧瓶（CCF）内部含有一个由 8 个等距间隔的矩形条带组成的径向圆盘，以提供更多的生物附着所需的表面（图 6.4），这种装置可以用作能够形成生物膜的潮间带 γ-Proteobacterium 的培养来生产蛋白酶。这种烧瓶设计可以用来比较在亲水性材料（玻璃）或疏水材料（PMMA）的表面培养时生产蛋白酶的不同。相比于 Erlenmeyer 锥形烧瓶，疏水表面的 CCF 具有更高的蛋白酶产量。这项研究打破了传统的摇瓶培养，通过形成生物膜来提高蛋白酶的产量（Sarkar et al.，2011）。

6.4.2　木聚糖酶和纤维素酶

热古细菌 *Pyrodictium abyssi* 在 97℃具有最佳生长效率，是高度热稳定木聚糖酶的一个潜在来源。Andrade 等（2001）系统地比较了 *P. abyssi* 的培养参数。在 97℃下将培养物在血清瓶中培养 48 h，并以此为接种物接种到 16 L 的生物反应器中，加入 13 L 培养基进行培养（生物工程，沃尔德（Wald），瑞士）。利用非线性方法评估了 Monod、Contois 和

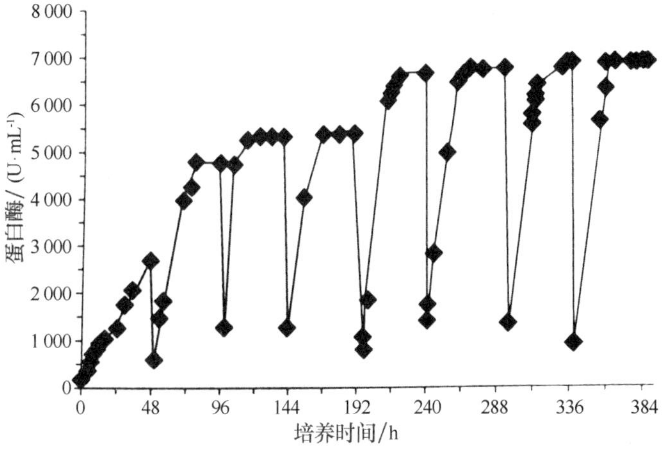

图 6.3　固定化的 *Teredinobacter turnirae* 分泌的碱性蛋白酶产物活性与时间关系（由 Elsevier 重新影印）

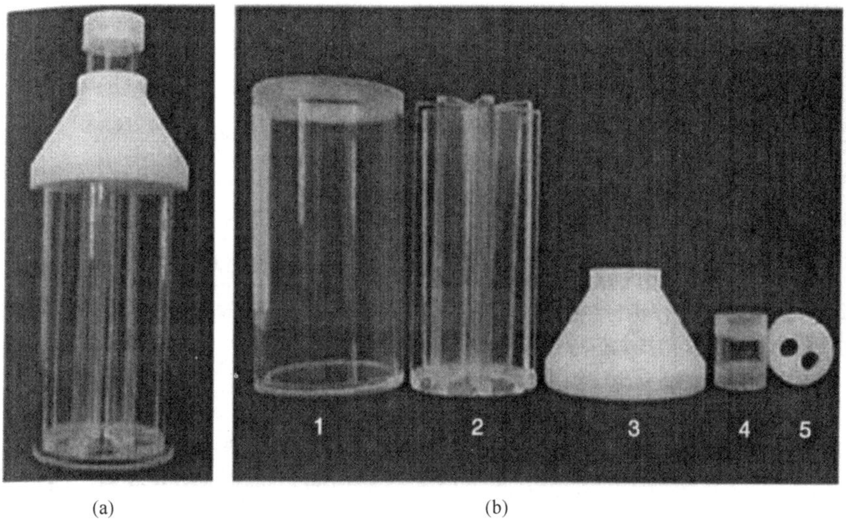

图 6.4　（a）锥形烧瓶（CCF）和（b）锥形烧瓶的组件

注：1：圆柱形底座；2：内部构造；3：上面漏斗部分；4：连接顶部的颈部；5：顶部通风盖.

Reprinted with permission from Elsevier

Tessier 模型的动力学参数。由于不能充分地利用底物，细胞产率非常低，但测定出了非常高的最大生长速率。由 *P. abyssi* 批次培养所产生的木聚糖酶单位酶活很低，其中 β-木糖苷酶和阿拉伯呋喃聚糖酶的活性低于 0.1 mU/L。这种低活性的出现可能是由于培养基中的酵母提取物浓度导致的，为获取更大的生长量需要在培养基中加入元素硫。发现基于 Contois 模型的可变性比 Monod 和 Tessier 模型要小。Contois 模型中各参数的物理意义和能够参数允许高细胞密度的培养，使得 Contois 模型成为预测 *P. abyssi* 利用木聚糖维持生长的首选模型。在不搅拌发酵罐中可以成功培养 *P. abyssi*。剪切力可能会破坏 *P. abyssi* 脆弱的微生物细胞网络，对木聚糖酶的生产造成显著影响。

测试耐碱性真菌 *Chaetomium* sp.（NIOCC 36）在 SSF 生物加工过程中，利用农业和工业废物作为底物，生产碱性纤维素酶（β-内切，β-外切葡聚糖酶，β-葡萄糖苷酶）的能力。在 SSF 培养条件下，棉花种子和高碱性 pH 值条件下，表现出最高酶产量。而碱性条件、棉花种子、SSF 培养条件，50℃条件下产生的纤维素酶比非碱性条件下产生的纤维素酶具有更高的稳定性和活性（Ravindran et al.，2010）。

之前提到的聚甲基丙烯酸甲酯圆锥圆柱瓶（Sarkar et al.，2011）被用于培养 *Chaetomium crispatum* 和 *Gliocladium viride* 分别生产纤维素酶和木聚糖酶。这些设计可以用来比较疏水表面 CCFs（PMMA-CCF），亲水玻璃表面（GS-CCF）和 500 mL 的锥形烧瓶（EF）之间产量的不同。在 PMMA-CCF 中，β-1，4-葡聚糖酶，FPase（滤纸降解）活性和 *C. crispatum* 的生长率最高。在 EF 中，木聚糖酶的产量和 *G. viride* 的生长率最高（图 6.5）。通过对细胞外聚合物分子的双通道荧光检测以及在共焦激光扫描显微镜下全细胞的观察，发现形成的生物膜在格局等结构参数上表现出一定的增长（Mitra et al.，2011）。

6.4.3 几丁质酶

装有 2.0 L 培养基的 3 L 夹套桌面搅拌式发酵罐（Fenice et al.，1998），通过微电脑控制器（Applikon Dependable Instruments，Schiedam，The Netherlands）控制，用于培养 *Penicillium janthinellum* 以生产几丁质酶，发现酶活性的出现与菌丝体生长相关。

Kao 等（2007a）利用 *Paenibacillus* sp. CHE-N1，在 5 L 搅拌发酵罐培养器（BTF-A5L，Bio-Top Inc.，Taiwan）中加入 3 L 培养基进行发酵生产几丁质酶。测定了单位体积传质系数、$K_L\alpha$，混合时间和培养黏度。在 100 r/min 的搅拌速率下几丁质酶水平明显低于在 200 r/min 下几丁质酶水平，这可能是由于在较低搅拌速度下的混合不完全或者氧传递受阻。然而，在 300 r/min 的搅拌速率条件下观察到几丁质酶的水平最低，这是高剪切应力的结果。在 100 r/min 条件下的 $K_L\alpha$ 值比较高搅拌速率下的值低。因此，有理由相信在较低的搅拌速度下，氧传递是主要的限制因素。每小时 35.5 的 $K_L\alpha$ 值被认为是适于细胞生长和几丁质酶生产的最基本的要求。如图 6.6 所示发酵培养的流变曲线图。发酵肉汤被视为非牛顿宾厄姆流体，即剪切应力与剪切速度是线性相关的，其斜率等于表观黏度（η，g/cm）和 τ_0 的截距（1.8 dyn/cm^2）。剪切应力计算为黏度和剪切速率加 τ_0 的产物。当搅拌速率增加至 200 r/min 时，在恒定细胞的水平下，孢子的数量不断增加，然而，在 300 r/min 时，孢子的数量增加而细胞的数量减少。结果表明，剪切应力高于 5.8 dyn/cm^2 不利于细胞的生长和几丁质酶的生产。

Liu 等（2003）在深层培养的参数对 *Verticillium lecanii* 在发酵罐中生产几丁质酶的影响进行了研究。在一个 5 L 的搅拌发酵罐（STR）中，在最佳的培养条件（图 6.7）下获得了较高的几丁质酶活性。

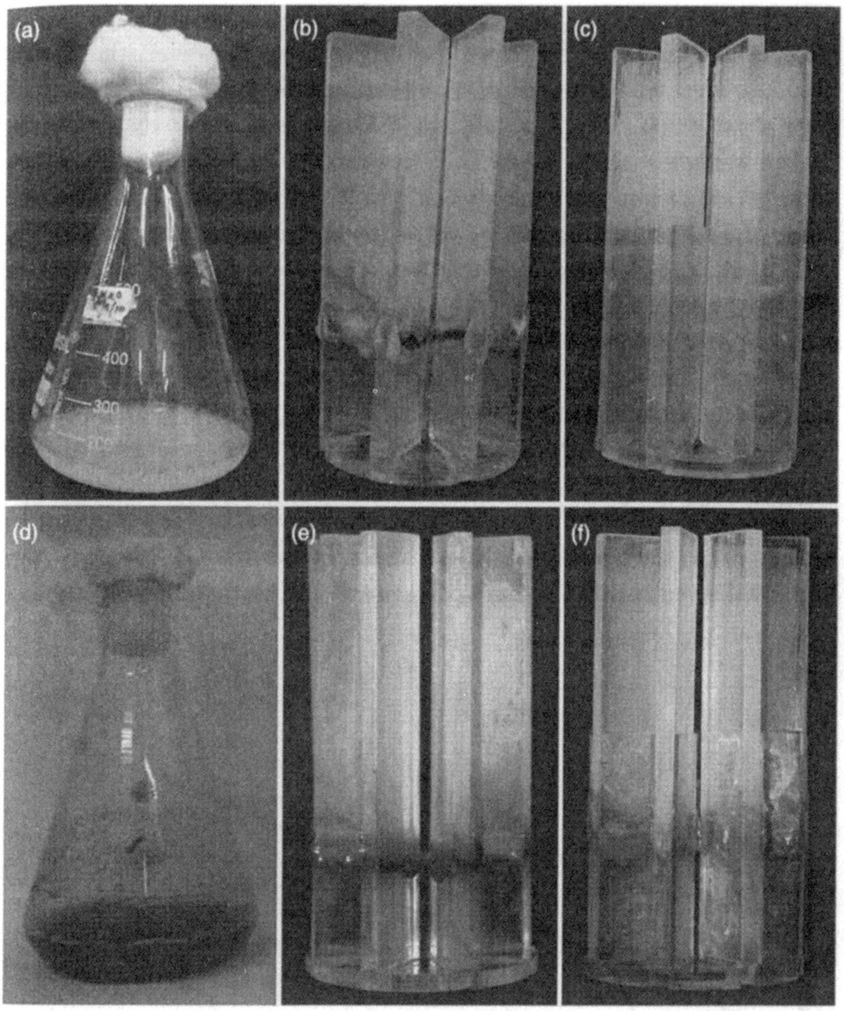

图 6.5（a）EF 的外观；（b）PMMA-CCF 的内部构造和（c）GS-CCF 在培养过 *C. crispatum* 7 d 之后的内部构造；（d）EF 的外观；（e）PMMA-CCF 的内部构造和（f）在培养过 *G. viride* 7 d 之后的内部构造

对发酵培养基 pH 值的影响也进行了研究。当培养基 pH 值从 2 上升至 9 时，几丁质酶的活性不断变化，最适 pH 值为 4.0。设制挡板的搅拌发酵罐 STR 能够提供更好的质量和气体转移效率。然而，挡板在培养过程中会引起的 *V. lecanii* 的聚集，真菌的生长和酶的合成都会受到限制。搅动也造成了生物培养器中形态的变化，从而导致目标酶的生产率不同。

对通气效果的影响也进行了研究，比较了在配有 24 目通风管网络的 30-L 通气生物培养器中，3 个通气速率的效率，即整合有 0.6 *V*/（*V*·min）、0.9 *V*/（*V*·min）和 1.2 *V*/

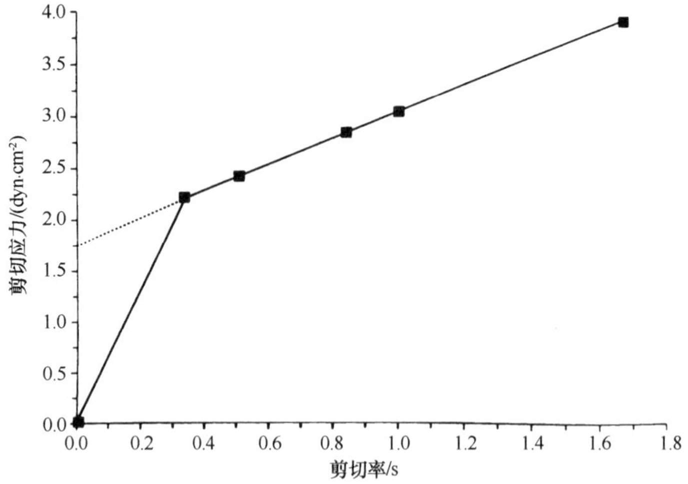

图 6.6　剪切应力下 *Paenibacillus* sp. CHE-N1 发酵的流变学

Reprinted with permission from Elsevier

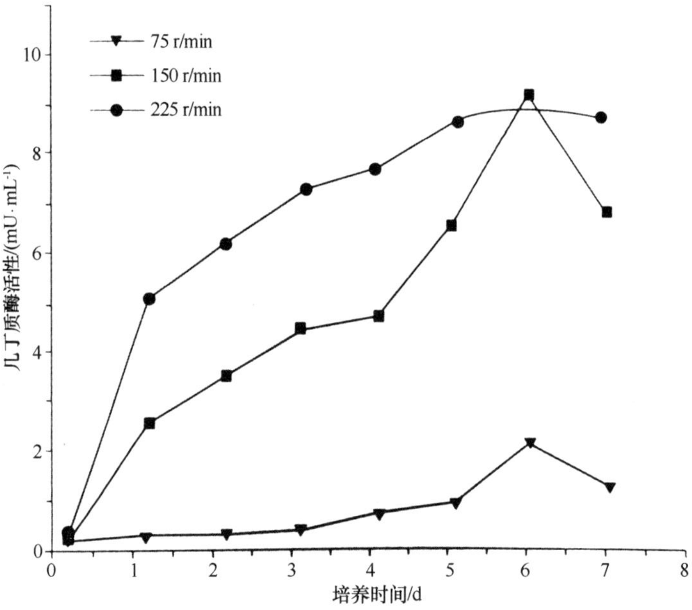

图 6.7　在不同的搅动频率下 *Verticillium lecanii* F091 儿丁质酶活性与时间的关系

Reprinted with permission from Elsevier

（*V*·min）（每分钟单位培养基的体积的空气体积）的 24 目通风管网络。图 6.8 显示 *V. lecanii* 30-L 发酵罐中的儿丁质酶活性和 DO（溶解氧）时间曲线。通风情况会略微影响溶解氧浓度，进而影响细胞生长、酶活性和底物利用。当溶解氧浓度低于临界水平，细胞呼吸从 DO 转换到气态形式。然而，这个现象只发生在高通风率情况时。否则，将会是线性

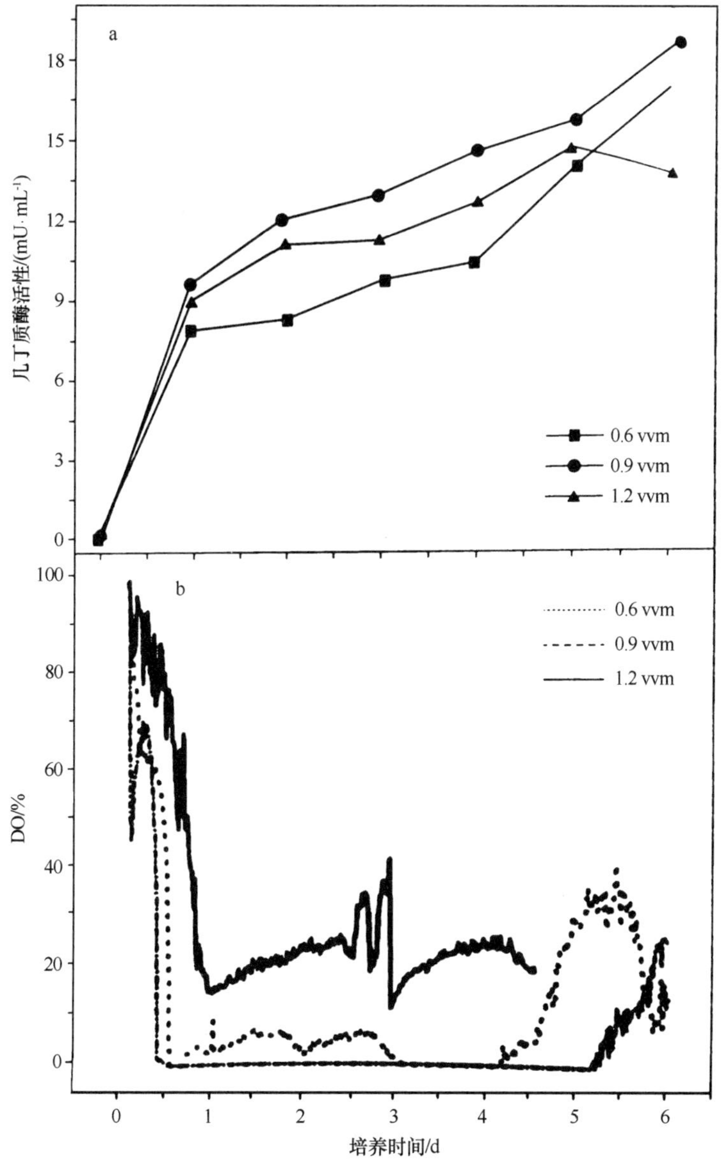

图 6.8 不同通风率下 *Verticillium lecanii*（a）几丁质酶活性与时间的关系和溶解氧的层级（b）

Reprinted with permission from Elsevier

增长状态，表明气态和 DO 的不足。这个概念表明溶解氧的消耗和与之相伴的二氧化碳分压的升高可以刺激呼吸转换和代谢或形态变化的开始。这项研究表明，发酵 pH 值和搅拌速度对发酵的影响比通风速率对发酵的影响更加显著。

Kao 等（2007b）研究了采用薄膜方式发酵操作以连续性生产几丁质酶的可行性，研究采用的是类芽孢杆菌属（*Paenibacillus* sp.）CHE-N1。膜外循环回路的生物培养器用来评价膜的孔径对细胞保留的效率、渗透溢流速率、积垢和在渗透物中几丁质酶回收的影响。结果表明，在 0.9 kg/cm^2 的跨膜压力、300 kDa 的孔径大小的 M 9 微滤柱表现出最好的微滤特性，可用于薄膜模式操作。在薄膜操作方式中，总几丁质酶的活性比在分批模式培养中得到的总几丁质酶的活性高出约 78%（图 6.9）。通过每 3~4 天补充几丁质能得到进一步提高。

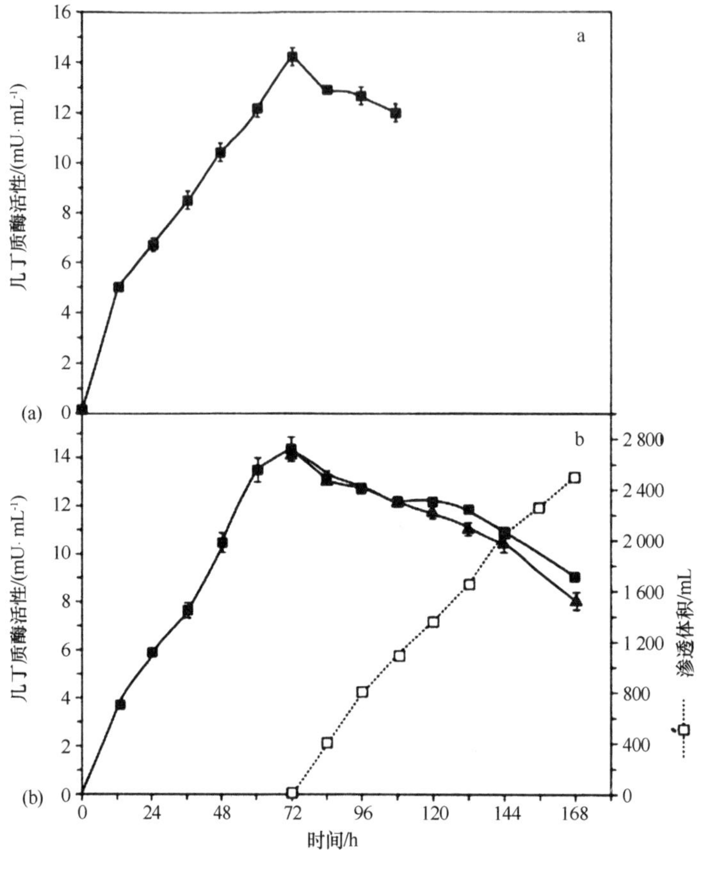

图 6.9　在批量处理模式下 *Paenibacillus* sp. 几丁质酶产率（a）与时间的关系（b）膜上操作模式

注：正方形为在培养基中的活性水平中的活性层级，实心三角形是渗透中活性水平，空心正方为在培养基中渗透物体积，反应器中工作体积保持在 2 L.

Reprinted with permission from Elsevier

6.4.4　谷氨酰胺酶

Sabu 等（2002）报道了在填充床培养器中培养海洋真菌 *Beauveria bassiana* BTMF S-10 连续生产细胞外 L-谷氨酰胺酶。对影响粒珠生产能力的各种参数和批次培养模式下的效果进行了优化。在较高藻酸盐浓度时，出现了扩散限制，因为众所周知，增加藻酸盐浓度导致更紧密的交联。对在分批模式下生产谷氨酰胺酶优化得到的参数纳入连续生产的研究范围，研究培养基的溢流速率、底物浓度、通风和床层高度等对连续生产 L-谷氨酰胺酶的影响。通过 Ca-褐藻酸固定化的孢子进行酶的连续生产，非常适合于 *B. bassiana*，因为在很短的时间内就获得了更高的酶产量。

海洋 *Pseudomonas* sp. BTMS-51，由 Ca-藻酸盐凝胶包埋固定化，被用于在重复培养模式和使用 PBR 的持续模式下生产胞外 L-谷氨酰胺酶（Kumar and Chandrasekaran，2003）。重复培养 20 个循环之后，固定化细胞再次使用时并没有表现出任何产量上的下降。酶的产量与生物质的含量密切相关（图 6.10）。

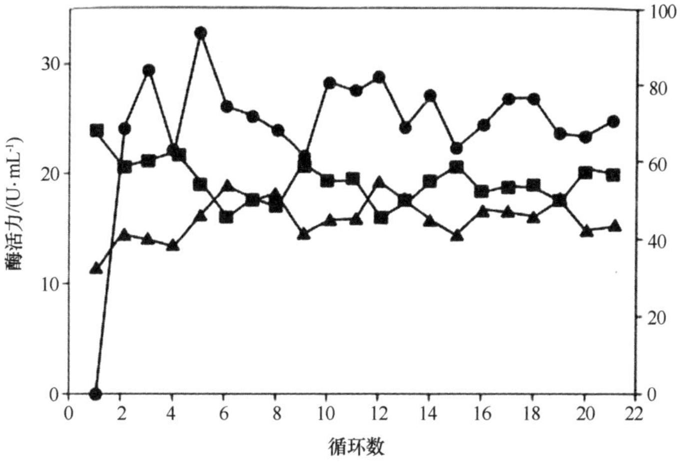

图 6.10　由 Ca-alginate 固定细胞的 *Pseudomonas* sp. BTMS-51 制备 L-谷氨酰胺酶实验

注：三角形为珠子中生物量，正方形为培养基中生物量，圆形为酶活力 .

Reprinted with permission from Elsevier

此外，对运用 PBR 技术中酶的连续生产进行了研究，研究不同的底物浓度和稀释速率。总的来讲，单位体积生产率随着稀释率，底物浓度的升高而升高，但底物转化效率下降。该系统可以连续运行 120 h，生产率没有任何下降。

6.4.5　菊糖酶

能够编码外切菊糖酶的 *INU*1 基因从 *Kluyveromyces marxianus* CBS 6556 中被克隆，并被连接至表达质粒 pINA1317，在 *Yarrowia lipolytica* 中被表达（Cui et al.，2011）。在装有挡板、搅拌器、碱泵、加热元件、氧传感器以及温度传感器的 Biastat B2 2-L 发酵罐中（贝朗，德国）进行发酵。种子培养物被转移到含有 4.0%（*w/v*）菊糖或 8.0%洋蓟结的生产培养基中。在 250 r/min 的搅拌速度，10 L/min 通气量，28 ℃的条件下进行发酵，发酵周期为 80 h。在 80 h 的培养后，分泌出的菊糖酶活性为（43.1±0.9）U/mL。当工程化的酵母细胞在含有 8.0%洋蓟结的 2 L 发酵罐中生长时，得到的粗蛋白和细胞质量值比在含有 4.0%菊粉的培养基中更高（图 6.11）。

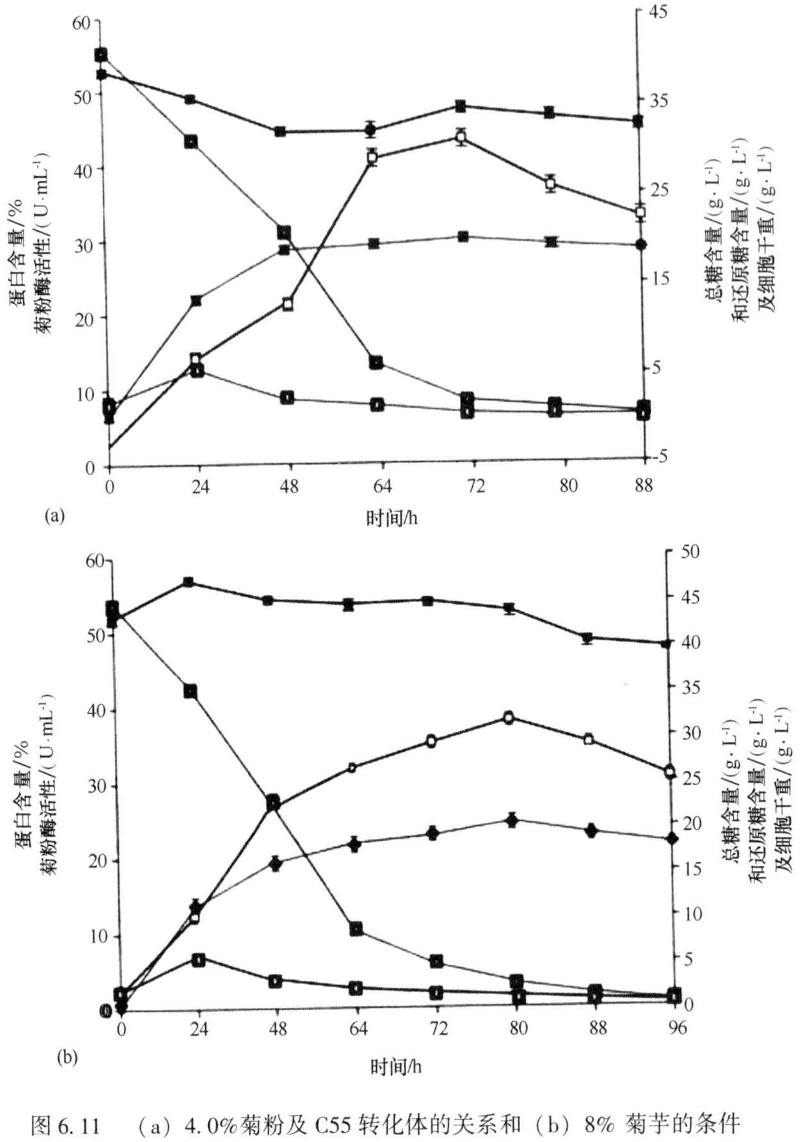

图 6.11　（a）4.0%菊粉及 C55 转化体的关系和（b）8% 菊芋的条件

注：不同时间细胞生长（菱形）单细胞蛋白（实心正方形）菊粉酶活性（空心正方形），总糖含量（交叉）和还原糖（点），2-1 培养中.

从海洋 *Pichia guilliermondii* 菌株 1 中克隆的菊糖酶的基因，在 *Pichia pastoris* X-33 中进行表达，对菊糖酶高表达的条件进行了优化。对携带质粒 pPICZaAINU1 的转化子 INU1 进行了高细胞密度的发酵培养，在 Biostat B2 2-L 发酵罐（B. Braun，Germany）中进行。搅拌速度，通气速度和温度分别为 160 r/min，4 L/min 和 28℃。发酵 48 h 后，加入 1.5%

的甲醇到培养基中，诱导表达重组菊糖酶。然后，每 24 h 将 1.5%的甲醇加入到培养基中。优化表达条件后，重组菊糖酶的表达效率可达到（13.04±0.4）mg/（protein）/（mL·d）(Zhang et al.，2009)。

6.4.6 琼脂糖酶

由弧菌（*Vibrio* sp.）菌株 JT0107 生产琼脂水解酶的最佳培养条件是由 Sugano 等于 1995 年，在含有 3 L 培养基的 5 L 发酵罐（Marubishi Eng. Co.，Tokyo）中测定的。研究了在通气和不通气的情况下，氧气浓度的改变对细胞生长和琼脂糖水解酶表达情况的影响(图 6.12)。

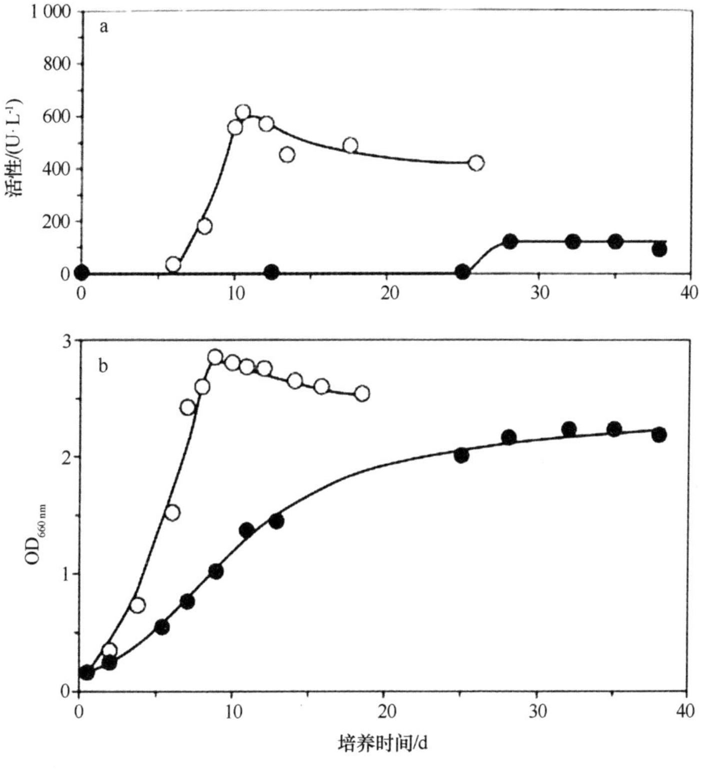

图 6.12　a：*Vibrio* sp. JT0107 和琼脂糖活性和孵化时间的关系

b：生长（OD_{660}）和孵化时间之间的关系

注：空心圆形为 0.5 vvm 通风量，实心圆形为没有通风．

Reprinted with permission from Elsevier Pyrococcus furiosus

充气培养条件下的最大琼脂糖酶活性是非充气培养条件下的 7 倍多。溶解氧的存在显著提高了细胞的生长速度和琼脂糖酶的生产。

6.4.7 淀粉酶

新设计的培养系统使得可以在温度接近 100℃下连续培养嗜热菌。使用该系统对生产糖化酶的激烈热球菌进行了连续培养，测定了其稀释率和产气情况（Brown and Kelly 1989；Kelly et al.，2002）。*Pyrococcus furiosus* 是一种专性厌氧异养菌，在有硫元素存在和没有硫元素的情况下都可以生长。当硫存在时，菌的生长伴随着 H_2S 和 CO_2 的产生，同时也会出现微量的 H_2。在没有硫时，只产生 CO_2，而 H_2 会对细胞的生长起到抑制作用。此外，发现加入适当的碳水化合物底物到 *P. furiosus* 的培养物中，能够同时诱导增强胞内外糖化酶的产生，通常是那些能够参与特定底物的降解代谢的酶的活性。*P. furiosus* 的细胞密度可超过 10^8 个/mL，这个密度相对于这类微生物是非常高的。

通过液面下发酵和 SSF 生长的 *Penicillium* sp. NIOM-02 能够产生色素和淀粉酶，这些代谢物的生产有所不同。与液面下发酵培养相比，在 SSF 过程中，淀粉酶活性较低，说明了溶解氧和连续搅拌的作用。氧传递是 SSF 的主要问题，缠结在底物表面和内部的菌丝体会阻碍氧气的扩散（Dhale and Vijay-Raj，2009）。

6.4.8 DNA 聚合酶

重组 Tma（*Thermotoga maritima*）的 DNA 聚合酶是从含有质粒 Tma12-3 的大肠杆菌菌株 DG116 提取纯化得到的（Gelfand et al.，1997）。接种到发酵罐的种子培养物的体积是这样计算出来的：使得细菌浓度为（干重）0.5 mg/L。通过加入丙二醇控制发酵过程中的泡沫。控制气体流速在 2 L/min。将培养物在 30℃条件下培养，然后将生长温度上升到 35℃以诱导重组 Tma DNA 聚合酶的表达。这种温度的转变增加了 Tma12-3 质粒的拷贝数并同时抑制了控制改变后 Tma DNA 聚合酶基因转录的启动子的活性。

6.4.9 酯酶和脂肪酶

Sun 等（2009）申请了一个有关制备钙-藻酸盐固定化海洋细菌 MP-2 酯酶的方法的专利。海藻酸钠和海洋细菌 MP-2 酯酶溶解于甘氨酸-NaOH 缓冲溶液，然后混合均匀；将混合物滴入氯化钙溶液成球，稳定，固定，洗涤和干燥，从而获得固定化海洋细菌 MP-2 酯酶。

Tetrasphaera sp. 产生的分子量为 32 kDa 或 40 kDa 的脂肪酶可以通过吸附被固定在阴离子交换树脂或疏水性树脂上，从而被用做固定化酶（Nakao et al.，2011）。当填充到柱子中时，固定化酶允许当原料通过该柱时，反应可以连续地进行。此外，固定化酶可以很容易地从反应液中被除去，实现再利用。

6.4.10　连接酶

Pyrococcus furiosus 具有发酵形式的代谢，能够产生有机酸，CO_2 和 H_2 作为最终产物。H_2 的生产能够抑制生长，因此培养物须用氩气喷射以除去 H_2。或者可以加入元素硫。原被用来生产 H_2 的还原剂可以被用来将元素硫还原至 H_2S。加入元素硫对于玻璃容器的小规模培养是比较方便的，但是在 500 L 不锈钢发酵罐中，不能用这一还原反应来除去抑制性的 H_2，因为 H_2S 具有一定的腐蚀性（Mathur et al.，1997）。在烧瓶（2 L）中接种两次 100 mL 的培养物，并用氩气喷射。向 20 L 的培养中接种 2 L 培养物进行培养，20 L 的培养物作为种子接种到 500 L 培养基中。培养温度为 88℃，通入氩气，搅拌速度约为 50 r/min。

6.4.11　氧化酶

Qureshi 等 1998 年研究了一种深海嗜压细菌，*Shewanella* sp. DB-172F 的呼吸链系统。从在加压容器中生长的细胞中获得一种膜结合 ccb-类型的苯二酚氧化酶。将细菌培养在预灭菌袋中，其中含有海水肉汤 2216，含有或者不包含氧化的氟化。将装有培养基的袋子置于钛压力容器中（manufactured by HiP，http：//www. highpressure. com），并保持大气压力（0.1 MPa），或加压至 60 MPa。在指数生长期的早期收集细胞。该酶可在高静水压力下被特异性诱导，在 60 MPa 条件下生长的细胞中表达水平最高。结果表明在深海细菌 DB-172F 中存在两种类型呼吸链以适应压力。在不同条件下培养的相同湿重的细胞中均部分纯化到苯二酚氧化酶。在 60 MPa 下生长的细胞中的苯二酚氧化酶的含量最高量，而在大气压（0.1 MPa）下生长的细胞中未检测到这种酶。

6.4.12　过氧化物酶

从海洋真菌 *Caldariomyces fumago* 中研究开发了一种生产木质素过氧化物酶的极好的方法（Irvine and Venkatadri，1994）。这种微生物生长在透氧的液体培养基表面的一侧，而该表面的另一侧用于氧气供给。发明者发现通过将微生物固定在透氧的表面，并通过透氧表面向生物体供给氧气，微生物中胞外酶的产量在八批培养周期后达到最高水平。当生物体悬浮培养，并通过向液体培养基中鼓泡来提供微生物所需的空气或氧气，得到的木质素过氧化物酶的活性降低。当生物体附着在氧渗透膜上，但氧的供给是通过反应器的顶部空间，而不是通过膜提供时，初始加入生产培养基后，检测不到酶的活性。必须要对 *C. fumago* 进行压力选择以刺激过氧化物酶的生产。通常情况下可以通过减少对生物体营养物的供给来实现。当营养供应减少，真菌开始饥饿时，就会启动利用一些自身的原生质以生产木质素过氧化物酶。当生产木质素过氧化物酶一段时间之后，有必要再次供给食物以刺激真菌的生长。因此，过氧化物酶的生产通常是通过生长和产酶周期的更替而不断进行，这可以通过供给真菌生长培养基，然后是营养物缺乏的产酶培养。几个循环后，微生物开始表现出老化的迹象，其酶的生产率下降。微生物的复壮可以通过冲洗膜片以除去老的真菌，让年轻的真菌能够生长得以实现（图 6.13）。

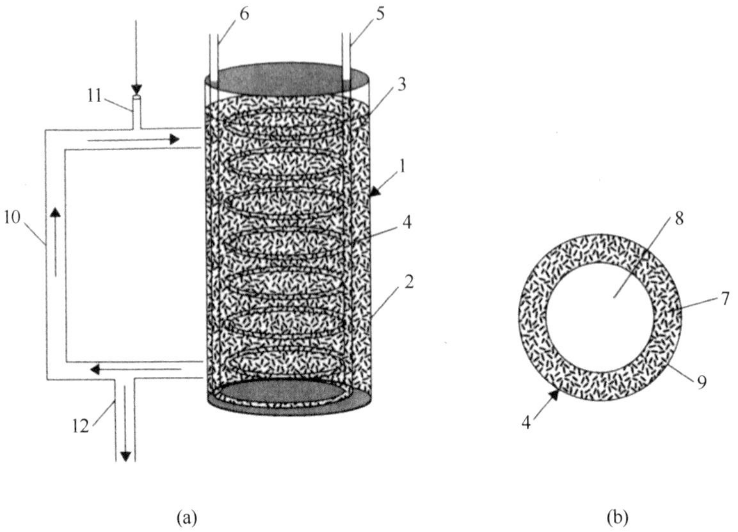

图 6.13 （a）图解说明发明仪器的过程（Irvine and Venkatadri，1994）

注：1. 反应器；2. 包括一个罐；3. 包含一个水性介质；4. 管道沉浸其中；有一个进口；5. 一个出气口；6. 氧气透过管的（b）展现在（a）中截面，管（4）由氧气透过材料组成；7. 氧气包含气体穿过流明；8. 生长在管子外部有氧微生物的薄片；9. 液体培养基穿过反应器（1）再通过环；10. 不断循环，营养由进口（11）供应到环内，酶产物从出口（12）移除 .

Reprinted with permission of the United States Patent and Trademark Office

6.4.13 丙酮酸羧化酶

从一种超嗜热，严格氢营养，自养海洋甲烷菌-*Methanococcus jannaschii* 中制备了丙酮酸羧化酶。该生物被培养在 STR 中（Mukhopadhyay et al.，1999）。反应器规模是实验的 16-L（12 L 工作容积）不锈钢连续搅拌釜中进行（型号 Microgen；New Brunswick Scientific Company，New Brunswick，N. J.）。为了达到厌氧和无菌增补（手动或自动）的培养条件，在容器上顶部板上的两个额外的端口分别安装了橡胶塞。所有气体通过容器底部的一个位于搅拌轴正下方的单孔喷射器向培养物供给。N_2、CO_2 和 H_2 流分别通过常见的加热的旋转铜板床除去氧气；氢流保证了氧化铜的不断再生，培养器内容物的加热和冷却通过使用蒸汽流来实现。灭菌培养基在氮气下冷却至 30℃，然后开启氢气和 CO_2 流，氮气流被中断。在这个阶段，含有盐营养成分的无氧无菌溶液被添加到培养基中。当培养基氧化还原电位读数稳定后，开启连续的 N_2 和 H_2S 的混合气体流。在这一阶段，搅拌速度和放气率调节到所需的值，培养基中接种 25 mL 对数生长期的培养物。整个培养过程中，将容器保持在正压，培养温度保持在 85℃。根据需要，培养物中的泡沫，通过加入 0.2 mL 无氧无菌的泡剂 289（Sigma）的水溶液进行抑制。

6.5 结论

本章描述了通过采用不同的传统生物培养器和特殊的反应器培养海洋微生物以进行酶的生产。包括深层培养（SDS 碱性稳定性蛋白酶，金属蛋白酶，几丁质酶，外切菊粉酶，琼脂糖，丙酮酸羧化酶，糖化酶等），全细胞固定化培养（碱性蛋白酶，胞外 L-谷氨酰胺酶，酯酶，脂肪酶，木质素过氧化物酶），深层培养和表面附着组合培养（蛋白酶，耐热木聚糖酶和纤维素酶）和固态发酵（糖化酶）。

海洋来源的生物酶已经进入了市场，一个例子是 Aquabeautine XL™，Aqua Biotechnology（http：//aquabiotechnology.com）的一款皮肤护理产品。Biotec Pharmacon（http：//www.biotec.no/）公司在生产最畅销产品虾碱性磷酸酶的基础上，也在扩大其对新型的海洋生物适冷酶产品的投放。该公司介绍了 COD-UNG（尿嘧啶 DNA -N-糖基化酶），这是一种用于 DNA/RNA 分析的理想的酶，已经被国际知名的检测公司所采用。Zymetech（http：//www.zymetech.com）成立于 1996 年，主要涉及海洋酶及其衍生产品的开发，生产和市场开发。它已经在制药和化妆品领域开发出了多项海洋酶专利技术。

在生物培养器的设计和操作方面，在过去 10~15 年所取得的进展为生产至今发现的许多有趣的海洋酶开辟了很多机会。我们希望在生物加工工程技术人员、海洋学家、生物学家和化学家的多学科团队的努力下，最终将成功建立海洋酶的发现与商业化生产之间的桥梁。

致谢

十分感谢 Sreyashi Sarkar（CSIR No.09/096-0549-2K8-EMR-I）对本研究的经济协助。

参考文献

Aertsen，A.，Meersman，F.，Hendrickx，M.E.G.，Vogel，R.F. and Michiels，C.W.（2009）Biotechnology under high pressure：applications and implications. *Trends Biotechnol.* 27，434-441.

Andrade，C.M.M.C.，Aguiar，W.B. and Antranikian，G.（2001）Physiological aspects involved in production of xylanolytic enzymes by deep-sea hyperthermophilic archaeon *Pyrodictium abyssi*. *Appl. Biochem. Biotechnol.* 91-93，655-669.

Beshay，U. and Moreira，A.（2003）Repeated batch production of alkaline protease using porous sintered glass as carriers. *Process Biochem.* 38，1463-1469.

Brown，S.H. and Kelly，R.M.（1989）Cultivation techniques for hyperthermophilic archaebacteria：continuous culture of *Pyrococcus furiosus* at temperatures near 100℃. *Appl. Environ. Microbiol.* 55, 2086-2088.

Clauwaert, P. and Verstraete, W. (2009) Methanogenesis in membraneless microbial electrolysis cells. *Appl. Microbiol. Biotechnol.* 82, 829-836.

Cui, W., Wang, Q., Zhang, F., Zhang, S.-C., Chi, Z.-M. and Madzak, C. (2011) Direct conversion of inulin into single cell protein by the engineered *Yarrowia lipolytica* carrying inulinase gene. *Process Biochem.* 46, 1442-1448.

Deusner, C., Meyer, V. and Ferdelman, T. G. (2010) High-pressure systems for gas-phase free continuous incubation of enriched marine microbial communities performing anaerobic oxidation of methane. *Biotechnol. Bioeng.* 105, 524-533.

Dhale, M. A. and Vijay-Raj, A. S. (2009) Pigment and amylase production in *Penicillium* sp. NIOM - 02 and its radical scavenging activity. *Int. J. Food Sci. Technol.* 44, 2424-2430.

Doran, P. M. (1995) 'Reactor engineering'. In *Bioprocess Engineering Principles*. 333-391. San Diego, CA: Elsevier.

El-Gendy, M. M. A. (2010) Keratinase production by endophytic *Penicillium* spp. Morsy1 under solid-state fermentation using rice straw. *Appl. Biochem. Biotechnol.* 162, 780-794.

Elibol, M. and Moreira, A. R. (2003) Production of extracellular alkaline protease by immobilization of the marine bacterium *Teredinobacter turnirae*. *Process Biochem.* 38, 1445-1450.

Estrada-Badillo, C. and Marquez-Rocha, F. J. (2003) Effect of agitation rate on biomass and protease production by a marine bacterium *Vibrioharveyi* cultured in a fermentor. *World J. Microbiol. Biotechnol.* 13, 129-133.

Fenice, M., Leuba, J. L. and Federici, F. (1998) Chitinolytic enzyme activity of *Penicillium janthinellum* P9 in bench-top bioreactor. *J. Ferment. Bioeng.* 86, 620-623.

Gelfand, D. H., Lawyer, F. C. and Stoffel, S. (1997) Purified thermostable nucleic acid polymerase enzyme from *Thermotoga maritime*. US Patent No. 5, 624, 833.

Ghosh, D., Saha, M., Sana, B. and Mukherjee, J. (2005) 'Marine enzymes'. In *Marine Biotechnology* I, *Advances in Biochemical Engineering/ Biotechnology*, Le Gal, Y. and Ulber, R. (volume eds), Scheper, T. (series ed.); 189-218.

Girguis, P. R., Orphan, V. J., Hallam, S. J. and DeLong, E. F. (2003) Growth and methane oxidation rates of anaerobic methanotrophic archaea in a continuous flow bioreactor. *Appl. Environ. Microbiol.* 69, 5472-5482.

Grossart, H.-P. and Gust, G. (2009) Hydrostatic pressure affects physiology and community structure of marine bacteria during settling to 4000 m: an experimental approach. *Mar. Ecol. Prog. Ser.* 390, 97-104.

Houghton, J. L., Seyfried Jr., W. E., Banta, A. B. and Reysenbach, A.-L. (2007) Continuous enrichment culturing of thermophiles under sulfate and nitratereducing conditions and at deep-sea hydrostatic pressures. *Extremophiles* 11, 371-382.

Imai, E.-I., Honda, H., Hatori, K., Brack, A. and Matsuno, K. (1999) Elongation of oligopeptides in a simulated submarine hydrothermal system. *Science* 283, 831-833.

Irvine, R. L. and Venkatadri, R. (1994) Method of producing extracellular products from aerobic microorganisms. US Patent No. 5, 342, 765.

Kao, P. M., Chen, C. I., Huang, S. C., Chang, Y. C. et al., (2007a) Effects of shear stress and mass transfer on chitinase production by *Paenibacillus* sp. CHEN1. *Biochem. Eng. J.* 34, 172-178.

Kao, P. M., Huang, S. C., Chang, Y. C. and Liu, Y. C. (2007b) Development of continuous chitinase production process in a membrane bioreactor by *Paenibacillus* sp. CHE-N1. *Process Biochem.* 42, 606-611.

Kelly, R. M., Brown, S. H. and Costantino, H. R. (2002) Saccharification enzymes from hyperthermophilic bacteria and processes for their production. US Patent No. 6, 355, 467.

Konopka, A. (2000) Microbial physiological state at low growth rate in natural and engineered ecosystems. *Curr. Opin. Microbiol.* 3, 244-247.

Kristensen, J. B., Meyer, R. L., Poulsen, C. H., Kragh, K. M., Besenbacher, F. and Laursen B. S. (2010) Biomimetic silica encapsulation of enzymes for replacement of biocides in antifouling coatings. *Green Chem.* 12, 387-394.

Kumar, C. G., Joo, H. S., Koo, Y. M., Paik, S. R. and Chang, C. S. (2004) Thermostable alkaline protease from a novel marine haloalkalophilic *Bacillus clausa* isolate. *World J. Microbiol. Biotechnol.* 20, 351-357.

Kumar, S. R. and Chandrasekaran, M. (2003) Continuous production of L-glutaminase by an immobilized marine *Pseudomonas* sp. BTMS-51 in a packed bed reactor. *Process Biochem.* 38, 1431-1436.

Lang, S., Hiiners, M. and Lurtz, V. (2005) Bioprocess engineering data on the cultivation of marine prokaryotes and fungi. *Adv. Biochem. Eng. Biotechnol.* 97, 29-62.

Liu, B. L., Kao, P. M., Tzeng, Y. M. and Feng, K. C. (2003) Production of chitinase from *Verticillium lecanii* F091 using submerged fermentation. *Enzyme Microb. Tech.* 33, 410-415.

Logan, B. E., Hamelers, B., Rozendal, R., Schröder, U., Keller, J. et al., (2006) Microbial fuel cells: methodology and technology. *Environ. Sci. Technol.* 40, 5181-5192.

Manyak, D. M., Weiner, R. M., Carlson, P. S. and Quintero, E. J. (2010) Preparation and use of biofilm-degrading, multiple-specificity, hydrolytic enzyme mixtures. US Patent Application No. 20, 100, 159, 563.

Mathur, E. J., Marsh, E. J. and Schoettlin, W. E. (1997) Purified thermostable *Pyrococcus furiousus* DNA ligase. US Patent No. 5, 700, 672.

Matsumoto, M., Yokouchi, H., Suzuki, N., Ohata, H. and Matsunaga, T. (2003) Saccharification of marine microalgae using marine bacteria for ethanol production. *Appl. Biochem. Biotech.* 105, 247-254.

Meulepas, R. J. W., Jagersma, C. G., Gieteling, J., Buisman, C. J. N., Stams, A. J. M. and Lens, P. N. L. (2009) Enrichment of anaerobic methanotrophs in sulfate-reducing membrane bioreactors. *Biotechnol. Bioeng.* 104, 458-470.

Mitra, S., Banerjee, P., Gachhui, R. and Mukherjee, J. (2011) Cellulase and xylanase activity in relation to biofilm formation by two intertidal filamentous fungi in a novel polymethylmethacrylateconico-cylindrical flask. *Bioprocess Biosyst. Eng.* 34, 1087-1101.

Muffler, K. and Ulber, R. (2008) Fed-batch cultivation of the marine bacterium *Sulfitobacter pontiacus* using immobilized substrate and purification of sulfite oxidase by application of membrane adsorber technology. *Biotechnol. Bioeng.* 99, 870-875.

Mukhopadhyay, B., Johnson, E. F. and Wolfe, R. S. (1999) Reactor-scale cultivation of the hyperthermophilic methanarchaeon *Methanococcus jannaschii* to high cell densities. *Appl. Environ. Microbiol.* 65, 5059-5065.

Nakao, M., Kanamori, M., Fukami, H., Kasai, H. and Ochiai, M. (2011) Lipase. US Patent No. 7, 893, 232.

Narinx, E., Baise, E. and Gerday, C. (1997) Subtilisin from psychrophilic antarctic bacteria: characterization and site-directed mutagenesis of residues possibly involved in the adaptation to cold. *Protein Eng.* 10, 1271-1279.

Nijland, R., Hall, M. J. and Grant Burgess, J. (2010) Dispersal of biofilms by secreted, matrix degrading, bacterial DNase. *PLoS ONE* 5, el 5668.

Parkes, R. J., Sellek, G., Webster, G., Martin, D., Anders, E., et al., (2009) Culturable prokaryotic diversity of deep, gas hydrate sediments: first use of a continuous high-pressure, anaerobic, enrichment and isolation system for subseafloor sediments (DeepIsoBUG). *Environ. Microbiol.* 11, 3140-3153.

Postec, A., Lesongeur, F., Pignet, P., Ollivier, B., Querellou, J. and Godfroy, A. (2007) Continuous enrichment cultures: insights into prokaryotic diversity and metabolic interactions in deep-sea vent chimneys. *Extremophiles* 11, 747-757.

Pradillon, F., Shillito, B., Chervin, J.-C., Hamel, G. and Gaill, F. (2004) Pressure vessels for *in vivo* studies of deep-sea fauna. *High Pressure Res.* 24, 237-246.

Qureshi, M. H., Kato, C. and Horikoshi, K. (1998) Purification of a ccb-type quinol oxidase specifically induced in a deep-sea barophilic bacterium, *Shewanella* sp. strain DB-172F. *Extremophiles* 2, 93-99.

Ravindran, C., Naveenan, T. and Varatharajan, G. R. (2010) Optimization of alkaline cellulase production by the marine derived fungi, *Chaetomium* sp. using agricultural and industrial wastes as substrates. *Bot. Mar.* 53, 275-282.

Sabu, A., Kumar, S. R. and Chandrasekaran, M. (2002) Continuous production of extracellular L-glutaminase by Ca-alginate-immobilized marine *Beauveria bassiana* BTMF S-10 in packed-bed reactor. *Appl. Biochem. Biotechnol.* 71, 71-79.

Sarkar, S., Roy, D. and Mukherjee, J. (2011) Enhanced protease production in a polymethylmethacrylate conico-cylindrical flask by two biofilm-forming bacteria. *Bioresour. Technol.* 102, 1849-1855.

Sarkar, S., Saha, M., Roy, D., Jaisankar, P., Das, S., et al., (2008) Enhanced production of antimicrobial compounds by three salt-tolerant actinobacterial strains isolated from the Sundarbans in a niche-mimic bioreactor. *Mar. Biotechnol.* 10, 518-526.

Shuler, M. L. and Kargi, F. (2003) 'Operating considerations for bioreactors for suspension and immobilized cultures'. In *Bioprocess Engineering: Basic concepts, 2nd edn.* 245-284. Upper Saddle River, NJ: Prentice-Hall.

Sugano, Y., Nagae, H., Inagaki, K., Yamamoto, T., Terada, I. and Yamazaki, Y. (1995) Production and characteristics of some new β-agarases from a marine bacterium, *Vibrio* sp. strain JT0107. *J. Ferment. Bioeng.* 79, 549-554.

Sun, M., Liu, J., Wang, H., Zheng, Y., Ping, R., et al., (2009) Method for preparing calcium-alginate-immobilized marine bacterium MP-2 esterase. European Patent Application No. CN2, 009, 108, 736, 520, 090, 619.

Weiner, R. M., Shi, J. and Coyne, V. E. (1996) Thermostable alkaline metalloprotease produced by a hyphomonas and preparation thereof. US Patent No. 5, 589, 373.

Wright, P. C., Westacott, R. E. and Burja, A. M. (2003) Piezotolerance as a metabolic engineering tool for the biosynthesis of natural products. *Biomol. Eng.* 20, 325-331.

Zeng, X., Birrien, J.-L., Fouquet, Y., Cherkashov, G., Jebbar, M., et al., (2009) *Pyrococcu* CHI, an obligate piezophilic hyperthermophile: extending the upper pressure-temperature limits for life. *ISME J.* 3, 873-876.

Zhang, C. and Kim, S. -K. (2010) Research and application of marine microbial enzymes: status and prospects. *Mar. Drugs* 8, 1920-1934.

Zhang, T., Gong, F., Peng, Y. and Chi, Z. (2009) Optimization for high-level expression of the *Pichia guilliermondii* recombinant inulinase in *Pichiapastoris* and characterization of the recombinant inulinase. *Process Biochem.* 44, 1335-1339.

Zhang, Y., Arends, J. B. A., Van de Wiele, T. and Boon, N. (2011) Bioreactor. technology in marine microbiology: from design to future application. *Biotechnol. Adv.* 29, 312-321.

Zhang, Y., Henriet, J. P., Bursens, J. and Boon, N. (2010) Stimulation of *in vitro* anaerobic oxidation of methane rate in a continuous high-pressure bioreactor. *Bioresour. Tech nol.* 101, 3132-3138.

Zuo, Y., Xing, D., Regan, J. M. and Logan, B. E. (2008) Isolation of the exoelectrogenic bacterium *Ochrobactrum anthropi* YZ-1 by using a U-tube microbial fuel cell. *Appl. Environ. Microbiol.* 74, 3130-3137.

第7章　盐碱生态系统中可培养与非可培养细菌群落的多样性，种群动态以及生物催化潜力

Vikram H. Raval, Megha K. Purohit and Satya P. Singh, Saurashtra University, Rajkot, India

DOI：10.1533/9781908818355.2.165

摘要：嗜盐碱生物主要是从盐碱湖和死海中发现的，而开发其他栖息地研究的相对比较少。探索这些生物以及它们独特的生物分子特征为其未来铺平了道路。最近几年，能够应用于工业方面的生物催化的研究增多。然而，迄今研究重点集中在实验室条件下可以生长的微生物。在极端环境栖息地中，盐碱地富含大量的细菌、放线菌和古生菌。嗜盐碱蛋白结构与功能的关系研究显示特定的结构域特性决定了酶在多种极端条件下具有催化能力和稳定性。在极端的pH值、温度、盐度条件下，可以稳定存在使得这些生物体更加吸引人。这些生物体的通性反映在它们的生化性质，催化能力和盐度依赖的稳定性。最近，应用非依赖培养的方法探索非可培养的微生物多样性取得了巨大的进展。对种群动态、多样性、系统发育和盐碱生态系统中整个微生物群体的生物催化潜力的分析具有重要意义。宏基因组的方法与体外生物催化能力进化的结合产生更新的应用。

关键词：嗜盐碱细菌，稳定性，碱性蛋白酶，宏基因组学

7.1　引言

海洋环境的生物与化学多样性推动了多个工业领域的发展，如药物、化妆品、营养添加剂、分子探针、酶学、精细化工以及农业。在众多具有附加值的海洋生命产品中，海洋生物酶由于具有不同于陆地生物酶的特点而引起了更多的关注。广义上来说，海洋酶为海洋环境的生物综合评价提供参考。除了微生物，许多其他的海洋生命体如鱼、虾、蟹类、蛇、植物、海绵类和藻类在某种程度上也可以作为具有生物催化能力的生物进行研究。其高耐盐性、热稳定性、耐压性以及易于规模化培养的特征是科学家重点研究的内容。这些特征在海洋生物中很好地被体现出来，因为海洋生物可以在热液口、洞穴以及那些高压和

缺光的环境中生存（Horikoshi, 1990; Niehaus et al., 1999）。

最近，海洋无脊椎动物与其共生细菌是天然产物研究的热点。海洋微生物通常与它们的宿主——脊椎动物和无脊椎动物具有共生关系。海岸附近沉积物，富盐土地和海水中已经有细菌多样性和酶类的报道（Dodia et al., 2006, 2008a; Joshi et al., 2008; Patel et al., 2005, 2006a, b）。在这些酶类中，蛋白酶类、淀粉酶和过氧化物酶是研究最多的，其中有的具有极大的商业开发前景（Rao et al., 1998）。迄今，自然界中的极端微生物研究集中在嗜热微生物，但是对嗜盐碱菌的关注却很少（Demirjian et al., 2001; Ladenstein and Antranikian, 1998; Niehaus et al., 1999）。海洋微生物来源的稳定性酶的发现会进一步推动海洋酶的应用（Rao et al., 1998）。

7.2 海洋栖息地

相对于为适应高盐环境微生物的新陈代谢多样性而言，高盐环境本身是一个相对简单的生态系统。然而，盐及高盐栖息地的多样性与当地微生物群体为适应该生活环境是息息相关的（Oren, 2002a, 2002b）。盐湖和其他拥有接近饱和盐浓度的系统为微生物生态学提供了便利的模型系统。含盐水生栖息环境通常盐浓度高于海水（3%~4%, *w/v*）。嗜盐的生物群体的栖息地是盐水湖，蒸发咸水湖沉淀物和沿海盐田。盐碱土和动物盐排泄物表面是极少研究的栖息地（Grant et al., 1998）。这些栖息地是新型微生物和它们生物催化剂的潜在来源。

7.2.1 非海盐型环境

非海洋型环境的离子组成与海水有极大不同。自然环境对于所有的耐盐生物体的耐受力有一个阈值（DasSarma and Arora, 2001）。死海、高碱湖、碳酸盐泉、盐场卤水和碱性土代表非海盐性环境。

苏打湖

苏打湖是一个稳定、高产的极端水环境系统，其 pH 值达 10 以上。在非洲、印度和中国的苏打湖（pH 值为 11，盐浓度超过 300 g/L）庇护着大量的微生物（Oren, 2002a, 2002b, 2002c）。稳定的高盐碱环境是由地质的、地理的和气候条件的异常组合造成的（Grant and Mwatha, 1989）。然而，只对少数高碱湖的微生物多样性和生态学方面开展了研究。已经研究了位于东非大裂谷的肯尼亚碱湖的生物多样性和地球化学特征（Grand and Horikoshi, 1992）。它们含有 5%~35%（*w/v*）的盐（Magadi and Natron）及其 pH 值在 8.5~11.5（Tindall et al., 1980; Duckworth et al., 1996; Rees et al., 2004）。位于莫诺湖和加利福尼亚东部的欧文斯湖，塞拉利昂、西藏、巴基斯坦和俄罗斯的几个苏打湖和盐碱土壤也拥有大量的原核生物群体（Satyanarayana et al., 2005）。

死海

死海的特征是 pH 值在 6 左右，二价阳离子（1.9 mol/L Mg^{2+}和 0.4 mol/L Ca^{2+}）相对于一价阳离子（1.6 mol/L Na^{+}和 0.14 mol/L K^{+}）占绝对优势。然而，这样的恶劣环境却养育了大量的微生物（Oren, 1998）。死海位于中东，其面积为 800 km^2，深度为 340 m，是迄今研究最详细的最大的高盐环境。它含有高浓度的镁离子，而镁离子对海洋生物体来源的酶稳定性具有很重要的作用（DasSarma and Arora, 2001）。

碳酸盐泉

富有碳酸盐的泉为不同的群体（异养生物主要是 *Bacillus* spp. 和 *Cynobacterial* spp.）提供有机物。蛋白的降解和尿素的水解引起了氨化作用，进而导致 pH 值增加，因此支持了嗜碱菌的生长。Horikoshi 与其他研究人员分离出了许多嗜碱菌并且充分研究了它们的应用（Grand and Horikoshi, 1992）。

7.2.2 海盐型环境

海水蒸发后形成的高盐栖息环境被称为海盐性环境。它的盐分组成与海水相似，钠离子氯离子占主要成分，其 pH 值在中性及稍微偏碱的范围。蒸发后，离子组成由于二水硫酸钙以及其他矿物的沉淀发生了改变（Oren, 2002a, 2002b）。

盐湖和碱性环境

自然界的碱性环境分为高 Ca^{2+}（地下水中含有 $Ca(OH)_2$）和低 Ca^{2+}环境（以碳酸钠为主的苏打湖和苏打荒漠）（Grant and Mwatha, 1989; Grand and Horikoshi, 1992）。富含 Ca^{2+}的地下水在世界上的几个地方被发现，包括加利福尼亚、阿曼、前南斯拉夫、塞浦路斯、约旦、土耳其（Barnes et al., 1982）。Grant（Grant and Mwatha, 1989; Grand and Horikoshi, 1992）和 Jones（Jones et al., 1994）已经研究了含高 Ca^{2+}碱泉。苏打湖和苏打荒漠代表了天然形成的稳定的碱性环境。在高盐环境中较普通的现象由于海水的蒸发造成了盐梯度的变化。

太阳能盐田

太阳能盐田代表盐梯度从海水到盐饱和的过渡。每个盐池的盐浓度是相对稳定且拥有主要的微生物群体。虽然盐田表面相似，但是它们在成分和水的滞留时间上是不同的（Javor, 1983）。饱和 NaCl 卤水由于有颜色的微生物的存在而呈亮红色（Oren, 2002a, 2002b）。中度嗜盐菌在盐池通常聚集形成海水中的菌群。大部分被分离出来的好氧异养菌可以在高达 20% NaCl 的条件下生长，有的甚至会增到 30%（Forsyth et al., 1971）。已经有亚得里亚海海岸盐场结晶池嗜盐古生菌多样性的报道（Pasic et al.，2005）。

海水

海水的盐分受许多因素的影响，如溶冰，河水的汇入、蒸发、降雨量、降雪量、风、波动，海洋气流引起的水平与竖直方向上盐水的混合。海水的某些成分例如钙、镁、碳酸氢盐和二氧化硅还会被生物体利用，以及化学沉淀，化学反应所利用，同时开放的海洋也含有溶解的机碳（0.1～1.0 mg/L）。许多嗜盐菌都是从海滩和海藻中分离出来的（Onishi et al.，1980）。因此，海水中包含从中度到极度耐盐的菌，这些菌是高度稳定的酶的来源。针对海洋微生物酶的分离、优化、纯化、生化表征研究有助于获得关于该酶的新的信息（Ventosa，1998；Oren，2002a，2002b；Joshi et al.，2008）。

7.2.3 盐碱化土壤

土壤生境是不均一的并且含有变化范围很广的不同浓度的盐（Grant，1991）。由于盐浓度的周期性变化，盐土壤中含有大量的耐盐微生物（Quesada et al.，1982，1983）。

7.3 极端微生物

极端的条件通常反映在一些理化参数例如温度、压力、辐射；碳源和氮源以及其他营养，地理化学参数例如盐度和 pH 值。极端条件下的微生物是多种多样的，能在极端的温度、压力、辐射、干燥、盐度、pH 值、活性氧、氧化还原电位、金属和气体的条件下存活（van den Burg，2003）。

7.3.1 极度嗜盐菌和嗜碱菌

极端微生物中，嗜盐菌又代表了一种更加独特的异养微生物群体（Oren，2002a，2002b），根据他们对盐的需求又分为极度嗜盐菌 15%～30%（*w/v*）和中度嗜盐菌 3%～15%（*w/v*）（Ventosa et al.，1998）。嗜碱菌可以在含盐和碱性条件下生长，已经在很多高盐环境例如苏打湖、太阳能盐田、卤水、碳酸盐泉、死海中都大量发现。几年前我们在索拉什特拉大学的研究发现嗜盐碱菌的生存远不止在苏打湖（Dodia et al.，2006；Nowlan et al.，2006）。最近，因为其在生物技术和生态方面的重要意义，具有不止一个极端环境的微生物引起了人们的关注。这些微生物在分类、分子系统进化和应用前景方面的研究正在开展。

7.3.2 海洋酶

从不同海洋栖息地来源的微生物已经被研究过，包括海绵、珊瑚等的共生微生物。位于极端环境例如热液口发现的微生物也引起了关注。微生物具有很大范围的生物催化能力，这些使得它们能在极端条件下生存（Debashish et al.，2005）。除此之外，海洋生物学家也研究了其他的海洋栖息环境，如近海沉淀物、深海沉淀物、海水。在这些酶中，蛋白酶、碳水化合物酶、氧化物酶都是研究最多的对象。酶的新颖性反映在它们在极高或极低温度、极端 pH 值、高盐浓度和高压力下的稳定性（Antranikian et al.，2005；Demirjian et al.，2001；Ferrer et al.，2007），而且，许多极端酶通过可培养或不可培养的途径被发现（Ferrer et al.，2007）。

蛋白酶

蛋白酶在生物体中广泛存在，其对细胞生长和分化是至关重要的。在细胞中，微生物蛋白酶在膜蛋白的加工与维护方面的重要性已经确定（Weski and Ehrmann，2012）。除此之外，蛋白酶是最具有商业开发价值的酶并且占据着世界酶市场的主导地位（Niehaus et al.，1999；Gupta et al.，2002）。在这些蛋白酶中，碱性丝氨酸蛋白酶被认为是非常有潜力的。蛋白酶在有机合成和许多工业上的应用已经在相关文献中被重点报道（Dodia et al.，2006，2008a，2008b；Purohit and Singh，2011；Rao et al.，1998）。海洋来源的蛋白酶在热稳定性上不同于陆地来源的蛋白酶。来自于中度嗜盐菌 *Bacillus* sp. Vel（Gupta et al.，2005），*Filobacillus* sp. RF2-5（Hiraga et al.，2005），*Halobacillus* sp. SR5-3，（Namwong et al.，2006），*Salinivibrio* sp. AF-2004（Herdari et al.，2007），*Haloalkaliphilic bacterium* sp. AH-6（Dodia et al.，2006，2008a，2008b）和 *Halobacillus karajensis*（Heidari et al.，2009）的蛋白酶被纯化和表征。海洋生物来源的嗜盐和嗜盐碱蛋白酶在洗涤行业中的应用已经被评价（Haddar et al.，2009；Maurer 2004；Oberoi et al.，2001），同时其在鱼的腌制，鱼味调味酱以及腌泡汁、调整鱼蛋白浓度、脱毛、蜕皮中的应用也被研究（Akolkar et al.，2001）。催化非水相条件下的反应时，蛋白酶的溶剂耐受性也已经被报道（Heidari et al.，2007；Karan and Khase 2011；Pandey et al.，2012）。最近几年，来自海洋栖息地耐盐碱细菌的蛋白酶已经有在酶的制备，纯化和酶的表征方面的报道（Dodia et al.，2006，2008a，2008b；Joshi et al.，2008；Patel et al.，2005，2006a，2006b；Nowlan et al.，2006）。

脂酶

脂酶催化水解吸附于油水界面的三酰甘油成甘油和脂肪酸（Kurtovic et al.，2009）。他们催化反应在脂水界面（Beisson et al.，2000；Reetz，2002）。酯酶和脂酶广泛应用于精细化工产品的制备（Demirjian et al.，2001），以及食品和其他工业应用。脂酶通常是细菌产生的，细菌来源的脂酶尤其在商业应用中扮演着关键角色（Gupta et al.，2004）。然而，脂酶巨大的商业市场是在清洁剂及洗衣行业。细菌来源的脂酶已经用深层发酵生产。许多来源于 *Pseudomonas* spp. 的脂酶的产品有巨大的潜力，如清洁行业的 Lumafast 与 Lipomax；有机合成中的 Chiro CLEC-PC，Chirazyme L-1 和 Amano P，P-30（Godfrey and West 1996；Gupta et al.，2004；Jaeger and Reetz，1998）。

淀粉酶

许多极端微生物产生 α-淀粉酶，α-淀粉酶在许多工业生产过程中都有应用，比如淀粉液化和纸浆生产过程（UpaDek and Kottwitz, 1997）。淀粉酶在多个领域都有应用，例如临床、制药、分析化学、纺织品、洗衣、食品、酿造、烘焙等产业。由于其广泛应用及在碱性和高温条件下的特殊活性，来自于耐盐碱细菌的淀粉酶获得了更多的关注（UpaDek Kottwitz, 1997; Pandey et al., 2000, 2012; Pandey and Singh, 2012）。*Bacillus amyloliquefaciens* 和 *B. stearothermophilus*, *B. subtilis* 和 *B. licheniformis* 都产生淀粉酶（Sajedi et al., 2005）。来源于海洋细菌的 α-淀粉酶的热稳定性，pH 值稳定性，有机溶剂耐受性已经被研究（Amoozegar et al., 2003; Carvalho et al., 2008; Coronado et al., 2010; Deutch, 2002; Onishi and Hidaka, 1978; Pandey and Singh, 2012）。同时，其细胞及酶的固定化研究探讨了酶的重复利用、pH 值及热稳定性，这也大大推动了酶的应用（Abdel-Naby et al., 1998; Akkaya et al., 2012; Riyaz et al., 2009）。

甲壳质酶

甲壳质酶降解甲壳质，即均一的由 β-1，4 糖苷键链接的 N-乙酰-β-D 氨基葡萄糖（GlcNAc）。大量的生物，例如细菌、真菌、昆虫、植物都产生甲壳质酶，其中很多生物是来源于海洋的。甲壳质酶 1911 年由 Bernard 从兰花纸浆中发现，更深入的研究由 Karrer 和 Hoffmann 进行（Flach et al., 1992）。甲壳质酶被分为 3 类：内切甲壳质酶，外切甲壳质酶以及甲壳二糖酶（Kupiec and Chet, 1998）。内切甲壳质酶与外切甲壳质酶联合降解不溶性的甲壳质成为可溶性的甲壳二糖，进而被进一步降解成为 GlcNAc。降解产物可以作为生物体的碳源和氮源被利用（Tsujibo et al., 2002）。

甲壳质是水环境中的重要成分，甲壳质降解的微生物在栖息地的生态环境中（Yu et al., 1991）扮演着核心角色，而甲壳质酶产生菌在生物控制植物病原的真菌和昆虫方面具有巨大的潜力（Freeman et al., 2004; Jung et al., 2005; Chuan, 2006）。许多细菌来源的甲壳质酶是由芽孢杆菌属 *Bacillus species* 产生的，并且它们的耐盐、pH 值的稳定性和热稳定性都已经被研究。比如，来源于 *B. circulans*，（Watanabe et al., 1990）*B. licheniformis*（Trachuck et al., 1996），*B. cereus*（Pleban et al., 1997），*B. subtilis* 和 *B. pumilus*（Wang et al., 2006; Ahmadian et al., 2007），*Virgibacillus marismortui* 以及 *Planococcus rifitoensis*（Essghaier et al., 2010, 2012）的甲壳质酶都已经被广泛的研究。一些具有高甲壳质酶活性的耐盐菌也有报道（Liaw and Mah, 1992）。这些微生物都具有水解甲壳质的活性酶和作为活性控制物质的潜力。

纤维素酶

Bolobova 及其同事报道了首个来源于耐盐菌的降解纤维素的微生物（*Halocella cellulo-*

lytica）（Bolobova et al.，1992）。另一种来源于地下盐内的利用纤维素的古生菌是由Vreeland等报道（1998）。最初的嗜盐微生物分泌的细胞外基质降解酶来源于纤维素存在的地下岩石（Cojoc et al.，2009）。表现出高纤维素酶活性的菌都是革兰氏阴性菌，不能在含抗生素（新霉素、青霉素、红霉素）的环境中生长，但能耐受的NaCl浓度高达3 mol/L（Cojoc et al.，2009）。产生耐盐碱的纤维素酶的海洋细菌 *Bacillus flexus* 是与其共生的绿藻（*Ulva lactuca*）中分离得到的。该酶的生化特征，及不同pH值、盐浓度和温度等方面的性质已经被研究（Trivedi et al.，2011）。

最近，纤维素酶已经应用在纺织品和服装产业中有砂洗工装裤产品的纤维织物的生物抛光方面（Aygan and Arikan，2008）。对纤维素酶产生生物乙醇再次有兴趣是因为其可以水解预先处理的纤维素材料，成为可发酵的糖类从而转化为乙醇（Wang et al.，2009a）。最近几年，来自许多海洋菌 *Bacillus* sp.（Aygan et al.，2008），*Salinivibrio* sp.（Wang et al.，2009b）和宏基因组文库（Voget et al.，2006）的耐盐碱纤维素酶的pH值、盐度和热稳定性都有研究报道。

木聚糖酶

碱性木聚糖酶已在烘焙领域有所应用，同时在环境友好型技术发展中有应用，比如生物漂洗纸以及纸浆形成过程中（Mamo et al.，2009）。然而，木聚糖酶在生物漂白过程中的高效应用需要它具有耐盐碱和高温环境的特点。已经有一些耐盐碱的木聚糖酶研究报道，这包括来自海洋像 *Glaciecola mesophila*（Guo et al.，2009），*Chromohalobacter* sp.（Prakash et al.，2009）以及 *Nesterenkonia* sp.（Govender et al.，2009）等的酶。其中有些酶在很广的pH值范围内（pH 6~11），温度高于60℃，表现出来很好的稳定性，当然这些酶都普遍需要含有NaCl的盐环境（Wejse et al.，2003；Guo et al.，2009；Prakash et al.，2009）。来自 *Bacillus* sp. 的木聚糖酶的制备、纯化及生化性质（Balakrishnan et al.，1992；Kulkarni et al.，1999；Chang et al.，2004；Kumar et al.，2004；Martinez et al.，2005）都已经有报道。Kulkarni及其同事详细开展了来自极端微生物的木聚糖酶的分子和生物技术应用方面的研究（Kulkarni et al.，1999）。因其在经济方面的巨大前景，一些综述已经被报道（Gilbert and Hazlewood，1993；Prade，1995）。

7.4 培养方法

7.4.1 基因克隆和重组酶

最近几年，来自极端生物的酶的基因克隆，及其在常温宿主中有利用价值的酶的表达已引起了极大关注。来自极端微生物的基因已经被过表达在常规的宿主中并且获得了大量酶（Carolina et al.，2009；Ni et al.，2009；Purohit，2012）。然而，重组蛋白的折叠及性质的研究是非常重要的（Huang et al.，2012）。到目前为止，只有一些来自耐盐碱菌的耐碱蛋白酶被纯化和性质表征。典型的嗜盐碱菌的活性和稳定性都需要高盐环境，而在体外环境中，这些条件的缺乏会导致了其失活。一些耐盐碱酶性质的研究将为阐明其适应机制及在极端 NaCl、pH 值和温度的稳定性提供重要信息（Kumar et al.，2012）。

重组 DNA 技术和其他分子技术被用于改善和发展具有特定目标的酶的改造，获得经基因改造的菌株（Battestein and Macedo，2007；Carolina et al.，2009）；同时，近几年重组 DNA 技术，高通量筛选及基因组学与蛋白质组学的发展已经极大地促进了新型生物催化技术的发展。基因克隆和定向进化已经成为蛋白质工程中强有力的工具，去设计具有新型特征的生物分子（Syed et al.，2012；Matsuo et al.，2001；Yan et al.，2009）。最近几年，一些细菌来源的耐盐碱蛋白酶的基因已经在 *E. coli* 和 *B. subtilis* 中克隆及表达（Fu et al.，2003；Wang et al.，2009a；Krishnaveni et al.，2012）。

基因克隆和表达的发展已经提高了来自极端微生物的蛋白酶在异源宿主中可溶性的表达。这些发展推动了大量的催化酶在化工、食品、药学及其他工业上的应用（Kim et al.，1998；Machida et al.，2000；Singh et al.，2002）。来自耐盐碱菌的耐碱蛋白酶基因在大肠杆菌中的克隆与表达与天然菌株产生的酶的性质比较工作已经进行（Purohit，2012；Purohit and Singh，2011；Singh et al.，2010a，2010b，2012）。同时，新型极端微生物的发现及其基因组序列的确定为具有新的应用前景的酶的研究提供了更大可能。分子克隆和蛋白质的表达，蛋白质工程和定向进化可以改善酶的稳定性并获得其在自然界中不存在的底物特异性（Colquhouna and Sorumb，2002）。

7.5 不可培养与宏基因组学方法

7.5.1 宏基因组学的基础

现在大家普遍广泛接受，常规的微生物方法只能获得特定生境下有限的微生物多样性信息。因此，对整个微生物多样性的探索将会引入一个全新的领域，这称为宏基因组研究。宏基因组的第一步是从一个给定的栖息地提取高质量的环境 DNA（Purohit and Singh，2008；Siddhpura et al.，2010；Wooley et al.，2010；Santosa et al.，2012；Engel et al.，2012）。随后，环境 DNA 被片段化，构建宏基因文库。最后基于搜索数据库的同源性，对分离出来的基因通过生物信息学的方法进行分析。

基于序列和功能的宏基因组学

基因序列的全部信息可由基于序列的方法获得（Glockner et al.，2010；Tringe and Rubin, 2005)。系统发育和插入大片段宏基因的方法可以提供获得基因信息的途径，以特异性系统发育标记基因序列的形式获取已知微生物种群的基因信息。

这里主要强调的是与海洋无脊椎动物相关的微生物的一些关键基因簇的确认，这些基因簇可以编码新的且在生物制药领域具有重要意义的化合物的生物合成途径（Raes et al., 2007)。一种新型的低温活性的脂酶就是从波罗的海的沉淀物菌的宏基因组库获得。该酶已经在 *E. coli* 获得表达，并且重组蛋白的生化性质已经被表征（Jeon et al., 2011)。同样的，来自宏基因组文库的烷烃羟化酶和酯酶基因也被克隆，同时这些新蛋白质的生物催化应用也已经被研究（Xu et al., 2008; Okamura et al., 2010)。

基于功能的宏基因组学探索来源于微生物群体的特定产物。宏基因组文库被用于对各种功能的筛选，例如生物催化、维生素和抗生素的生产（Raes et al., 2007)。同样的，最近技术的进步使得我们可以直接从某一特定栖息地的微生物群落提取和鉴定蛋白质和代谢物（Craig et al., 2010)。宏基因组结合体外进化和高通量筛选技术给探索来自未开发的非培养性群体的新型分子提供了空前的机会（Eamonn et al., 2012)。

宏基因组文库的筛选

正如之前介绍中重点指出的，理论上不同的生物催化剂和其他功能分子可以通过一个给定的环境样品中 DNA 的提取而获得。耐盐碱海洋样品的总环境 DNA 克隆到表达载体上筛选脂酶，最终发现 120 种新酶，它们分属 21 个蛋白质家族（Miller et al., 2008)。近几年里，来自宏基因组的基因克隆已经为发现来自非可培养型环境的新酶铺平了道路（Kennedy et al., 2008; Purohit, 2012)。

分子工具和技术

过去的 20 年里，分子工具的发展推动了基因与基因组的鉴定、克隆、筛选和测序。这些发展帮助微生物生态学家从整体上去研究和评估微生物的多样性（Gilbert 2010; Glockner et al., 2010)。依靠特殊的遗传标记，利用基于 PCR 的技术开展的微生物分类工作，有助于复杂的微生物群落的分析（Bach et al., 2001)。相同长度的 DNA 片段甚至是单碱基可以被鉴定（Ercolini, 2004)。在此背景下，来源于印度的 Gujarat Coast 的嗜盐生物的 16S rRNA 基因序列已经利用 DGCE 技术进行分析（Purohit and Singh, 2008; Siddhpura et al., 2010)。

宏基因组的比较和特定栖息地的快速鉴定为捕捉栖息地特定的宏基因组标签提供了快速、准确的方法（Xu, 2006)。水生高盐度环境中丰富的微生物通过宏基因组学用焦磷酸测序进行鉴定，发现 *Haloquadratum walsbyi* 占优势。此外，它也发现了新型、丰富和先前未知的微生物群体。缺氧沉积物微生物群落的系统发育分析和功能结构研究表明了栖息地的固有的物理和化学特性（Ferrer ct al., 2011)。这些意外的发现揭示了宏基因组和单细胞基因组学方法的重要性（Ghai et al., 2011)。

海洋微生物群落的宏基因组含有典型的生物活性天然产物生物合成的基因和基因簇（Kauffmann et al.，2004；Kennedy et al.，2008）。基因 cassettes 里没有任何与数据库条目相似的序列（Glockner et al.，2010）。因此，在大多数情况下，使用简并引物、杂交与保守区相似的目标基因的末端（Liles et al.，2008；Ni et al.，2009；Purohit and Singh，2011），已经建成了为所有对宏基因组有兴趣的研究人员提供高性能计算的高通量研究策略（Morgan et al.，2010）。这个策略里，通过与已知蛋白质和碱基序列比较，系统可以自动化完成宏基因组序列的功能注释。它的发展是基于水和土壤样品的序列和功能分析（Grant and Heaphy，2012）。总之，宏基因组学可以获得大部分的基因组，提供大部分通路上的基因。现在的研究说明宏基因组方法结合异源表达为探索海洋生态系统多样性带来更大希望。

由此可知，宏基因组方法曾被用来探索来自印度古吉拉特邦沿海的不可培养的盐碱生境中的蛋白酶。蛋白酶序列通过利用生物信息学设计的简并引物，利用克隆技术获得，后续开展了重组表达和重组酶的性质研究（Purohit 2012；Purohit and Singh，2011，2012；Singh et al.，2010a，2010b，2012）。然而，这种方法并不是特别适合去获得那些与已知序列共享序列的功能特点（Craig et al.，2010）。更多地关注以表达为基础的生物催化剂的鉴定工作，大规模鸟枪测序和在酶编码区的筛选正在进行中。蒙特雷湾滨海海洋微生物天文台就是这样一个例子（Nakamura et al.，2009）。当前关注点都集中在从潜在的栖息地提取大量的酶基因核酸，最大化地覆盖给定的生境的整个遗传资源。

7.6 结论

极端微生物分泌了多种人类可以应用的胞外水解酶。这些酶主要包括蛋白酶、脂酶、淀粉酶、纤维素酶、甲壳质酶和木聚糖酶。这些酶的 pH 值、盐度和热稳定性使得它们适合在工业上应用。为获得具有多种极端环境的微生物，新物种从之前未研究过的栖息地被分离出来。近期重组 DNA 技术、基因组学和蛋白质组学的发展更加推动了寻找微生物来源生物催化剂的研究步伐。基因克隆和定向进化为开发和设计具有在多种极端条件下具有生物催化剂作用下的酶提供了可能。因此，获得来源于新环境，包括海洋环境的酶的序列分析，基因表达和性质表征的相关知识是必要的。宏基因组学方法为研究未知栖息地和它们大量不可培养的生物群体提供了可能。通过宏基因组，分子克隆和表达，蛋白质工程和定向进化技术寻找新的栖息地，获得大多数不可培养微生物种群将为新的生物催化剂领域提供一个大平台。

致谢

我们研究小组的研究得到 Saurashtra 大学与 UGC 大学的支持。海岸古吉拉特帮嗜极端生物多样性，分子进化，酶分子性质的研究得到多个研究计划支持，包括 DBT 和印度政府。Megha Purohit 与 Vikram Raval 分别感谢来自科学与工业研究会与 DBT 的支持。感谢 DBT 的支持，对 Rajesh K. Patel, Mital Dodia, Rupal Joshi 与 Sandeep Pandey 博士的工作也一并表示感谢。

参考文献

Abdel-Naby M. A., Hashem, A. M., Esawy, M. A. and Abdel-Fattah, A. F. (1998) Immobilization of *Bacillus subtilis* α-amylase and characterization of its enzymatic properties. *Microbiol. Res.* 153, 1-7.

Ahmadian, G., Degrassi, G., Venturi, V., Zeigler, D. R., Soudi, M. and Zanguinejad, P. (2007) *Bacillus pumilus* SG2 isolated from saline conditions produces and secretes two chitinases. *J. Appl. Microbiol.* 103, 1081-1089.

Akkaya B., Yenidunya, A. F. and Akkaya R. (2012) Production and immobilization of a novel thermoalkalophilic extracellular amylase from *Bacilli isolate. IJBIOMAC* 50 (4), 991-995.

Akolkar, A. V., Durai, D. and Desai A. J. (2010) *Halobacterium* sp. SP1 (1) as a starter culture for accelerating fish sauce fermentation. *J. Appl. Microbiol.* 109 (1), 44-53.

Amoozegar, M. A., Malekzadeh, F. and Khursheed, A. M. (2003) Production of amylase by newly isolated moderate halophile, *Halobacillus* sp. strain MA-2. *J. Microbiol. Methods* 52, 353-359.

Antranikian, G., Vorgias, C. E. and Bertoldo, C. (2005) Extreme environments as a resource for microorganisms and novel biocatalysts. Adv. Biochem. Eng. Biotechnol. 96, 219-262.

Aygan, A. and Arikan, B. (2008) A new halo-alkaliphilic, thermostable endoglucanase from moderately halophilic *Bacillus* sp. Cl4 isolated from Van Soda Lake. *Int. J. Agric. Biol.* 10, 369-374.

Aygan, A., Arikan, B., Korkmaz, H., Dincer, S. and Colak O. (2008) Highly thermostable and alkaline α-amylase from a halotolerant alkaliphilic Bacillus sp. AB68. *Braz. J. Microbiol.* 39, 547-553.

Bach, H. J., Hartmann, A., Schloter, M. and Munch, J. C. (2001) PCR primers and functional probes for amplification and detection of bacterial genes for extracellular peptidases in single strains and in soil. *J. Microbiol. Methods* 44, 173-182.

Balakrishnan, H., Dutta-Choudhary, N., Srinivasan, M. C. and Rele, M. V. (1992) Cellulase-free xylanase production from an alkalophilic *Bacillus* sp. NCL-87-6-10. *World J. Microbiol. Biotechnol.* 8, 627-631.

Barnes, I., Presses T. S., Saines, M., Dickson, P. and Van Goos, A. F. K. (1982) Geochemistry of highly basis calcium hydroxide groundwater in Jordan. *Chem. Geol.* 35, 147-154.

Battestein, V. and Macedo, G. A. (2007) Effects of temperature, pH and additives on the activity of tannase produced by *Paecilomyces variotii. Elect. J. Biotechnol.* 10 (2), 9.

Beisson, F., Tiss, A., Riviere, C. and Verger, R. (2000) Methods for lipase detection and assay: a critical review. *Eur. J. Lipid. Sci. Technol.* 133-153.

Bolobova, A. V., Simankova, M. C. and Markovich, N. A. (1992) Cellulase complex of a new halophilic bacterium Halocella cellulolytica. *Mikrobiologiya* 61, 804-811.

Carolina, P. M., Augusto, G., Castro-Ochoa, D. and Farres, A. (2009) Purification and biochemical characterization of a broad substrate specificity thermostable alkaline protease from *Aspergillus nidulans. Appl. Microbi-*

ol. Biotechnol. 78 603-612.

Carvalho, R. V., Correa, T. L. R., Matos da Silva, J. C., Mansur, L. R. C. and Martins, M. L. L. (2008) Properties of an amylase from thermophilic *Bacillus* sp. *Braz. / · Microbiol.* 39 (1), 102-107.

Chang, P., Tsai, W. S., Tsai, C. L. and Tseng, M. J. (2004) Cloning and characterization of two thermostable xylanases from an alkaliphilic *Bacillus firmus*. *Biochem. Biophys. Res. Commun.* 319, 1017-1025.

Chuan, L. D. (2006) Review of fungal chitinases. *Mycopathology* 161, 345-360.

Cojoc, R., Merciu, S., Popescu, G., Dumitru, L., Kamekura, M. and Enache, M. (2009) Extracellular hydrolytic enzymes of halophilic bacteria isolated from a subterranean rock salt crystal. *Rom. Biotechnol. Lett.* 14, 4658-4664.

Colquhouna, D. and Sorumb, H. (2002) Cloning, characterization and phylogenetic analysis of the fur gene in *Vibrio salmonicida* and *Vibrio logei*. *Gene* 296, 213-220.

Coronado, M. J., Vargas, C., Hofemeister, J., Ventosa, A. and Nieto, J. J. (2000) Production and biochemical characterization of an α-amylase from the moderate halophile *Halomonas meridiana*. *FEMS Microbiol. Lett.* 183, 67-71.

Craig J. W., Chang, W., Kim, J. H., Obiajulu, S. C. and Brady S. F. (2010) Expanding small-molecule functional metagenomics through parallel screening of broad-host-range cosmid environmental DNA libraries in diverse Proteobacteria. *Appl. Environ. Microbiol.* 76 (5), 1633-1641.

DasSarma, S. and Arora, P. (2001) 'Halophiles'. In: *Encyclopedia of Life Sciences*. Macmillan Press, Nature Publishing Group, London 1-9. *www. els. net*.

Debashish, G., Malay, S., Barindra, S. and Joydeep, M. (2005) Marine enzymes. *Adv. Biochem. Eng. Biotechnol.* 96, 189-218.

Demirjian, D. C., Morís-Varas, F. and Cassidy, C. S. (2001) Enzymes from extremophiles. *Curr. Opin. Chem. Biol.* 5, 144-151.

Deutch, C. E. (2002) Characterization of a salt-tolerant extracellular α-amylase from *Bacillus dipsosauri*. *Lett. Appl. Microbiol.* 35 (1), 78-84.

Dodia, M. S., Bhimani, H. G., Rawal, C. M., Joshi, R. H. and Singh S. P. (2008a) Salt dependent resistance against chemical denaturation of alkaline protease from a newly isolated haloalkaliphilic *Bacillus* sp. *Bioresour. Techttol.* 99, 6223-6227.

Dodia, M. S., Joshi, R. H., Patel, R. K. and Singh, S. P. (2006) Characterization and stability of extracellular alkaline proteases from moderately halophilic and alkaliphilic bacteria isolated from saline habitat of coastal Gujarat, India. *Braz. J. Microbiol.* 37, 244-252.

Dodia, M. S., Rawal, C. M., Bhimani, H. G., Joshi, R. H., Khare, S. K. and Singh S. P. (2008b) Purification and stability characteristics of an alkaline serine protease from a newly isolated *Haloalkaliphilic bacterium* sp. AH-6. *J. Ind. Microbiol. Biotechnol.* 35, 121-131.

Duckworth, A. W., Grant, W. D., Jones, B. E. and Steenbergen, R. V. (1996) Phylogenetic diversity of soda lake alkaliphiles. *FEMS Microbiol. Ecol.* 19, 181-191.

Eamonn, P. C., Roy, D. S., Julian, R. M. and Colin, H. (2012) Functional metagenomics reveals novel salt tolerance loci from the human gut microbiome. *ISME J.* 6 (10), 1916—1925. doi: 10.1038/ismej.2012.38.

Engel, K., Pinnell, L., Cheng, J., Charles, T. C. and Neufeld, J. D. (2012) Nonlinear electrophoresis for purification of soil DNA for metagenomics. *J. Microbiol. Methods.* 88 (1), 35-40.

Ercolini, D. (2004) PCR-DGGE fingerprinting: novel strategies for detection of microbes in food. *J. Microbiol. Meth.* 56, 297-314.

Eriksen, N. (1996) Detergents, In: Industrial Enzymology, Godfrey, T. and West S. (eds) Stockton Press, Boston and New York, 187-200.

Essghaier, B., Hedi, A., Bejji, M., Jijakli, H., Boudabous, A., Sadfi-Zouaoui, N. (2012) Characterization of a novel chitinase from a moderately halophilic bacterium, *Virgibacillus marismortui* strain M3-23. *Ann. Microbiol.* 62, 835-841.

Essghaier, B., Rouaissi, M., Boudabous, A., Jijakli, H. and Sadfi-Zouaoui, N. (2010) Production and partial characterization of chitinase from a halotolerant Planococcus rifitoensis strain M2-26. *World J. Microbiol. Biotechnol.* 26 (6), 977-984.

Ferrer, M., Golyshina, O., Beloqui, A. and Golyshin, P N. (2007) Mining enzymes from extreme environments. *Curr. Opin. Microbiol.* 10, 207-214.

Ferrer, M., Guazzaroni, M. E., Richter, M., García-Salamanca, A., Yarza, P., et al., (2011) Taxonomic and functional metagenomic profiling of the microbial community in the anoxic sediment of a sub-saline shallow lake (Laguna de Carrizo, Central Spain). *Microb. Ecol.* 62 (4), 824-837.

Flach, J., Pilet, P. E. and Jolles, P. (1992) What is new in chitinase research? *Experientia* 48, 701-713.

Forsyth, M. P., Shindler, D. B., Gochnauer, M. B. and Kushner, D. J. (1971) Salt tolerance of intertidal marine bacteria. *Can. J. Microbiol.* 17, 825-828.

Freeman, S., Minzm, O., Kolesnik, I., Barbul, O., Zveibil, A., et al., (2004) Trichoderma biocontrol of *Colletotrichum acutatum* and *Botrytis cinerea* and survival in strawberry. *Eur. J. Plant. Pathol.* 110, 361-370.

Fu, Z., Hamid, S. B. A., Razak, C. A. N., Basri, M., Salleh, A. B. and Zaliha Abd, R. N. (2003) Secretory expression in *Escherichia coli* and single-step purification of a heat-stable alkaline protease. *Prot. Exp. Purif.* 28, 63-68.

Ghai, R., Pasic, L., Fernandez, A., Martin-Cuadrado, A., Mizuno, C., et al., (2011) New abundant microbial groups in aquatic hypersaline environments. *Scientific Rep.* 1, 135. doi: 10.1038/srep00135.

Gilbert J. A. (2010) 'Aquatic metagenome library (archive; expression) generation and analysis'. In: *Handbook of Hydrocarbon and Lipid Microbiology*, Timmis, K. (ed). Springer-Verlag, Berlin and Heidelberg, 4347-4352. Doi: 10.1007/978-3-540-77587-4.340.

Gilbert, H. J. and Hazlewood, G. P. (1993) Bacterial cellulases and xylanases. *J. Gen. Microbiol.* 139, 187-194.

Glockner, J., Kube, K., Shrestha, P., Weber, M., Glockner, F. O., et al., (2010) Identification of novel catalyst by cassette construction from sponges through metagenomic approaches. *Env. Microbiol.* 12 (5), 1218-1229.

Godfrey, T. and West, S. (1996) 'The application of enzymes in industry'. In: Industrial enzymology, 2nd edn, Godfrey, T. and Reichelt, J. (eds). The Nature Press, New York, 512.

Govender, L., Naidoo, L. and Setati, M. E. (2009) Isolation of hydrolases producing bacteria from Sua pan solar salterns and the production of endo-1, 4-b-xylanase from a newly isolated haloalkaliphilic *Nester enkonia* sp. Afr. *J. Biotechnol.* 8, 5458-5466.

Grant, W. D. (1991) 'General view of halophiles'. In: *Superbugs: microorganisms in extreme environments*, Horikoshi, K. and Grant, W. D. (eds). Japan Scientific Societies Press, Tokyo, 15–37.

Grant, W. D. and Heaphy, S. (2012) Metagenomics and recovery of enzyme genes from alkaline saline environments. *Env. Technol.* 31, 10–16.

Grant, W. D. and Horikoshi, K. (1992) 'Alkaliphilcs; ecology and biotechnological applications'. In: *Molecular Biology and Biotechnology of Extremophiles*, Herbert, R. A. and Sharpe, R. J. (eds). Blackie 8c Sons, Glasgow, 143–162.

Grant, W. D. and Mwatha, W. E. (1989) Bacteria from alkaline, saline environments. In: *Recent advances in microbial ecology*, Hattori, T., Ishida, Y., Maruyama, Y., Morita, R. Y. and Uchida, A. (eds) Japan Scientific Societies Press, Tokyo, 64–67.

Grant, W. D., Gemmell, R. T. and McGenity, T. J. (1998) 'Halophiles'. In: *Extremophiles: microbial life in extreme environments*, Horikoshi, K. and Grant, W. D. (eds). Wiley–Liss, Inc. New York, 93–132.

Guo, B., Chen, X. L., Sun, C. Y., Zhou, B. C. and Zhang, Y. Z. (2009) Gene cloning, expression and characterization of a new cold–active and salt–tolerant endo–xylanase from marine Glaciecola mesophila KMM 241. *Appl. Microbiol. Biotechnol.* 84, 1107–1115.

Gupta, A., Roy, I., Patel, R. K., Singh, S. P., Khare, S. K. and Gupta, M. N. (2005) One–step purification and characterization of an alkaline protease from haloalkaliphilic *Bacillus* sp. *J. Chromatog*r. A. 1075, 103–108.

Gupta, R., Beg, Q. K. and Lorenz, P. (2002) Bacterial alkaline proteases: molecular approaches and industrial applications. *Appl. Microbiol. Biotechnol.* 59, 15–32.

Gupta, R., Gupta, N. and Rathi, P. (2004) Bacterial lipases: an overview of production, purification and biochemical properties. *Appl. Microbiol. Biotechnol.* 64 (6), 763–781.

Haddar, A., Agrebi, R., Bougatef, A., Hmidet, N., Sellami–Kamoun, A. and Nasri, M. (2009) Two detergent stabile alkaline proteases from *Bacillus mojavensis A*21: purification, characterization and potential application as laundry detergent additive. *Bioresour. Technol.* 100, 3366–3373.

Heidari, H. R. K., Amoozegar, M. A., Hajighasemi, M., Ziaee, A. A. and Ventosa, A. (2009) Production, optimization and purification of a novel extracellular protease from the moderately halophilic bacterium *Halobacillus karajensis*. *J. bid. Microbiol. Biotechnol.* 36, 21–27.

Heidari, H. R. K., Ziaee, A. A. and Amoozegar, M. A. (2007) Purification and biochemical characterization of a protease secreted by the *Salinivibrio* sp. strain AF–2004 and its behavior in organic solvents. *Extremophiles* 11, 237–243.

Hiraga, K., Nishikata, Y., Namwong, S., Tanasupawat, S., Takada, K. and Oda, K. (2005) Purification and characterization of serine proteinase from a halophilic bacterium, *Filobacillus* sp. RF2–5. *Biosci. Biotechnol. Biochem.* 69, 38–44.

Horikoshi, K. (1999) Alkaliphiles: some applications of their products for biotechnology. *Microbiol. Mol. Biol. Rev.* 63, 735–750.

Huang, W., Zuo, Z., Shen, W., Singh, S., Chen, X. Z., et al., (2012) High–level expression of alkaline protease using recombinant Bacillus timyloliquefaciens Afr. *J. Biotechnol.* 11 (14), 3358–3362.

Jaeger K. E. and Reetz M. T. (1998) Microbial lipases form versatile tools for biotechnology. *Trends Biotechnol.*

16, 396-403.

Javor, B. J. (1983) Planktonic standing crop and nutrients in a saltern ecosystem. *Limnol. Oceanogr.* 28, 153-159.

Jeon, J., Kim, J., Lee, H., Kim, S., Kang, S., et al., (2011) Novel lipolytic enzymes identified from metagenomic library of deep-sea sediment. *Evidence-Based Complementary and Alternative Medicine*, Article 9D 271419, 9 pages, doi: 10.1155/2011/271419.

Jones, B. E., Grant, W. D., Collins, N. C. and Mwatha, W. E. (1994) 'Alkaliphiles: diversity and identification'. In: *Bacterial Diversity and Systematics*, Priest F. G. (ed.) Plenum Press, New York, 195-230.

Joshi, R. H., Dodia, M. S. and Singh, S. P. (2008) Production and optimization of a commercially viable alkaline protease from a haloalkaliphilic bacterium. *Biotechnol. Bioproc. Engg.* 13, 552-559.

Jung, W. J., Kuk, J. H., Kim, K. Y., Kim, T. H. and Park, R. D. (2005) Purification and characterization of chitinase from *Paenibacillus illinoisensis* KJA-424. *J. Microbiol. Biotechnol.* 15, 274-280.

Karan, R. and Khare, S. K. (2011) Stability of haloalkaliphilic *Geomicrobium* sp. protease modulated by salt. *Biochem.* (*Moscow*) 76, 686-693.

Kauffmann, I. M., Schmitt, J. and Schmid, R. D. (2004) DNA isolation form soil sample for cloning in different host. *Metagenomics* 64, 665-670.

Kennedy, J., Marchesi, J. R. and Dobson, D. W. (2008) Direct metagenomic detection of viral pathogens in nasal and fecal specimens using unbiased high-throughput sequencing approach. *Microb. Cell Fact.* 7, 27-33.

Kim, D., Singh, S., Machida, S., Chika, Y., Kawata, Y. and Hayashi, K. (1998) Importance of five amino acid residues at C-terminal region for the folding and stability of β-Glucosidase of Cellvibrio gilvus. *J. Ferment. Bioengg.* 85, 433-435.

Krishnaveni, K., Mukeshkumar, D. J., Balakumaran, M. D., Ramesh, S. and Kalaichelvan, P. T. (2012) Production and optimization of extracellular alkaline protease from *Bacillus subtilis* isolated from dairy effluent. *Der. Pharmacia. Lett.* 4 (1), 98-109.

Kulkarni, N., Shendye, A. and Rao, M. (1999) Molecular and biotechnological aspects of xylanases. *FEMS Microbiol. Rev.* 23, 411-456.

Kumar, K. B., Balakrishnan, H. and Rele, M. V. (2004) Compatibility of alkaline xylanases from an alkaliphilic *Bacillus* NCL (87-6-10) with commercial detergents and proteases. *J. Ind. Microbiol. Biotechnol.* 31 (2), 83-87.

Kumar, V., Morya, S., Kim, E. and Yadav, D. (2012) *In-silico* characterization of alkaline proteases from different species of *Aspergillus*. *Appl. Biochem. Biotechnol.* 166 (1), 243-257.

Kupiec, R. C. and Chet I. (1998) The molecular biology of chitin digestion. *Curr. Opin. Biotech.* 9, 270-277.

Kurtovic, I., Marshall, S. N., Zhao, X. and Simpson, B. K. (2009) Lipases from mammals and fishes. *Rev. Fisheries Sci.* 17 (1), 18-40.

Ladenstein, R. and Antranikian, G. (1998) Proteins from hyperthermophiles: stability and enzymatic catalysis close to the boiling point of water. *Adv. Biochem. Eng. Biotechnol.* 61, 37-85.

Liaw, H. J. and Mah, R. A. (1992) Isolation and characterization of *Haloanaerobacter chitinovorans* gen. nov., sp. nov., a halophilic, anaerobic, chitinolytic bacterium from a solar saltern. *Appl. Env. Microbiol.* 58 (1), 260-266.

Liles, M., Williamson, L., Rodbumrer, J., Torsvik, V., Goodman, R. and Handelsman, J. (2008) Recovery, purification, and cloning of high-molecular-weight DNA from soil microorganisms. *Appl. Env. Microbiol.* 7 (10), 3302-3305.

Machida, S., Ogawa, S., Xiaohua, S., Takaha, T., Fujii, K. and Hayashi K. (2000) Cycloamylose as an efficient artificial chaperone for protein refolding. *FEBS Lett.* 486, 131-135.

Mamo, G., Thunnissen, M., Hatti-Kaul, R. and Mattiasson, B. (2009) An alkaline active xylanase: insights into mechanisms of high pH catalytic adaptation. *Biochimie.* 91, 1187-1196.

Martinez, M. A., Delgado, O. D., Baigori, M. D. and Sineriz, F. (2005) Sequence analysis, cloning and over-expression of an endoxylanase from the alkaliphilic *Bacillus halodurans. Biotechnol. Lett.* 27 (8), 545-550.

Matsuo, T., Ikeda, A., Seki, H., Ichimata, T., Sugimori, D. and Nakamura, S. (2001) Cloning and expression of the ferredoxin gene from extremely halophilic archaeon *Haloarcula japonica* strain TR-1. *BioMetals* 14, 135-142.

Maurer, K. H. (2004) Detergent proteases. *Curr. Opin. Biotechnol.* 15, 330-334.

Miller J. P., Reyes, F., Parra, L. P., Salazar, O., Andrews, B. A. and Asenjo, J. A. (2008) Cloning of complete genes for novel hydrolytic enzymes from Antarctic seawater bacteria by use of an improved genome walking technique. *J. Biotechnol.* 33, 277-286.

Morgan, J., Darling, A. and Eisen, J. (2010) High throughput screening of metagenomic library. *PLoS ONE* 5 (4), el0209. doi: 10.1371/journal. pone.0010209.

Nakamura, S., Yang, C., Sakon, N., Ueda, M., Tougan, T., et al., (2009) The Monterey Bay Coastal Ocean Microbial Observatory on marine picoplancton. *PLoS ONE* 4 (1), 4219-4225.

Namwong, S., Hiraga, K., Takada, K., Tsunemi, M., Tanasupawat, S. and Oda, K. (2006) A halophilic serine proteinase from *Halobacillus* sp. SR5-3 isolated from fish sauce: purification and characterization. *Biosci. Biotechnol. Biochem.* 70 (6), 1395-1401.

Ni, X., Yue, L., Chi, Z., Li, Z., Wang, X. and Madzak, C. (2009) Alkaline protease gene cloning from the marine yeast *Aureobasidium pullulans* HN2-3 and the protease surface display on *Yarrowia lipolytica* for bioactive peptide production. *Mar. Biotechnol.* 11, 81-89.

Niehaus, F., Bertoldo, C., Kahler, M. and Antranikian, G. (1999) Extremophiles as a source of novel enzymes for industrial application. *Appl. Microbiol. Biotechnol.* 51, 711-729.

Nowlan, B., Dodia, M. S., Singh, S. P. and Patel, B. K. C. (2006) Bacillus okhensis sp. nov., a halotolerant and alkalitolerant bacterium from an Indian saltpan. *Int. J. Syst. Evol. Microbiol.* 56, 1073-1077.

Oberoi, R., Beg, Q. K., Puri, S., Sazena, R. K and Gupta, R. (2001) Characterization and wash performance analysis of an SDS-Stable alkaline protease from *Bacillus* sp. *World J. Microbiol. Biotechonl.* 17, 493-497.

Okamura, Y., Kimura, T., Yokouchi, H., Meneses-Osorio, M., Katoh, M., et al., (2010) Isolation and characterization of a GDSL esterase from the metagenome of a marine sponge-associated bacteria. *Mar. Biotechnol.* 12 (4), 395-402.

Onishi, H., Fuchi, H., Konomi, K., Hidaka, O. and Kamekura, M. (1980) Isolation and distribution of a variety of halophiiic bacteria and their classification by salt response. *Agric. Biol. Chem.* 44, 1253-1258.

Onishi, H. and Hidaka, O. (1978) Purification and properties of amylase produced by a moderately halophiiic

Acinetobacter sp. *Can. J. Microbiol.* 24, 1017–1023.

Oren, A. (1988) The microbial ecology of the Dead Sea. *Adv. Microb. Ecol.* 10, 193–229.

Oren, A. (2002a) Diversity of halophilic microorganisms: environments, phylogeny, physiology, applications. *J. Ind. Microbiol. Biotechnol.* 28, 56–63.

Oren, A. (2002b) *Halophilic Microorganisms and their Environments*, Kluwer Academic Publishers, Dordrecht. DOI: 10.1007/0-306-48053-0.

Oren, A. (2002c) Molecular ecology of extremely halophilic archaea and bacteria. *FEMS Microbiol. Ecol.* 39, 1–7.

Pandey, A., Nigam, P., Soceol, C. R., Soccol, V. T., Singh, D. and Mohan, R. (2000) Advances in microbial amylases. *Biotechnol. Appl. Biochem.* 31, 135–152.

Pandey, S., Rakholiya, K., Raval, V. H. and Singh, S. P. (2012) Catalysis and stability of an alkaline protease from a haloalkaliphilic bacterium under non-aqueous conditions as a function of pH, salt and temperature. *J. Biosci. Bioengg.* 114 (3), 251–256, doi: 10.1016/J. JIbiosc, 2012.03.003.

Pandey, S. and Singh, S. P. (2012) Organic solvent tolerance of an α-amylase from haloalkaliphilic bacteria as a function of pH, temperature, and salt concentrations. *Appl. Biochem. Biotechnol.* 166, 1747–1757.

Pasic, L., Bartual, S. G., Ulrih, N. P., Grabnar, M. and Velikonja, B. H. (2005) Diversity of halophiiic archaea in the crystallizers of an Adriatic solar saltern. *FEMS Microbiol. Ecol.* 54, 491–498.

Patel, R. K., Dodia, M. S., Joshi, R. H. and Singh, S. P. (2006a) Production of extracellular halo-alkaline protease from a newly isolated haloalkaliphilic *Bacillus* sp. isolated from seawater in Western India. *World J. Microbiol. Biotechnol.* 22 (4), 375–382.

Patel, R. K., Dodia, M. S., Joshi, R. H. and Singh, S. P. (2006b) Purification and characterization of alkaline protease from a newly isolated haloalkaliphilic *Bacillus* sp. *Process Biochem.* 41, 2002–2009.

Patel, R. K., Dodia, M. S. and Singh S. P. (2005) Extracellular alkaline protease from a newly. isolated haloalkaliphilic *Bacillus* sp.: Production and optimization. *Process Biochem.* 40, 3569–3575.

Pleban, S., Chernin, L. and Chet, I. (1997) Chitinolytic activity of endophytic strain of *Bacillus cereus*. *Lett. Appl. Microbiol.* 25, 284–288.

Prade, R. A. (1995) Xylanases: from biology to biotechnology. *Biotechnol. Genet. Eng. Rev.* 13, 101–131.

Prakash, B., Vidyasagar, M., Madhukumar, M. S., Muralikrishna, G. and Sreeramulu, K. (2009) Production, purification, and characterization of two vtremely halotolerant, thermostable, and alkali-stable α-amylase from *Chrornohalobacter* sp. TVSP 101. *Process Biochem.* 44, 210–215.

Purohit, M. (2012) PhD thesis, Saurashtra University, Rajkot, India.

Purohit, M. K. and Singh, S. P. (2008) Assessment of various methods for extraction of metagenomic DNA from saline habitats of Coastal Gujarat (India) to explore molecular diversity. *Lett. Appl. Microbiol.* 49 (3), 338–344.

Purohit, M. K. and Singh, S. P. (2011) Comparative analysis of enzymatic stability and amino acid sequences of thermostable alkaline proteases from two haloalkaliphilic bacteria isolated from coastal region of Gujarat, India. *IJBIOMAC*49, 103–112.

Purohit, M. K. and Singh, S. P. (2012) 'Metagenomics of saline habitats with respect to bacterial phylogeny and biocatalytic potential'. In: *Microorganisms in Sustainable Agriculture and Biotechnology*, Satyanarayana,

T. , Johri, B. N. and Prakash, A. (eds) . Springer Science & Business Media B. V. , 295-308.

Quesada, E. , Ventosa, A. , Rodriguez-Valera, F. and Ramos-Cormenzana, A. (1982) Types and properties of some bacteria isolated from hypersaline soils. *J. Appl. Microbiol.* 53, 155-161.

Quesada, E. , Ventosa, A. , Rodriguez-Valera, F. , Megias, L. and Ramos-Cormenzana, A. (1983) Numerical taxonomy of moderately halophilic gram negative bacteria from hypersaline soils. *J. Gen. Microbiol.* 129, 2649-2657.

Raes, J. , Husenholts, P. , Tringe, S. G. , Doerks, T. , Jensen, L. J. , et al. , (2007) Qualitative phylogeny assessment of microbial communities in diverse environment. *Sci. Exp.* 1-2/10, 1126.

Rao, M. B. , Tanksale, A. M. , Ghatge, M. S. and Deshpande, V. V. (1998) Molecular and biotechnological aspects of microbial proteases. *Microbiol. Mol. Biol. Rev.* 62, 597-635.

Rees, H. C. , Grant, W. D. and Jones, B. E. (2004) Diversity of Kenyan Soda Lake alkaliphiles assessed by molecular methods. *Extremophiles* 8, 63-71.

Reetz, M. T. (2002) Lipases as practical biocatalysts. *Curr. Opin. Chem. Biol.* 6, 145-150.

Riyaz, A. , Qader, S. A. , Anwar, A. and Iqbal, S. (2009) Immobilization of a thermostable a-amylase on calcium alginate beads from *Bacillus subtilis* KIBGE-HAR. *Aus. J. Basic Appl. Sci.* 3 (3), 2883-2887.

Sajedi, R. H. , Naderi-Manesh, H. , Khajeh, K. , Ahmadvand, R. , Ranjbar, B. , et al. , (2005) A Ca-independent α-amylase that is active and stable at low pH from the *Bacillus* sp. KR-8104. *Enzyme Microb. Technol.* 36, 666-671.

Santosa, F. , Yarzal, P. , Victor, P. , Inmaculada, M. , Ramon, R. and Josefa, A. (2012) Viruses from hypersaline environments: a culture-independent approach. *Appl. Environ. Microbiol.* 78, 1635-1643.

Satyanarayana, T. , Raghukumar, C. and Shivaji, S. (2005) Extremophilic microbes: diversity and perspectives. *Curr. Sci.* 89 (1), 78-90.

Siddhpura, P. K. , Vanparia, S. , Purohit, M. K. and Singh, S. P. (2010) Comparative studies on the extraction of metagenomic DNA from the saline habitats of Coastal Gujarat and Sambhar Lake, Rajasthan (India) in prospect of molecular diversity and search for novel biocatalysts. *IJBIOMAC* 47, 375-379.

Singh, S. P. , Kim, J. D. , Machida, S. and Hayashi, K. (2002) Over-expression and protein folding of a chimeric β- glucosidase constructed from *Agrobacterium tumefaciens* and *Cellvibrio gilvus. lrtd. J. Biochem. Biophy.* 39, 235-239.

Singh, S. P. , Purohit, M. K. , Raval, V. H. , Pandey, S. , Akbari V. G. and kaw C. M. (2010a) 'Capturing the potential of haloalkaliphilic bacteria from the saline habitats through culture dependent and metagenomic approaches' . In: *Current Research Technology and Education Topics in Applied Microbiolol and Microbial Biotechnology*, Mendez-Vilas, A. (ed) . Formatex Research Centre, Badajoz, Spain, 81-87.

Singh, S. P. , Raval, V. H. , Purohit, M. K. , Pandey, S. , Thumar, J. T. et al. , (2012) 'Haloalkaliphilc bacteria and actinobacteria from the saline habitats: new opportunities for biocatalysis and bioremediation' . In: *Microorganisms in Environmental Management: Microbes and Environment*, Satyanarayana T. , Johri, B. N. and Prakash, A. (eds) . Springer, New York, 415-429.

Singh, S. P. , Thumar J. T. , Gohel S. D. and Purohit, M. K. (2010b) 'Molecular diversity and enzymatic potential of salt-tolerant alkaliphilic actinomycetes' . In: *Current research technology* and *Education Topics in Applied Microbiology and Microbial Biotechnology.* Mendez-Vilas, A. (ed) . Formatex Research Centre, Badajoz, Spain, 280–286.

Syed, R., Rani, R., Sabeena., Masoodi, A., Shafi, G. and Alharbi, K. (2012) Functional analysis and structure determination of alkaline protease from *Aspergillus flavus*. *Bioinformation* 8 (4), 175–180.

Tindall, B. J., Mills, A. and Grant, W. D. (1980) An alkalophilic red halophilic bacterium with a low magnesium requirement from a Kenyan soda lake. *J. Gen. Microbiol.* 116, 257–260.

Trachuck, L. A., Revina, L. P., Shemyakina, T. M., Chestukhina, G. G. and Stepanov, V. M. (1996) Chitinases of Bacillus licheniformis B6839: isolation and properties. *Can. Microbiol.* 42, 307–315.

Tringe, S. G. and Rubin, E. M. (2005) Metagenomics DNA sequencing of environmental samples. *Nat. Rev. Genet.* 805–814.

Trivedi, N., Gupta, V., Kumar, M., Kumari, P., Reddy, C. R. K. and Jha, B. (2011) An alkali-halotolerant cellulase from *Bacillus flexus* isolated from green seaweed *Ulva lactuca*. *Carbohydr. Poly.* 83 (2), 891–897.

Tsujibo, H., Orikoshi, H., Baba, N., Miyahara, M., Miyamoto, K., et al., (2002) Identification and characterization of the gene cluster involved in chitin degradation in a marine bacterium, Alteromonas sp. strain O-7. *Appl. Environ. Microbiol.* 68, 263–270.

UpaDek, H. and Kottwitz, B. (1997) 'Application of analyses in detergents'. In: *Enzymes in Detergency*, van Ec, J. H., Misset, O. and Baas, E. J. (eds). Marcel Dekker, Inc, New York, 203–212.

van den Burg, B. (2003) Extremophiles as a source for novel enzymes. *Curr. Opin. Microbiol.* 6, 213–218.

Ventosa, A., Nieto J. J. and Oren, A. (1998) Biology of moderately halophilic aerobic bacteria. Microbiol. *Mol. Biol. Rev.* 62 (2), 504–544.

Voget, S., Steele, H. L. and Streit, W. R. (2006) Characterization of a metagenome-derived halotolerant cellulase. *J. Biotechnol.* 126, 26–36.

Vreeland, R. H., Piselli, Jr. A. F., McDonnough, S. and Meyers S. S. (1998) Distribution and diversity of halophilic bacteria in a subsurface salt formation *Extremophiles* 2, 321–331.

Wang, F., Hao, J., Yang, C. and Sun, M. (2009a) Cloning, expression, and identification of a novel extracellular cold-adapted alkaline protease gene of the marine bacterium strain YS-80-122. *Appl. Biochem. Biotechnol.* 162 (5), 1497–1505.

Wang, C. Y., Hsieh, Y. R., Ng, C. C., Chan, H., Lin, H. T., et al., (2009b) purification and characterization of a novel halostable cellulase from *Salinivibrio* sp. strain NTU-05. *Enzyme Microb. Technol.* 44, 373–379.

Wang, S. L., Lin, T. Y., Yen, Y. H., Liao, H. F. and Chen, Y. J. (2006) Bioconversion of shellfish chitin wastes for the production of Bacillus subtilis W-118 chitinase. *Carbohydr. Res.* 341, 2507–2515.

Watanabe, T., Oyanagi, W., Suzuki, K. and Tanaka, H. (1990) Chitinase system *Bacillus circulans* WL-12 and importance of chitinase Al in chitin degradation. *J. Bacteriol.* 172, 4017–4022.

Wejse, P. L., Ingvorsen, K. and Mortensen, K. K. (2003) Purification and characterization of two extremely halotolerant xylanases from a novel halophilic bacterium. *Extremophiles* 7, 423–431.

Weski, J. and Ehrmann, M. (2012) Genetic analysis of 15 Protein folding factors and proteases of the *Escherichia coli* cell envelope. *J. Bacteriol.* 194 (12), 3225–3233.

Wooley, C., Godzik, A. and Friedberg, I. (2010) Gleaning information from metagenomic data. *PLoS Comput.*

Biol. 6 (2), el000667.

Xu, J. (2006) Microbial ecology in the age of genomics and metagenomics: concepts, tools, and recent advances. *Mol. Ecol.* 15, 1713-1731.

Xu, M., Xiao, X. and Wang, F. (2008) Isolation and characterization of alkane hydroxylases from a metagenomic library of pacific deep-sea sediment. *Extremophiles* 12 (2), 255-262, doi: 10.1007/S00792-007-0122-X.

Yan, B., Chen, X., Hou, X., He, H., Zhou, B. and Zhang, Y. (2009) Molecular analysis of the gene encoding a cold-adapted halophilic subtilase from deep-sea psychrotolerant bacterium *Pseudoalteromonas* sp. SM9913: cloning, expression, characterization and function analysis of the C-terminal PPC domains. *Extremophiles.* 13, 725-733.

Yu, C., Lee, A. M., Bassler, B. L. and Roseman, S. (1991) Chitin utilization by marine bacteria. A physiological function for bacterial adhesion to immobilized carbohydrate. *J. Biol. Chem.* 266, 24260-24267.

第8章 来源于海洋栖息地的放线菌及其催化潜能

Satya P. Singh, Saurashtra University, Rajkot, India Jignasha T. Thumar, M. & N. Virani Science College, Rajkot, India Sangeeta D. Gohel, Bhavtosh Kikani, Rushit Shukla, Amit Sharma and Kruti Dangar, Saurashtra University, Rajkot, India

DOI：10.1533/9781908818355.2.191

摘要：海洋已经被视为地球上生命的起源，并且占据了大部分有助于各式生物生活的栖息地。海洋环境中，微生物之间对空间和营养的竞争已经成为导致进化的主要选择力量，这也促使海洋微生物去获得独特酶系统以适应复杂的环境。海洋占超过70%的地球总面积，海洋栖息地包括无数为适应生存环境而具有生化和分子奥秘的微生物。因此，海洋微生物可以提供新颖的生物催化剂和其他有附加值的分子。虽然已经从微生物、动物、植物中分离纯化得到了酶，但由于具有普遍的生化多样性，易于大规模培养和遗传操作简单等优势，微生物依旧是酶最主要的来源。为获得有价值的次级代谢产物，通常是从土壤中发现放线菌。然而，具体到多样性和有附加值分子的研究，水生生态系统中的放线菌只开展了有限研究。近几年来，许多海洋栖息地中的新型放线菌已经被分离，并且这些海洋放线菌产生不同类型的新酶。针对海洋放线菌的出现，多样性和酶潜力的有限研究已突出显示了其巨大潜能。随着对极端环境下放线菌研究兴趣的迅速增长，对其多样性、酶谱、生理学和适应性的研究将具有重要意义。

关键词：海洋放线菌，海洋微生物酶，极端酶，生物催化剂

8.1 引言

微生物约占地球上所有已知生命类型的90%。通常认为，生命起源于海洋，因此，海洋微生物在地球上扮演着重要的角色。微生物是海洋中的主要初级生产者，并且决定着海洋能量和营养的流动。微生物的生命具有高度多样性并且可以从不同温度、压力、盐浓度、pH值等条件的环境中发现。其多样的海洋栖息地从冰雪覆盖的南北两极地区，直到

深海热液喷孔，包括海水、近海地和很多人造含盐栖息地。根据最近宏组学分析的研究，盐度是微生物群落动态的主要环境决定因素（Lozupone and Knight, 2007）。微生物的生命涵盖了整个海洋生物圈，微生物可以适应不同的环境，此外，在多数海洋动物、脊椎动物和无脊椎动物寄主的细胞内或细胞外，也发现了与寄主共生的微生物，这些共生微生物拥有系列酶和通路以满足宿主的需要（Debashish et al., 2005）。近年来，这些具有在可变温度、极端 pH 值、高盐和高压条件下生长良好的独特微生物受到了极大关注。其酶的稳定性和代谢机制是它们生存的主要原因（Demirjian et al., 2001, Antranikian et al., 2005; Ferrer et al., 2007）。从极端微生物中鉴定出来一些酶，他们与不同的乙醇脱氢酶一起作用于糖、蛋白质和脂质。

放线菌在形态学和系统发生学上具有多样性，是革兰氏阳性鞭毛细菌且具有高 G+C 含量（>55 mol%）。他们很少从海洋栖息地中被研究，只有少量文献报道了其的存在、多样性和酶的催化潜能。因此，关于海洋放线菌的研究将具有重大意义。近来一些基于可培养的和非可培养方法的发现，证明了放线菌在海洋生态系统的广泛分布。海洋放线菌拥有大量的未被开发的新奇的初级代谢产物。

在近期关于生物催化的研究中，海洋生态系统拥有的巨大生物多样性物种库被认为是酶资源开发最有利的宝库。此外，新兴化学和立体化学特性进一步增加了它们在研究和产业上的重要意义。

8.2 海洋栖息地

地球表面超过 70%的面积被海洋覆盖。海洋生态系统是地球水生生态系统的一部分，包括从有生产力的近海地区到几乎贫瘠的海底。每当我们想到海底的生活，通常想到的就是肉眼可见的鱼和鲸。然而，令我们惊讶的是最多的海洋生命代表是微生物。微生物群落约占海洋总生物量的 90%，组成一个隐藏的却在海洋环境中繁荣的生命群落。然而，还有很多种微生物由于在实验室环境中不能培养而仍未被了解。因此，对海洋微生物生命的探索，我们仍然处于初期，需要大量的工作去探索它们的可培养性和宏基因组的多样性。

8.2.1 海洋沉淀物

海洋沉淀物是沉积在海底松散的有机和无机粒子。其源于生物活动、风化、大陆侵蚀、火山爆发和海洋内的化学过程。海洋沉淀物中存在的多样的微生物菌落一直被忽视。然而，现在当科学家们评估海洋微生物的多样性及其产生的天然产物的回收时，海洋沉淀物被认为是产孢子微生物的最好来源，如放射菌类。Montalavo 等通过 16S rDNA 菌落分析分离到了独特的放线菌亚类，并建立将其作为与 *Xestospongia* 属相关细菌种群的主要组分。海洋沉淀物中占主导的微生物主要是 *Micromonospora* sp.、*Rhodococcus* sp.、*Streptomyces* sp.

（Maldonado et al.，2005）。然而，利用非培养方法依赖的手段证明，先前对于从海洋环境分离出的放线菌微生物物种多样性差的描述，没有反映出实际的物种的多样性。这说明通过选择性分离方法可以发现新的分类群。Magarvey 与合作者从巴布亚新几内亚沿岸的俾斯麦海和所罗门海的亚潮海洋沉积物分离鉴定了 102 种放线菌。综合生理参数、化学分类特征、16S rRNA 基因序列和系统发育分析都充分地证明其为 *Micromonosporaceae* 属家族的两个新属。

8.2.2 高盐湖

苏打湖跟其他高盐环境之间的主要不同是在于由 pH 值和盐水的离子组成。除苏打湖之外，最值得注意的高盐生物栖息地是大的内陆盐湖（Gilmour，1990；Grant，1993），如大盐湖和死海都是高盐环境。这些湖都在亚热带和热带地区，这些地方的高温和强烈光照导致了更高的蒸发量。当蒸发量超过来源于河水和降水等新鲜水的流入时，就会形成高盐湖。同样，高盐栖息地也会在快速蒸发的海滨自然形成。在相似地区，许多地方在短时间内通过人类在盐田的活动或海岸边的蒸发池形成。依赖于 Ca^{2+} 与 Mg^{2+} 相对含量，这些湖的 pH 值与碱湖非常相似。大盐湖的离子组成与海水类似，呈微碱性（pH 7.7），低 Ca^{2+} 高 Mg^{2+} 含量。另一方面，死海在自然环境下呈弱酸性（pH 5.9~6.3），天然状态下有高含量的 Ca^{2+}（0.4 mol/L）与 Mg^{2+}（0.4 mol/L）。在这种栖息地生长的生物体通常在实验室条件下培养时需要高 Ca^{2+} 和 Mg^{2+} 含量。多年来，高盐环境的微生物多样性的信息一直在稳步增长（Ventosa，1989；Singh et al.，2012）。这些物种有多样性要求并且可以容忍 NaCl 上升至饱和水平。已经从高盐湖沉积物中分离出许多产烷嗜盐菌、发酵嗜盐菌、硫酸还原剂（Nakatsugwa，1991；Lowe et al.，1993）。

通常认为，在饱和 NaCl 含量超高盐环境中占主导的主要是许多嗜盐古细菌，与细菌和放线菌并没有很大关系。16S rRNA 方法已经被用来表征太阳能盐田的结晶池中的极端嗜盐细菌和放线菌（Wang et al.，2012；Todkar et al.，2012）。一种嗜盐细菌从卡尔斯巴德的一种埋在地下超过 1 500 英尺的盐结晶中分离出来，推测其已经存在 2.5 亿年了。从盐洞中提取的古老的孢子可以生长，其细菌像是 *Bacillus* 物种（Vreelan et al.，2000）。一种从中国西部的高盐湖中分离的放射菌类 *Nesterenkonia halotolerans* sp. 可在 20% NaCl 条件下生长（Li et al.，2004）。

8.2.3 太阳能盐田

盐田有很多高盐水，通常有大量嗜盐微生物，主要是嗜盐古菌。然而，其他形式，例如藻类和细菌也同样存在。盐田通常由海水作为最初源头，但有的也利用地下水或者其他水源。这些水分经过一系列盐池而蒸发浓缩，随后获得 NaCl 或其他盐沉积在池塘。太阳能盐田由代表一种从海水到岩盐饱和度浓度梯度，但是具有高密度微生物种群的池塘的盐浓度维持相对稳定。尽管世界上不同地方的盐田表面上相似，但在营养和保水时间上是不同的，这取决于其处于的气候环境。Maturramo 与同事研究了马拉什盐田的微生物多样性，该高盐环境是坐落于秘鲁安第斯山海平面以上 3 380 m。这些盐田有超过 3 000 个小的独立的作为盐结晶器的池塘组成，这些池塘由富含钠和氯化物组成的高盐泉水组成。通过荧光原位杂交检测，16S rRNA 基因克隆库分析和培养技术对栖息于这些盐田的微生物菌落进行了研究。

NaCl 饱和盐水，例如盐田晶体池塘呈现出一种明亮的红色，这是因为大量的有颜色微生物存在（Oren et al.，2002）。海水来源的盐田池塘中的中度嗜盐放线菌菌落，可能大部分是来源于海水的菌落，并且与海洋菌落比较相似。一个支持该观点的事实是，许多海洋微生物有广泛的耐盐性。在过去的 10 年里，许多嗜盐放线菌已被分离，其可以在 20% NaCl 中生长，其中 *Nocardiopsis kunsanensis* 是从韩国的一个太阳能盐田中分离的（Chun et al.，2000）。

8.3 海洋栖息地中的放线菌

不同进化路线中的放线菌研究显示出明显的化学和形态多样性（Goodfellow and Donnell，1989）。它们在各种极端的栖息地中被发现，如南极土壤（Schumann et al.，1997）、沙漠（Dobrovolskaya et al.，1994；Zenova et al.，1996）、盐田旁的富盐土壤（Mehta et al.，2006；Vasavada et al.，2006；Thumar and Singh，2007a，b，2009，2011；Gohel and Singh，2012；Borad et al.，2012）、温泉（Carreto et al.，1996，Bhadreshwara et al.，2012；Borad et al.，2012）。最近，放线菌也在独特的海洋环境中被发现，例如河口（Dhanasekara et al.，2009）、红树林土壤及植物（Hong et al.，2009）、海洋有机集合体和深海汽油水合物储层，在这些地方，放线菌是主要的微生物菌落（Lam，2006）。在墨西哥湾（Lanoil et al.，2001）和近日本（Colwell et al.，2004；Reed et al.，2002）的 Nanki Trough 栖息地的水合沉积物克隆库中，放线菌占 30% ~ 40%。最近，一种名叫 *Amycolatopsis marina* sp. nov. 的放线菌从海洋沉积物中被分离（Bian et al.，2009）。

从瓦登海的海洋有机聚合体中分离出的放线菌发现对微生物菌落有高的拮抗活力。由于海洋聚合体被密集的细菌侵占，其内在独特的相互影响对细菌繁衍至关重要（Grossart et al.，2004），抑制活性反映了每株特异性的模式，即使是一些分离获得的 16S rRNA 基因序列分析发现关系密切的微生物。黄杆菌/鞘脂类（35%）的拮抗活力最低，而放线菌（80%）的最高。

一些新型线菌已经从大堡礁海绵中被分离：*Rhopaloeides odorabile*，*Pseudoceratina clavata*，*Candidaspongia flabellate* 和地中海海绵（Kim et al.，2005）。某些不寻常的属于 *Micrococceae*，*Dermatophilaceae* 和 *Gordoniaceae* 也已经从海绵中分离出（Hill，2004）。海绵中获得的放线菌具有合成新型生物活性代谢物的能力（Hill，2004）。Jensen 等报道了来源于藻

类和海绵样本的放线菌，并描述在样品处理过程中使用的不同方法，获得不同程度复苏的放线菌。

8.4 生长在高盐浓度中的适应性

适应性是有机体维持生命循环和在自然选择中进化的一个特性，它有利于个体的健康和生存。海水中含有3%~4% NaCl，然而，嗜盐有机体可以在这种限制以上生长，一些高度嗜盐菌的最优生长条件几乎是饱和盐溶液（大约35%的盐浓度）。嗜盐菌已进化出对抗细胞内与高盐环境的高渗透压的适应机制。一个有意思的极端嗜盐菌种群利用离子泵主动的维持细胞质内高浓度的 K^+。嗜盐菌生活在约4 mol/L NaCl而有较低的 Mg^{2+}，K^+ 的环境中，它们会主动地积累 K^+ 而泵出 Na^+。Na^+ 的泵出跟ATP依靠 H^+ 梯度的形成来完成。嗜盐菌生物体内参与内部生物过程的酶都具有很高的盐耐受性，并且在低盐的时候会几乎完全失活（Muller and Oren，2003）。

某些海洋放线菌通过积累可溶性溶质维持一个高浓度的内部环境，例如细胞中的糖类、乙醇、甘氨酸、甜菜碱、海藻酶、甘油和氨基酸。在生理pH值下，这些溶质是极性的易溶分子，不带电荷或者是两性离子。其能形成强的水结构，也可从蛋白质水化壳中排除。可溶性溶质通过防止蛋白质在加热或冷冻的条件变性，因而起到稳定剂的效果。大多数嗜盐菌，例如长盐单胞菌 *Halomonas elongata* 和专一性嗜盐古细菌 *Methanohalophilus portucalensis*，从环境中积累甘氨酸甜菜碱（Lai et al.，1999）。然而，某些嗜盐或者中度嗜盐生物能够利用甘氨酸甜菜碱作为碳源和能源（Madern et al.，2004），也将其作为渗透稳定剂（Nyyssola and Leisola，2001）。极端嗜盐性放线菌 *Actinopolyspora halophila* 代表了可以从简单的碳源中产生甜菜碱的异养微生物，它积累细胞内高甜菜碱浓度，在24% NaCl的情况下大约占细胞干重的33%，也可以合成可溶性溶质海藻糖，它占细胞干重的9.7%。甜菜碱浓度随NaCl浓度的上升而上升。然而，海藻糖在最低15%（*w/v*）NaCl浓度时取得最高水平。*A. halophila* 从培养基积累甜菜碱，并且分泌甜菜碱回到培养基中。除了可以合成全新甜菜碱，*A. halophila* 能氧化培养基中胆碱转化成甜菜碱。

8.5 嗜碱放线菌

嗜碱菌是生活在高碱性环境中，如苏打湖和高盐浓度的碱性湖泊（Groth et al.，1997；Jones et al.，1998；Duckworth et al.，1998）。Mikami等首先报道了嗜碱放线菌，接着又报道了关于它们分类地位和应用的研究（Groth et al.，1997；Duckworth et al.，1998；）嗜碱放线菌能够产生大量碱性酶和生物活性物质（Gohel and Singh，2012a；Horikoshi，1999），如抗生素（Gulve and Deshmukh，2012；Kumar et al.，2011；Thumar et al.，2010；Olano et al.，2009；Imada et al.，2007；Fiedler et al.，2005）。这些生物能产生

一些经典的适用于工业应用的代谢物。尽管有一些关于嗜碱微生物的生理机能和活性的报道（Krulwich et al.，2001；Yumoto，2002），但几乎没有对嗜碱放线菌的报道。因此，在微生物资源中需要更多地探究关于嗜碱放线菌的发现、多样性、生理机能和适应机制。

8.6 微生物酶和海洋环境

我们主要研究海洋酶的代谢作用，然而这些研究只关注了海洋酶的有限方面。海洋微生物积极地利用它们的代谢途径，分解和矿化复杂的有机体复合物，同时产生次级代谢产物，海洋环境中的复杂大分子，包括多糖、蛋白质、纤维素、木质素、胶质、木聚糖、淀粉、果糖或尿素。海洋微生物在分解这些有机体复合物的作用远大于其他分解体。

8.7 海洋放线菌中的酶

寻找海洋酶可以发现其具有独特特性的新型生物催化剂，例如高耐盐性、热稳定性、嗜压特性和冷适应性的酶。据近期一些报道，海洋生态系统的巨大多样性现状可为有用的生物催化剂开发提供潜在的自然储存库（Naveena et al.，2012；Senthil et al.，2012）。海洋生物催化剂的新型化学和立体化学特性将会增加它们的研究价值。

随着生物科技的进步，关键酶的种类和应用领域已经从传统工业领域过渡到一些新领域，如临床、药物和分析化学。对某些经典酶的持续需求要求寻找这些酶的新来源，例如淀粉酶和蛋白酶。目前工业酶的全球化市场约20亿美元（Godfrey and West，1996）。尽管蛋白酶和淀粉酶都是来源于霉菌和真菌，但最近也有研究来探讨利用放线菌来生产这些酶。许多放线菌的成员，例如 *Streptomycetes*，*Nocardiopsis*，*Thermomonospora* 和 *Thermoactinomyces* 已经被研究用于淀粉酶和蛋白酶的生产（Gohel and Singh，2012a；Andrews and Ward，1988；Bakhtiar et al.，2003）。来自海洋栖息地的 *Streptomycetes* sp.，和 *Nocardiopsis* sp.，分泌的细胞外的蛋白酶最近已经被深入研究，研究涉及酶的纯化，酶稳定性的热力学分析，克隆以及在常温寄主的基因克隆与表达（Gohel and Singh，2012a，b）。

8.7.1 蛋白酶

蛋白酶是有广泛应用的蛋白水解酶，例如废物处理、生物修复、木材质量改进、肉制品的嫩化作用、食品工业（Ara et al.，2012；Guravaiah et al.，2012）、皮革制品、药物和洗涤剂工业（Pandey and Singh，2012；Gulve and Deshmukh，2011；Najafi et al；2005）。因此，蛋白酶是重要的工业酶，约占全球酶总销量的60%。在不同来源的蛋白酶中，细菌来源的比动物和霉菌占有更大比重，约占20%世界市场（Dodia et al.，2008；Moreira et al.，2002；Ward，1985）。细菌碱性蛋白酶的主要应用是在配方洗涤剂上（Boguslawski and

Shultz, 1992; Wolff et al., 1996; Kwon et al., 1998)。碱性、嗜盐、碱性嗜盐和嗜热蛋白酶是最受欢迎的，因为它们的高活性、高稳定性和快速的反应性。适合于多种应用的碱性蛋白酶现已从多种海洋栖息地分离的嗜盐碱性细菌和放线菌被发现（Gohel and Singh, 2012a, b; Pandey et al., 2012; Purohit and Singh, 2011; Halpern, 1981; Hagiwara et al., 1958）。然而，总的来说，来源于放线菌中蛋白酶的信息很少，如 *Nocardiopsis* sp.（Gohel et al., 2012b; Kim, et al., 1993）*Streptomyces albidflarithavu*（Mohamed et al., 2011）和 *Streptomyces carpaticus*（Haritha et al., 2012）。

链球菌是生产蛋白水解酶优良菌种（Rao et al., 1998）。放线菌的细胞外蛋白酶可能参与了细胞外蛋白质的氮源吸收（Shapiro, 1989）。Jain 等已经研究了 305 种放线菌的蛋白水解活力，并且报道这些放线菌在工业废物处理上的重要性。根据化学分类学、培养、形态学，一株被鉴定为 *Streptomyes albidflarithavus* 的海洋放线菌，当培养在酪蛋白和蛋白胨的混合物中时会产生碱性蛋白酶（Shafei et al., 2010）。同样，也研究了不同碳源和氮源条件下，来自一种海洋碱性放线菌 MA1-1 可以产生碱性蛋白酶（Hames and Uzel, 2007）。一个丝氨酸蛋白酶在 *Streptomyces peucetius* 平台生长期的后期产生，说明了其对固态底物的细胞转运作用（Gibb and Strohl, 1988）。*Streptomyces clavuligerus* 中碱性蛋白酶的制备和纯化已被研究（Moreira et al., 2002），同时，一个来自碱性放线菌的分泌的细胞外丝氨酸蛋白酶也被报道（Mehta et al., 2006）。

8.7.2 淀粉酶

淀粉酶是最重要的酶之一并且在众多生物科技中有着举足轻重的应用（Pandey and Singh, 2012; Kikani and Singh, 2011; Kikani et al., 2010; Gupta et al., 2003）。其在淀粉降解中起到重要作用，并且在世界工业酶市场中占到 25%~33%（Saxena et al., 2000）。淀粉酶在许多领域有着广泛的应用，例如食品、淀粉糖化、纺织品、烘焙、酿造和蒸馏工业（Pandey et al., 2000）。随着海洋科技的发展，研究报道了许多海洋栖息地微生物具有产生淀粉酶的能力（Chakraborty et al., 2011; Rao et al., 2009）：从太平洋的深海沉积物中分离获得了一株海洋酵母（Li et al., 2007），*Aureobasidium pullulans* N13d，它可以产生一种细胞外的淀粉酶；一株被命名为 *Bacillus amyloliquificiences* 的嗜热细菌从热泉中分离出来，它可以产生在广泛 pH 值条件下高耐热性 α-淀粉酶（Kikani and Singh, 2011）；一种新型的来自海洋 *Streptomyces* sp. D1 的 α-淀粉酶最近被报道（Chakraborty et al., 2009）。终产物分析说明其有 α-1，4 与 α-1，6 的水解活性。当盐缺少时，这种酶的活性大大降低，在 7%（*w/v*）和 10%（*w/v*）NaCl 浓度时，酶仍然保持 100%和 50%的活性。在 pH 值 7~11 范围内，淀粉酶可以稳定的存在 48 h。这是首次报道链球菌酶可适应这么大范围的 pH 值条件（Trincone, 2011）。对其多样性、抗菌活性、酶的制备和其他高附加值分子兴趣的增加推动了对新栖息地的探索。

8.7.3 木聚糖酶

木聚糖是被子植物细胞壁半纤维素的主要成分，可能是植物中第二大丰富的碳水化合物（Timell, 1967）。木聚糖是一种杂多糖，其主链由 D-吡喃木糖链接组成（Whistler and Richards, 1970）。木聚糖酶在不同植物之间，不同组织之间都不尽相同（Jeffries, 1990）。可去除侧链的阿拉伯呋喃糖酶、乙酰酯酶（Biely et al., 1985）、甲基化葡萄糖苷酸酶（MacKenzie et al., 1987; Puls et al., 1987）和阿魏酸酯酶（Hatfield et al., 1991）都参与了木聚糖酶对低聚木聚糖的作用。降解木聚糖酶已被关注，并且其也有一些新的应用，包括牛皮纸浆的预漂白和发酵果糖的回收（Viikari et al., 1986）。

8.7.4 纤维素酶

纤维素是生物圈中固定碳最丰富的形式。它不溶于水，只能被特定的能产生多种多样的纤维素酶的微生物群体降解（Lynd et al., 2002）。许多微生物降解非晶体纤维素，但只有一部分能够完全降解高结晶度纤维素。在陆生环境中，纤维素高度木质化，这增加了降解难度（McCarthy, 1987）。真菌和放线菌由于其菌丝生长能够伸入到木质部中的纤维素。与陆地微生物相比，水生生态系统中只有部分群体可以参与纤维素降解（Arunachalam et al., 2010）。在水生放线菌中，小单孢子菌属 *Micromonosporas* 通常大量出现在土壤中，但是能很好地适应水扩散，因而容易从水生环境中复苏，特别是湖底沉积物中（Cross, 1981）。它们因为生长的慢而在培养研究中是很难被发现的（Wohl and McArthur, 1998）。然而，小单孢子菌属降解复杂多糖的能力是很常见的，例如纤维素和甲壳质（Kawamoto, 1984; Park et al., 1993; Damude et al., 1993）。来源于嗜碱 *Streptomyces* 家族中 KSM-9 的碱性纤维素酶已被报道（Dasilva et al., 1993）。类似地，来源于嗜碱链球菌 *Streptomyces* sp. 家族中 S36-2 的一种纤维素酶被发现在 pH 值 8~9 范围内有活性。Singh 等（2004）报道了 *Bacillus sphaericus* JS1 的碱性纤维素酶的制备和表征。从非洲碱湖中分离出的新型嗜碱链球菌中内切纤维素酶的基因克隆和表达增加了分子水平对海洋放线菌纤维素酶的认识（Van et al., 2001）。

8.7.5 脂肪酶

脂肪酶可以水解脂肪和石油（triacylglycerol acylhydrolases E. C. 3. 1. 1. 3），从而释放游离脂肪酸、甘油二酯、单酰甘油和丙三醇。除此之外，脂肪酶也参与酯化作用、酯基转移作用和氨解作用，它具有相当重要的生理功能和工业应用潜能（Schmid and Verger, 1998; Babu et al., 2008）。当培养在疏水培养基上时，不同的细菌、酵母和真菌能分泌脂肪酶。细菌脂肪酶在有机溶剂中表现出高易变性、反应性和稳定性（Haba et al., 2000; Gao et al., 2000）。来源于 *Pseudomonas* sp. 和其他石油降解菌的脂肪酶已经被深入研究（Deb et al.，2006）。在这种背景下，放线菌的研究却被忽略。当然也有一些来自 SAP1089 *Streptomyces*（Jain et al.，2003）和 Z94-2 *Streptomyces* 脂肪酶（Zhou et al.，2000）的研究。Lescic 跟他的同事（2004）对来源于 *Streptomyces rimosus* 的细胞外脂肪酶进行了结构特性表征，并且利用质谱方法证明了二硫键的存在方式。一种新型的来源于 *Streptomyces rimosus* R6-554W 中的细胞外碱性脂肪酶在完成基因克隆、序列和表达后，开展了其纯化和生化特性研究。该脂肪酶与数据库中其他脂肪酶的氨基酸相似性低。通过一种高拷贝的载体，脂肪酶基因转移到一种脂肪酶缺陷的 *Streptomyces rimosus* 家族中，其表达量高于原始菌株 22 倍（Vujaklija et al.，2002）。

8.7.6 甲壳质酶

甲壳质是 N-乙酰氨基葡萄糖组成的不溶性线性 β-1，4 聚合物（Glc NAc），是自然界中第二大丰富的聚合物。这种多聚糖出现在真菌的细胞壁以及昆虫和甲壳纲动物的外骨骼中。甲壳质酶扮演着重要的生理学和生态学角色（EC 3.2.1.14），它可以被多种生物分泌产生，包括病毒、细菌、高等植物和动物（Gooday，1990）。甲壳质酶水解 β-1，4 链，主要产生二乙酰壳寡糖，再进一步通过 N-乙酰葡糖氨糖苷酶降解成 GlcNAc 单体。

在放线菌家族内，*Streptomyces* 是主要的甲壳质分解菌株。来源于多种 *Streptomyces* 菌种的甲壳质酶已经被纯化和表征，且基因也被克隆和测序，已经有多种甲壳质酶基因从 *Streptomyces* sp. 中被克隆（Gupta et al.，1995）。然而，迄今没有文献报道其他种群来源的甲壳质酶的类似的分子克隆和分析工作。Tsujibo 等（1992）报道了从嗜碱放线菌（一种 *Nocardiopsis albus* sub sp. *prasina* OPC-131）中分离出了两种甲壳质酶。A、B 两种甲壳质酶的最适 pH 值分别是 8 和 7。甲壳质酶基因也在一些研究中被表征（Tsujibo et al.，2003）。表观分子量 55 kDa 的细胞外甲壳质酶从 *Streptomyces halstedii* AJ-7（Joo and Chang，2005）*Streptomyces* sp. 中被鉴定和纯化（Kim et al.，2003）。早年间，通过离子交换色谱法纯化了一种 *Streptomyces* M-20 的甲壳质酶。甲壳质酶 C，第一个来源于细菌的 19 家族的甲壳质酶从 *Streptomyces griseus* HUT6037 中被发现（Itoh et al.，2002），与来源于植物的 19 家族的甲壳质酶在催化结构域上有很大的相似性。然而，N-末端甲壳质结合区的功能分析确认了新型 19-家族甲壳质酶的关联性。来源于 *Streptomyces thermoviolaceus* OPC-520 链球菌中的热稳定性甲壳质酶（Chi，40）的过表达、纯化和特性研究也已经开展（Christodoulou et al.，2001）。

8.7.7 琼胶水解酶

琼脂是一种高异源化的杂多聚糖。中性琼脂是一种 D-半乳糖和 3，6-酐-L-半乳糖交替出现的聚合物，通过交替的 β-4→1 和 α-1→3 键的链接。琼脂广泛应用在食品工业中，例如饮料、面包和低卡路里食品的生产。在日本，琼脂寡糖作为一种保湿化妆品添加剂和头发保养剂使用（John and John，1981；Oren，2004；Rasmussen and Morrissey，2007）。琼脂的酸降解正逐步被酶解法代替。琼脂水解酶是在琼胶降解微生物中被发现（Parro and Mellado，1994）。琼胶降解微生物分为两种：一种软化琼胶的细菌；另一种是直接溶解琼脂的细菌。琼脂水解酶是在下面的微生物中被发现：*Cytophaga*，*Bacillus*，*Vibrio*，*Alteromonas*，*Pseudoalteromonas* 和 *Streptomyces*（Aoki et al.，1990；Leon et al.，1992；Hosoda et al.，2003）。

8.7.8 DNA 聚合酶

一种热稳定性的 DNA 聚合酶从 *Thermotoga maritima* 中获得，它是可以在 55~90℃之间生长的一种超嗜热真菌。*Thermotoga* 的 3 个菌株从意大利和亚马孙的地热海底被分离出来。这种细菌最初是从地热海洋沉积物和热泉中分离出来。Slater 等获得了一个来源于超嗜热真菌-*Thermotoga neapolitana*（Tne）的热稳定性 DNA 聚合酶的美国专利。目标基因克隆到了大肠杆菌菌株。Tne 聚合酶通过含有 Tne 表达载体而产生。IPTG 诱导后，规模化培养在 37℃下 5 h。然而，这个过程中的最大缺点是 IPTG 价格昂贵。

8.8 结论

海洋生态系统中巨大的生物多样性为新型又有价值的生物催化剂提供了巨大的资源库。海洋酶被认为拥有与栖息地有关的特性，例如耐盐性、热稳定性、亲水性和冷适应性。在这篇文献中讨论了与栖息地有关的特性。这些生物催化剂的新的化学和立体化学特点被重点介绍，它们来源于海洋生物，并且在化学和制药工业中具有重要应用前景。海洋酶有关的文献和专利显示了它们巨大潜能依然未被开发。目前，只有小部分海洋栖息地的微生物被研究。培养极端微生物和制备其新型分子产物是需要重点解决的问题。此外，海洋放线菌中基因的异源克隆与表达的研究将增加对基于酶的生物转化的发展。

致谢

作者感谢印度拨款委员会（UGC）新德里，印度以及索拉什特拉大学，拉杰果德，印度的财政和基础设施的支持。JT 感谢 M & N. Virani 理学院，拉杰果德，印度，并感谢 DBT，新德里印度的财政支持。SDG 和 AS 非常感谢 UGC 新德里，印度，授予的优秀奖（Meritorious Fellowships）。BK 感谢科学与工业研究委员会，新德里，印度授予的高级研究成果奖。KD 作为女性科学家感谢 DST 新德里，印度的财政支持。

参考文献

Andrews，L. and Ward，J.（1988）Extracellular amylases from *Streptomyces aureofaciens*-purification and properties. *Starch/Starke* 30, 338-341.

Antranikian, G., Vorgias, C. E. and Bertoldo, C. (2005) Extreme environments as a resource for microorganisms and novel biocatalysts. *Adv. Biochem. Eng. Biotechnol.* 96, 219-262.

Aoki, T., Araki, T. and Kitamikado, M. (1990) Purification and characterization of a novel β-agarase from *Vibrio* sp. AP-2. *Eur. J. Biochem.* 187, 461-465.

Arunachalam, R., Wesely, E. G., George, J. and Annadurai, G. (2010) Novel approaches for identification of *Streptomyces noboritoensis* TBG-V20 with cellulase production. Curr. Res. Bacteriol. 3, 15-26.

Babu, J., Pramod, W. R. and George, T. (2008) Cold active microbial lipases: some hot issues and recent developments. *Biotechnol. Adv.* 26, 457-470.

Bakhtiar, S. J., Vevodova, R., Hatti-Kaul R. and Su, X. D. (2003) Crystallization and preliminary X-ray analysis of an alkaline serine protease from *Nesterenkonia* sp. *Acta Cryst.* 59, 529-531.

Bhadreshwara, A., Hedpara, R., Kanjariya, H., Shukla, S. and Singh, S. P. (2012) Optimization of xylanase enzyme production from thermophilic/ thermo tolerant actinomycetes TSI14 isolated from hot spring reservoir Tulsi Shyam, Gujarat, India. M. Sc. dissertation thesis, Saurashtra University, Rajkot, India.

Bian, J., Li, Y., Wang, J., Song, F. H., Liu, M., et al., (2009) *Amycolatopsis marina* sp. nov., an actinomycete isolated from an ocean sediment. *Int. J. Syst. Evol. Microbiol.* 59, 477-481.

Biely, P., Puls, J. and Schneider, H. (1985) Acetyl xylan esterases in fungal cellulolytic systems. *FEBS Lett.* 186, 80-84.

Boguslawski, G. and Shultz, J. W. (1992) US Patent No. 5 (118): 623.

Borad, P., Agravat, D., Sindhav, P., Makwana, P., Shukla, S. and Singh, S. P. (2012) Statistical optimization of amylase enzyme production from thermophilic/ thermo tolerant actinomycetes TSI14 isolated from hot spring reservoir Tulsi Shyam, Gujarat, India. M. Sc. dissertation thesis, Saurashtra University, Rajkot, India.

Carreto, L., Moore, E., Nobre, M. F., Wait, R., Riley P. W., et al., (1996) *Rubrobacter nylanophilus* sp. nov. a new Thermophilic species isolated from a thermally polluted effluent. *Int. J. Syst. Bacteriol.* 46, 460-465.

Chakraborty, S., Khopade, A., Biao, R., Jian, W., Liu, X., et al., (2011) Characterization and stability studies on surfactant, detergent and oxidant stable α-amylase from marine haloalkaliphilic Saccharopolyspora sp. A9. *J. Mol. Cat. B: Enz.* 68 (1), 52-58.

Chakrabortya, S., Khopadea, A., Kokarea, C., Mahadika, K. and Chopadeb, B. (2009) Isolation and characterization of novel α-amylase from marine *Streptomyces* sp. Dl. *Mol. Cat. B: Enz.* 58, 17-23.

Christodoulou, E., Duffner, F. and Vorgias, E. (2001) Overexpression, purification, and characterization of a thermostable chitinase (Chi40) from *Streptomyces themtoviolaceus* OPC-520. *Prot. Expression and Purification* 23, 97-105.

Chun, J., Bae, K. S., Moon, E. Y., Jung, S. O., Lee, H. K. and Kim, S. J. (2000) *Nocardiopsis kunsanensis* sp. nov., a moderately halophilic actinomycete isolated from a saltern. *Int. J. Syst. Evol. Microbiol.* 50 (5), 1909-1913.

Chun, J., Lee, J. H., Jung, Y., Kin, M., Kim, S., et al., (2007) EzTaxon: a web-based tool for the identification of prokaryotes based on 16S ribosomal RNA gene sequences. *Int. J. Syst. Evol. Microbiol.* 57, 2259-2261.

Colwell, F., Matsumoto, R. and Reed, D. (2004) A review of gas hydrates, geology, and biology of Nankai Trough. *Chem. Geol.* 205, 391-404.

Cross, T. (1981) Aquatic actinomycetes-a critical survey of the occurrence, growth and role of actinomycetes in aquatic habitats. *J. Appl. Bacteriol.* 50, 397-423.

Damude, H. G., Gilkes, N. R., Kilburn, D. G., Mille, R. C., Antony, R. and Warren, J. (1993) Endoglucanase CasA from alkalophilic Streptomyces strain KSM-9 is a typical member of family B of β-1, 4-glucanases. *Gene* 123, 105-107.

Dasilva, R., Yim, D. K., Asquieri, E. R. and Park, Y. K. (1993) Production of microbial alkaline cellulase and studies of their characteristics. *Rev. Microbiol.* 24, 269-274.

Deb, C., Daniel, J., Sirakova, T. D., Abomoelak, B., Dubey, V. S. and Kolattukudy, P. E. (2006) A novel lipase belonging to the hormone-sensitive lipase family induced under starvation to utilize stored triacylglycerol in *Mycobacterium tuberculosis*. *J. Biol. Ghent.* 281 (7), 3866-3875.

Debashish, G., Malay, S., Barindra, S. and Joydeep, M. (2005) Marine enzymes. *Adv. Biochem. Eng. Biotechnol.* 96, 189-218.

Demirjian, D. C., Moris-Varas, F. and Cassidy, C. S. (2001) Enzymes from extremophiles. *Curr. Opin. Chem. Biol.* 5, 144-151.

Dhanasekara, D., Selvamani, S., Panneerselvam, A. and Thajuddin, N. (2009) Isolation and characterization of actinomycetes in Vellar Estury, Annagkoil, Tamilnadu. *Afric. J. Biotechnol.* 8, 4159-4162.

Dobrovolskaya, T. G., Chernov, I. Y., Lysak, L. V., Zenova, G. M., Gracheva, T. A. and Zvyagintsev, D. G. (1994) Bacterial communities of the Kyzyl Kum desert: spatial distribution and taxonomic composition. *Microbiologica* 63, 188-192.

Dodia, M. S., Bhimani, H. G., Rawal, C. M., Joshi, R. H. and Singh, S. P. (2008) Salt dependent resistance against chemical denaturation of alkaline protease from a newly isolated haloalkaliphilic Bacillus sp. *Bioresource Technol.* 99, 6223-6227.

Duckworth, A. W., Grant, S., Grant, W. D., Jones, B. E. and Meijer, D. (1998) *Dietzia Natronolimnaios* sp. nov., a new member of the genus Dietzia isolated from an East African soda lake. *Extremophiles* 2, 359-366.

Ferrer, M., Golyshina, O., Beloqui, A. and Golyshin, P. N. (2007) Mining enzymes from extreme environments. *Curr. Opin. Microbiol.* 10, 207-214.

Fiedler, H. P., Bruntner, C., Bull A. T., Ward, A. C., Goodfellow, M., et al., (2005) Marine actinomycetes as a source of novel secondary metabolites. *Antoni Van Leeuwenhoek* 87, 37-42.

Gao, X. G., Cao, S. G. and Zhang, K. (2000) Production, properties and application to non aqueous enzymatic catalysis of lipase from a newly isolated *Pseudomonas* strain. *Enzym. Microb. Technol.* 27, 74-82.

Gibb, G. D. and Strohl, W. R. (1988) Physiological regulation of protease activity in *Streptomyces peucetius*. *Can. J. Microbiol.* 34, 187-190.

Gilmour, D. (1990) 'Halotolerant and halophilic microorganisms'. In *Microbiology of Extreme Environments*, Edwards, C. (ed.). Milton Keynes; Open University Press, pp. 147-177.

Godfrey, T. and West, S. (1996) *Industrial Enzymology*. McMillan Publishers Inc., New York.

Gohel, S. and Singh, S. P. (2012a) Cloning and expression of alkaline protease genes from two salt-tolerant alkaliphilic actinomycetes in E. *coli*. *IJBIOMAC* 50, 664-671.

Gohel, S. and Singh, S. P. (2012b) Single step purification, characteristics and thermodynamic analysis of a highly thermostable alkaline protease from a salt-tolerant alkaliph*ilic actinomycete*, *Nocardiopsis alba OK*-5. *J. Chrom. B* 889-890, 61-68.

Gooday, G. W. (1990) The ecology of chitin decomposition. *Adv. Microb. Ecol.* 11, 387-430.

Goodfellow, M. and O'Donnell, A. G. (1989) 'Search and discovery of significant actinomycetes'. In *Microbial Products*, Baumberg, S., Hunter, I. and Rhodes, M. C. (eds). New approach 44 the symposium of the society for General Microbiology, 343-383.

Grant, W. D. (1993) 'Hypersaline environments'. In *Trends in Microbial Ecology*, Guerrero, R. and Pedros-Alio, C. (Eds). Socieded Espanola de Microbiologica (Spanish Society for Microbiology, pp. 13-17.

Grossart, H. P., Schlingloff, A., Bernhard, M., Simon, M. and Brinkhoff, T. (2004) Antagonistic activity of bacteria isolated from organic aggregates of the German Wadden Sea. *FEMS Microbiol. Ecol.* 47, 387-396.

Groth, I., Schumann, P., Rainey, F. A., Martin, K., Schuetze, B. and Augsten, K. (1997) *Bogoriella caseilytica* gen. nov., sp. nov., a New Alkaliphilic Actinomycete from a Soda Lake in Africa, *Int. J. Syst. Bacteriol.* 47, 788-794.

Gulve, R. M. and Deshmukh, A. M. (2011) Enzymatic activity of actinomycetes isolated from marine sediments. *Recent Res. Sci. Technol.* 3 (5), 80-83.

Gulve, R. M. and Deshmukh, A. M. (2012) Antimicrobial activity of the marine actinomycetes. *Int. Multidisciplin. Res. J.* 2 (3), 16-22.

Gupta, R., Gigras, P., Mohapatra, H., Goswami, V. K. and Chauhan, B. (2003) Microbial #αOγα-amylases: a biotechnological perspective. *Process Biochem.* 38, 1599-1616.

Gupta, R., Saxena, R. K., Chaturvedi, P. and Virdi, J. S. (1995) Chitinase production by Streptomyces viridificans: it's potential on fungal cell wall lysis. *J. Appl. Bacteriol.* 78, 378-383.

Haba, E., Bresco, D., Ferror, C., Margues, A., Busquets, M. and Manresa, A. (2000) Isolation of lipase-secreting bacteria by deploying used frying oil as selective substrate. *Enzym. Microb. Technol.* 26, 40-44.

Hagiwarara B., Matsubara, H., Nakai, M. and Okunuki, K. J. (1958) Crystalline proteinase. I. Preparation of crystalline proteinase of B*acillus subtilis*. *J. Biocbem.* 45, 188.

Halpern, M. G. (1981) Industrial enzymes from microbial sources. *Recent Advances* 51-75.

Hames, K. E. and Uzel, A. (2007) Alkaline protease production by an actinomycete MA1-1 isolated from marine sediments. *Annals of Microbiol.* 57 (1), 71-75.

Haritha, R., Sivakumar, K., Swati, A., Jagan Mohan, Y. S. and Ramana, T. (2012) Characterization of marine *Streptomyces carpaticus*. And optimization of onditions for production of extracellular protease. *Microbiol. J.* 2 (1), 23-35.

Hatfield, R-D., Helm, R. F. and Ralph, J. (1991) Synthesis of methyl 5-O-trans-feruloyl-alfa-L-arabinofuranoside and its use as a substrate to assess feruloy esterase activity. *Anal. Biochem.* 194, 25-33.

Hill R. T. (2004) 'Microbes from marine sponges: a trove of biodiversity of natural products discovery'. In *Microbial Diversity and Bioprospecting*, Bull, A. T. (ed.). Washington: ASM Press, pp. 177-190.

Hong, K., Gao, A. H., Xie, Q. Y., Gao, H., Zhuang, L., et al., (2009) Actinomycetes for marine drug discovery isolated from mangrove soils and plants in China. *Mar. Drugs* 7, 24-44.

Horikoshi, K. (1999) Microbiol. Alkaliphiles: some applications of their products for biotechnology. *Mol. Biol.*

Rev. 63, 4735–4750.

Hosoda, A., Sakai, M. and Kanazawa, S. (2003) Isolation and characterization of agar-degrading *Paenibacillus* spp. associated with the rhizosphere of spinach. *Biosci. Biotechnol. Biochem.* 67, 1048–1055.

Imada, C., Koseki, N., Kamata, M., Kobayashi, T. and Hamada-Sato, N. (2007) Isolation and characterization of antibacterial substances produced by marine actinomycetes in the presence of seawater. *Actinomycetologica* 21, 27–31.

Itoh, Y., Kawase, T., Nikaidou, N., Fukada, H., Mitsutomi, M. and Watanabe, T. (2002) Functional analysis of the chitin-binding domain of a family 19 chitinase from *Streptomyces griseus* HUT6037 substrate-bonmding affinity and cis-dominant increase of antifungal function. *Biosci. Biotechnol. Biochem.* 66 (5), 1084–1092.

Jain, P. K., Jain, R. and Jain, P. C. (2003) Production of industrially important enzymes by some actinomycetes producing antifungal compounds. *Hindustan Antibiot. Bull.* 46 (1–4), 29–33.

Jeffries, T. W. (1990) Biodegradation of lignin-carbohydrate complexes. *Biodegradation* 1, 163–176.

Jensen, P. R., Gontang, E., Mafnas, C., Mincer, T. J. and Fenical, W. (2005) Culturable marine actinomycete diversity from tropical Pacific Ocean sediments. Environ. *Microbiol.* 7, 1039–1048.

John, N. C. W. and John, R. E. (1981) The agar component of the red seaweed *Gelidium purpurascens. Phytochemistry* 20, 237–240.

Jones, B. E., Grant, W. D., Duckworth, A. W. and Owenson, G. G. (1998) Microbial diversity of soda lakes. *Extremophiles* 2, 191–200.

Joo, H. S. and Chang, C. S. (2005) Oxidant and SDS stable alkaline protease from a halo-tolerant *Bacillus clausii* I–52: enhanced production and simple Purification. J. *Appl. Microbiol.* 98 (2), 491–497.

Kawamoto, L (1984) 'Genus Micromonospora'. In *Bergey's Manual of Systemati Bacteriology*, vol. 4, Williams, S. T., Sharpe, M. E. and Holt, J. G. (eds), 2442. 2450. Williams & Wilkins, Baltimore, M. D.

Kikani, B. and Singh, S. P. (2011) Single step purification and characterization of a highly thermostable and calcium independent α–amylase from *bacillus amyloliquifaciens* TSWK1–1 isolated from a Hot spring Reservoir, Gujarat (India). *IJBIOMAC* 48, 676–681.

Kikani, B. A., Shukla, R. J. and Singh, S. R (2010) 'Biocatalytic potential of thermophilic bacteria and actinomycetes'. In *Current Research, Technology and Education topics in applied microbiology and microbial biotechnology*, Mendez. V. (ed.). Formatex Publishers, Spain, pp. 1000–1007.

Kim, K. J., Yang, Y. J. and Kim, J. G. (2003) Purification and characterization of chitinase from *Streptomyces* sp. M–20. *J. Biochem. Mol. Biol.* 36 (2), 185–189.

Kim, M. J., Chung, H. S. and Park, S. J. (1993) Properties of alkaline protease isolated from *Nocardiopsis dassonvillei. Korean Biochem. J.* 26, 81–85.

Kim, T. K., Garson, M. J. and Fuerst, J. A. (2005) Marine actinomycetes related to the Salinospora group from the Great Barrier Reef sponge *Pseudoceratina clavata. Environ. Microbiol.* 7, 509–518.

Krulwich, T. A., Masahiro, I. and Guffanti, A. A. (2001) The Na^+-dependence of alkaliphily in *Bacillus. Biochimica et Biophysica Acta (BBA) -Bioenergetics* 1505, 158–168.

Kumar, K. S., Haritha, R., Mohan, J. and Ramana, T. (2011) Screening of marine actinobacteria for antimicrobial compounds. *Res. J. Microbiol.* 6, 385–393.

Kwon, S. T. , Terada, T. , Matsuzawa, H. and Ohta, T. (1988) Determination of the positions of the disulfide bonds in Aqualysin I (a thermophilic alkaline serine protease) of *Thermus aquaticus YT*-1. *J. Biochem.* 173, 491.

Lai, M. C. , Yang, D. R. and Chuang, M. J. (1999) Regulatory factors associated with synthesis of the osmoly te glycine betaine in the halophilic methanoarchaeon *Methanohalophilus portucalensis. Appl. Environ. Microbiol.* 65, 828-833.

Lam, K. S. (2006) Discovery of novel metabolites from marine actinomycetes. *Curr. Opinion Microbiol.* 9 (3), 245-251.

Lanoil, B. D. , Sassen, R. , La Due, M. T. , Sweet, S. T. and Nealson, K. H. (2001) Bacteria and archaea physically associated with Gulf of Mexico gas hydrates. *Appl. Environ. Microbiol.* 67, 5143-5153.

Leon, O. , Quintana, L. , Peruzzo, G. and Slebe, J. C. (1992) Purification and properties of an extracellular agarase from *Alteromonas* sp. *strain C-l. Appl. Environ. Microbiol.* 58, 4060-4063.

Lescic, I. , Zehl, M. , Muller, R. , Vukelic, B. , Abramic, M. , et al. , (2004) Structural characterization of extracellular lipase from Streptomyces rimosus: assignment of disulfide bridge pattern by mass spectrometry. *Biol. Chem.* 385 (12), 1147-1156.

Li, H. F. , Chi, Z. M. , Wang, X. H. and Ma, C. I. (2007) Amylase production by the marine yeast *Aureobasidium pullulans N*13*d. J. Ocean Univ. Chin.* 6, 61-66.

Li, W. J. , Chen, H. H. , Zhang, Y. Q. , Schumann, P. , Xu, L. H. and Jiang, C. L. (2004) *Nesterenkonia halotolerans* sp. nov. and *Nesterenkonia xinjiangensis* sp. nov. , actinobacteria from saline soils in the west of China. *Int. J, Syst Evol. Microbiol.* 54, 837-841.

Lowe, S. E Jain, M. K. and Zeikus, J. G. (1993) Biology, ecology and biotechnological application of anaerobic bacteria adapted to environmental stresses in temperature, pH, sflbnity or substrates. *Microbiol. Rev.* 57, 451-509.

Lozupine, A. C. and Knight, R. (2007) Global patterns in bacterial diversity. *PANS* 104 (27), 11436-11440.

Lynd, L. R. , Weimer, P. J. , Zyl, W. H. V. and Pretorius, I. S. (2002) Microbial cellulose utilization: fundamentals and biotechnology. *Microbiol. Mol. Biol. Rev.* 66, 506-577.

MacKenzie, C. R. , Bilous, D. , Schneider, H. and Johnson, K. G. (1987) Induction of cellulolytic and xylanolytic enzyme systems in *Streptomyces* spp. *Appl. Environ. Microbiol.* 53, 2835-2839.

Madem, D. , Camacho, M. , Rodriguez-Arnedo, A. , Bonete, M. J. and Zaccai, G. (2004) Salt-dependent studies of NADP-dependent isocitrate dehydrogense from the halophilic archaeon *Haloferax volcanii. Extremophiles* 8, 377-384.

Magarvey, N. A. , Keller, J. M. , Bernan, V. , Dworkin, M. and Sherman, D. H. (2004) Isolation and characterization of novel marine-derived actinomycete taxa rich in bioactive metabolites. *Appl. Environ. Microbiol.* 70 (12), 7520-7529.

Maldonado, L. A. , Starch, J. E. , Pathom-aree, W. , Ward, A. C. , Bull, A. T. and Goodfellow, M. (2005) Diversity of cultivabe actinobacterial in geographically widespread marine sediments. *Antonie Van Leeuwenhoek* 87 (1), 11-18.

Maturrano, L. , Santos, F. , Rossello-Mora, R. and Anton, J. (2006) Microbial diversity in Maras salterns, a hypersaline environment in the peruvian andes. *Appl. Environ. Microbiol.* 72 (6), 3887-3895.

McCarthy, A. J. (1987) Lignocellulose-degradingactinomycetes. *FEMS Microbiol. Rev.* 46, 145-163.

Mehta, V. J., Thumar J. T. and Singh S. P. (2006) Production of alkaline protease from an alkaliphilic actinomycete. *Bioresource Technol.* 97 (14), 1650-1654.

Mikami, Y., Miyashita, K. and Arai, T. (1986) Alkaliphilic actinomycetes. *Actinomycetes* 19, 176-191.

Mohamed, S., Abdel-Aziz, M. S., Hanaa, A., El-Shafei, Mohamed, F. G. and Hamed, A. A. (2011) Alkaline protease from marine *Streptomyces albidflarithavus* and its probable applications. *J. App. Sci. Res.* 7 (6), 897-906.

Montalavo, N. F., Mohamed, N. M., Enticknap, J. J. and Hill, R. T. (2005) Novel actinobacterial from marine sponges. *Antonie Van Leeuwenhoek* 87 (1), 29-36.

Moreira, K. A., Albuquerque, B. F., Teixeira, M. F. S., Porto, H. L. F. and Filho, J. L. L. (2002) Application of protease from *Nocardiopsis* sp. as a laundry detergent additive. *World J. Microbiol. Biotechnol.* 18, 307-312.

Muller, V. and Oren, A. (2003) Metabolism of chloride in halophilic prokaryotes. *Extremophiles* 7, 261-266.

Najafi, M. F., Deobagkar, D. and Deobagkar, D. (2005) Potential application of protease isolated from *Pseudomonas aeruginosa* PD100. *Electronic J. Biotechnol.* 8, 197-203.

Nakatsugwa, N. (1991) 'Novel methanogenic archaebacteria which grow in extreme environments'. In *Superbugs, Microorganism in Extreme Environments*, Horikoshi, K. and Grant, W. D. (eds), Springer-Verlag, pp. 212-220.

Naveena, B., Sakthiselvan, P., Elaiyaraju, P. and Partha, N. (2012) Ultrasound induced production of thrombinase by marine actinomycetes: kinetic and optimization studies. *Biochem. Eng. J.* 61, 34-42.

Nyyssola, A. and Leisola, M. (2001) *Actinopolyspora halophila* has twosepa pathways for betaine synthesis. *Arch. Microbiol.* 176 (4), 294-300.

Olano, C., Mendez, C. and Salas, J. A. (2009) Antitumor compounds from marin actinomycetes. *Mar. Drugs* 7, 210-48.

Oren, A. (2004) Prokaryote diversity and taxonomy: current status and future challenges. *Philos. Trans. R. Soc. B-Biol. Sci.* 359, 623-638.

Oren, A., Heldal, M., Norland, S. and Galinski, E. A. (2002) Intracellular ion and organic solute concentrations of the extremely Halophilic Bacterium *Salinibacter rubber. Extremophiles* 6, 491-498.

Pandey, A., Nigam, P., Soccol, C. R., Soccol, V. T., Singh, D. and Mohan, R. (2000) Advances in microbial amylases. *Biotechnol. Appl. Biochem.* 31, 135-152.

Pandey, S., Rakholiya, K., Raval, V. H. and Singh, S. P. (2012) Catalysis and Stability of an alkaline protease from a Haloalkaliphilic Bacterium under non-aqueous conditions as a function of pH, salt and temperature. *J. Biosci. Bioeng.* 114, 251-256.

Pandey, S. and Singh, S. P. (2012) Organic solvent tolerance of an amylase from haloalkaliphilic bacteria as a function of pH, temperature and salt concentrations, *Appl. Biochem. Biotechnol.* 166, 1747-1757.

Park, J. S., Hitomi, J., Horinouchi, S. and Beppu, T. (1993) Identification of two amino acids contributing the high enzyme activity in the alkaline pH range of an alkaline endoglucanase from a *Bacillus* sp. *Protein Eng.* 6, 921-926.

Parro, V. and Mellado, R. P. (1994). Effect of glucose on agarase overproduction in *Streptomyces. Gene* 145,

49-55.
Puls, J., Schmidt O. and Granzow, C. (1987) α-Glucuronidase in two microbial xylanolytic systems. *Enzyme Microbio. Technol.* 9, 83-88.
Purohit, M. and Singh, S. P. (2011) Comparative analysis of enzymatic stability and amino acid sequences of thermostable alkaline proteases from two Haloalkaliphilic bacteria isolated from Coastal region of Gujarat, India. *IJBIOMAC* 49, 103-112.
Rao, K., Jayasri, M. A. and Kannabiran, K. K. (2009) α-Glucosidase and α-amylase inhibitory activity of *Micromonospora* sp. VITSDK3 (EU551238). *Int. J. Integrat. Biol.* 6 (3), 115-120.
Rao, M. B., Tanksale, A. M., Ghatge, M. N. and Deshpande, V. V. (1998) Molecular and biotechnological aspects of microbial proteases. *Microbiol. Mol. Biol. Rev.* 62, 597-635.
Rasmussen, R. S. and Morrissey, M. T. (2007) Marine biotechnology for production of food ingredients. *Adv. Food Nutr. Res.* 52, 237-292.
Reed, D. W., Fujita, Y., Delwiche, M. E., Blackwelder, D. B., Sheridian, P. P., et al., (2002) Microbial communities from methane hydrate-bearing deep marine sediments in a forearc basin. *Appl. Environ. Microbiol.* 68, 3759-3770.
Saxena, R. K., Dutt, K., Agarwal, L. and Nayyar, P. A. (2007) Highly thermostable and alkaline amylase from a Bacillus sp. PN5. *Bioresource Technol.* 98, 260-265.
Schmid, R. D. and Verger, R. (1998) Lipases: interfacial enzymes with attractive applications. *Angew Chem. bit. Ed. Engl.* 37, 1608-1633.
Schumann, P., Prauser, H., Rainey, F. A., Stackebrandt, E. and Hirsch, P. (1997) *Friedmanniella antarctica* gen. nov., sp. nov., an LL-diaminopimelic acid-containing actinomycete from antarctic sandstone. *Int. J. Syst. Bacterial* 42, 278-283.
Senthil, K M. j Selvam, K. and Singaravel. (2012) Statistical assessment of medium components by factorial design and surface methodology of L-asparaginase production by isolated *Streptomyces radiopugnans* MSI in submerged Lmentotion using tapioca effluent. *Asian J. Appl. Sci.* 5, 252-265.
Shafei H. A. E., Abdel-Aziz, M. S., Ghaly, M. F. and Abdalla, A. A. H. (2010) Optimizing some factors affecting alkaline protease production by a marine bacterium *Streptomyces albidoflavus*. Proceedings of fifth scientific environmental conference, Zagazig University, Egypt, pp. 125-142.
Shapjro, H. S. (1989) Nitrogen assimilation in actimomycctes and the influence of nitrogen nutrition on actinomycete secondary metabolism. In *Regulation of Secondary Metabolism in Actinomycetes*, Shapiro, H. S. (ed.). Boca Raton, FL: CRC Press, pp. 149-153.
Singh, J., Batra, N. and Sobti, C. (2004a) Purification and characterization of alkaline cellulase produced by a novel isolate, *Bacillus sphaericus JS*1. *J. Ind. Microbiol. Biotechnol.* 31, 51-56.
Singh, S. P., Raval, V. H., Purohit, M. K., Thumar, J. T. and Gohel, S. D., et al., (2012) In *Microorganisms in Environmental Management*, Satyanarayan, T., Johri, B. N. and Prakash, A. (Eds). Springer Publisher, pp. 415-429.
Slater, M. R., Huang, E, Hartnett, J. R., Bolchakova, E., Storts, D. R., et al., (2000). US Patent no. 6, 077, 664.
Thumar, J. T. and Singh, S. P. (2007a) Secretion of an alkaline protease from salt-tolerant and alkaliphilic,

Streptomyces clavuligerus strain Mit-1. *Braz. J. Microbiol.* 38, 1-9.

Thumar, J. T. and Singh, S. P. (2007b) Two-step purification of a highly thermostable alkaline protease from salt-tolerant alkaliphilic *Streptomyces clavuligerus* strain Mit-1. *J. Chrom. B* 854, 198-203.

Thumar, J. T. and Singh, S. P. (2009) Organic solvent tolerance of an alkaline protease from salt-tolerant alkaliphilic *Streptomyces clavuligerus* strain Mit-1. *J. Ind. Microbiol. Biotechnol.* 36, 211-218.

Thumar, J. T. and Singh, S. P. (2011) Repression of alkaline protease in salt-tolerant alkaliphilic *Streptomyces clavuligerus* strain Mit-1 under the influence of amino acids in minimal medium. *Biotechnol. Biopr. Eng.* 16 (6), 1180-1186.

Thumar, J. T., Dhulia, K. and Singh, S. P. (2010) Isolation and partial purification of an antimicrobial agent from halotolerant alkaliphilic *Streptomyces aburaviensis* strain Kut-8. *World J. Microbiol. Biotechnol.* 26, 2081-2087.

Timell, T. E. (1967) Recent progress in chrysosporium. *Methods Enzymol.* 161, 238-249.

Todkar, S., Todkar, R., Kowale, L., Karmarkar, K. and Kulkarni, A. (2012) Isolation and screening of antibiotic producing halophiles from Ratnagri coastal area, State of Maharahstra. *Int. J. Scient. Res. Pub.* 2 (4), 82-84.

Trincone, A. (2011) Review: Marine biocatalysts: enzymatic features and applications. *Marine Drugs* 9, 478-499.

Tsujibo, H., Kubota, T., Yamamoto, M., Miyamoto, K. and Inamori, Y. (2003) Characterization of chitinase genes from an alkaliphilic actinomycete, *Nocardiopsis prastna* OPC-131. *Appl. Environ. Microbiol.* 69 (2), 894-900.

Tsujibo, H., Yoshida, Y., Miyamoto, K., Hasegawa, T. and Inamori, Y. (1992) Purification and properties of two types of chitinases produced by an alkalophilic actinomycete. *Biosci. Biotechnol. Biochem.* 56, 1304-1305.

Van, S. P., Meijer, D., Van der Kleij, W. A., Barnett, C, Bolle, R., et al., (2001) Cloning and expression of an endocellulase gene from a novel *Streptomycete* isolated from an East African soda lake. *Extremophiles* 5 (5), 333-341.

Vasavada, S. H., Thumar, J. T. and Singh, S. P. (2006) Secretion of a potent antibiotic by salt-tolerant and alkaliphilic actinomycete *Streptomyces sannanensis* strain RJT-1. *Current Science* 91 (4), 1393-1397.

Ventosa A. (1989) 'Taxonomy of halophilic bacteria'. In *Microbiology of Extreme Environments and its potential for Biotechnology*, De Costa, MS., Duarte, J. C. and Williams, R. A. D. (eds), New York: Elsevier Applied Science, pp. 262-279.

Viikari, L., Pauna, M., Kantelinen, A., Sundquist, J. and Linko. M. (1986) 'Bleaching with enzymes'. In Proceedings of the Third International Conference on Biotechnology in the Pulp and Paper Industry, Stockholm, 67-69.

Vreelan, A., Rosenzweig, W. D. and Powers, W. D. (2000) Isolation of a 250 million years old *Halotolerant bacterium* from a primary salt crystal. *Nature* 407, 897-900.

Vujaklija, D., Schroder, W. and Abramic, M. (2002) A novel *Streptomycete* lipase: cloning, sequencing and high-level expression of the *Streptomyces rimosus* GDS (L) -lipase gene. *Arch. Microbiol.* 178 (2), 124-130.

Wang, Y. X., Liu, J. H., Xiao, W., Zhang, X. X, Li, Y. Q. et al., (2012) *Fodinibius salinus* gen. nov., sp. nov., a moderately halophilic bacterium isolated from a salt mine. *Int. J. Syst. Evol. Microbiol.* 62 (2), 390-396.

Ward, O. P. (1985) 'Proteolytic enzymes'. In *Comprehensive Biotechnology. The Principles, Application and Regulations of Biotechnology in Industry, Agriculture and Medicine*, Moo-Young, M. (ed.). VCH Verlagsgesellschaft mbH, Germany, pp. 819-835.

Whistler, R. L. and Richards, E. L. (1970) Hemicelluloses. In *The carbohydrates Hemicelluloses*, Pigman, W. and Horton, D. (eds). New York: Academic Press, pp. 447-469.

Wohl, D. L. and McArthur. J. V. (1998) Actinomycete-flora associated with submersed freshwater macrophytes. *FEMS Microbiol. Ecol.* 26, 135-140.

Wolff, A. M., Showed, M. S., Venegas, M. G., Barnett, B. L. and Werts, W. C. (1996) Laundry performance of subtilisin proteases. In *Subtilisin Enzymes: Proc. Protein Engineering*, Bott, R. and Betzd, C. (Eds). New York: Plenum Press, pp. 113-120.

Yumoto, I., Nakamura, A., Iwata, H., Kojima, K., Kusumoto, K., et al., (2002) *Dietzia psychralcaliphila* sp. nov., a novel, facultatively psychrophilic alkaliphile that grows on hydrocarbons. *Int. J. Syst. Evol. Microbiol.* 52, 85-90.

Zenova, G. M., Chernov, I. Y., Gracheva, T. A. and Zvyagintsev, D. G. (1996) Structure of actinomycete complexes in deserts. *Microbiologica* 65, 616-620.

Zhou, X., Huang, J., Ou, Z., Wang, H. and Wang R. (2000) Conditions of enzyme production and properties of alkaline lipase by *Streptomyces* Z94-2. *Wei Sheng Wm Xue Bao.* 40, 75-79.

第9章 日本普通乌贼太平洋褶柔鱼肝胰腺中3种蛋白水解酶对鱼肌肉蛋白降解的研究

Kunihiko Konno, Hokkaido University, Hokkaido, Japan and Yuanyong Tian, Dalian Polytechnic University, Dalian, China

DOI: 10.1533/9781908818355.3.217

摘要: 在乌贼的肝胰腺提取物中3种蛋白酶（半胱氨酸蛋白酶、金属蛋白酶、丝氨酸蛋白酶）活性被检测出来。金属蛋白酶有选择地裂解肌球蛋白，而半胱氨酸蛋白酶没有选择性。在这3种酶中，丝氨酸蛋白酶的活性最低。半胱氨酸蛋白酶的热稳定性最好，在50℃时仍表现出较高活性。变性的肌原纤维更容易被半胱氨酸蛋白酶进行随机降解。利用合成多肽作为底物，半胱氨酸蛋白酶被认为是组织蛋白酶L。在这三者中，由于组织蛋白酶L在pH值5和50℃下展现出来的最佳的反应条件，有助于酸溶性多肽的制备。为了长期保存酶，可以将乌贼肝脏利用丙酮干燥成粉末。将粉末与肉沫共同孵育是生产多肽的最佳方法。

关键词: 半胱氨酸蛋白酶，组织蛋白酶L，肌球蛋白，变性，肝胰腺，乌贼，蛋白酶，肝脏，TCA可溶性多肽

9.1 引言

蛋白酶是一种被广泛深入研究的酶。例如，胃蛋白酶是胃液中主要的酶，早在1930年就获得了晶体。已经在不同生物的不同组织中发现了各种不同类型的蛋白酶。根据催化类型蛋白酶可以分为两大类：从多肽内部水解肽键而产生相对较长肽段的肽链内切酶；从多肽链的N末端或C末端连续释放氨基酸的肽链外切酶。还可以根据活性位点的相似性将蛋白酶分为不同的种类；包括催化必需的基本氨基酸和其他组分。有丝氨酸蛋白酶、半胱氨酸蛋白酶、金属酶和天冬氨酸蛋白酶。因此，这些酶可以通过阻断其催化的必需氨基酸或者移除必要的金属进行特异性抑制。

在日本尤其是北海道，日本常见乌贼太平洋褶柔鱼是海洋渔业最重要的海洋资源之

一。套膜、鳍和腿可以作为食品，但是内脏（肝胰腺），尤其是肝脏，通常被作为工业废物来丢弃。众所周知，乌贼肝胰腺中含有几种类型的蛋白水解活性（Hatate et al.，2000；Makinodan et al.，1993）。乌贼肝脏是生产传统发酵乌贼肉“*Ika-Shiokara*”（盐乌贼罐头）的主要成分，这种食物通过将含有乌贼肉片、约10%的肝脏、10%~20%的盐的混合物腌制几个月来制成。普遍认为，肝脏中的蛋白水解酶活性会影响产品的口味（Hatate et al.，2000）。同时有报道称组织蛋白酶是盐乌贼罐头的肉成熟过程中改变肉质的酶（Makinodan et al.，1993）。同理，加入乌贼肝脏的目的就是缩短鱼酱的腌制时间（Raksakulthai et al.，1986）。乌贼肝脏中已有多种蛋白酶活性被报道：金属酶（Okamoto et al.，1993；Tajima et al.，1998；Tsujioka et al.，2005；Tamori et al.，1999），半胱氨酸蛋白酶（Sakai-Suzuki et al.，1986；Cardenas Lopez and Haard 2005，2009），丝氨酸蛋白酶（Ebina et al.，1995）和组织蛋白酶D类似的半胱氨酸蛋白酶（Komai et al.，2004）。尽管乌贼肝脏中的蛋白酶活性已经被很好的研究，但是它们在整体提取物中的作用还没有被很好的研究。多数研究重点关注单一目标蛋白酶的特征描述上。因此，为了研究酶的活性，酪蛋白或合成多肽通常被用来作为底物，酸溶性的多肽产物作为检测活性的常用指标。一旦酶被纯化，应用酶特异性的合成底物可以方便的检测。

当我们试图使用乌贼肝脏中的酶作为一种生产功能性多肽的工具的时候，关注作为底物的蛋白和产物的生成是十分重要的。我们的目的是利用附着在渔业废物像骨头、鱼头上的肉来作为底物生产多肽。因此，我们利用鱼肌原纤维作为研究乌贼肝脏中蛋白酶的模式底物。我们将注意力集中于肌原纤维中的蛋白尤其是肌球蛋白是如何被酶消化的。首先，乌贼肝脏粗提液中的酶的种类被鉴别出来，对它们的功能做了比较；即，那种类型的蛋白酶活性在乌贼肝脏中发挥主导作用。然后，我们研究了肌原纤维的变性对其作为底物被消化的影响。最后，利用乌贼肝脏本身直接将鱼肉进行消化，以找到利用鱼肉提高多肽产量的条件。

9.2 乌贼肝脏中蛋白酶活性的鉴定

从日本常见乌贼的完整肝脏（肝胰腺）中提取粗酶，利用普通鲤鱼的肌原纤维作为酶的模式底物。通过SDS-PAGE检测肌原纤维蛋白的降解情况，对蛋白酶的活性进行研究。在SDS-PAGE电泳图上观察到的由于肌球蛋白的降解而导致的肌球蛋白重链（MHC）的消失速率可以代表酶的活性。

我们可以合理地认为乌贼肝脏中不同的酶可以通过酶对肌原纤维特异性的消化模式来进行表征，这些都可以很容易地通过SDS-PAGE来进行分析。抑制谱的研究是鉴定乌贼肝脏中存在的蛋白酶的一种方法。当肌原纤维与粗酶在不含有任何抑制剂的条件下进行共同孵育时，肌球蛋白重链与肌动蛋白会降解成碎片（图9.1a），表明提取物中含有很高的蛋白水解酶活性。将混合物与不同类型的蛋白酶抑制剂共同孵育。在这些抑制剂实验中，E-64、亮抑酶肽、p-chrolomerculic苯酸盐、N-丁烯二酰亚胺、大豆胰蛋白酶抑制剂、苯甲基磺酰氟（PMSF）、乙二胺四乙酸（EDTA）和 o-邻二氮杂菲展示出了不同程度的抑制作用。结果表明在提取物中存在半胱氨酸蛋白酶，金属蛋白酶和丝氨酸蛋白酶。当 3 种抑制剂（E-64，o-邻二氮杂菲和 PMSF）共同加入时，几乎能够完全抑制 MHC 和肌动蛋白的降解（图 9. 1e）。因此，通过减去目标酶的特异抑制剂，能够很容易地分析各种酶的消化模式。金属蛋白酶（从抑制剂混合物中去除 o-邻二氮杂菲）将肌球蛋白降解成一些迁移率在 MHC 和肌动蛋白条带之间的片段（图 9. 1b）。正如 Tamori 等（1999）报道的一样，降解是非常有选择性的。据他们报道，150 kDa，140 kDa 和 110 kDa 是重链酶解肌球蛋白类似物，其他的 3 个（50 kDa，60 kDa，90 kDa）是它们相对应的轻链肌球蛋白类似片段。半胱氨酸蛋白酶在 135 kDa 处产生条带和分散于整条泳道中的一些条带（图 9. 1c）。半胱氨酸蛋白酶降解肌动蛋白同时也降解肌球蛋白，这与金属蛋白酶的模式不同。产生的条带迁移率在肌球蛋白轻链以下，表明肌球蛋白和肌动蛋白被裂解成了非常短的片段。丝氨酸蛋白酶的消化活性非常低，在 170 kDa 处产生了模糊的条带。

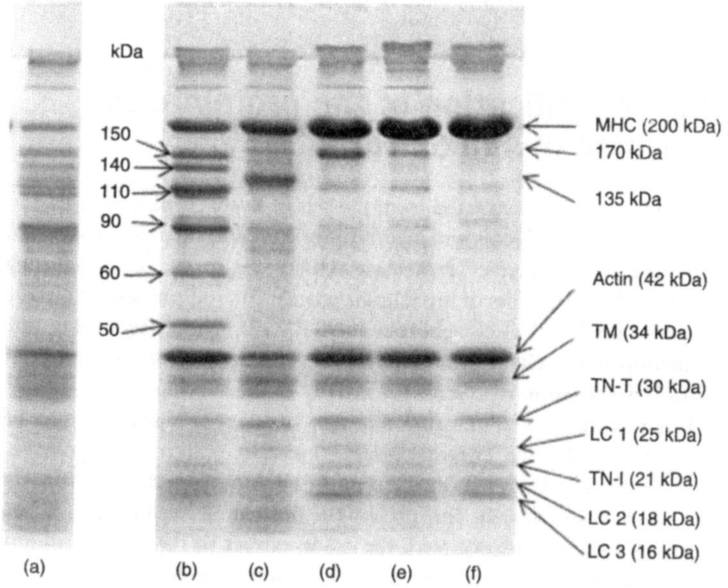

图 9. 1　用 3 种类型乌贼肝中蛋白酶消化鲤鱼肌原纤维模式

注：在 20℃消化鲤鱼肌原纤维：（a）没有抑制剂；（b）金属蛋白酶；（c）半胱氨酸蛋白酶；（d）丝氨酸蛋白酶；（e）有 3 种抑制剂；（f）消化前肌原纤维．MHC，Actin，TM，TN-T，I，C 和 LC1，2，3 分别代表肌球蛋白重链，肌动蛋白，原肌球蛋白，肌球蛋白亚基和轻链元件，片段单位为 kDa.

为了表征各种酶，检测了它们温度依赖性的酶活性（图 9. 2）。所有酶的活性都随着温度的增加而升高。在低温下例如 20℃，3 种酶中的金属蛋白酶活性最高，而半胱氨酸蛋白酶在 50℃时表现出了最高的活性，因为不同酶的最高反应温度不同；半胱氨酸蛋白酶在高于 50℃时有最适的活性，而其他两种酶在 35~40℃时表现出最高活性。造成这些差异的原因是各种酶的热稳定性不同，半胱氨酸蛋白酶是最稳定的。利用粗提取物的热失活实验

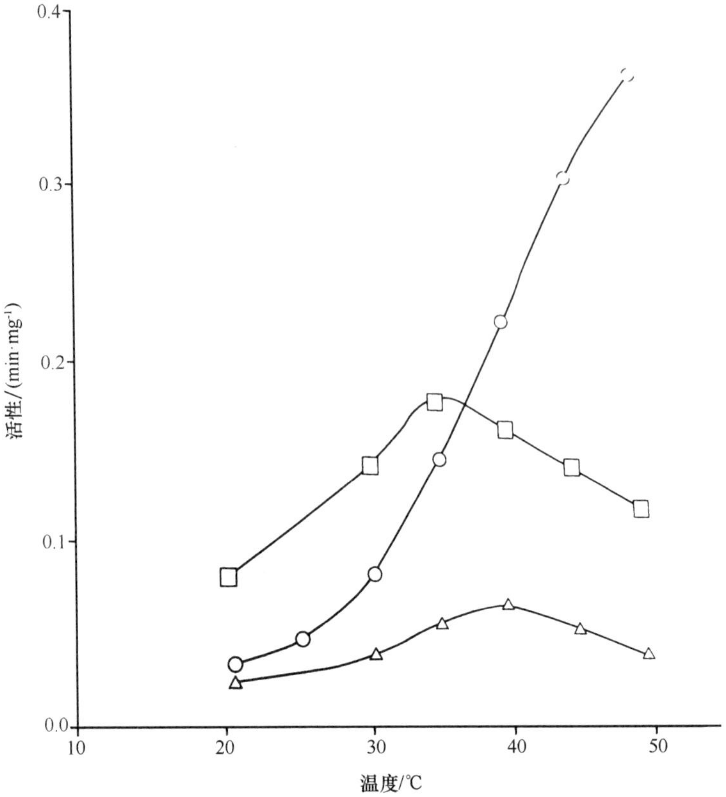

图 9.2　温度依赖的 3 种蛋白酶效率

注：金属蛋白酶（正方形），半胱氨酸蛋白酶（圆形），丝氨酸蛋白酶（三角形）在不同温度下的测试.

证实了半胱氨酸蛋白酶的稳定性。

为了表征这 3 种酶，测定了其在 20℃下的酶活性的 pH 值的依赖性（图 9.3）。金属蛋白酶在 pH 值 7 的条件下显示出了最大的酶活，将 pH 值调成酸性或碱性范围，酶活急剧下降。半胱氨酸蛋白酶在 pH 值 6 时有最大活性。丝氨酸蛋白酶在 pH 值 9 时表现出最佳活性但是其本来活性非常低。

9.3　肌球蛋白的变性对可消化性的影响

在高温下测定酶活性会影响肌原纤维、底物和酶的活性。为了了解肌原纤维的变性对酶活性的影响，利用在不同温度下加热过的肌原纤维作为底物（图 9.4）。

将肌原纤维在 30℃加热 30 min，对 3 种酶的活性没有任何影响，但是在 20℃，pH 值 7 的条件下测量时，金属蛋白酶表现出了最高的酶活。但是，将肌原纤维在 30～50℃之间加热时，改变了酶的活性。当用 50℃以上的温度加热过的肌原纤维做底物时，金属蛋白酶的活性降低到了一半，然而半胱氨酸蛋白酶的活性增加了 8 倍。丝氨酸蛋白酶的活性也增加了，但是活性仍然很低。这些结果表明，相同的肌球蛋白变性对各种酶引起了不同的影

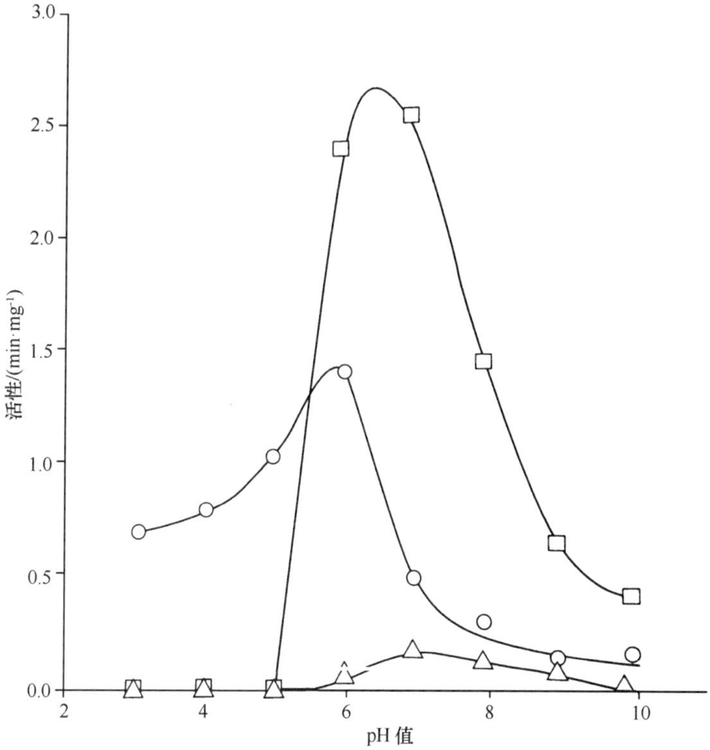

图 9.3　不同 pH 值条件下使用肌原纤维作为底物的活性，测试保持在 20℃

注：金属蛋白酶（正方形），半胱氨酸蛋白酶（圆形），丝氨酸（三角形）蛋白酶.

响。对加热后的肌原纤维的消化模式与未加热的肌原纤维相比较，在金属蛋白酶的消化产物中发现的片段与以未加热的肌原纤维做底物的相比，并没有明显不同，但是它们的量都减少了。或许，造成对变性后肌原纤维活性降低的原因是蛋白质通过聚集体的形成，屏蔽了肌球蛋白分子上的选择性裂解位点。半胱氨酸蛋白酶显示出了一种更加复杂的消化模式，表明是由于变性后增加了肌球蛋白分子上的裂解位点。

9.4　乌贼肝脏中半胱氨酸蛋白酶的鉴定

乌贼肝脏中的金属蛋白酶被鉴定为肌球蛋白特异性蛋白酶，又被 Tamori 等（1999）命名为肌球蛋白酶 I 和肌球蛋白酶 II。在不同组织中报道了各种类型的半胱氨酸蛋白酶活性。它们是组织蛋白酶 B，H，L，S，K 和钙蛋白酶（Barrett et al.，1981；Katsunuma and Kominami，1983；Pangkey et al.，2000；Aibe et al.，1996；Sakai-Suzuki et al.，1983）。我们利用商业化人工合成的酶特异性底物从乌贼肝脏中鉴定出半胱氨酸蛋白酶。试验的底物是 Z-Phe-Arg-MCA（组织蛋白酶 B/L 特异性），Z-Val-Val-Arg-MCA（组织蛋白酶 S 特异性），Z-Arg-Arg-MCA（组织蛋白酶 B 特异性），Arg-MCA（组织蛋白酶 H 特异性），Z-Gly-Pro-Arg-MCA

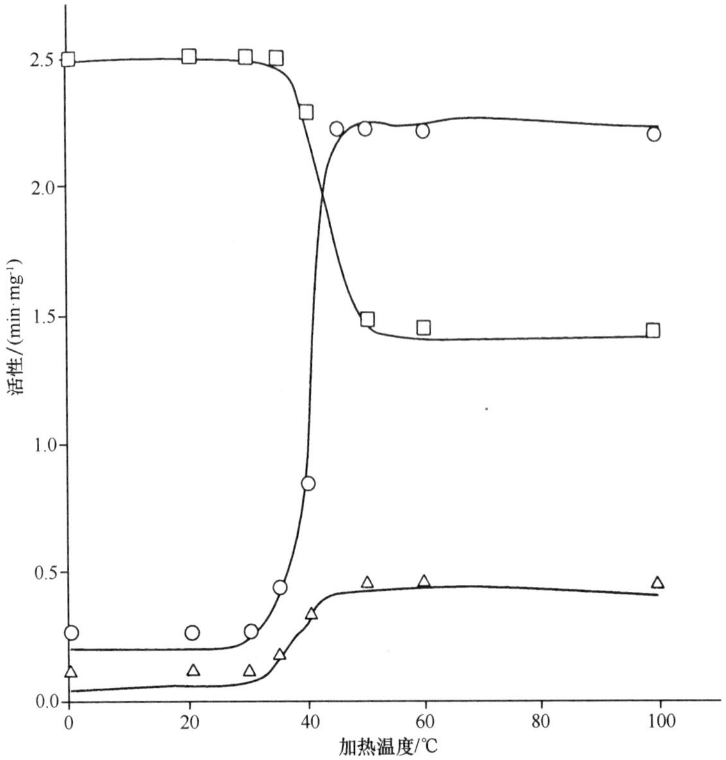

图 9.4　针对加热过的肌原纤维 3 种类型酶活性

注：加热 30 min 后的肌原纤维作为底物在 20℃进行测试，金属蛋白酶（正方形），半胱氨酸蛋白酶（圆形），丝氨酸（三角形）蛋白酶 .

（组织蛋白酶 K 特异性）。在这些底物中，Z-Phe-Arg-MCA 是被水解的底物。由于酶并不能水解 Z-Arg-Arg-MCA（组织蛋白酶 B 特异性），相对于酶对 Z-Phe-Arg-MCA 的水解，酶活只有 0.53%，因此，该酶被鉴定为组织蛋白酶 L。

我们利用丙酮干燥后的乌贼肝脏粉末代替未加工的肝脏作为分离酶的起始材料。这是因为提取物中油脂的污染会干扰后续的纯化步骤。来源于干燥粉末的粗提物中的油脂是可以忽略。丙酮处理不会改变任一酶的活性。通过利用几种纯化步骤将组织蛋白酶 L 进行纯化。Aranishi 等（1997）利用 S-琼脂糖凝胶从鲤鱼肝胰腺中纯化出组织蛋白酶 L。但是乌贼肝脏中的半胱氨酸蛋白酶不能与阳离子交换柱（Toyopearl SP-650M，CM-650M）结合，而是结合在阴离子交换柱（Toyopearl DE-650M）上，表明酶的表面电荷不一样。在 0.1～0.35 mol/L NaCl 洗脱下来的活性组分进一步用 Toyopearl HW-55F 凝胶柱纯化。活性组分在 SDS-PAGE 上显示单一条带，分子量大约为 24 kDa。分子量大小也与之前报道的从兔子脾脏（Maciewicz and Etherington，1998），巨型乌贼肝脏（Cardenas-Lopez et al.，2009），兔子肌肉（Okitani et al.，1980），老鼠肾脏（Reddy and Dhar，1992），鲤鱼肝脏（Aranishi et al.，1997），鲑鱼肌肉（Yamashita and Konagaya，1992）中分离的酶的大小相近。

9.5 用于短肽生产的酶

为了从蛋白质材料像渔业废物中生产功能性的多肽，人们已经做了很多尝试。为了达到这个目的，有两个途径可以实现：第一个就是利用商业的酶（Benjakul and Morrissey, 1997; Shahidi et al., 1995., Je et al., 2007）；另一种方法就是利用组织中的内源性的酶（Kubota and Sakai, 1978; Gildberg et al., 1984; Makinodan et al., 1993）。

我们的计划属于第二种。我们利用 MHC 降解作为蛋白水解活性测量的指标。然而，蛋白水解活性通常是检测底物蛋白释放的三氯乙酸（TCA）可溶性多肽来进行测定的。为了检测不同的检测方法是否将导致不同的结果，我们用含有不同底物组合和不同检测方法进行酶活的检测。分别是：①加热后的肌原纤维和 MHC 的降解；②加热后的肌原纤维和 TCA 可溶多肽；③含氮酪蛋白和 TCA 可溶性多肽；④Z-Phe-Arg-MCA。最显著的不同点体现在 pH 值依赖性（图 9.5）。检测 MHC 降解时，最适 pH 值在 6，但是当利用 TCA 可溶性多肽作为检测指标时，最佳 pH 值为 5。利用含氮酪蛋白中的 TCA 可溶性多肽检测最佳 pH 值也是 5. Z-Phe-Arg-MCA 检测时的最佳 pH 值为 7。结果表明讨论酶的性能时应该十分谨慎。

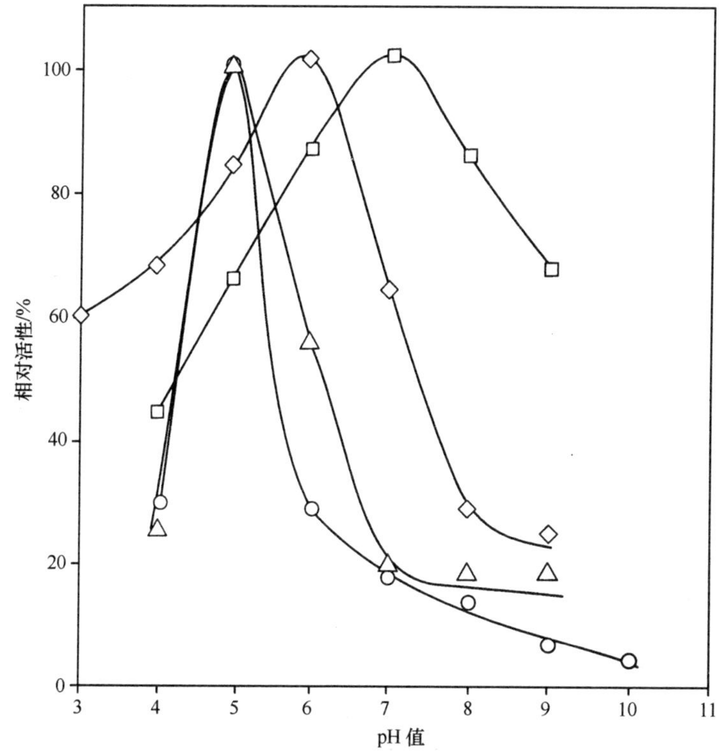

图 9.5 使用不同底物和分析方法 pH 值依赖的活性

注：用不同底物和方法的组合在 20℃下检测，肌原纤维和 MHC 消化（菱形），肌原纤维和可溶性 TCA 多肽（圆形,）偶氮酪蛋白和可溶性 TCA 多肽（三角形），Z-Phe-Arg-MCA（正方形）.

我们进一步研究了哪一种酶在多肽生产中的贡献最大。我们也好奇 TCA 可溶性多肽的生产需要肌原纤维降解到什么程度。TCA 可溶性多肽的产量可以通过检测 TCA 可溶性部分的吸光度进行检测，降解程度可通过 SDS-PAGE 分析消化后的不可溶部分来进行研究(图 9.6)。通过仅用半胱氨酸蛋白酶获得的多肽的产量与使用粗酶本身获得的产量几乎相同。金属蛋白酶与丝氨酸蛋白酶对 TCA 可溶性多肽的产率的贡献可忽略不计（图 9.6（A））。对剩余物进行 SDS-PAGE 电泳显示，在粗酶提取物消化后，肌球蛋白和肌动蛋白有着显著的降解（图 9.6（B）b）。值得注意的是，在这种模式下可以检测到非常浓的与前沿染料迁移到一起的条带，表明 TCA 不溶性部分仍然含有相当多的被降解的短片段。这部分的分子量应该小于 5 000 Da，换句话说，TCA 可溶部分多肽的回收并不是由分子量决定的。短肽的作用阻止了 TCA 可溶性部分的回收。半胱氨酸类蛋白酶消化残余物的模式与不含抑制剂几乎完全相同，表明粗酶提取物中对降解有贡献的酶是半胱氨酸蛋白酶(组织蛋白酶 L)。尽管用金属蛋白酶（泳道 d）与丝氨酸蛋白酶（泳道 e）对肌球蛋白的降解也是十分显著的，但是这些蛋白的降解并没有产生 TCA 可溶性肽。假设 TCA 可溶性多肽的产生是作为检测蛋白酶活性的指数，可以推断出乌贼肝脏中只含有半胱氨酸蛋白酶。

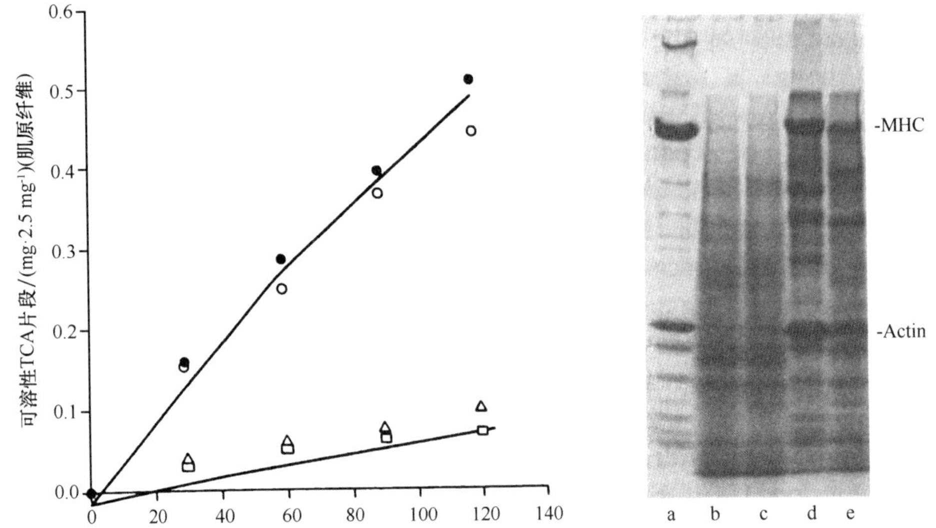

图 9.6 TCA 可溶性多肽产物 SDS-PAGE 对不溶性 TCA 片段分析，加热过的肌原纤维在 pH 值 5 和 50℃下经粗酶孵育

（A）孵育后可溶性 TCA 多肽的数量被测定；（B）可溶性 TCA 片段包含的组分在 120 min 内由 SDS-PAGE 检测，没有抑制剂（A 中封闭圆圈和 B 中 b），半胱氨酸（A 中开放圆圈和 B 中 c），金属（A 中正方形和 B 中 d），丝氨酸（A 中三角形和 B 中 e）蛋白酶，B 中 a 为肌原纤维空白.

在图 9.6（A）中，通过计算，在 120 min 中，TCA 可溶性多肽的回收率大约为

0. 5 mg/2. 5 mg 肌原纤维。这是通过读取肌原纤维溶液和 TCA 可溶性部分在 290 nm 处的吸光度计算出来的。TCA 沉淀的肌原纤维重新溶解于 0. 6 mol/L NaOH 中，并且在 290 nm 处有吸收峰。因此，TCA 可溶性组分也加入相同浓度的 NaOH 进行吸光度的测量。

9. 6 TCA 可溶性多肽的制备

由于半胱氨酸蛋白酶（组织蛋白酶 L）是与多肽制备有关的酶，反应物体系的 pH 值要设置在最佳的 pH 值 5，并且利用加热后的肌原纤维作为底物（图 9. 7）。

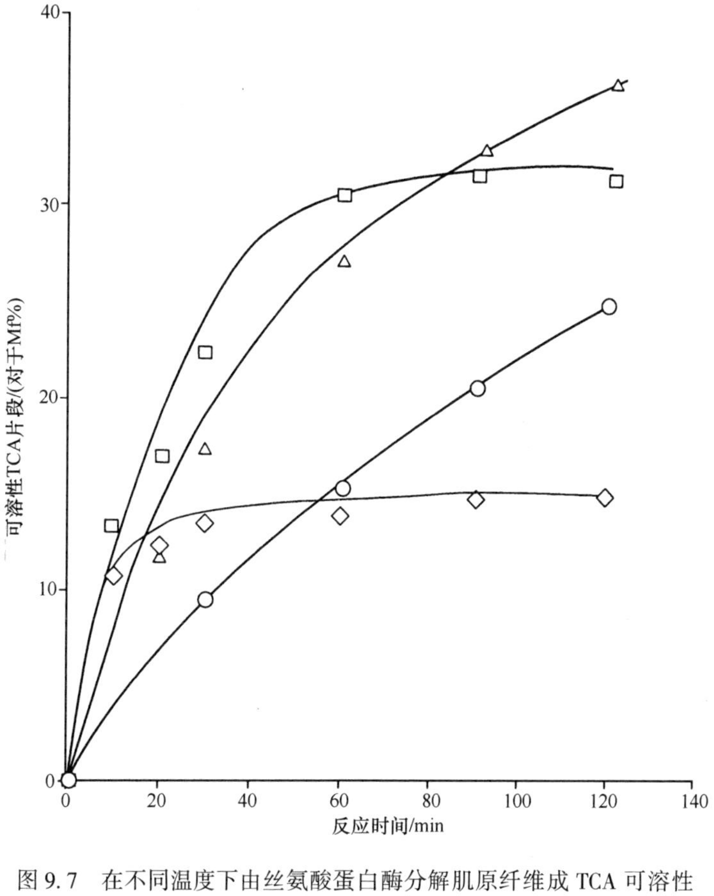

图 9. 7 在不同温度下由丝氨酸蛋白酶分解肌原纤维成 TCA 可溶性多肽产物

注：加热后的鲤鱼肌原纤维由粗酶（10%，与肌原纤维关系为质量分数）在 pH 5 不同温度下消化，回收的 TCA 可溶性多肽的数量作为使用肌原纤维的重量比例的关系，反应温度代表为 40℃（圆形）；50（三角形）；60℃（正方形）；70℃（菱形）.

通过将反应温度从40℃提高到70℃来生产短肽。在40℃时，会发现肽不断产生的过程，当酶粗提物占肌原纤维的10%，在120 min时，大约转化了25%。在50℃时，生产率变得更快，显示出了相似的进程，在120 min产量达到了约36%。在60℃时的情形与50℃略有不同。在60 min时，多肽的生产达到快速饱和，表明了在这个温度下酶失活，最大的转化率约为32%。在70℃时，酶在15~20 min就失活。这些特征表明最佳温度为50℃。我们注意到，即使在40℃，反应的进展也不是呈线性的。弯曲的原因并不是因为酶的失活，因为在50℃时与40℃时的情形是一样的。另一原因可能是底物浓度的降低。当肌原纤维的浓度是2.5 mg/mL，在40℃，120 min时的转化率是25%，表明反应系统中仍然含有2 mg/mL的TCA可溶性蛋白。我们怀疑底物浓度对酶是不是不够充足。为了阐明蛋白浓度对活性的影响，通过将肌原纤维的浓度从0.5 mg/mL改变到6 mg/mL来研究TCA可溶性多肽的产量。将酶活性与底物浓度进行双倒数法作图，得到表观K_m值在5 mg/mL，表明酶对肌原纤维的亲和力不是很高。研究报道了利用细菌来源的商业化的酶，如碱性蛋白酶或中性蛋白酶，进行相似的多肽生产过程，报道中利用的是不同来源的蛋白，鳕鱼肉（Benjakul and Morrissey，1997），小龙虾废物（Beak and Cadwallader，1995）、香鱼（Shahidi et al.，1995）、沙丁鱼（Quaglia and Orban，1987）和牛肉（Linder et al.，1996）。因此，利用含有高浓度蛋白的肉本身有利于多肽的生产。

9.7 通过直接加入乌贼肝脏粉末从鱼肉中生产TCA可溶性多肽

以上所有试验都是利用从肝脏提取的粗酶和肌原纤维进行的。考虑到乌贼肝脏中蛋白酶的实际应用，开展了利用乌贼肝脏本身和鱼肉的研究。然而，肝脏中的油脂含量较高，限制了它的储存和应用，因为多元不饱和脂肪酸的快速氧化很有可能会破坏酶的功能，尤其是含有巯基基团半胱氨酸类蛋白酶。有报道称过氧化的鱼脂质会很容易地与肌球蛋白的SH基团进行反应，导致ATP酶活性的改变（Kawasaki et al.，1991）。

为了避免这种可能性，肝脏中的油脂和水用丙酮处理进行除去。丙酮干燥的组织粉末是酶法提取常用的起始材料。我们证实了利用丙酮处理不会破坏任何蛋白酶的活性，换句话说，在粉末的提取物再次检测到了3种类型的蛋白酶。乌贼肝脏粉末的另一个优点就是其中的酶十分稳定。尽管半胱氨酸蛋白酶是稳定的，将肝脏的提取物在60℃进行加热30 min，能够完全失活。然而，将干粉在60℃加热30 min，一点也不会失活。在70℃、80℃下加热30 min，活性仍然还分别保留70%和40%。将粉末保存在冰箱中，能够存放很长时间而不会损失活性。

将粉碎的鲤鱼肉和乌贼肝脏干粉混合，在50℃下孵育12 h，随着增加粉末的量，TCA可溶性多肽的产量也会增加，但是粉末的量与可溶性多肽的产量之间并没有线性关系（图9.8）。在粉末量为1%时，产量是1 g湿重的肉（大约120 mg蛋白）产出46 mg，这远远高于10%时的产量（75 mg/g 肉的湿重）。当使用乌贼肝脏时，要考虑到重金属的污染，尤其是镉、铜和锌。不同物种的乌贼肝脏中镉的含量是不同的，大的乌贼累积镉的含量多达 200 μg/g 肝脏干粉（Martin and Flegal，1975；Tanaka et al.，1983；Miramand and Bentley，1992）。假设乌贼肝脏中镉的含量大约是 100 μg/g 肝脏干粉（常见镉含量），利用1%的干粉加到湿肉里，消化物中镉的含量大约是1 mg/kg，这将近是FAO/WHO 允许标准（2 mg/kg）含量的一半。报道称乌贼肝脏中的金属，包括镉，是以与蛋白结合的形式存在的，被定义为金属硫因（Margoshes and Valle，1957；Tanaka et al.，1983）。当消化程度不是很高时，可以将镉去除在 TCA 不溶性沉淀物中。

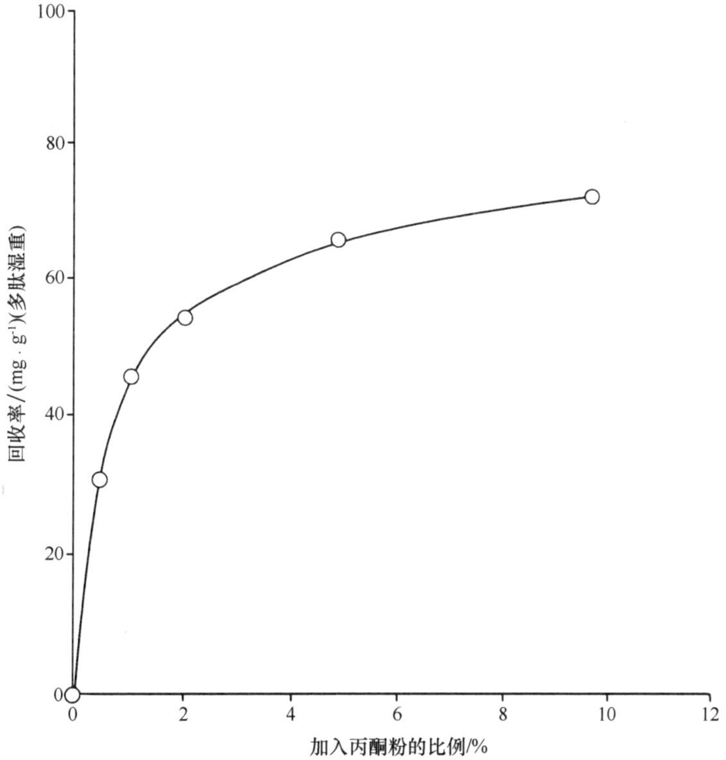

图 9.8　从鲤鱼肌肉通过直接添加丙酮干燥粉末得到的 TCA 可溶性多肽产物

注：在 pH 5 和 12 h 洗后的鱼肉糊（1 g）中添加不同数量的丙酮干粉所消化.

很显然 TCA 可溶性组分回收的短肽是不适合作为食品添加剂的。一种简单又方便的代替方法是煮沸消化产物再回收短肽。这样既简单又方便。实际上，利用 TCA 溶解和沸水溶解的组分在多肽回收的量上和多肽的分子大小分布上没什么不同。从煮沸后上清液中回收的可溶性部分应用于葡聚糖 G-25 凝胶过滤（图 9.9）。多肽图显示有很宽的峰形，说明是一种不同大小的片段的混合物，大多数多肽的大小在 3 000～5 000 kDa。总体来说，鱼肉与丙酮干燥的粉末在 50℃ 条件下孵育过夜可以转化为短肽。通过将混合物进行煮沸，可以很容易地从可溶性组分中对短肽进行回收。通过喷雾干燥的方法可以很容易地收集多肽。

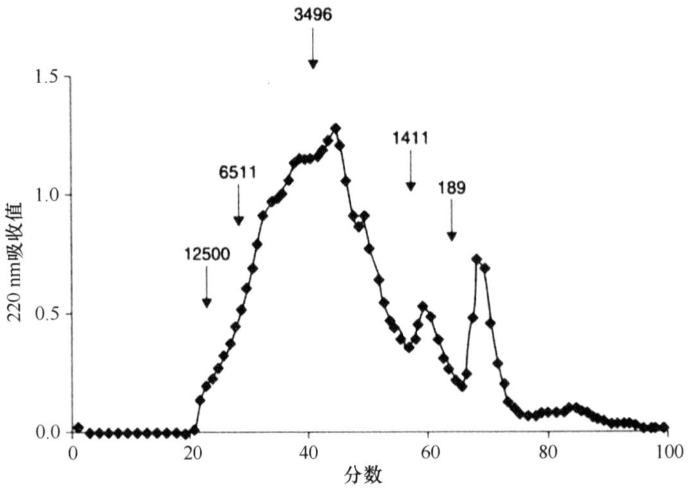

图 9.9　煮后多肽降解的分子量

注：煮后水溶片段由 50℃，pH 5 经 Sephadex G-25 作用 12 h 孵育产生，使用的标准组分为细胞色素 C（M. W. 12500），抑肽酶（M. W. 6511），胰岛素 B 链（M. W. 3496），杆菌肽（M. W. 1411），Gly-Gly-Gly（M. W. 189）.

参考文献

Aibe, K., Yazawa, H., Abe, K., Teramura, K., Kumegawa, M., et al., (1996) K. Substrate specificity of recombinant osteoclast-specific cathepsin K from rabbit. *Biol. Farm. Bull.* 19, 1026-1031.

Aranishi, F., Ogata, H., Hara, K., Osatomi, K. and Ishihara, T. (1997) Purification and characterization of cathepsin L from hepatopancreas of carp *Cyprinuscapio*. *Comp. Biochem. Physiol.* 118*B*, 531-537.

Barrett, AJ. and Kirschke, H. (1981) Cathepsin B, cathepsin H, and cathepsin L. *Methods Enzymol.* 80, 535-561.

Beak, H. H. and Cadwallader, K. R. (1995) Enzymatic hydrolysis of crayfish processing by-products. *J. Food Sci.* 60, 929-935.

Benjakul, S. and Morrissey, M. T. (1997) Protein hydrolysates from Pacific whiting solid wastes. *J. Agric. Food Chem.* 45, 3423-3430.

Cardenas-Lopez, J. L. and Haard, N. F. (2005) Cysteine proteinase activity in Jumbo squid *Dosidicus gigas* hepatopancreas extracts. *J. Food Biochem.* 29, 171-186.

Cardenas-Lopez, J. L. and Haard, N. F. (2009) Identification of a cysteine proteinase from Jumbo squid *Dosidicus gigas* hepatopancreas as cathepsin L. *Food Chem.* 12, 442-447.

Ebina, H., Nagasbima, Y., Ishizaki, S. and Taguchi, T. (1995) Myosin heavy chain-degrading proteinase from spear squid muscle. *Food Res. Int.* 28, 31-36.

Gildberg, A., Jasmin, E. and Florian, M. O. (1984) Acceleration of autolysis during fish sauce fermentation by adding acid and reducing the salt content. *J. Sci. Food Agric.* 36, 1363-1369.

Hatate, H. , Tanaka, R. , Suzuki, N. and Hama, Y. (2000) Comparison of proteinase activity in liver among several species of squid and cuttlefish. *Fish. Sci.* 66, 182-183.

Je, J. Y. , Qian, Z. J. , Byun, H. G. and Kim, S. K. (2007) Purification and characterization of an antioxidant peptide obtained tuna backbone protein by enzymatic hydrolysis. *Process Biochem.* 42, 840-846.

Katsunuma, N. and Kominami, E. (1983) Structures and functions of lysosomal thiol proteinases and their endogenous inhibitor. *Curr. Top Cell Reg.* 22, 71-101.

Kawasaki, K. , Ooizumi, T. and Konno, K. (1991) Effect of peroxidized lipid on the ATPase of carp myofibrils. *Nippon Suisan Gakkaishi* 57, 1185-1191.

Komai, T. , Kawabata, C. , Amanob, M. , Lee, B. R. and Ichishima E. (2004) Todarepsin, a new cathepsin D from hepatopancreas of Japanese common squid *Todarodes pacificus*. *Comp. Biochem. Physiol.* 137*B*, 373-382.

Kubota, M. and Sakai, K. (1978) Autolysis of Antarctic krill protein and its inactivation by combined effects of temperature and pH. Trans. *Tokyo Univ. fisheries* 2, 53-63.

Linder, M. , Fanni, J. and Parmenter, M. (1996) Functional properties of veal bone hydrolysates. *J. Food Sci.* 61, 712-720.

Maciewicz, R. A. and Etherington, D. J. (1988) A composition of four cathepsin (B, L, N, and S) with collageolytic activity from rabbit spleen. *Biochem J.* 256, 433-440.

Makinodan, Y. , Nakagawa, T. and Hujita, M. (1993) Effect of cathepsins on textural change during ripening of Ika-shiokara (salted squid preserves) . *Nippon Suisan Gakkaishi* 59, 1625-1629.

Margoshes, M. and Valle, B. L. (1957) Cadmium protein from equine kidney cortex. *J. Amer. Chem. Soc.* 79, 4813-4814.

Martin, J. and Flegal, A. (1975) High copper concentrations in squid livers in association with elevated levels of silver, cadmium, and zinc. *Mar. Biol.* 30, 51-55.

Miramand, P. and Bently, D. (1992) Concentration and distribution of heavy metals in tissues of two cephalopods, Eledone cirrhosa and Sepia officinalis, from the French coast of the English Channel. *Mar. Biot.* 114, 407-414.

Okamoto, Y. , Otauka-Fuchino, H. , Horiuchi, S. , Tamiya, T. , Matsuraoto, J. J. and Tsuchiya, T. (1993) Purification and characterization of two metalloproteinases from squid mantle muscle, myosinase Ⅰ and myosinase Ⅱ, Biochim. *Biophys. Acta* 1161, 97-104.

Okitani, A. , Matsumura, U. , Kato, H. and Fujimaki, M. (1980) Purification and some properties of a myofibrillar protein degrading protease, cathepsin L, from rabbit skeletal muscle. *J. Biochem.* 87, 1133-1143.

Pangkey, H. , Hara, K. , Tachibana, K. , Cao, M. , Osatomi, K. and Ishihara, T, (2000) Purification and characterization of cathepsin S from hepatopancreas of carp, *Cyprinuscarpio*. *Fish. Sci.* 66, 1130-1137.

Quaglia, G. and Orban, E. (1987) Enzymic solubilization of proteins of sardine (*Sardina pilchardus*) by commercial proteases. *J. Sci. Food Agric.* 38, 263-269.

Raksakulthai, N. , Lee, Y. Z. and Haard, N. F. (1986) Effect of enzyme supplements on the production of fish sauce from male capelin *Mallotus villosus*. *Food Sci. Technol. J.* 19, 28-33.

Reddy, C. K. and Dhar, S. C. (1992) Purification and characterization of a collagenolytic property of renal cathepsin L form arthritic rat. *Int. J. Biochem.* 24, 1465-1473.

Sakai-Suzuki, J. , Sakaguchi, Y. , Hoshino, S. and Matsumoto, J. J. (1983) Separation of cathepsin D-like proteinase and acid thiol proteinase of squid mantle muscle. *Comp. Biochem. Physiol.* 75*B*, 409-412.

Sakai-Suzuki, J. , Tobe, M. , Tsuchiya, T. and Matsumoto, J. J. (1986) Purification and characterization of acid cysteine proteinase from squid mantle muscle. *Comp. Biochem. Physiol.* 85*B*, 887-893.

Shahidi, F. , Han, X. and Synowiecki, J. (1995) Production and characteristics of protein hydrolysates from capelin (*Mallotus villosus*) . *Food Chem.* 53, 285-293.

Tajima, T. , Tamori, J. , Kanzawa, N. , Tamiya, T. and Tsuchiya, T. (1998) Distribution of myosinase I and myosinase II in tissues of Coleoidea. *Fish Sci* 64, 808-811.

Tamori, J. , Kanzawa, N. , Tajima, T. , Tamiya, T. and Tsuchiya, T. (1999) Purification and characterization of a novel isoform of myosinase from spear squid liver. *J. Biochem.* 126, 969-974.

Tanaka, T. , Hayashi, Y. and Ishizawa, M. (1983) Subcellular distribution and binding of heavy metals in the untreated liver of the squid; comparison with data from the livers of cadmium and silver exposed to rats. *Experientia* 39, 746-748.

Tsujioka, E. , Rhara, T. , Kanzawa, N. , Noguchi, S. and Tsuchiya, T. (2005) Effects of additives on the thermal gelation of Japanese common squid natural actomyosin. *Fish Sci.* 71, 688-690.

Yamashita, M. and Konagaya, S. (1992) Differentiation and localization of catheptic proteinases responsible for extensive autolysis of mature chum salmon muscle (*Oncorhyncjus keta*) . *Comp. Biochem. Physiol.* 103*B*, 999-1003.

第 10 章　基于海洋生物酶的立体选择性合成

Dunming Zhu, Jianjiong Li and Qiaqing Wu, Tianjin Institute of Industrial Biotechnology, Chinese Academy of Sciences, Tianjin, China

DOI：10. 1533/9781908818355. 3. 237

摘要：极度丰富的海洋微生物提供了大量的生物催化资源，这些资源具有独特的性能和重要的应用。本章总结了近期海洋生物催化剂在非对称化学合成中的应用进展，包括羟基化作用、环氧化作用、缩合作用、还原胺化作用，环氧化和酯的水解作用以及转糖基作用。

关键词：海洋酶，立体选择性，合成，生物催化，微生物

10.1　引言

利用生物催化剂进行化学合成中手性化合物的合成在精细化合物制备过程中的作用越来越重要，如制药和农用化合物的合成。发现合适的酶依然是完成目标反应或者合成一个靶标分子的关键。来源于极端海洋环境下的高度多样化的海洋微生物资源，提供了大量的具有新型化学与立体异构化学催化性质的酶，如耐盐度、耐热性、适冷性的一些微生物（Bornscheuer，2005；Carlos，2006；Sogin et al.，2006；Trincone，2010，2011）。同样的，已经发现海洋生物催化剂在大量的有机反应中有很好的应用。包括氢过氧化反应、羟基化反应、环氧化反应、缩合反应、还原反应、胺化反应、环氧化合物和酯的水解作用、转糖基作用。在这一章按照反应类型，主要讨论涉及海洋生物酶催化的化学手性反应研究的前沿进展，激发大家对这一蓬勃发展研究领域的更大兴趣。

10.2　氧化反应

以区域和立体选择性的成链方式，脂肪氧合酶催化长链脂肪酸发生氢过氧化物反应，

产生相应的氢过氧化物，随后转变成长链的醛类，这是一种精油的成分（Akakabe and Nyuugaku，2007）。以前报道过长链的饱和烃和非饱和烃脂肪酸，比如肉豆蔻酸、十五烷酸、软脂酸、油酸、亚油酸、亚麻酸，这些脂肪酸能在 *Ulva pertusa* 绿藻含有的一种的酶的催化下发生 2-过氧羟基取代反应，使相对应的（R）-过氧羟酸旋光异构化，右旋效率达到 99%（Akakabe et al.，1999，2000，2001）。过氧醛基化反应也能发生在脂肪酸其他位置。当加入亚麻酸和亚油酸与海洋绿藻 *Ulva conglobata* 的粗酶孵育时，（R）-9-HPODEA 和 HPOTrE 与各自对映异构体的比例达到了 99%以上（Akakabe et al.，2002）。当外源添加花生四烯酸到具有一个（R）-11-HPITE 的海洋绿藻天然酶中，它将进一步转化成 2，4-癸二烯醛，这充分说明了脂肪氧化酶和羟醛氧化酶的存在（Akakabe et al.，2003）。

芳香烃的羟基化不仅仅是这些化合物代谢的第一步，同时也能产生单或多羟基化产物，为其开发利用提供了一个新的途径，因为他们是各种精细化合物制备的关键中间体（Hollmann et al.，2011）。在很多海洋生物中发现了作用于芳香类物质的羟化酶活性。例如，在海洋多毛类环虫 *Nereis virens* 的肠道组织暴露 5 d 后，83%的芘被羟化，其中产生 65%葡糖苷酸芘，12%的硫酸芘，2%萘葡萄糖苷和 4%的 1-羟基芘（图 10.1）。这表明海洋多毛类环虫 *Nereis virens* 具有很高的多环芳烃生物转化能力。芘能够诱导芘羟化酶活性（Hollmsnn et al.，2011）。针对从 *N. virens* 的肠道组织中分离出来两种新型的 CYP 酶（CYP342A1 和 CYP4BB1）的表征发现，CYP342A1 和 CYP4BB1 有单氧化活性和催化芘羟化（1）产生 1-羟基芘（2）（Jørgrnsen et al.，2005b，2008）。

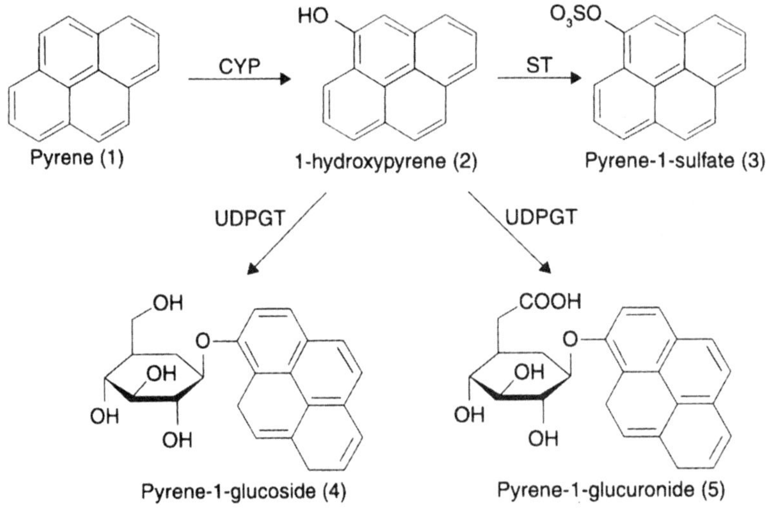

图 10.1 由海洋多毛类 *Nereis virens* 产生的氢氧化芘

各种各样的单分子或者双分子取代萘如（双甲基萘）都可以被重组大肠杆菌表达的芳香族双加氧酶（phnA1A2A3A4 和 phdABCD）双羟基化。目的基因（phnA1A2A3A4 和

phdABCD）分别来源于稠环芳烃海洋细菌 *Nocardioides* sp. KP7 和 *Cycloclasticus* sp. A5。然而双氧酶 PhnA1A2A3A4 有广泛的底物特异性，当 phdABCD 作为催化剂的时候，在所有的反应中，该酶经常倾向于芳香族环的双羟基化反应。例如，在大肠杆菌表达的 phnA1A2A3A4 基因存在时，二甲基乙酰胺比如 1，4-二甲基亚硝酸（6）和 1，6-二甲基亚硝酸（9）大部分转换成二羟基萘（7，10）；用 PhdABCD 作催化剂时，3，8-双甲基-1，2-萘-1，2-二醇（12）和 3，7 双甲基-1，2-萘-1，2-二醇（14）分别是 1，6-二甲基萘和 2，6-二甲基萘生物转化的唯一产物（13）见图 10. 2（Shindo et al.，2007）。

加双氧化酶 PhdABCD 催化 1，6-二甲基萘和 2，6-二甲基萘的生物转化见图 10. 2（Shindo et al.，2007）。

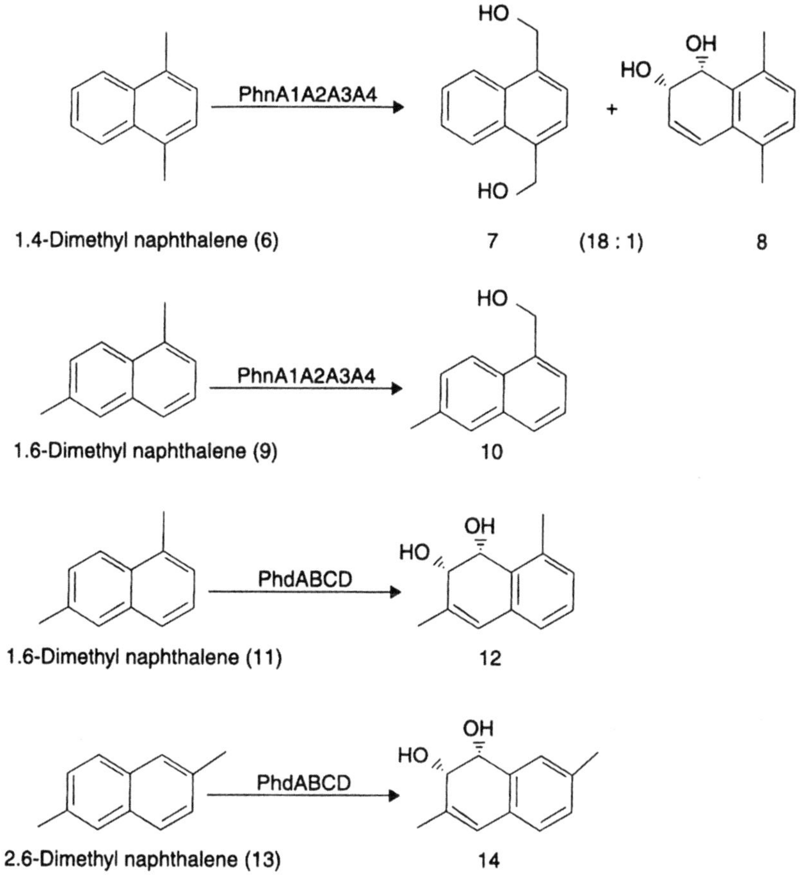

图 10. 2　由芳香双加氧酶氢氧化二甲基萘（phnA1A2A3A4 和 phdABCD）

据报道一种溴化倍半萜烯（aplysistatin）（15）通过一种海洋真菌 *R. atrovirens* NRBC 32362 可以转化成 3 种新的成分（图 10. 3）。5α-hydroxyaplysistatin（16）和 5α-hydroxyisoaplysistatin（17）和 9β-hydroxyaplysistatin（18）。生物催化的时序性提示了 aplysistatin（15）在 5-位羟基化产生了 5α-hydroxyaplysistatin（16），随后溴的丢失和甲基化生成了 5αhydroxyisoaplysistatin（17）（Koshimura et al. , 2009）。

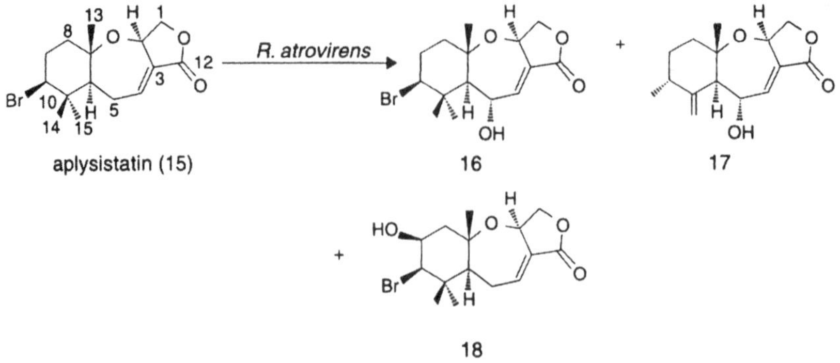

图 10.3　由海洋真菌 *R. atrovirens* NRBC 32362 产生的 aplysistatin 生物转化

在海洋藻类中发现了 V-BrPO 的存在（Butler and Sandy, 2009）。从红藻中克隆了 V-BrPO（*Plocamium cartilagineum*, *Laurencia pacifica*, *Corallina officinalis*），它们可以产生卤代化合物成分。这个酶催化橙花醇（19）的溴化和环化，在溴离子和过氧化氢存在的情况下修饰一溴八环乙醚（20），伴随着末端的溴醇反应成二溴化物和环氧化产物（Carter-Franklin et al. , 2003; Carter-Franklin and Butler, 2004）。然而，与香叶醇（21）反应产生了两个单独的溴化六元环产物，非环化的溴醇和环氧化物（22，23）。溴化及环化反应见图 10.4。来源于海洋红藻（*Delisea pulchra*）的含钒的溴化过氧化物酶（24）已经被证实可以在溴及过氧化氢存在的情况下，催化戊炔酸的溴化内酯化，生成 5E-溴化-亚甲基四氢-呋喃酮（25）（图 10.5），这种成分可以扰乱工程菌（*Agrobacterium tumefaciens*）（NTL4）群体感应（Sandy et al. , 2011）。

从丝状真菌 *Caldariomyces fumago* 来源的 CPO 是一种特别的生物催化剂，比起传统的氧化酶它更像 P-450 细胞色素。CPO 催化很多合成中有用的氧化反应，例如烯烃非对称的环氧化作用，羟基化作用和非对称的磺化氧化作用（Conesa et al. , 2001）。具有 99%选择性的 2-位吲哚衍生物氧化为 2-羟吲哚。苯硫基甲烷发生非对称氧化反应产生了对映 R-亚砜异构体效率超过 98%。单帖也是从 *C. fumago* 分离出来的氯化物过氧化物酶氧化和羟基化的底物。在卤化钠存在下，CPO 催化（1*S*）-(+)-3-carene 转化成（1*S*, 3*R*, 4*R*, 6*R*)-4-halo-3, 7, 7-trimethyl-bicyclo-heptane-3-ols（Kaup et al. , 2007）。在没有氯离子的情况下，CPO 立体选择氧化 R-(+)-limonene（26）成为（1*S*, 2*S*, 4*R*)-limonene-1, 2-diol（29），非对映产率超过了 99%。当反应在氯离子存在的情况下，反应的速度得到提高并且保留了反应的立体选择性，反应产物是（1*S*, 2*S*, 4*R*)-limonene-1, 2-diol（29）和（1*R*, 2*R*, 4*R*)-limonene-1, 2-diol（30）混合物（图 10.6）。这就证明在没有氯离子参与的情况下，CPO 酶活性位点对柠檬烯的立体选择性控制着反应的立体异构选择性。在氯化钾存在的情况下，柠檬烯通过次氯酸的氧化反应没有立体选择性（Aguila et al. , 2008）。

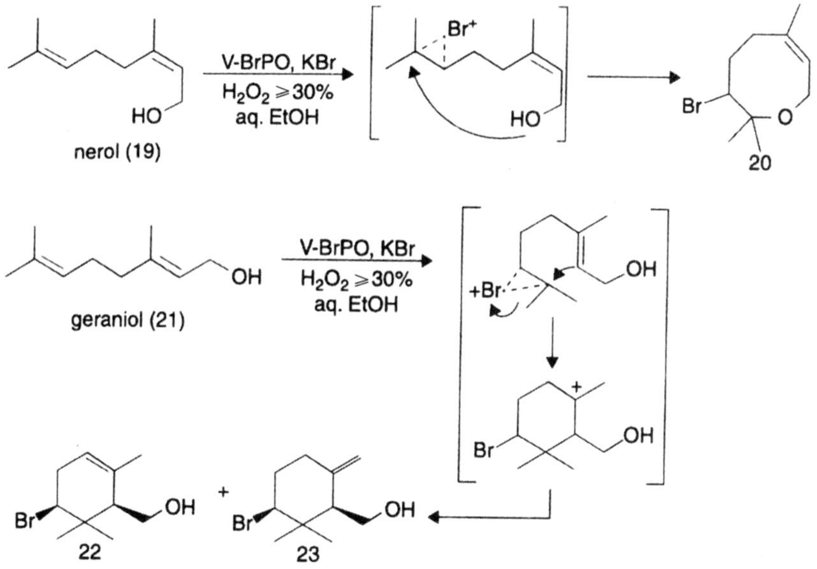

图 10.4　钒溴过氧化物酶对橙花醇和香叶醇的生物转化

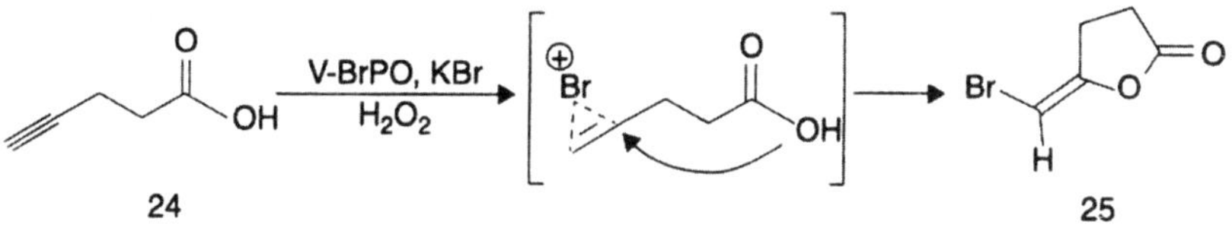

图 10.5　钒溴过氧化物酶对 4-戊炔酸的溴代内酯化

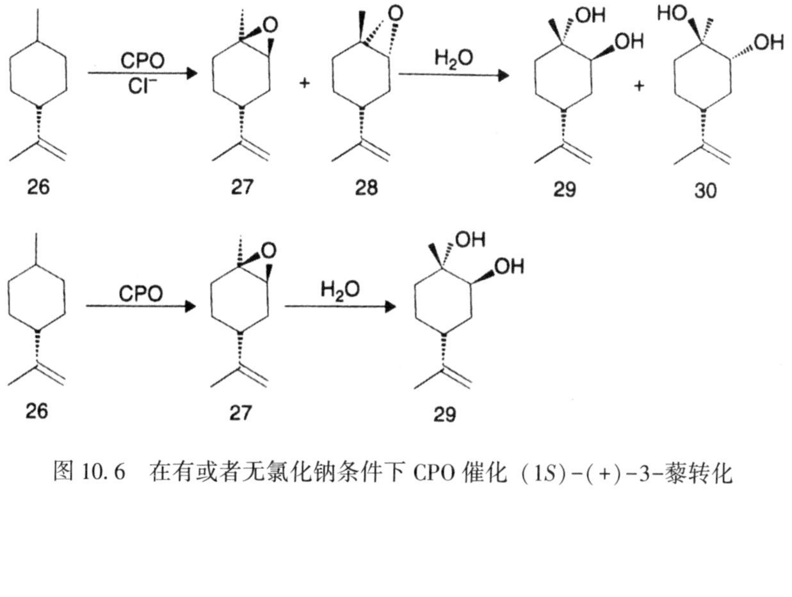

图 10.6　在有或者无氯化钠条件下 CPO 催化（1*S*）-（+）-3-蒈转化

10.3 还原反应

手性乙醇一直是制药行业和其他精细化工合成中重要的中间体。直接的合成途径是前手性的酮还原为手性乙醇。因为生物催化还原合成酮的反应具有很多优势，比如无毒、环境友好和温和的反应条件，高的化学，区域及立体选择性，产品中低的金属残留等（Patel，2001；Nakamura et al.，2003；Kroutil et al.，2004）。因此，分离出的酮还原酶的数量也越来越多，并且它们优于其他酮的还原反应。事实上，De Wildeman 与其合作伙伴已经宣称生物还原反应已经从“实验室”到工作的“首选”（De Wildeman et al.，2007；Moore et al.，2007）。在近几年，海洋生物来源的酮还原酶在合成手性乙醇的过程中的作用越来越多地被人们证实。

来源于极端嗜热古细菌 *P. furiosus* 的一种乙醇脱氢酶的基因 *adh*D 已经被克隆，并在大肠杆菌表达，重组酶（PFADH）也已经被纯化（Machielsen et al.，2006）。该酶的底物特异性及对映异构体的选择性也已经被评价，主要包括各种各样的酮、芳香族酮、α 和 β 型酮酯在包括含有 D-型葡糖糖和 D 型葡萄糖脱氢酶的 NADH 再生系统（图 10.7）（Zhu et al.，2006，2009）。芳香酮（表 10.1，1~7 条目），一种苯基取代的 α 和 β 酮还原成相应的具有很高的光学纯度的手性醇（32），然而，无苯基取代的还原反应呈现出一个中等的对映异构体选择性（表 10.1，条目 13、14、16~20），这表明靠近羰基的苯基对对映选择性是非常重要的。

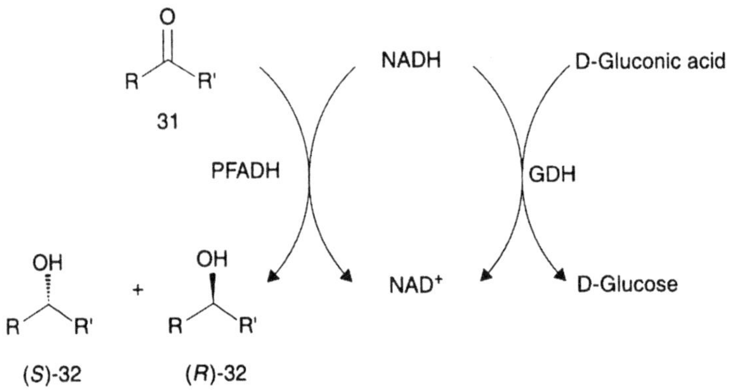

图 10.7　来自超嗜热古细菌 *P. furiosus* 的醇脱氢酶对酮的不对称还原

表 10.1　PFADH 催化不同酮的还原

Entry	R	R′	Relative activity[a]	ee/%
1	CH_3	C_6H_5	100	99（S）
2	CH_3	4′-Cl C_6H_4	131	99（S）

续表

Entry	R	R′	Relative activity[a]	ee/%
3	CH_3	4′-CH_3 C_6H_4	108	99 (S)
4	CH_3	2′-Cl C_6H_4	231	99 (S)
5	CH_3	3′-Cl C_6H_4	285	99 (S)
6	CH_3	3′, 5′- $(CF_3)_2$ C_6H_4	131	99 (S)
7	CH_2Cl	C_6H_5	1 331	99 (R)
8	$CO_2C_2H_5$	C_6H_6	754	99 (R)
9	$CO_2C_2H_5$	4′-Cl C_6H_4	915	99 (R)
10	$CO_2C_2H_5$	4′-CH_3 C_6H_4	523	99 (R)
11	$CO_2C_2H_5$	4′-CN C_6H_4	1 469	99 (R)
12	$CO_2C_2H_5$	3′, 5′-F_2 C_6H_4	646	99 (R)
13	$CO_2C_2H_5$	*iso*-C_3H_7	946	44 (S)
14	$CO_2C_2H_5$	*tert*-C_4H_9	654	71 (S)
15	$CH_2CO_2C_2H_5$	C_6H_5	154	99 (S)
16	$CH_2CO_2C_2H_5$	CH_3	285	95 (S)
17	$CH_2CO_2C_2H_5$	C_2H_5	162	60 (S)
18	$CH_2CO_2C_2H_5$	C_3H_7	215	24 (R)
19	$CH_2CO_2C_2H_5$	*iso*-C_3H_7	108	16 (S)
20	$CH_2CO_2C_2H_5$	CH_2Cl	1 092	4 (R)

a. 苯乙酮特有的活性在 13 $nmol \cdot min^{-1} \cdot mg^{-1}$在 37℃，它的相对活性定义为 100.

PFADH 催化的酮还原反应，NADH 作为一种辅因子既可以通过包括含有 D-型葡糖糖和 D 型葡萄糖脱氢酶共轭酶体系再生，也可以通过用一种用异丙醇作为供氢体的共轭底物体系再生。以氢转移模式，通过该酶完成了各种各样右旋 α-氯乙醇（33）还原成高纯度 R-构型的 α-氯苯乙酮（34）。这种模型，接近于异丙醇作为氢供体时辅因子 NAPH 可以循环利用（图 10.8 和表 10.2）（Zhu et al.，2009）。在该反应中，NADH 通过用异丙醇作为供氢体的共轭底物体系再生。既然氢转移是一个更加经济有效的手段，因此，该酶是一种非常有开发价值的催化获得手性氯乙醇的生物催化剂，它在医药领域具有重要应用前景，可以规模化地通过对 2-氯-4 对氯苯乙酮，2，2，4-三氯苯乙酮进行催化，获得纯化效率达到 83%～92% 的氯乙醇，纯化后的 ee 值 ≥98% 的氯乙醇（Zhu et al.，2009）。PFADH 也表现出来很好的有机溶剂耐受性，比如二甲基亚砜、异丙醇、甲基叔丁基醚和乙烷，这种特征对于在水中不溶的反应中是非常重要的（Zhu et al.，2006，2009）。

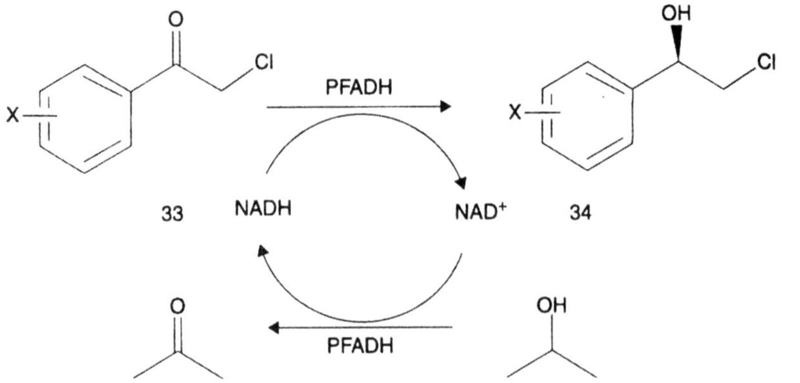

图 10.8　由超嗜热古细菌 *P. furiosus* 的醇脱氢酶在基质偶联模式下对 α-氯苯乙酮的还原

表 10.2　在基质偶联模式下 PFADH 催化 α-氯苯乙酮的还原

Entry	X	Yield/%	Ee/%
1	4′-H	99	99
2	4′-F	97	99
3	4′-Cl	75	99
4	4′-NO_2	96	98
5	4′-CH_3CONH	97	99
6	3′-Cl	75	99
7	2′, 4′-($Cl)_2$	90	99
8	3′, 4′-($Cl)_2$	100	99
9	3′-NO_2-4′-CH_3CONH	87	99

很多嗜热酶在室温下通常表现出低活性。为了得到一种在低温条件下有活性的乙醇脱氢酶催化生成（2S，5S）-已二醇，针对从嗜热古细菌 *Pyrococcus furiosus* 中获得的 AdhA 进行实验室的进化研究，并且获得了一种 S-选择的突变酶（R11L/A180V），它在 30℃时，表现比野生菌高 10 倍的催化 2，5-已二酮的活性。突变酶的两个突变的氨基酸残基位于高度保守的 NADP（H）结合区，并且靠近 183 位的高度保守的催化活性位点（Thr）(Machielsen et al.，2008)。

近来，许多来源于海洋蓝藻细菌的酮还原酶被克隆，并在大肠杆菌中获得了表达。这些重组酶研究了针对 4 种底物的活性，包括 2′，3′，4′，5′，6′-五氯苯乙酮、4′-氯乙酰苯、4-氯乙酰乙酸乙酯、苯甲酰乙酸乙酯，它们作为底物的组成单元（Hölsch and Weuster-Botz，2010）。其中，藻青菌里 *Synechococcus* sp. RCC 307 里的 3-酰基酮还原酶对上述 4 种酶底物表现出了很高的对映选择性。

虽然从海洋生物中获得酮还原酶是最近开始的工作，这些酶在合成光学纯乙醇也表现出来了潜力（Hölsch and Weuster-Botz，2010；Timpson et al.，2012）。同时，大量的酮被报道会被海洋真菌的全细胞（Rocha et al.，2009，2010，2012）、酵母、细菌（Hölsch and Weuster-Botz，2010；Mouad et al.，2011）、单孢子和藻类（Hook et al.，1999，2003；Ishihara et al.，2001；Mouad et al.，2011）还原。各种乙酰苯被海洋真菌还原成手性乙醇。当1-（4-甲酰基苯基）-乙酮被 *A. sydowii* Ce15，*A. sclerotiorum* CBMAI 849 Ce15 和 *Bionectria* sp. Ce5 不对称的生物还原，产生具有光学活性的（*R*）-1-（4-甲氧基苯基）-乙醇（>99%ee），这是依据反 Prelog 法则。当 1-(4-甲酰基苯基）-乙酮被 *B. felina* CBMAI 738 被不对称生物还原，产生具有光学活性的（*S*)-1-(4-甲氧基苯基）-乙醇（>99%ee），这是依据 Prelog 法则（Rocha et al.，2012）。7 种海洋真菌已经表现出可以催化 α-氯乙酰苯不对称还原反应成（*S*)-(-)-2-氯-1-苯基乙醇，纯度在 17%～66%，转化率在 23%～99%（Rocha et al.，2009）。α-溴苯乙酮的还原反应比利用来源于海洋真菌 *Aspergillus sydowii* Ce19 全细胞的生物转化更复杂。它会产生 2-溴化-1 苯甲醇（56%），同时含有 9%的 α-氯醇、26%1-苯基乙烷-1，2-二醇、4%苯乙酮、5 苯甲醇（Rocha et al.，2010）。原因是通过电子转移，不稳定的溴被来源于培养基中羟基或者氯离子所替换，失去电子转移给乙酰苯，它同时被还原。在相同的条件下，针对 *p*-溴-α-溴-乙酰苯和 *p*-硝基-α-溴苯乙酮，虽然大量的生物催化产物被鉴定到，但是没有观察到卤代化合物的形成（Rocha et al.，2010）。

海洋微藻也能还原乙酰苯衍生物。邻位、间位、对位的碘酸酯苯酮也可以被来源于 *B. tenella* 与 *B. radicans* 提取物进行生物还原反应，产生 *S*-构象的乙醇（>99% ee）。乙酰苯衍生物的邻位取代物由于空间位阻而难于被还原。然而，发现 *B. tenella* 与 *B. radicans* 提取物可以高效地将邻苯碘乙酮进行生物转化还原为碘苯乙酮。*B. tenella* 藻物质因此被用于进行乙酰苯的不同的邻位取代物的还原反应，并且发现（*S*）-构位乙醇的 ee 值在 95%～99%，转化效率会伴随着邻位取代基团的变化而减少（Cl>Br>NO_2）（Mouad et al.，2011）。相反的，5 种光合微藻培养基 *C. minutissima*，*N. atomus*，*D. parva*，*P. purpureum*，*I. gallbana* 对于乙酰苯表现出非常低的对映异构体选择性（Hook et al.，2003）。这些培养物能够还原环己酮和 1*S*，4*R*-(-)-薄荷酮成对应的乙醇、并不是环己烯酮共轭的羰基。然而，当 α，β-不饱和酮 4*R*-(-)-香芹酮被转化成 1*R*，4*R*-(+)-二氢黄篙萜酮和 1*R*，2*S*，4*R*-(+)-二氢香芹酮，4*S*-(+)-香芹酮被转化成 1*S*，4*S*-(-)-二氢黄篙萜酮、1*S*，2*R*，4*S*-(-)-二氢香芹酮，1*R*，4*S*-(+)-香芹酮和 1*R*，2*R*，4*S*-(-)-香芹酮，这说明成 C=C 和 C=O 被海洋微藻的培养基还原了（Hook et al.，2003）。

α-和 β-酮酯也能被一种海洋微藻还原成相应的羟化酯（图 10.9）。对来源于 4 种海洋藻类 *Chaetoceros gracilis*，*Chaetoceros* sp.，*Nannochloropsis* sp. 和（*Pavlova lutheri*），的 7 种对 α-酮酯的反应进行了测试。在所有的实验中，相应的 α-羟基酯的 ee 在 3%～98%。一些添加剂，如葡萄糖和乳糖，对乙基、3-甲基-2-氧代丁酸丁酯（37）对还原反应的立体选择性有很大影响。在 L-乳酸存在的情况下，*Chaetoceros gracilis* 完成 99% 的 ee 率。乙基 2-甲基-3-氧代丁酸丁酯会被 *Nannochloropsis* sp. 还原，产生具有很好非对映性（*syn/*

anti=1∶99）和高光学选择性的（*anti*>99%，*syn* 98%）的反向羟基酯（38），而其他藻类还原为高纯度反向羟基酯的能力比较弱（Ishihara et al.，2001）。

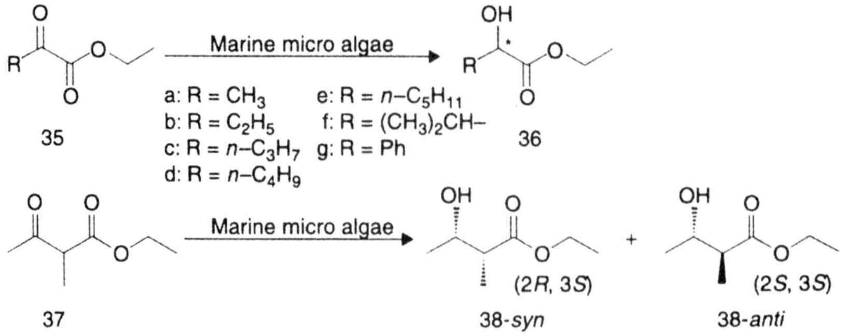

图 10.9 由海洋微藻对 α–和 β–酮酯的还原

海藻（*B. tenella* 与 *B. radicans*）的共生微生物也被发现可以提高邻位、间位、对位的碘酸酯苯酮的还原反应，且有较高纯度的对映异构体产生（>99% ee）（Mouad et al.，2011）。

除了羰基还原酶的活性，从海洋微生物中还发现了烯还原酶的活性。给海洋真菌 *Rhinocladiella* sp. K–001 注射他汀类药物（aplysistatin（15）），在他汀类的 C–3 位置的 C–C 双键被立体性还原（图 10.10），然而，用 *Rhinocladiella* sp. K–001 进行立体相似结构的生物转化过程中，palisadin A（40）在 C–9—C–10 脱溴化来得到相应的脱氢溴酸产物（41），12–hydroxypalisadin（42）的溴醇基团被优先脱溴化产生 palisadin A（40），palisadin A 可以在 C–9—C–10 脱溴化来提供脱氢溴酸产物（41）（路线图 10）（Koshimura et al.，2009）。

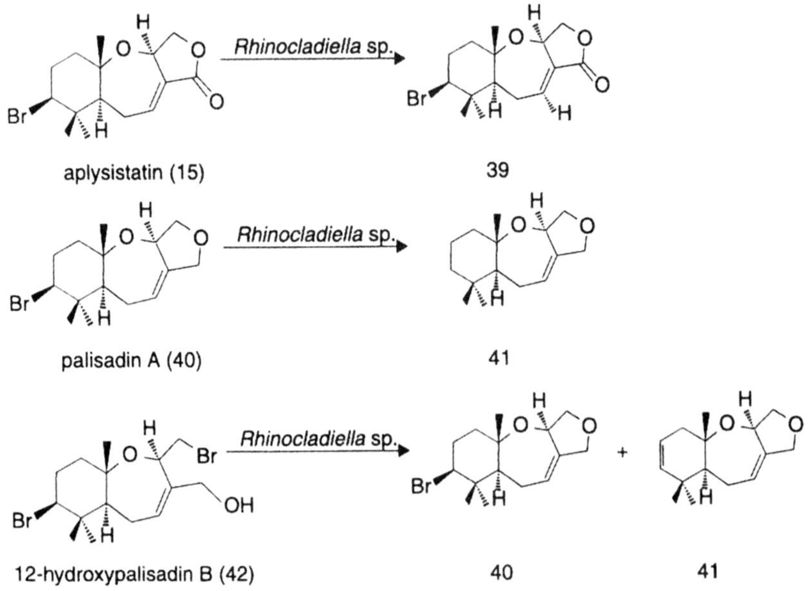

图 10.10 由海洋真菌 *Rhinocladiella* sp. K–001 对 aplysistatin，palisadin A 和 12–羟基尿素 B 的生物转化

10.4 还原氨化反应

氨基酸脱氢酶在 NAD^+存在下可催化 L-氨基酸的可逆脱氨反应生成相应的 α-酮酸和氨。由于氨基酸脱氢酶通常是高度对映特异性的，α-酮酸的逆向还原氨化为合成纯的 L-氨基酸是一个非常好的方法（Groeger et al.，2006）。来源于 *Vibrio proteolyticus* DSM30189 的丙氨酸脱氢酶基因在大肠杆菌 TG1 菌株中进行克隆和表达。重组酶对丙酮酸盐、β 氟代丙酮酸、β 羟基丙酮酸表现出高还原氨化活性，并被用来合成相应的 L-氨基酸。酶在 2 mol/L NaCl 浓度中仍保留高的酶热稳定性，这可以大大提高酶的稳定性（Kato et al.，2003）。

10.5 环氧衍生物的水解

环氧化物水解酶（EHase）催化外消旋环氧化物的对映选择性水解产生手性环氧化物和二醇，是手性化合物合成的一个很好的方法，此两类化合物是合成手性化合物中有用的中间体（Archelas and Furstoss，2001；Kotik et al.，2012）。据报道第一个具有环氧化物水解酶活性海洋微生物：*Sphingomonas echinoides* EH-983 是从海水中分离的。它优先水解 *R*-苯乙烯氧化物生成的 21.3%光学对映体 *S*-氧化苯乙烯（Kim et al.，2006）。通过对从各种海洋环境中分离出降解苯乙烯能力菌株的筛选，其中一株 *Erythrobacter* sp. JCS358，鉴定出对氧化苯乙烯和环氧丙基苯基醚有对映选择性水解活性。利用 *Erythrobacter* sp. JCS358 全细胞在 15 h 孵育后产生的 S-氧化苯乙烯占 99%以上，以此实现外消旋环氧苯乙烷的动力学拆分（Hwang et al.，2008）。

缩水甘油苯基醚（GPE）是合成手性氨基醇和生物活性化合物，如 β-阻滞剂的有用的中间体。从海洋细菌 *R. bacterium* HTCC2653 提取的高对映选择活性 EHase 被应用于 GPE 的动力学拆分，其中 S-GPE 比 R-GPE 优先水解，对映体过 ee 大于 99%，生成量为 38.4%（Woo et al.，2010）。南大西洋海洋真菌（*Aspergillus sydowii*）和（*Trichoderma* sp.）全细胞也能介导外消旋 GPE 的动力学拆分，但对映选择性较低（Martins et al.，2011）。

斑马鱼（*Danio rerio*）的微粒体 EHase 进行了确认和在大肠杆菌中进行了异源表达。该酶对 *R*-苯乙烯氧化物表现出对映优先性，*S*-苯乙烯氧化物可以制备 ee 为 99%，产量为 23.5%（Kim et al.，2005）。然而斑马鱼可溶性烯烃氧化物水解酶及其突变体催化水解 *S*-苯乙烯氧化物生成 ee 99%的 *R*-苯乙烯氧化物（Woo et al.，2012）。

动力学拆分过程本身限制的最大收率为 50%。外消旋环氧化物对映水解能产生二元醇高达 100%的 ee 和 100%的收率。例如，*R*-苯基-1，2-乙二醇（44）通过两种重组环氧化物水解酶催化水解苯乙烯氧化物（43）可获得收率达 94%，ee 为 90%，一种来源自（*Caulobacter crescentus*）；另一种来自 *Mugil cephalus*（图 10.11）（Kim et al.，2008）。

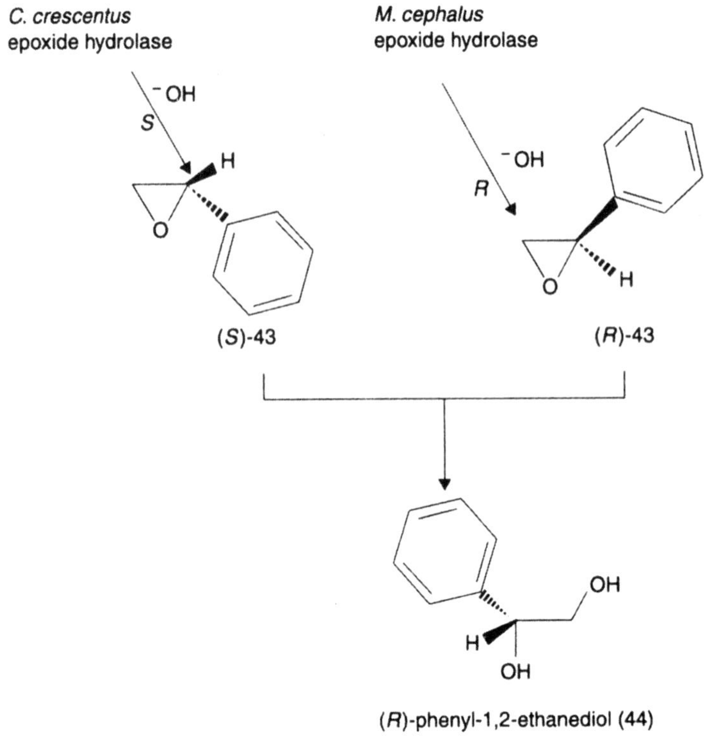

图 10.11　由一种细菌 *Caulobacter crescentus* 和一种海洋鱼类 *Mugil cephalus* 的环氧化物水解酶催化外消旋环氧化物的对映体水解

10.6　水解作用

从海洋环境中分离出一种具有酯酶活性的海洋微生物—（*Yarrowia lipolytica*）CL180。脂肪酶/脂酶基因（yl180）在大肠杆菌进行克隆和表达。重组酶优先水解少于 10 个碳链长度的对硝基苯酯类脂肪酸，属于 B1 羧酸酯酶/脂肪酶家族。它优先水解 S-对映体消旋氧氟沙星酯（45），显示出其潜在的作为水解催化剂的可能性（图 10.12）（Kim et al.，2007）。来自北极沉积物宏基因组或海兔卵分离的海洋细菌（*Vibrio* sp. GMR509）的酯酶都优先催化（S）-氧氟沙星丁基酯的水解（Park et al.，2007；Jeon et al.，2009）。

近年来，通过对来自海洋环境包括潮滩沉积物（Jeon et al.，2012）、表层海水（Chu et al.，2008）和深海沉积物宏基因组文库筛选，陆续发现了各种脂肪酶和酯酶（Jiang et al.，2012）。其中，一种来源于韩国潮滩沉积物的宏基因组文库的酯酶（EstEH112）能在 4 h 内有效水解一组庞大的叔醇乙酸酯，但立体选择性较低（Oh et al.，2012）。

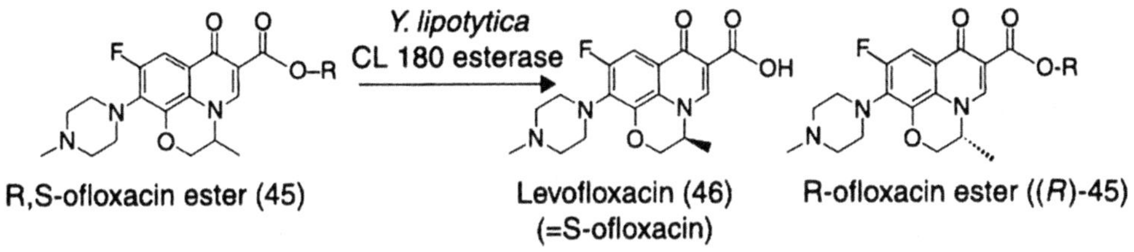

图 10.12 来自 *Yarrowia lipolytica* CL180 的海洋酯酶对外消旋氧氟沙星酯的动力学

10.7 转糖基作用

糖基水解酶，如β-甘露糖苷酶、α-葡萄糖苷酶和β-半乳糖苷酶不仅水解糖苷键，也催化这种键的立体特异性的形成，从而发现其在低聚糖合成中的应用。对来自海洋软体动物海兔（*Aplysia fasciata*）的α-葡萄糖苷酶的水解底物特异性和转糖苷反应都进行了研究（Andreotti et al.，2006）。这种酶可以催化吡哆醇糖基化，合成有色二糖和三糖以及纤维二糖和蔗糖的糖基化。吡哆醇被海洋酶催化在 5 端区域选择性 α 糖化，葡萄糖苷和二糖的摩尔收率为 80%（吡哆醇单糖为 24 g/L。吡哆醇二糖为 35 g/L）（Tramice et al.，2006）。源自海兔（*Aplysia fasciata*）的β-半乳糖苷酶也被纯化并应用于抗病毒和抗癌核苷的β半乳糖基衍生物的常规合成。所有的反应具有非常严格的异构和立体选择性，只有端基异构体纯的 5′-*O*-半乳糖缀合物生成。例如，合成了 60%收率的 5′-*O*-β-半乳糖-5-氟尿苷和 43%收率的 5′-*O*-β-半乳糖-3′-叠氮脱氧胸苷是一种抗艾滋病药物的衍生物（Andreotti et al.，2007）。

10.8 结论

虽然海洋酶已经逐渐被大量地用作生物催化剂进行不对称氢过氧化、羟基化、环氧化、还原、还原胺化、环氧化合物和酯水解、转糖基化等，但这一领域还有待进一步探索。不对称合成往往涉及疏水底物或产物，因此需要具有高稳定性和耐有机溶剂的生物催化剂。极端生活环境的特性，如耐盐、高温稳定性、嗜压性和海洋微生物低温适应性可以提供多样化的生物催化剂。应该可以预期的是，未来几年新的化学和海洋生物催化剂的立体化学特性肯定会引发更多的兴趣，对新型海洋酶及其在化学合成中的应用进行更深的探索研究。

感谢

感谢中国国家自然科学基金（No. 21072151）与中国国家基础研究计划（“973”项

目，No. 2011CB710801）的资金支持。

参考文献

Aguila, S., Vazquez-Duhalt, R., Tinoco, R., Rivera, M., Pecchi, G. and Alder J. B. (2008) Stereoselective oxidation of R- (+) -limonene by chloroperoxidas from *Caldariomyces fumago*. *Green Ghent.* 10, 647–653.

Akakabe, Y., Matsui, K. and Kajiwara, T. (1999) Enantioselective α-hydroperoxylation of long-chain fatty acids with crude enzyme of marine green alga *Ulva pertusa*. *Tetrahedron Lett.* 40, 1137–1140.

Akakabe, Y., Matsui, K. and Kajiwara, T. (2000) Alpha-oxidation of long-chain unsaturated fatty acids in the marine green alga *Ulva pertusa*. *Biosci. Biotechnol.*, *Biochem.* 64, 2680–2681.

Akakabe, Y., Matsui, K. and Kajiwara, T. (2001) Enantioselective 2-hydroperoxylation of long-chain fatty acids in marine algae. *Fisheries Sci.* 67, 328–332.

Akakabe, Y., Matsui, K. and Kajiwara, T. (2002) Enantioselective formation of (R) -9-HPODE and (R) -9-HPOTRE in marine green alga *Ulva conglobata*. *Bioorg. Med. Ghem.* 10, 3171–3173.

Akakabe, Y., Matsui, K. and Kajiwara, T. (2003) 2, 4-Decadienals are produced via (JR) -l 1-HPITE from arachidonic acid in marine green alga *Ulva conglobata*. *Bioorg. Med. Ghem.* 11, 3607–3609.

Akakabe, Y. and Nyuugaku, T. (2000) An efficient conversion of carboxylic acids to one-carbon degraded aldehydes via 2-hydroperoxy acids. *Biosci.*, *Biotechnol Biochem.* 71, 1370–1371.

Andreotti, G., Giordano, A., Tramice, A., Mollo, E. and Trincone, A. (2006) Hydrolyses and transglycosylations performed by purified a-d-glucosidase of the marine mollusc *Aplysia fasciata*. *J. Biotechnol.* 122, 274–284.

Andreotti, G., Trincone, A. and Giordano, A. (2007) Convenient synthesis of *β*-galactosyl nucleosides using the marine *β*-galactosidase from *Aplysia fasciata*. *J. Mol. Catal. B-Enzym.* 47, 28–32.

Archelas, A. and Furstoss, R. (2001) Synthetic applications of epoxide hydrolases. *Curr. Opin. Chem. Biol.* 5, 112–119.

Bornscheuer, U. T. (2005) Deep sea mining for unique biocatalysts. *Chem. Biol.* 12, 859–860.

Butler, A. and Sandy, M. (2009) Mechanistic considerations of halogenating enzymes. *Nature* 460, 848–854.

Carlos, P.-A. (2006) Marine microbial diversity: Can it be determined? *Trends Microbiol.* 14, 257–263.

Carter-Franklin, J. N. and Butler, A. (2004) Vanadium bromoperoxidase-catalyzed biosynthesis of halogenated marine natural products. *J. Am. Chem. Soc.* 126, 15060–15066.

Carter-Franklin, J. N., Parrish, J. D., Tschirret-Guth, R. A., Little, R. D. and Butler, A. (2003) Vanadium haloperoxidase-catalyzed bromination and cyclization of terpenes. *J. Am. Chem. Soc.* 125, 3688–3689.

Chu, X., He, H., Guo, C. and Sun, B. (2008) Identification of two novel esterases from a marine metagenomic library derived from south china sea. *Appl Microbiol. Biotechnol.* 80, 615–625.

Conesa, A., van de Velde, F., van Rantwijk, F., Sheldon, R. A., van den Hondel, C. A. M. J. J. and Punt, P. J. (2001) Expression of the Caldariomyces fumago chloroperoxidase in Aspergillus niger and characterization of the recombinant enzyme. *J. Biol. Chem.* 276, 17635–17640.

De Wildeman, S. M. A., Sonke, T., Schoemaker, H. E. and May, O. (2007) Biocatalytic reductions: From lab curiosity to 'first choice'. *Acc. Chem. Res.* 40, 1260–1266.

Groeger, H. , May, O. , Werner, H. , Menzel, A. and Altenbuchner, J. (2006) A 'second-generation process' for the synthesis of l-neopentylgiycine: asymmetric reductive amination using a recombinant whole cell catalyst. *Org. Proc. Res. Dev.* 10, 666-669.

Hölsch, K. and Weuster-Botz, D. (2010) New oxidoreductases from cyanobacteria: exploring nature's diversity. *Enzyme Microb. Technol.* 47, 228-235.

Hollmann, F. , Arends, I. W. C. E. , Buehler, K. , Schallmey, A. and Buhlei; B. (2011) Enzyme-mediated oxidations for the chemist. *Green Chem.* 13, 226-265.

Hook, I. L. , Ryan, S. and Sheridan, H. (1999) Biotransformation of aromatic aldehydes by five species of marine microalgae. *Phytochem.* 51, 621-627.

Hook, I. L. , Ryan, S. and Sheridan, H. (2003) Biotransformation of aliphatic and aromatic ketones, including several monoterpenoid ketones and their derivatives by five species of marine microalgae. *Phytochem.* 63, 31-36.

Hwang, Y. -O. , Kang, S. , Woo, J. -H. , Kwon, K. , Sato, T. , et al. , (2008) Screening enantioselective epoxide hydrolase activities from marine microorganisms: detection of activities in *Erythrobacter spp. Mar. Biotechnol.* 10, 366-373.

Ishihara, K. , Nakajima, N. , Yamaguchi, H. , Hamada, H. and Uchimura, Y. -S. (2001) Stereoselective reduction of keto esters with marine micro algae. *J. Mol. Catal. B-Enzym.* 15, 101-104.

Jørgensen, A. , Giessing, A. M. B. , Rasmussen, L. J. and Andersen, O. (2005a) Biotransformation of the polycyclic aromatic hydrocarbon pyrene in the marine polychaete *Nereis virens. Environ. Toxicol. Chem.* 24, 2796-2805.

Jørgensen, A. , Giessing, A. M. B. , Rasmussen, L. J. and Andersen, O. (2008) Biotransformation of polycyclic aromatic hydrocarbons in marine polychaetes. *Mar. Environ. Res.* 65, 171-186.

Jørgensen, A. , Rasmussen, L. J. and Andersen, O. (2005b) Characterisation of two novel cyp4 genes from the marine polychaete Nereis virens and their involvement in pyrene hydroxylase activity. *Biochem. Biophys. Res. Commun.* 336, 890-897.

Jeon, J. , Kim, J. -T. , Kang, S. , Lee, J. -H. and Kim, S. -J. (2009) Characterization and its potential application of two esterases derived from the arctic sediment metagenome. *Mar. Biotechnol.* 11, 307-316.

Jeon, J. , Lee, H. , Kim, J. , Kim, S. -J. , Choi, S. , et al. , (2012) Identification of a new subfamily of salt-tolerant esterases from a metagenomic library of tidal flat sediment. *Appl. Microbiol. Biotechnol.* 93, 623-631.

Jiang, X. , Xu, X. , Huo, Y. , Wu, Y. , Zhu, X. , et al. , (2012) Identification and characterization of novel esterases from a deep-sea sediment metagenome. *Arch. Microbiol.* 194, 207-214.

Kato, S. -I. , Ohshima, T. , Galkin, A. , Kulakova, L. , Yoshimura, T. and Esaki, N. (2003) Purification and characterization of alanine dehydrogenase from a marine bacterium, Vibrio proteolyticus. *J. Mol. Catal. B-Enzym.* 23, 373-378.

Kaup, B. -A. , Piantini, U. , Wiist, M. and Schrader, J. (2007) Monoterpe as novel substrates for oxidation and halo - hydroxylation with chloroperoxidase from Caldariomyces fumago. *Appl. Microbiol. Biotechnol*, 73, 1087-1096.

Kim, H. , Lee, O. , Hwang, S. , Kim, B. and Lee, E. (2008) Biosynthesis of (R) -phenyl-1, 2-ethanediol from racemic styrene oxide by using bacterial marine fish epoxide hydrolases. *Biotechnol. Lett.* 30, 127-133.

Kim, H. S. , Lee, O. K. , Lee, S. J. , Hwang, S. , Kim, S. J. , et al. , (2006) Enantio-selective epoxide hydrolase activity of a newly isolated microorganism Sphingomonas echinoides EH-983, from seawater. *J. Mol. Catal. B-Enzym.* 41, 130-135.

Kim, H. S. , Lee, S. J. , Lee, E. J. , Hwang, J. W. , Park, S. , et al. , (2005) Cloning and characterization of a fish microsomal epoxide hydrolase of Datiiorerio and application to kinetic resolution of racemic styrene oxide. *J. Mol. Catal B-Enzym.* 37, 30-35.

Kim, J. -T. , Kang, S. , Woo, J. -H. , Lee, J. -H. , Jeong, B. and Kim, S. -J. (2007) Screening and its potential application of lipolytic activity from a marine environment: characterization of a novel esterase from Yarrowia lipolytica cll80. *Appl. Microbiol. Biotechnol.* 74, 820-828.

Koshimura, M. , Utsukihara, T. , Kawamoto, M. , Saito, M. , Horiuchi, C. L and Kuniyoshi, M. (2009) Biotransformation of bromosesquiterpenes by marine fungi. *Phytochem.* 70, 2023-2026.

Kotik, M. , Archelas, A. and Wohlgemuth, R. (2012) Epoxide hydrolases and their application in organic synthesis. *Curr. Org. Chem.* 16, 451-482.

Kroutil, W. , Mang, H. , Edegger, K. and Faber, K. (2004) Recent advances in the biocatalytic reduction of ketones and oxidation of sec-alcohols. *Curr. Opin. Chem. Biol.* 8, 120-126.

Machielsen, R. , Leferink, N. , Hendriks, A. , Brouns, S. , Hennemann, H. -G. , et al. , (2008) Laboratory evolution of Pyrococcusfuriosus alcohol dehydrogenase to improve the production of (2S, 5S) -hexanediol at moderate temperatures. *Extremophiles* 12, 587-594.

Machielsen, R. , Uria, A. R. , Kengen, S. W. M. and van der Oost, J. (2006) Production and characterization of a thermostable alcohol dehydrogenase that belongs to the aldo-keto reductase superfamily. *Appl. Environ. Microbiol.* 72, 233-238.

Martins, M. , Mouad, A. , Boschini, L. , RegaliSeleghim, M. , Sette, L. and Meleiro Porto, A. (2011) Marine fungi Aspergillussydowii and Trichoderm sp. Catalyze the hydrolysis of benzyl glycidyl ether. *Mar. Biotechnol.* 13, 314-320.

Moore, J. C. , Pollard, D. J. , Kosjek, B. and Devine, P. N. (2007) Advances in the enzymatic reduction of ketones. Acc. Chem. Res. 40, 1412-1419.

Mouad, A. M. , Martins, M. P. , Debonsi, H. M. , de Oliveira, A. L. L. , de Felicio, R. , et al. , (2011) Bioreduction of acetophenone derivatives by red marine algae Bostrychia radicans and B. tenella, and marine bacteria associated. *Helv. Chim. Acta* 94, 1506-1514.

Nakamura, K. , Yamanaka, R. , Matsuda, T. and Harada, T. (2003) Recent developments in asymmetric reduction of ketones with biocatalysts. *Tetrahedron: Asymmetry* 14, 2659-2681.

Oh, K. -H. , Nguyen, G. -S. , Kim, E. -Y. , Kourist, R. , Bornscheuer, U. , et al. , (2012) Characterization of a novel esterase isolated from intertidal flat metagenome and its tertiary alcohols synthesis. *J. Mol. Catal. B-Enzym.* 80, 67-63.

Park, S. -Y. , Kim, J. -T. , Kang, S. , Woo, J. -H. , Lee, J. -H. , et al. , (2007) A new esterase showing similarity to putative dienelactone hydrolase from a strict marine bacterium, *Vibrio* sp. GMD509. *Appl. Microbiol. Biotechnol.* 77, 107-115.

Patel, R. N. (2001) Biocatalytic synthesis of intermediates for the synthesis of chiral drug substances. *Curr. Opin. Biotechnol.* 12, 587-604.

Rocha, L. , Ferreira, H. , Luiz, R. , Sette, L. and Porto, A. (2012) Stereoselectivebioreduction of 1- (4-methoxyphenyl) ethanone by whole cells of marine-derived fungi. *Mar. Biotechnol.* 14, 358-362.

Rocha, L. , Ferreira, H. , Pimenta, E. , Berlinck, R. , Seleghim, M. , et al. , (2009) Bioreduction of α-chloroacetophenone by whole cells of marine fungi. *Biotechnol. Lett.* 31, 1559-1563.

Rocha, L. C. , Ferreira, H. V. , Pimenta, E. F. , Berlinck, R. G. S. , Rezende, M. O. O. , et al. , (2010) Biotransformation of α-bromoacetophenones by the marine fungus *Aspergillus sydowii*. *Mar. Biotechnol.* 12, 552-557.

Sandy, M. , Carter-Franklin, J. N. , Martin, J. D. and Butler, A. (2011) Vanadium bromoperoxidase from Deliseapulchra: enzyme-catalyzed formation of bromofuranone and attendant disruption of quorum sensing. *Chem. Commun.* 47, 12086-12088.

Shindo, K. , Osawa, A. , Kasai, Y. , Iba, N. , Saotome, A. and Misawa, N. (2007) Hydroxylations of substituted naphthalenes by Escherichia coli expressing aromatic dihydroxylating dioxygenase genes from polycyclic aromatic hydrocarbon-utilizing marine bacteria. *J. Mol. Catal. B-Enzym.* 48, 77-83.

Sogin, M. L. , Morrison, H. G. , Huber, J. A. , Welch, D. M. , Huse, S. M. , et al. , (2006) Microbial diversity in the deep sea and the underexplored 'rare biosphere'. *Proc. Natl. Acad. Sci. USA* 103, 12115-12120.

Timpson, L. , Alsafadi, D. , Mac Donnchadha, C. , Liddell, S. , Sharkey, M. and Paradisi, F. (2012) Characterization of alcohol dehydrogenase (ADH12) from Haloarcula marismortui, an extreme halophile from the dead sea. *Extremophiles* 16, 57-66.

Tramice, A. , Giordano, A. , Andreotti, G. , Mollo, E. and Trincone, A. (2006) High-yielding enzymatic α-glucosylation of pyridoxine by marine α-glucosidase from *Aplysia fasciata*. *Mar. Biotechnol.* 8, 448-452.

Trincone, A. (2010) Potential biocatalysts originating from sea environments. *J. Mol. Catal. B-Enzym.* 66, 241-256.

Trincone, A. (2011) Marine biocatalysts: Enzymatic features and applications. *Marine Drugs* 9, 478-499.

Woo, J.-H. , Kang, J.-H. , Hwang, Y.-O. , Cho, J.-C. , Kim, S.-J. and Kang, S. G. (2010) Biocatalytic resolution of glycidyl phenyl ether using a novel epoxide hydrolase from a marine bacterium, *Rhodobacterales bacterium* HTCC2654. *J. Boiosci. Bioeng.* 109, 539-544.

Woo, M. H. , Kim, H. S. and Lee, E. Y. (2012) Development and characterization of recombinant whole cells expressing the soluble epoxide hydrolase of Daniorerio and its variant for enantioselective resolution of racemic styrene oxides. *J. Ind. Eng. Chem.* 18, 384-391.

Zhu, D. M. , Malik, H. T. and Hua, L. (2006) Asymmetric ketone reduction by a hyperthermophilic alcohol dehydrogenase. The substrate specificity, enantioselectivity and tolerance of organic solvents. *Tetrahedron: Asymmetry* 17, 3010-3014.

Zhu, D. M. , Hyatt, B. A. and Hua, L. (2009) Enzymatic hydrogen transfer reduction of alpha-chloro aromatic ketones catalyzed by a hyperthermophilic alcohol dehydrogenase. *J. Mol. Catal. B-Enzym.* 56, 272-276.

第 11 章　单宁酸酶：来源、生物催化特点以及生产的生物过程

M. Chandrasekaran, Cochin University of Science and Technology, India and P. S. Beena, SciGenom Labs Pvt Ltd. Cochin, India

DOI：10. 1533/9781908818355. 3. 259

摘要：单宁酸酶在食品和饮料工业、制药、皮肤护理、生物治疗等领域都有应用。用于生产单宁酸酶的微生物通常是陆地来源的，其中主要的是曲霉属真菌（*Aspergilli* sp.）。这一章涉及一种新型的从海水中分离的泡盛曲霉（*Aspergillus awamori*）BTMFW032 产生的海洋单宁酸酶，并将它的生物催化活性与陆源单宁酸酶做了比较。此酶的分子量为 230 kDa，由 6 个相同的亚基组成，pI=4. 4，糖含量 8. 02%，最适催化温度为 30℃，在 5~80℃的范围内具有活性，有 pH 2 和 pH 8 两个最适 pH 值，在 pH 2 时 24 h 内稳定。对没食子酸甲酯表现出最强的亲和力，$K_m = 1.9\times10^{-3}$ mol/L，$V_{max} = 830$ μmol/min。此酶可以通过深层发酵（SmF）、浆态发酵（SLF）以及固态发酵（SSF）的胞外产物生产而来。然而，由于在发酵过程中孢子的不良水平极高，SSF 发现不是一种理想途径。响应面法被认为是一种有价值和可信赖的工具，用来优化通过 SmF 从海洋 *A. awamori* 生产单宁酸酶的同时生产没食子酸的工艺。我们的研究提示了从海洋环境中开发新酶的潜力。

关键词：海洋泡盛曲霉，单宁酸酶，特性，发酵生产

11.1　引言

单宁酸酶是一种多功能的生物催化剂，被广泛应用于食品和饮料工业、制药、皮肤护理甚至生物修复领域。由于它在蛋白质沉淀条件下可以完成一系列的生物转化反应，因而可以在众多领域中发挥重要作用，同时它在合适的溶液系统中具有水解和合成能力。因此，单宁酸酶最近被看做是一种重要的“工业酶”。单宁酸酶复杂的催化性质也提高了其商业价值。目前，单宁酸酶已被 Biocon（印度）、Kikkoman（日本）、ASA special enzyme GmbH（德国）、Julich Chiral Solutions GmbH（德国）、Wako Pure Chemical Industries，Ltd.

（日本）、Novo Nordisk（丹麦）以及 Sigma-Aldrich Co.（美国）等一些公司商品化。它们是主要的供应商，提供不同纯度以及催化活性的单宁酸酶制品（Aguilar et al.，2007）。然而，较高的生产成本，对其基本特性、理化性质、催化特性、调节机制以及潜在的应用认识不充分，这限制了单宁酸酶的工业化应用。

单宁酸酰基水解酶（TAH）通常被称为单宁酸酶（EC. 3. 1. 1. 20），参与了单宁酸的生物降解，在众多工业领域，特别是食品和制药行业有重要的应用。TAH 催化单宁酸中酯键的水解和缩酚酸键的水解，生成葡萄糖和没食子酸。TAH 也可以催化没食子酸酯中酯键的水解。

单宁是单宁酸酶的天然底物，广泛分布在植物中，在许多可食用的水果和蔬菜中被发现。由于它会与蛋白质、淀粉以及消化酶结合形成复合体，从而降低食品的营养价值（Chung et al.，1998）。可水解的单宁是由没食子酸（没食子单宁）或者鞣花酸（鞣花单宁）与一个糖核心（通常是葡萄糖）形成的酯组成（图 11.1）（Bhat et al.，1998）。由单宁酸酶催化的单宁酸水解会产生葡萄糖、没食子酸以及没食子酸酯化葡萄糖（Van de Lagemaat and Pyle，2006）。从单宁酸到没食子酸（3，4，5-三羟基苯甲酸）的工业生物转化由单宁酸酶完成。

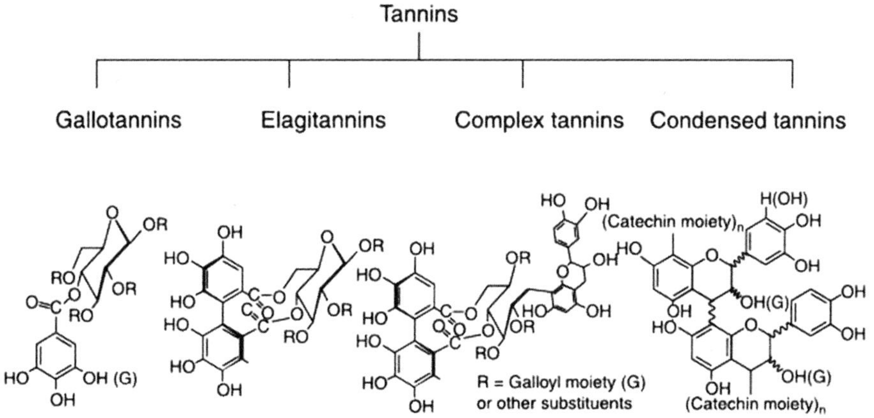

图 11.1　单宁酸的结构

资料来源：Augilar et al.，2007.

根据 Augilar 等（2007）的研究，单宁可以被分为 4 种类型：没食子单宁、鞣花单宁、缩合类单宁以及络合单宁（图 11.1）。没食子单宁中存在一些有机酸分子，比如没食子酸、双没食子酸和诃子次酸，它们与一分子葡萄糖发生酯化。另一方面，鞣花单宁具有与葡糖苷连接的鞣花酸结构单元。为了保持结合能力，没食子单宁和鞣花单宁必须具有超过两个的酸性部位与葡萄糖核心发生酯化。事实上，人们研究这种酶已经超过 100 年了。一些研究者已经研究了单宁酸酶的性质、来源、应用、反应机制以及特异性。研究结果表明单宁酸酶是一种可诱导酶，能够由丝状真菌如曲霉属（*Aspergillus*）和青霉属（*Penicillium*）通过固态发酵合成（Belur and Mugeraya，2011）。1970—1990 年间，人们申请了一些关于单宁酸酶在食品和饮料工业中潜在应用的专利（Van de Lagemaat and Pyle, 2006），并且出现了许多关于单宁酸酶的来源、分析、应用、固定化、纯化以及特性方面的研究。后来的研究指出，除了丝状真菌外，动物（Lekha and Lonsane, 1997）和细菌（Deschamps et al. , 1983）也能够生产单宁酸酶。回顾大量关于单宁酸酶的著作，特别是单宁酸酶研究的进展，会发现许多有关单宁酸酶生产和应用的报道（Lekha and Lonsane, 1997; Aguilar, et al. , 2007; Belur and Mugeraya, 2011）。最近，Chávez-González 等综述了微生物单宁酸酶在现代生物技术中的应用（2012）。表 11.1 中列出了一些可查询到的关于单宁酸酶的专利，这些专利表明了它在工业方面的重要性和潜力。

表 11.1 关于单宁酸酶生产和应用的公开的专利

年份	题目	专利号
1974	Conversion of green tea and natural tea leaves using tannase	USP3812266
1975	Production of tannase by *Aspergillus*	JP7225786
1975	Tea soluble in cold water	UKP1280135
1976	Extraction of tea in cold water	GP2610533
1976	Enzymatic solubilization of tea cream	USP3959497
1985	Gallic acid ester（s）preparation	EP-137601
1985	Enzymatic treatment of black tea leaf	EP135222
1987	Preparation of tannase	JP62272973
1987	Manufacturing of tannase with *Aspergillus*	JP62272973
1988	Production of tannase by *Aspergillus oryzae*	JP63304981
1988	Elaboration of tannase by fermentation	JP63304981
1989	Preparation of spray-concrete coating in mining shaft	SUP1514947
1989	Antioxidant catechin and gallic acid preparation	JP01268683
1989	Tannase production by culture of *Aspergillus tamari*	EP-339011
1989	New *Aspergillus niger* B1 strain	EP307071
1989	Tannase production process by *Aspergillus* and its application to obtain gallic acid	EP339011
1992	Tannase preparation method	JP4360684
1995	Enzymatic clarification of tea extracts	USP5445836
1997	DNA fragment containing a tannases gene, a recombinant plasmid, a process for producing tannases, and a promoter	USP5665584
2000	Tea concentrate prepared by enzymatic extraction and containing xanthan gum that is stable at ambient temperature	USP6024991

续表

年份	题目	专利号
2000	Producing theaflavin	USP6113965
2004	Compositions based on vanilloid catechin synergies	USP6759064
2006	Diagnostic agent and test method for colon cancer using tannases as index	USP7090997
2006	Isolation of a dimmer di-gallate a potent endothelium-dependent vasorelaxing compound	USP7132446
2007	Diagnostic agent and test method for colon cancer using tannase as index	USP7090997
2008	Anon tea based packaged beverage with a green tea extract	USP11845356
2008	A process for preparing a theaflavin-enhanced tea product	USP 11998613
2009	Novel tannase gene and protein thereof	US0239216A1 P
2009	Antiplaque oral composition containing enzymes and cyclodextrins	USP 7，601，338
2009	Methods for enhancing the degradation or conversion of cellulosic material	USP 7，608，689
2010	Process for producing purified tea extract	EP 2225952
2010	Methods for lowering blood pressure in prehypertensive individuals and/or individuals with metabolic syndrome	USP 7，651，707
2010	High-cleaning silica materials and dentifrice containing such ones	USP 7，670，593
2010	Polypeptides having cellulolytic enhancing activity and nucleic acids encoding the same.	USP 7，741，466
2010	Grape extract，dietary supplement thereof，and processes therefore	USP 7，767，235
2011	Composition for inhibiting thrombosis	USP 7，914，830
2011	Polyol oxidases	USP 7，919，295

资料来源：Augilar et al.，2007；Belur and Mugeraya，2011.

到目前为止，几乎所有报道的单宁酸酶都是分离自陆地生物的，而海洋来源单宁酸酶的相关报道非常有限。在这篇文章中，我们将对已知的微生物单宁酸酶的催化特点和海洋单宁酸酶生产的生物过程做一个简短讨论。文章中包含陆地单宁酸酶的基本信息是为了使读者对单宁酸酶有一个全面的认识并与海洋单宁酸酶做对比。

11.2 单宁酸酶的应用

11.2.1 没食子酸（3，4，5-三羟基苯甲酸）

单宁酸酶的主要应用是用来生产没食子酸（3，4，5-三羟基苯甲酸）。没食子酸以游离分子形式或者作为单宁酸分子的一部分存在于许多植物中。它可以化学合成并应用于制药业，用来生产一种广谱抗菌药物三甲氧苄二氨嘧啶（Bajpai and Patil，1996；Lekha and Lonsane，1997），该药物可以与磺胺进行联合用药（Kar and Banerjee，2000）。三甲氧苄二氨嘧啶与磺胺的联合用药能够有效地抑制许多耐药性细菌。尽管技术的进步促使大量的抗生素不断出现，但三甲氧苄二氨嘧啶依旧是非常有效的抗生素。没食子酸还是合成丙基没食子酸的重要底物，丙基没食子酸在食品及饮料业被用作油脂的抗氧化剂（Van de Lagemaat and Pyle，2006）。没食子酸还在化工领域中用于作为化学或酶法合成丙基没食子酸以及其他抗氧化成分、化妆品、美发产品、黏合剂以及润滑剂的底物。没食子酸还被用于半导体、染料的生产以及摄影胶片处理过程（Chávez-González et al.，2012）。它还被用于墨水和染料业。每年对没食子酸的需求总量为 8 000 t，其中 75%是用来制备三甲氧苄二氨嘧啶。没食子酸是由单宁酸水解生产而获得，这涉及单宁酸与硫酸在高温下发生酸化反应。这一技术有着成本高、产量低、纯度低以及之后的杂质会形成干扰并与产品一起沉淀等缺点。用这种方法生产的没食子酸已经被证实不适合用来作为生产药物的中间体。作为一个替代的方法，没食子酸可以通过使用单宁酸酶对单宁酸进行微生物水解而得到，因为没食子酸是一种单宁酸水解释放出的产物。（Iibuchi et al.，1972）。单宁酸的工业化生物转化通常是由单宁酸酶完成的，用于生产没食子酸。一个没食子酸分子中同时含有羟基和一个羧酸基团，两分子没食子酸之间能够相互形成一个酯，双没食子酸。当被加热到 220℃以上，没食子酸会失去二氧化碳形成焦性没食子酸，或者 1，2，3－三羟基苯酚，$C_6H_3(OH)_3$，可以用于偶氮染料的生产以及摄影的显影剂制备。用于水解单宁酸生产没食子酸的主要是曲霉属（*Aspergilli*）的真菌（Mondal et al.，2001；Seth and Chand，2000）。尽管没食子酸在商业上的重要性还没有被大量报道，但针对没食子酸在发酵罐水平生产过程相关研究的论文依然可以被找到（Pouratt et al.，1987）。

11.2.2 丙基没食子酸

没食子酸可用作焦性没食子酸和没食子酸酯制备的中间体。单宁的水解产生没食子酸以及没食子酸脱羧基最终形成焦性没食子酸。

n－丙基和戊醇没食子酸酯已经利用单宁酸酶在有机溶剂中通过酶法合成而获得。丙基没食子酸很贵，在油脂中被当做抗氧化剂，在食品中可以用作防腐剂，以及应用于化妆品、美发用品、黏合剂以及润滑剂等产品中（Gaathon et al.，1989；Hadi et al.，1994；Lekha and Lonsane，1997；Yamada and Tanaka，1972）。

在低水条件下使用丙醇本身作为有机反应溶剂，单宁酸酶通过单宁酸的直接转酯作用合成丙基没食子酸的方法已经被建立。由海洋泡盛曲霉 *A. awamori* BTMFW032 制备的单宁酸酶可以作为通过转酯作用进行丙基没食子酸生产酶是可行的（Beena et al.，2011a）。

11.2.3 葡萄酒

在葡萄酒中，单宁主要以儿茶素和表儿茶素的形式存在，它们能够与半乳糖-儿茶素以及其他没食子酰基衍生物形成复合物。白葡萄酒中的儿茶素含量在10~50 mg/mL，在其他葡萄酒中能达到800 mg/mL（Ribereau-Gayon，1973）。葡萄酒中50%的颜色成分都是由于单宁的存在；然而，如果这些成分由于与空气接触而被氧化成奎宁，就会形成一种不能溶解的混浊物，造成严重的质量问题。单宁酸酶能够水解麦芽酚，麦芽酚在啤酒混合物中与其他成分结合导致形成浑浊（Giovanelli，1989）。化学方法处理葡萄酒来去除非风味酚类的方法可能会被利用单宁酸酶，替代将氯原酸水解成咖啡酸和奎尼酸的方法（图11.2），这种方法对保持葡萄酒的风味更有利（Chae and Yu，1983）。

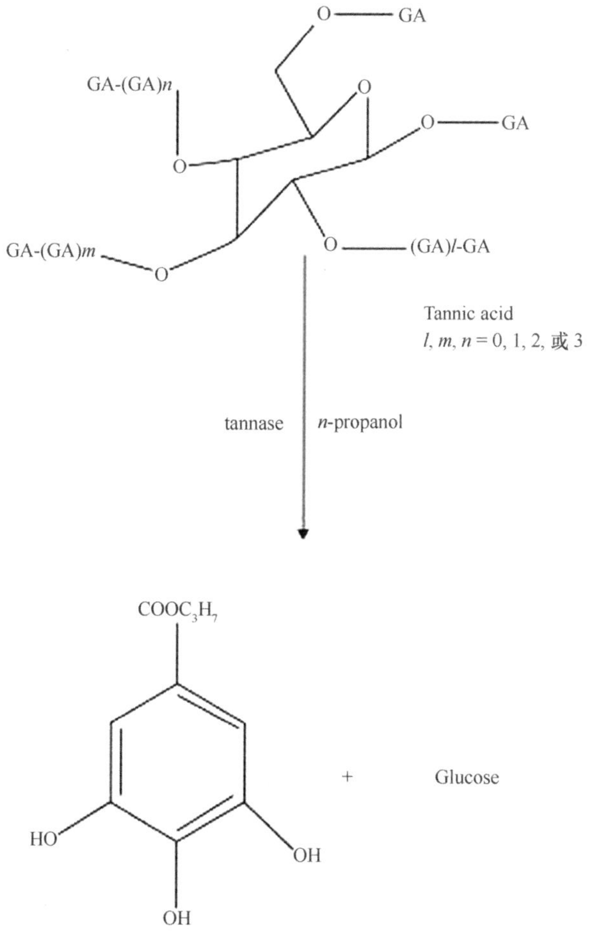

图 11.2 *n*-丙醇存在时单宁酸酶催化的单宁酸到丙基没食子酸的转酯反应

资料来源：Sharma and Gupta，2003.

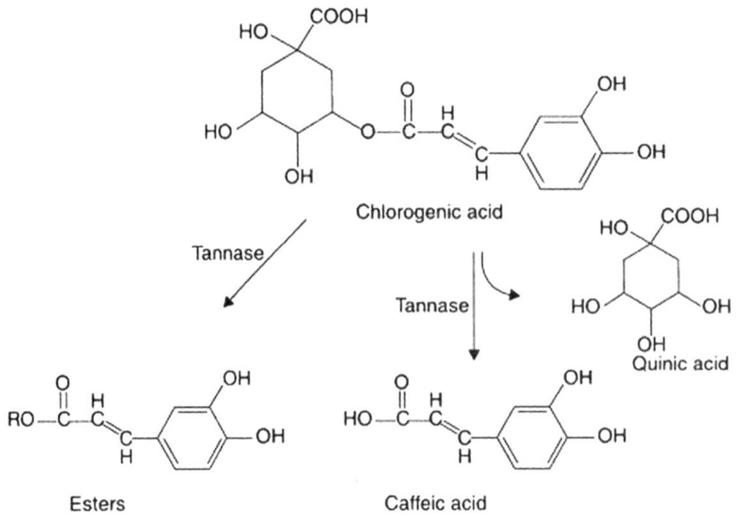

图 11.3　单宁酸酶催化的氯原酸到咖啡酸和奎尼酸的转变

11.2.4　啤酒

在啤酒的制造过程中，单宁酸被添加而形成啤酒花。啤酒中的蛋白质含量过高会导致不可溶的浑浊物。这是由于蛋白质和啤酒花单宁形成了复合物，单宁酸酶可以解决这一问题（Belmares et al.，2004）。

11.2.5　茶

单宁酸酶在冷水可溶的速溶茶产品的制造中被广泛应用（Lekha and Lonsane，1997；Garcia-Conesa et al.，2001）。茶中雾状物质的形成是由于茶黄酮类化合物的凝聚，黄酮类化合物主要是由表儿茶素、表儿茶素没食子酸酯、表没食子儿茶素和表没食子儿茶素没食子酸酯组成。茶多酚与咖啡因以氢键结合导致膏状物的形成。顾客更加喜欢清凉的茶产品，因此应该加入一些能使产品变澄清的物质，使能形成雾状物的成分被去除，避免产品出现浑浊。单宁酸酶具有催化茶提取物中单宁和多酚去除其中没食子酸部分的活性，从而得到冷水可溶的产品。而且，被单宁酸酶处理过的绿茶中的抗氧化能力比正常绿茶和红茶中的更强。单宁酸酶处理后的绿茶能够有效地抑制 N-亚硝基胺类，这类物质存在于大多数腌制肉中，具有致癌、致突变、致畸性等作用（Lu and Chen，2007）。对分离来自海洋泡盛曲霉（*A. awamori*）的部分纯化后的单宁酸酶催化效率进行评估，针对冲泡茶，发现 699 U 的酶在最适反应温度、孵育温度以及酶浓度情况下能够在 1 h 内增加溶解度 60%（Beena et al.，2011a）。

11.2.6 动物饲料

由于单宁是一类可溶于水的分子量大小不同的酚类化合物，其与蛋白质会形成依靠氢键连接的不溶性单宁-蛋白质复合物。许多饲料原料中的单宁会与膳食或内源性蛋白质和消化酶结合从而干扰正常的消化。单宁也能够妨碍铁的吸收，损害消化道的黏膜内层。将单宁酸酶作为动物饲料的一种成分能够改善饲料的可消化性（Lekha and Lonsane，1997）。来自米曲霉（*Aspergillus oryzae*）的单宁酸酶催化 Diethyl diferulates 的水解能够改善动物饲料（Garcia-Conesa et al.，2001）。

11.2.7 废水处理

单宁的抗菌作用会降低土壤有机质的生物降解率（Scalbert，1991）。单宁底物结合部位的多酚类物质与来自生物降解组织的胞外和胞内酶形成复合物。这种复合物会导致对生物降解酶的抑制，从而抑制微生物的生长，延长土壤有机质分解的生物转化时间。在这个背景下单宁酸酶能够降低土壤有机质分解的生物转化时间（Albertse，2002）。

皮革业的废水中单宁含量很高，其中主要是多酚类，它们是危险的污染物，会造成严重的环境问题（Van de Lagemaat and Pyle，2001）。单宁酸酶可以潜在地降解皮革业废水中单宁物质，它能够以相对较低的成本去除这些有害物质。它还被用于清洁硬度大、酸性强的含有单宁成分的工业废水（Banerjee，2005）。

11.2.8 其他应用

单宁酸酶在食品和饮料产品中被用于消除单宁的不良影响。使用酶法处理果汁能够减少苦味，由于能产生更少的雾状物质而不影响品质，使用这一做法能够提高果汁质量。由于对健康有好处，新型的果汁（石榴、小红莓、覆盆子、冷茶等）最近广受好评，特别是由于它们具有抗病，抗氧化功能。这些果汁中高单宁含量会导致雾状物和沉淀的形成，同样在果汁的储藏过程中会影响其颜色、产生苦味以及收敛作用。由于传统饮料生产过程中很难有效去除苦味物质，使用酶法去除苦味物质应该是首选的方法。根据 Rout 和 Banerjee（2006）针对石榴汁早期的研究发现，单宁酸酶处理后能够降解 25%的单宁，单宁酸酶和明胶 1∶1 结合后能够降解 49%的单宁。

单宁酸酶被广泛应用于果汁的生产、咖啡风味软饮料以及改善葡萄酒的风味等方面，还作为一种分析探针来确定天然存在的没食子酸酯的结构（Haslam and Tanner，1970；Seth and Chand，2000）以及在高级皮革制品制造中的应用（Barthomeuf et al.，1994）。在美白牙齿表面的过程中，单宁酸酶有利于去除牙齿外在污点（Laurence Du-Thumm et al.，2005）。

目前，从农业废弃物中生产生物乙醇的方法特别受欢迎。当对原料进行脱木质化预处理时，单体或低聚化的酚类或其衍生物会从木质素中产生。单宁酸酶能够被用来降解这些低聚化的酚类进而提高这一过程的效率（Tejirian and Xu，2011）。单宁酸酶基因和单宁酸酶活性能够被利用来进行人金黄色葡萄球菌（*Staphylococcus lugdunensis*）的鉴定，还能够被当作结肠癌的指示物（Noguchi et al.，2007）。单宁酸酶已经被用来从 prunioside A 中生产一些具有抗炎症活性的酯类衍生物（Jun et al.，2007），还应用于切割多酚类，比如植物细胞壁中的脱氢二聚体交联，它对植物细胞壁的可消化性来说是必需的（Garcia-Conesa et al.，2001）。

11.3 单宁酸酶的来源

单宁酸酶广泛存在于微生物、植物和动物中，然而，只有微生物主要用来进行单宁酸酶商业化生产。Chávez-González 等（2012）提供了一份完整的生产单宁酸酶的微生物表格。从这份表格中列出的物种我们可以看出真菌、酵母和细菌是其中的主要种类。尽管已知有大约 20 种真菌能够生产单宁酸酶，但是其中 *Aspergilli* 和 *Penicillia* 是主要的种类。到目前为止，27 种 *Aspergilli* 真菌、24 种 *Penicillia* 真菌、4 种 *Trichoderma* 真菌以及 3 种 *Fusarium* 真菌被报道能够生产单宁酸酶。已知大约有 21 种不同的细菌能够生产单宁酸酶，其中 *Lactobacilli* 是最主要的种类（13 种），接下来是 *Pediococcus*（4 种）、*Serratia*（3 种）、*Leuconostoc*（2 种）、*Pantonea*（2 种）和 *Streptococcus*（2 种）。

早期的研究着重于从菌种保藏中心寻找可用的微生物，其中主要的来源是土壤。然而，后来的研究者设计专门的筛选培养基，从自然环境中，例如森林凋落物、人类粪便、发酵食品、羊排泄物、皮革废水、橄榄油厂废水等，筛选能够生产单宁酸酶的菌株（表 11.2）。

之前的论文中提到的几乎所有的来源都是陆地环境。然而，最近人们发现海水也是生产单宁酸酶的重要来源，以其中的 *Aspergillus awamori* 和一种藻青菌 *Phormidium valderianum* BDU 140441 为代表。

在众多微生物来源中，*Aspergillus* 的真菌是最主要的单宁酸酶生产者。尽管有许多其他微生物也被报道具有生产单宁酸酶的潜力，*Aspergillus* sp. 是商业上用于制备这种酶最有效的生产者。

在微生物单宁酸酶中，真菌单宁酸酶被证明具有更高的水解单宁的活性，而酵母单宁酸酶被报道对天然存在的单宁的选择性更低，能更高效地降解单宁酸（Deschamps et al.，1983）。细菌单宁酸酶被报道能够很有效地降解和水解天然单宁和单宁酸（Deschamps et al.，1983；Lewis and Starkey，1969）。然而，尽管发现了一些微生物，但用于工业的新的更好的生产单宁酸酶菌株仍然是急需的。

表 11.2 从不同环境样品中分离单宁酸酶

Source	Name of organism	Reference
	Bacteria	
Terrestrial		
Forest litter Decaying bark samples of oak and pine	*Bacillus pumilus* *Bacillus polymyxa* *Corynebacterium* sp. *Klebsiella pneumoniae* *Pseudomonas solanaceaum*	Deschamps et al., 1983
Tannery effluent	*Citrobacter freundi*	Kumar et al., 1999
Human faeces	*Lactobacillus* *Lactobacillus plantarum,* *L. paraplantarum,* *L. pentosus*	Osawa et al., 2000
Fermented food	*Lactobacillus*	Osawa et al., 2000
Forest soil	*Bacillus licheniformis* KBR6	Mondal et al., 2000
Forest soil	*Bacillus cereus* KBR9	Mondal et al., 2001
Stored olive wastes	*Lactobacillus plantarum*	Ayed and Hamdi, 2002
Human faeces	*Lactobacillus*	Nishitani et al., 2004
Fermented foods	*Lactobacillus*	Nishitani et al., 2004
Sheep excreta	*Lactobacillus* sp. *ASR-S1*	Sabu et al., 2006
Effluent pit of a tea processing factory Rotting skin of grapes	*Serratia ficaria* *Serratia marcescens* *Microbacterium terregens* *Providencia retgeri*	Belur et al., 2010
Terrestrial	*Fungi*	
Olive oil mill wastes	*Aspergillus nigermHA37*	Aissam et al., 2005
Tannery effluent	*Aspergillus niger*	Murugan et al., 2007.
Tannery effluent	*Aspergilli and Penicilli*	Batra and Saxena, 2005.
Marine Environment		
Seawater	*Aspergillus awamori* BTMFW032	Soorej, 2008.
Seawater	*Phormidium valderium* BDU 140441 (cyanobacterium)	Palanisami et al., 2011

11.4 作为酶来源的海洋微生物

海洋环境覆盖了地球表面积大约 71%，其中不同栖息地的微生物千差万别，包括漂浮生物、浮游生物、自游生物、漂浮物和残存的、寄生的、远洋的以及深海的不同环境。这些不同的栖息环境造就了不同的微生物，包括古细菌、蓝细菌、放线菌、酵母、丝状真菌、微藻、藻类和原生动物等。几乎所有种类的海洋微生物都是有用的酶的潜在来源。根据不同的生存环境和生态功能，细菌和真菌能分泌多种不同的酶，包括蛋白酶、淀粉酶、脂肪酶、几丁质酶、纤维素酶、木质酶、果胶酶、木聚糖酶、核酸酶（DNA 酶、RNA 酶、限制性酶等）。海洋微生物依然没有被作为工业酶来源而充分利用和开发。一般认为，只要能够付出一些努力，我们就能从它们身上得到一些工业酶（Chandrasekaran，1997；Chandrasekaran and Rajeevkumar，2002）。

尽管单宁酸酶成为一个研究课题已经有很长一段时间了，但是除了对分离自海水的海洋真菌泡盛曲霉 *Aspergillus awamori* BTMFW032 的后续研究（Soorej，2008；Beena et al.，2010）以及一种能够在 0.25 mmol/L 单宁酸中生长的海洋藻青菌 *Phormidium valderianum* BDU 140441（Palanisamy et al.，2011），没有其他更多的研究报道。

11.5 单宁酸酶的生物催化特征

11.5.1 通过比较陆地和海洋单宁酸酶的生物催化特征说明单宁酸酶的生物物理和结构特征

单宁酸酶是一种膜结合或细胞内的酶，它的性质更多地取决于来源和培养条件，所有经过鉴定的来自酵母和真菌的单宁酸酶都是糖蛋白。然而，细菌单宁酸酶并没有这样的翻译后修饰。同样的微生物在不同培养系统中生产的单宁酸酶在糖基化方面也有明显的不同（Renovato et al.，2011）。

分子量

单宁酸酶的分子量在 50 ~ 320 kDa（Iwamoto et al.，2008；Boer et al.，2009）。*Aspergillus* 的单宁酸酶的分子量在 150 ~ 350 kDa（Yamada et al.，1968；Pourrat et al.，1982；Haslam and Tanner，1970）。对于纯化后的单宁酸酶进行 SDS-PAGE 分析，证明了两种酶形式的存在。通过 Con A-Ultrogel 亲和层析对一种来自 *Aspergillus oryzae* 的商业化单宁酸酶进行纯化，可以将两种单宁酸酶（单宁酸酶 Ⅰ 和单宁酸酶 Ⅱ）分离（Beverini and Metche，1990）。研究报道单宁酸酶含有两个或更多亚基。Hatamoto 等（1996）得出 *Aspergillus oryzae* 天然单宁酸酶含有 4 对亚基，每对含有两个通过二硫键连接的亚单位（30 kDa 和 34 kDa），形成一个 300 kDa 的杂合八聚体的结论。来自 *Candida* sp. 的单宁酸酶 K.1 也是由两个 120 kDa 的亚基组成，并能通过 SDS 和 β-巯基乙醇处理后分开（Aoki et al.，1976）。

通过固态发酵生产的单宁酸酶以单体和二聚体形式存在，分子量分别为 90 kDa 和 180 kDa（Ramirez-Coronel et al.，2003）。单宁酸酶等电点为 3.8，最适反应温度 60 ~ 70℃，最适 pH 值为 6.0。*A. falvus* 含有一种分子量 192 kDa 的单宁酸酶，糖含量为 25.4%（Yamada et al.，1968）。*A. niger* 含有分子量 186 kDa 的单宁酸酶，含糖量 43%（Barthomeuf et al.，1994；Parthasarathy and Bose，1976），而 *A. oryzae* 含有分子量 300 kDa

的单宁酸酶，含糖量22.7%（Hatamoto et al.，1996；Abdel-Naby et al.，1999）。

所有已经报道的真菌单宁酸酶都是糖蛋白，一般含有甘露糖、半乳糖、己糖胺等中性糖（Aoki et al.，1976；Piater，1999）。圆二色性光谱分析纯化后的单宁酸酶显示β片层结构是主要的二级结构，提示其整体呈球形构象（Mahapatra et al.，2005）。不同物种的单宁酸酶的多肽部分差别很小，例如，来自*A. falvus*的氮含量12.5%的单宁酸酶，而来自*Candida* sp. K-1的单宁酸酶含有38 %糖蛋白。如此高糖含量的生物学意义仍然是未知。然而，因为单宁酸底物中存在大量的酚羟基团，糖基部分的作用很可能是防止蛋白质肽键的羰基与其形成氢键（Lekha and Lonsane，1994）。

单宁酸酶的催化活性

在一项针对来自*A. kawachii*的单宁酸酶的研究中，从凝胶纯化和胰蛋白消化方式得到了单宁酸酶的内部序列。这两种方式获得的肽段序列与来自*A. kawachii*的β-葡糖苷酶的序列相同。检测纯化后的单宁酸酶的β-葡糖苷酶活性，发现其能够高效地水解纤维二糖。然而，当单宁酸存在时，并没有检测出此酶具有β-葡糖苷酶活性（Ramirez-Coronel et al.，2003）。

纯化后的*A. tamari*单宁酸酶具有一些对于工业应用来说很有吸引力的特点，比如在宽pH值范围和最高40℃条件下具有低K_m值活性，在有机溶剂和表面活性剂中具有抗变性能力。这些特点在食品加工工业中是非常有利的（Costa et al.，2012）。

苯甲基磺酰氟（PMSF）是一种丝氨酸蛋白酶抑制剂，能够明显抑制海洋单宁酸酶的活性，使其只保留4.85%的活性，这表明*A. awamori*单宁酸酶的催化位点存在一个丝氨酸和半胱氨酸残基（Beena et al.，2010）。稍早时候相似的研究发现PMSF抑制*A. niger* GH1单宁酸酶的活性（Marco et al.，2009）。PMSF被证明能够完全抑制二型酵母*Arxula adeninivorans*单宁酸酶的活性，推测单宁酸酶是一种需要一个丝氨酸残基的丝氨酸水解酶（Boer et al.，2009）。

菲咯啉能够螯合Fe-S键上的一个铁原子从而影响其结构。通常Fe与1，10-O-菲咯啉形成一个复合物，抑制单宁酸酶的活性（Boumans et al.，1997）。硫酸亚铁被发现能够明显提高海洋单宁酸酶的活性，而菲咯啉强烈抑制其活性，这说明酶的活性位点中存在一个Fe基团（Beena et al.，2010）。

β-巯基乙醇对海洋*Aspergillus awamori*单宁酸酶活性有明显的抑制作用，这也表明它是一种丝氨酸水解酶。二硫苏糖醇（DTT）作为一种还原剂能够抑制单宁酸酶的活性，说明它的结构中存在二硫键（Beena et al.，2010）。巯基乙酸盐能够通过维持蛋白的巯基处于还原态从而保护单宁酸酶不被灭活。随着巯基乙酸钠浓度的升高，单宁酸酶的酶活性会提高，这可能是由于在高浓度时巯基会保持在还原状态（Beena et al.，2010）。

11.5.2 影响单宁酸酶活性的参数

温度

被报道的大多数单宁酸酶的最适温度在 30～40℃，但是也存在更低和更高的最适温度，例如 20℃（Kasieczka-Burnecka et al.，2007）和 70℃（Ramírez-Coronel et al.，2003；Battestin and Macedo，2007）。单宁酸酶在 30℃能稳定存在几个月。在 47.5℃以下时，单宁酸酶的活性随着温度的升高而提高，而超过这一值后，由于发生了可逆的生物催化剂失活，酶活性会快速下降。考虑到单宁酸酶的活性和稳定性，单宁酸酶催化的酯化反应的最适反应温度应该是 40℃（Farias et al.，1994）。*Aspergillus* 的单宁酸酶的最适反应温度为 35～40℃（Pourrat et al.，1982）。*A. oryzae*、*Aspergillus* sp. 和 *Penicillium chrysogenum* 单宁酸酶的最适反应温度为 30℃（Iibuchi et al.，1968；Lekha and Lonsane，1997）。随着温度的进一步升高，酶活性会下降。一种陆地 *A. awamori* nakazawa 单宁酸酶的最适反应温度是 35℃（Mahapatra et al.，2005）。由海洋 *A. awamori* 生产的单宁酸酶表现出宽范围的反应温度，在 80℃加热 1 h 和 60℃加热 24 h 后分别保留有 43% 和 25%的活性（Beena et al.，2010）。

pH 值

关于最适 pH 值，所有的已经研究过的单宁酸酶最适 pH 值为酸性（4.3～6.5），大部分单宁酸酶的等电点在 4.3～5.1，在宽范围的 pH 值条件下稳定（3～7）（Albertse，2002；Barthomeuf et al.，1994；Farías et al.，1994；Ramírez-Coronel et al.，2003）。pH 对于酶活的影响取决于催化位点的氨基酸的特点，pH 通过质子化和去质子化以及氨基酸电离引起的构象变化对酶活性产生影响。*Aspergillus* 的单宁酸酶最适 pH 5～6，在 pH 3.5～8 时稳定（Pourrat et al.，1982）。真菌单宁酸酶通常是酸性蛋白质（Mahapatra et al.，2005）。来自 *A. awamori* 的海洋单宁酸酶在酸性和碱性环境中都表现出活性。而且，这种酶只有在 pH 2 时稳定。这种性质可能与它在高盐的海水中的适应能力有关。这种海洋酶在强酸性环境仍然具有活性，这说明它在茶膏增溶等工业领域可能具有较大的开发潜力（Beena，et al.，2010）。

金属辅助因子

大多数单宁酸酶需要金属离子的存在才能表现出完全的活性，因此酶促反应中金属离子活化在工业生物催化、追求最大催化效率方面起着重要作用。在低浓度时，金属离子作为许多酶的辅助因子起作用，因而可以增加酶的催化活性；反之高金属离子浓度会降低催化活性。这可能是由于过多自由离子的存在造成酶的部分变性。针对来自 *A. awamori* 的海洋单宁酸酶的研究表明，单宁酸酶的活性被 Fe^{3+}、Na^{+}和低浓度的 K^{+}提高后，Mg^{2+}、Zn^{2+}、

Cu^{2+}、Hg^{2+}、Ba^{2+}、Li^{+}、Cd_2^{2+}和Al^{3+}等离子能够抑制酶活性（Beena et al.，2010）。Hg^{2+}能与蛋白质的-SH 和 S-S 基团相互作用从而造成蛋白质构象的变化。Kar 等（2003）研究了金属离子对一种 *Rhizopus oryzae* 单宁酸酶的影响，他们发现 1 mmol/L 的 Mg^{2+}或 Hg^{+}可以使单宁酸酶的活性增强，而 Ba^{2+}、Ca^{2+}、Zn^{2+}、Hg^{2+}和 Ag^{+}抑制单宁酸酶的活性，Fe^{3+}和 Co^{2+}完全抑制单宁酸酶的活性。另一方面，来自 *A. niger* GH1 的单宁酸酶被 Fe^{3+}强烈抑制，Cu^{2+}和 Zn^{2+}只能有温和的抑制活性，而 Co^{2+}能够提高酶活性（Mata-Gómez et al.，2009）。另外，Mg^{2+}、Mn^{2+}、Ca^{2+}、Na^{+}和 K^{+}刺激 *Aspergillus awamori* 单宁酸酶的活性，而 Cu^{2+}、Fe^{3+}和 Co^{2+}能够作为这种酶的抑制剂（Chhokar et al.，2010）。

底物特异性

酶动力学常数（K_m 和 V_{max}）取决于特定的底物和酶，而来自一些不同真菌的单宁酸酶对底物的亲和力也不相同。当单宁酸作为底物，反应条件为 30℃，pH 5～6 时，来自 *A. niger*、*Cryphonectria parasitica*、*Verticillium* sp. 和 *Pencillium chrysogenum* 的单宁酸酶的 K_m 分别为 0.28、0.95、1.05 以及 0.048（Bhardwaj et al.，2003；Farías et al.，1994；Kasieczka-Burnecka et al.，2007；Rajakumar and Nandy，1983）。然而，当比较这些值时要注意，在定量分析反应产物的的特性时，采用了不同底物的量和不同方法。没食子酸甲酯为底物时 K_m 和 V_{max}分别为 19 mmol/L 和 830 μmol/min（Beena et al.，2010）。*Aspergillus niger* GH1 单宁酸酶（以没食子酸甲酯为底物）的 K_m 和 V_{max} 分别为 0.41×10^{-4} mol/L 和 11.03 μmol/min（Marco et al.，2009）。Kasieczka-Burnecka 等（2007）报道了分离于南极菌 *Verticillium* sp. p9 的两种适冷胞外单宁酸酶，当没食子酸甲酯为底物时 K_m值分别为 3.65×10^{-3} mol/L 和 2.43×10^{-3} mol/L；当 TA 为底物时 K_m值分别为 5×10^{-4} mol/L 和 3.88×10^{-3} mol/L。从 *A. niger* ATCC 16620 中得到一种 K_m值 1.03×10^{-3} mol/L 的单宁酸酶。尽管对没食子酸甲酯的亲和力很高，但从 *Aspergillus awamori* 中得到的单宁酸酶表现出与已经报道的微生物来源的单宁酸酶具有相似的动力学常数（Beena et al.，2010）。K_m值越小表示亲和力越高，具有更小 K_m值的底物能够更快地达到 V_{max}要达到 V_{max}需要很高的底物浓度，意味着只有当底物使酶饱和时才能达到 V_{max}（Tropp and Freifelder，2007）。通过研究发现，与其他单宁酸酶底物（例如丙基没食子酸和单宁酸）相比，没食子酸甲酯在摩尔水平明显表现出更高的特异性（Beena et al.，2010）。

表面活性剂

表面活性剂是能够在分界面改变其特性的物质，这主要是由于它们具有两亲性质。它们具有在分界面聚集，并吸附在表面的倾向。它们可以通过将蛋白质分散成带有疏水末端的肽段，并通过另一个末端与水相介质相互作用而改变表面张力。表面活性剂能够使酶变性，因此研究表面活性剂对酶的作用就十分重要（Marco et al.，2009）。海洋 *Aspergillus awamori* 生产的单宁酸酶在 0.4% Triton X、低浓度的吐温 80 以及任何浓度的吐温 20 存在时活性会提高，而 Brij-35 存在时会抑制其活性（Beena et al.，2010）。一种嗜干真菌 *A. niger GH*1 生产的单宁酸酶也被报道具有类似的结果（Marco et al.，2009）。这种抑制作用可能是由于一些因素的共同作用，例如疏水相互作用的降低在保持蛋白质三级结构方面发挥关键作用，以及表面活性剂直接与蛋白质分子相互作用。

过氧化氢（H_2O_2）

过氧化氢（H_2O_2）作为氧化剂，在浓度达到 10%（*v/v*）时能够大幅提高海洋 *Aspergillus awamori* 单宁酸酶的活性（Beena，2010）。在过氧化氢存在的条件下，单宁酸酶前处理 dhool（浸泡的茶叶）在固态发酵条件下会出现抗氧化性的增强和酶活性的提高。这样生产的干叶产品放入冷水中就会产生良好的风味和颜色（Us Patent 6482450，2003）。

乙二胺四乙酸（EDTA）

乙二胺四乙酸（EDTA）对金属依赖型酶来说是一种有效地抑制剂，它通常被用来作为商业化的蛋白酶的抑制剂。它是通过与铅和锌等金属螯合而起作用的。在常规测试条件下，它能够抑制单宁酸酶 30%以上的活性（Beena et al.，2010）。EDTA 对 *A. oryzae* 单宁酸酶有强烈的抑制作用（Iibuchi et al.，1972），EDTA 对来自其他 *Aspergillus* sp. 的单宁酸酶不具有调节作用（Bhardwaj et al.，2003）。

二甲基亚砜（DMSO）

如果酶可以在有机溶剂中应用，而不是仅在水相反应体系中具有活性，其在工业上的应用价值将会大大提高。有机溶剂中，甚至超临界流体以及气相中的酶促反应有众多潜在应用，其中一些已经商业化。DMSO 的激活作用与酶结构的微小变化有关，这会导致其构象灵活性的增加。因此，在水中添加 DMSO 能够增加疏水性底物的溶解度，混合溶剂也能在应用中发挥提高催化能力的作用（Amitabh et al.，2002）。参考海洋 *A. awamori* 单宁酸酶在不同溶剂中的研究，发现只有 DMSO 能够在一定程度上提高它的活性（Beena et al.，2010）。

11.5.3 海洋单宁酸酶的新特性

一种海洋藻青菌 *Phormidium valderium* BDU140441（Palanisami et al.，2011）能够在 0.25 mmol/L 单宁酸环境中生长，通过活性染色和活性诱导实验确定了它的单宁酸酶活性定位。诱导后，多酚氧化酶活性增强，同时通过活性染色确定了一些新的酯酶亚型。但是在特性研究水平上，海洋单宁酸酶不如 *A. awamori* 开发的充分（Beena et al.，2010，2011a，2011b）。研究中提到的单宁酸酶能够在极端酸性环境中发挥作用，表明了它有潜力应用在许多领域。而且，宽温度范围下的稳定性以及表现出来的其他特性增加了它能够应用于工业化生产的潜力。另外还有关于茶膏增溶以及丙基没食子酸合成的研究，这也证明了此酶的开发潜力。从表 11.3 中的数据中可以看出，来自海水的 *A. awamori* 单宁酸酶同陆源单宁酸酶相比明显不同，具有一些新的特性，这进一步证明了它在多个工业领域的应用潜力。

表 11.3 来自海洋泡盛曲霉 *Aspergillus awamori* BTMFW032 和陆地泡盛曲霉的单宁酸酶的特性对比

Properties	Marine-*Aspergillus awamori* (Beena et al., 2010)	Terrestrial *Aspergillus awamori* (Chhokar et al., 2010)
Culture system	SMF	SMF
MW/kDa	230	101±2
Enzyme sub units	6×37.8	1
Glycosylation/%	8	Not determined
Substrate specificity	Methyl gallate	Not determined
Optimum temperature/℃	30	30
Temperature stability/℃	30~80	25~70
pH optimum	2.0, 8.0	5.5
pH stability	2.0	5~7
pl	4.4	Not determined
K_m	1.9	Not determined
V_{maxx}	830	Not determined

对于真菌来说，采用液体深层发酵时单宁酸酶是胞内酶，而固态发酵时是胞外酶。根据一些研究者的报道，当采用液体深层发酵（SmF）产 *Penicillium chrysogenum*（Rajkumar and Nandy, 1983）、*Aspergillus niger*（Pourrat et al., 1982）和 *A. flavus*（Yamada et al., 1968）时，生产的单宁酸酶是胞内酶。*A. niger* PKL 104 液体深层发酵生长最初的 48 h 生产的单宁酸酶完全是胞内酶。之后，随着发酵时间的延长，单宁酸酶会分泌到培养基中。然而，用相同培养基采用固态发酵法生产的单宁酸酶是完全的胞外酶（Lekha and Lonsane, 1994）。Seth 和 Chand（2000）发现 *A. awamori* 在实验室发酵 60 h 后所生产的单宁酸酶完全是胞内酶。

11.6 海洋单宁酸酶生产的生物过程

一些研究者已经实现了通过液体深层发酵、固态发酵以及优化的固态发酵 *Aspergillus* 真菌生产单宁酸酶。由海洋 *A. awamori* BTMFW032 生产的单宁酸酶，在不同的发酵条件下

（浆态发酵、液体深层发酵以及固态发酵）以最大的酶产量为目标进行了优化（Beena，2010）。以下的章节将对海洋真菌单宁酸酶生产的不同生物过程进行讨论。

11.6.1 培养基

Czapek Dox 培养基通常用于陆地 *Aspergillus* sp. 单宁酸酶的生产，因此这种培养基作为一种基础培养基被用于海洋 *A. awamori* BTMFW032。此外，由于这种真菌最初是分离自海洋环境，海水也被用于做酶生产的基础培养基成分（Beena，2010）。研究发现虽然海水能提高酶的产量，但当补充1%的单宁酸作为唯一的碳源时，对两种基础培养基都促进了单宁酸酶的产生。而且，制备培养基时联合使用海水和 Czapek Dox minimal 培养基以及不同浓度的单宁酸也能够提高酶的产量。

11.6.2 酶的诱导/调节

根据不同的菌株和培养条件，诱导和表达出的酶具有不同的活性，具有不同的生产模式。酚类化合物，例如没食子酸、焦性没食子酸、没食子酸甲酯以及单宁酸能够诱导单宁酸酶的合成（Bajpai and Patil，1997）。然而，诱导的机理还没有被明确地证明，一些可水解的单宁成分在围绕与单宁酸酶合成中的作用还存在一些争论（Deschamps et al.，1983；Aguilar er al.，2001a）。单宁酸酶被单宁酸或一些它的衍生物所诱导，但是它的合成调控机制仍然未知。Huang 等（2005）将单宁酸酶作为一种模式系统来实验考察两种培养系统中酶调节机制的不同，并建立了 *A. niger* Aa-20 生产单宁酸酶的诱导和抑制模式。他们使用单宁酸和葡萄糖作为碳源进行固态培养基（SSC）和深层培养基（SmC）培养黑曲霉。在他们的研究中，诱导和抑制率是分别通过单宁酸和葡萄糖的不同浓度得到。

11.6.3 液体深层发酵

尽管在液体深层发酵时，真菌能够在培养基中含有最高浓度为12.5%（*w/v*）的单宁酸时存活。由50%（*v/v*）海水和50%（*v/v*）Czapeks Dox minimal 培养基配制的含有7.5%（*w/v*）单宁酸的生产培养基能够用于生产单宁酸酶，单宁酸的浓度在4%~5%时似乎是高酶产量（423 U/mL）的最适浓度。海洋 *A. awamori* 能够在很短时间内生产单宁酸酶，接下来对发酵过程中影响单宁酸酶和没食子酸生产的参数进行了优化。数据也表明培养基中的海水在提高单宁酸酶产量中发挥了积极的作用。显而易见的是，海水中多种离子和无机盐的存在能够满足它提高单宁酸酶合成对微量元素的需求，从而对真菌产生积极影响。当然，需要详细的研究来证明这一事实。

11.6.4 固态培养基发酵

固态发酵（SSF）具有一些有吸引力的优势，包括高产量（最高达到 SmC 的5.5倍）、胞外酶的产生、在宽 pH 值范围及温度下稳定（Lekha and Lonsane，1994）。由于 SSF 能够以较低的代价提取出高度浓缩的粗酶，因此它在酶的生产中受到青睐（Tao et al.，1997）。一些研究报道了与深层培养基（SmC）相比，固态培养基生产的单宁酸酶具有一些明显的优势（Lekha and Lonsane，1997；Aguilar et al.，2001a，2001b，2002）。以不同的富含单宁的农业废物作为底物，在固态发酵条件下，采用 *A. ruber* 生产的酶被研究。其中 Jamun 叶子是 SSF 条件下生产酶的最适底物，它生产单宁酸酶在 30.1℃经过 96 h 培养后达到最大产量。向培养基中添加碳源和氮源不能提高单宁酸酶的产量（Kumar et al.，2007）。Reddy 和 Rathod（2012）研究了利用分离得到的 *P. purpurogenum* BVG 7 菌株，通过固态发酵的方式，利用富含单宁的农业废物（合欢的荚、红革壳、高粱壳以及废茶粉）研究了 pH 值对底物和温度对生产单宁酸酶过程和没食子酸产量的影响。通常来说，具有高单宁含量的底物能够通过 SSF 进行单宁酸酶的生产。底物经过矿物溶液的湿润并接种上筛选过的菌种。自然界中被用于单宁酸酶生产的有甘蔗渣、小麦糠、罗望子粉、木馏灌木叶（*Larrea tridentata*）、栗单宁（*Caesalpinia spinosa*）、橡树瘿（*Quercus infectoria*）、漆树（*Rhus coriaria*）叶、余甘子（*Terminalia chebula*）果实、高粱（*Sorghum vulgaris*）叶以及印度鹅莓（*Phyllanthus emblica*）叶（Aguilar et al.，2007）。然而，近些年，营养培养基中添加惰性载体（例如聚氨酯泡沫）逐渐增多（Cruz-Hernández et al.，2006；Mata-Gómez et al.，2009；Renovato et al.，2011）。惰性载体的添加和固定成分的培养基能够帮助监控和控制 SSF 过程中的参数（Zhu et al.，1994）；Kar 等（1999）利用一株 *R. oryzae*，使用了一种优化的固态发酵（MSSF）用于同时生产没食子酸和单宁酸酶。发现与传统的 SSF 系统相比，MSSF 分别提高了单宁酸酶和没食子酸 1.7 倍和 3 倍的产量（Kar and Banerjee，2000）。

关于海洋 *A. awamori*，在椰子纤维和椰子中果皮作为惰性载体进行固态发酵过程会导致非常低的单宁酸酶产量，因此，并没有更多的研究开展（Beena，2010）。

11.6.5 浆态发酵

与固态发酵和液体深层发酵相比，浆态发酵有许多优势。尽管它的生产培养基非常简单，只需要某些特定农业废物作为底物，但是它还没有引起人们足够的重视。当海洋真菌 *A. awamori* BTMFW032 被选作研究对象，在固态发酵的早期产生巨大芽孢和缺乏胞外单宁酸酶活性，浆态发酵生产单宁酸酶的前景开始被研究（Beena，2010）。使用海水作为培养基，利用含有单宁酸的天然底物作为单宁酸酶生产的诱导剂和底物。具有药用价值的藤黄叶，首次在浆态发酵下被用作生产单宁酸酶的底物（Beena et al.，2011b）。Beena 等（2011b）评估了藤黄（*Garcinia gummi gutta*）作为没食子酸酶生产的底物，并发现了它同样具有很好的作用。根据 Folin-denins 的方法，500 mg 叶子中含有 5 mg 单宁酸（即藤黄叶中含有 1%的单宁酸），能够诱导单宁酸酶和后续的没食子酸生产。应该注意到，这是之前被认为具有药用价值的 *Garcinia gummi gutta* 叶首次被发现能够成功诱导单宁酸酶和没食子酸合成（Beena et al.，2011b）。事实上，所有其他尝试作为底物的天然底物都有一定含量的单宁，因此尝试其诱导单宁酸酶生产。但是与对照组相比，那些底物不能提高酶的合成和提高酶活。关于天然碳源，最高的单宁酸酶活性是使用藤黄叶和海水，而不需要诱导剂（26.2090 U/mL），接下来是使用藤黄叶和 Czapek Doxz 培养基（22.059 U/mL）发酵 48 h 后。研究发现只使用藤黄叶就可以在 Czapek Doxz 培养基和海水中诱导单宁酸酶的生产，表明它作为单宁酸酶生产底物的潜力。大概藤黄中的单宁能够提高 *A. awamori* 生产单宁酸酶的产量。进一步的数据表明尽管天然底物能够诱导单宁酸酶的生产，但生物体依然需要培养基中单宁酸来提高酶的合成，证明了单宁酸诱导剂的作用（Beena et al.，2011b）。

11.6.6 单宁酸酶生产生物过程的统计建模

Plackett-Bruman 设计提供了一个有效的筛选过程，用以计算一个有大量因素的实验中各因素的重要性，是一种省时的方法并在各个部分保持令人信服的信息（Sharma and Satyanarayana，2006）。Plackett-Burman 设计实验结果证明推断出 18 个变量值中只有藤黄叶、单宁酸、葡萄糖、温度和培养液 5 个变量是最重要的变量（Beena et al.，2011b）。Plackett-Burman 设计中单独参数研究的结果证明，单宁酸和接种量浓度对提高酶的产量具有积极的作用，呈现浓度依赖性。相反的，葡萄糖、温度和藤黄叶变量的升高会对酶产量造成负面影响。关于 5 种筛选的理化因素的实验数据一共进行了 46 个实验，显示了对这些因素的强烈依赖性，因为在实验条件下，酶的产量在 37.4~76.79 U/mL。在最适条件下生产单宁酸酶时，通过测定反应表面确定这些因素的两两相互作用。这 5 个变量的最适值为藤黄叶 26%，葡萄糖 3.2 mmol/L，单宁酸 1%，温度 40℃，以及接种量 3%。最终在最适条件下培养 *A. awamori* 的时间过程研究表明，在发酵的初始阶段单宁酸酶的产量迅速上升，在 24 h 达到最大酶活 75.23 U/mL。然而，随着发酵过程的进行，酶活性会降低。在早期的研究中，单宁酸的浓度、震荡速度和 pH 被证明是影响细胞生长和酶合成的重要参数。这些参数在一个实验室生物反应器中通过反应表面法优化，使用 Box 和 Behnken 因子设计来确定生产酶和没食子酸积累的最适条件（Seth and Chand，2000）。Jamun 叶和余柑子叶被用于单宁酸酶的生产（Kumar et al.，2007），经过 96 h 的培养后得到酶活性的最大产量是 69 U/g 干重底物。同时棕榈仁饼和罗望子种子粉作为底物分别得到酶活性 13.03 U/g 干燥底物和 6.44 U/g 干燥底物（Sabu et al.，2005），而用添加 0.8%单宁酸的小麦糠能提高酶活性，达到 67.5 U/g 干重底物（Gustavo et al.，2001）。

从已发表的论文中可以看出，液体表面和固态发酵过程分别主要被用于生产胞内单宁酸酶和胞外单宁酸酶（Lekha and Lonsane，1994）。为了挖掘开发这种多用途酶的应用潜力，急需探索更多有效的能够提高产量的方法（Rana and Bhat，2005）。因此为了最大酶和没食子酸的产量，优化了过程参数，在液体深层发酵下通过 Box-Behnken 设计，使用一种能够接受 Plackett-Burman（PB）和 Response Surface Methodology（RSM）的一种单一培养基（Beena et al.，2011a）。在评估的 11 种过程变量包含单宁酸、氯化钠、硝酸钠、氯化钾、硫酸镁、硫酸亚铁、磷酸氢二钾、pH、接种物、孵育期以及震荡，其中只有单宁酸、氯化钠、硫酸亚铁、磷酸氢二钾、孵育以及震荡证明是影响酶和没食子酸产量的最关键参数。pH 被认为是非关键因素，因为在 RSM 中更低的 pH 和高水平的单宁酸抑制菌体的生长，因此 pH 在求最大产量时被软件当作一个常数因子。模型预测经过 48 h 培养后单宁酸酶活性达到 4 824.61 U/mL，没食子酸达到 136.206 μg/mL。经过时间过程实验之后，在培养第 36 小时得到 5 085 U/mL 单宁酸酶活性，第 84 小时得到 372.6 μg/mL 没食子酸（Beena et al.，2011a）。在最适条件下酶和没食子酸的产量比非最适条件下的产量高出接近 15 倍。作为诱导剂，单宁酸很明显对酶和没食子酸的产量有很大的影响，它的高浓度保障了高的酶产量。即使在 Plackett-Burman 设计中，伴随着单宁酸浓度单独对生物过程的影响，单宁酸与其他培养基组分和过程参数（例如硫酸亚铁、氯化钠、震荡和孵育）的相互作用显示了在响应面法中单宁酸在培养基中的最适浓度。在液体深层发酵中，2%单宁酸使生物生产单宁酸酶的产量达到最大（Banerjee et al.，2001）。根据 Mahapatra 和 Banerjee（2009）的研究，通过 SSF，尽管有记录显示在 15%单宁酸浓度，仍然有酶的产生。但 2.5%是使用 *Hyalopus* sp. 属生产单宁酸酶达到最大合成时最适的单宁酸浓度，随着浓度的升高，酶的产量会下降。然而，Sabu 等（2005）报道了 5%单宁酸对通过 SSF 和 *A. niger* 生产单宁酸酶是合适的。在高浓度的单宁酸下单宁酸酶活性也更高，相反的在液体深层发酵时活性会随单宁酸浓度的升高而受到抑制。

Box-Behnken 实验的结果证明单宁酸、孵育期、硫酸亚铁、磷酸氢二钾、震荡和氯化钠基于 Plackett-Burman 设计实验对系统有明显的线性影响。研究发现单宁酸对单宁酸酶生产具有最明显的影响。目前的研究发现培养基中 2.6%（*w/v*）浓度的单宁酸比其他浓度效果更好。曾经也有报道类似的发现。Hadi 等（1994）报道了通过 *R. oryzae* 生产单宁酸酶，在 2%单宁酸时酶产量达到最大的 6.12 U/mL，而 Bradoo 等（1997）报道了通过 *A. japonicas* 生产单宁酸酶的最适单宁酸浓度是 2%。Aguilar 等（2001b）报道了 *A. niger* Aa-20 分泌到液体深层发酵培养基中的单宁酸酶最初支持浓度为 50 g/L。更早的研究者报道了 5%的单宁酸是最适浓度（Aoki et al.，1976；Lekha and Lonsane，1997；Sharma et al.，2007）。还有报道指出单宁酸酶在微生物生长的早期被生产，之后会出现下降（Rajakumar and Nandy，1983；Sharma et al.，2007）。Kar 和 Banerjee（2000）也报道了对于单宁酸酶生产来说 48 h 是最适的培养时间。酶产量的提高被认为是与特定的活性平行的，这显示了在工业生产中设计培养基的作用。因此目前的工作证明了反应曲面分类研究法能够作为一种有价值且可靠的工具，用于优化从 *A. awamori* 中同时生产单宁酸酶和没食子酸。

11.7 结论

海洋微生物是开发新型工业化酶有潜力的来源，但是它们仍然未被开发利用。本章强调了海洋酶的重要性，特别是微生物酶，例如最近因其多种应用领域而收到越来越多关注的单宁酸酶。与陆地酶相比，有许多强有力的证据表明海洋微生物酶在许多新特性上是独一无二的。当然，还需要开展大量的研究和开发工作，以利用不同的海洋微生物群体来生产新型的工业生物催化剂。

参考文献

Abdel-Naby, M. A., Sherif, A. A., El-Tanash, A. B. and Mankarios, A. T. (1999) Immobilization of *Aspergillus oryzae* tannase and properties of the immobilized enzyme. *J. Appl. Microbiol.* 87, 108-114.

Aguilar, C. N., Augtu; C., Favela-Torres, E. and Viniegra-Gonzúlez, G. (2001a) Induction and repression patterns of fungal tannase in solid-state and submerged cultures. *Proc. Biochem* 36, 565-570.

Aguilar, C. N., Augur, C., Favela-Torres, E. and Viniegra-Gonzúlez, G. (2001b) Production of tannase by *Aspergillus niger* Aa-20 in submerged and solid state fermentation: influence of glucose and tannic acid. *J. bid. Microbiol. Biotechnol.* 26, 296-302.

Aguilar, C. N., Favela-Torres, E., Viniegra-Gonzúlez, G. and Augur, C. (2002) Culture conditions dictate protease and tannase production in submerged and solid-state cultures by *Aspergillus niger* Aa-20. *Appl. Biochem. Biotechnol.* 102, 407-414.

Aguilar, C. N., Rodriguez, R., Gutierrez-Sanchez, G., Augur, C., Favela-Torres, E. et al., (2007) Microbial tannases: advances and perspectives. *Appl. Microbiol. Biotechnol.* 76, 47-59.

Aissam, H., Errachidi, F., Penninckx, M. J., Merzouki, M. and Benlemlih, M. (2005) Production of tannase by *Aspergillus niger* HA37 growing on tannic acid and Olive Mill Waste Waters. *World J. Microbiol. Biotechnol.* 21, 609-614.

Albertse, E. K. (2002) Cloning, expression and characterization of tannase from *Aspergillus* species. M. Sc. thesis, Faculty of Natural and Agricultural Sciences, Department of Microbiology and Biotechnology, University of the Free State Bloemfontein, South Africa.

Amitabh, C. S., Richele, T., John, C. and Robert, M. K. (2002) Structural and catalytic response to temperature and cosolvents of carboxylesterase EST1 from the extremely thermoacidophilic archaeon *sulfolobus solfataricus* P1. *Biotechnol Bioertg.* 80, 784-793.

Aoki, K., Shinke, R. and Nishira, H. (1976) Purification and some properties of yeast tannase. *Agric. Biol. Chem.* 40, 79-85.

Ayed, L. and Hamdi, M. (2002) Culture conditions of tannase production by *Lactobacillus plantarum*. *Biotechnol. Lett.* 24, 1763-1765.

Bajpai, B. and Patil, S. (1996) Tannin acyl hydrolase activity of *Aspergillus*, *Penicillium*, *Fusarium* and *Trichoderma*. *World J. Microbiol. Biotechnol.* 12, 217-220.

Bajpai, B. and Patil, S. (1997) Introduction of tannin acyl hydrolanc (EC 3. 1. 1. 20) activity in some members of fungi imperfecti. *Enzyme, Microb. Technol.* 20, 612-614.

Banerjee, D., Mondal, K. C. and Bikas, R. (2001) Production and characterization of extracellular and intracellular tannase from newly isolated *Aspergillus aculeatus* DBF9. *J. Basic. Microbiol.* 6, 313-318.

Banerjee, S. D. (2005) Microbial production of tannase and gallic acid. Ph. D. thesis, Vidyasagar University, India.

Barthomeuf, C., Regerat, F. and Pourrat, H. (1994) Production, purification and characterization of a tannase from *Aspergillus niger* LCF8. *J. Ferment*, *Bioeng*, 77, 320-323.

Batra, A. and Saxena, R. K. (2005) Potential tannase producers from the genera *Aspergillus* and *Penicillium*. *Proc. Biochem.* 40, 1553-1557.

Battestin, V. and Macedo, G. A. (2007) Tannase production by *Paecilomyces variotii*. *Biores. Technol.* 98, 1832-1837.

Beena P. S. (2010) Production, purification, genetic characterization and application studies of tannase enzyme from marine fungus *Aspergillus awamori*. Ph. D. thesis, Cochin University of Science and Technology, India.

Beena, P. S., Soorej, M. B., Elyas, K. K., Bhat, S. G. and Chandrasekaran, M. (2010) Acidophilic tannase from marine *Aspergillus awamori* BTMFW032. *J. Microbiol. Biotechnol.* 20 (10), 1403-14.

Beena, P. S., Soorej, M. B., Sarita, G., Bhat, S. G., Bahkali, A. H. and Chandrasekaran, M. (2011a) Propyl gallate synthesis using acidophilic tannase and simultaneous production of tannase and gallic acid by marine *Aspergillus awamori* BTMFW032, *Appl. Biochem. Biotechnol.* 164 (5), 612-628.

Beena et al., (2011b) *Garcina cambogia* leaf and seawater for Tannase production by marine *Aspergillus awamori* BTMFW032 under slurry state fermentation. *Natural Product Communications* 6 (12), 1933-1938.

Belmares, R., Contreras-Esquivel, J. C., Rodriguez-Herrera, R., Coronel, A. R. and Aguilar, C. N. (2004) Microbial production of tannase: an enzyme with potential use in food industry. *Lebensmittel-Wissenschaft und-Technologie-Food Sci. Technol.* 37, 857-864.

Belur, P. D. and Mugeraya, G. (2011) Microbial production of tannase: state of the art, *Res. J. Microbiol*, 6 (1), 25-40.

Belur, P. D., Gopal, M., Nirmala, K. R. and Basavaraj, N. (2010) Production of novel cell-associated tannase from newly isolated *Serratia ficaria* DTC, *J. Microbiol. Biotechnol.* 20 (4), 732-736.

Beverini, M. and Metche, M. (1990) Identification, purification and physiochemical properties of tannase of *Aspergillus oryzae*. *Scinces des Aliments* 10, 807-816.

Bhardwaj, R., Bhat, T. K. and Singh, B. (2003) Purification and characterization of tannin acyl hydrolase from *A. niger* MTCC-2425. *J. Basic Microbiol.* 43, 449-461.

Bhat, T. K., Singh, B. and Sharma, O. P. (1998) Microbial degradation of tannins-current perspective. *Biodegrad.* 25, 343-357.

Boer E., Bode, R., Mock, H. P., Piontek, M. and Kunze, G. (2009) Atan1p-an extracellular tannase from the dimorphic yeast *Arxula adeninivorans*: molecular cloning of the *ATAN1* gene and characterization of the recombinant enzyme. *Yeast* 26, 323-337.

Boumans, H., van Gaalen, M. C., Grivell, L. A. and Berden, J. A. (1997) Differential inhibition of the yeast bc1 complex by phenanthrolines and ferroin. Implications for structure and catalytic mechanism. *J. Biol. Chem.* 272, 16753-16760.

Bradoo, S., Gupta, R. and Saxena, R. K. (1997) Parametric optimization and biochemical regulation of extracellular tannase from *Aspergillus japonicus*. *Proc. Biochem.* 32, 135-139.

Chae, S. and Yu, T. (1983) Experimental manufacture of a corn wine by fungal tannase. *Hanguk Sipkum Kwahakoechi* 15, 326-332.

Chandrasekaran, M. (1997) Industrial enzymes from marine microorganisms: the Indian scenario. *J. Mar. Biotechnol.* 5, 86-89.

Chandrasekaran, M. and Rajeevkumar, S. (2002) ‘Marine microbial enzymes’. In: *Encyclopedia of Life Support Systems (EOLSS-UNESCO publication)*. First hard copy, CD ROM and online form published in 2002. EOLSS Publishers, Oxford, UK. *http: //www. eolss. com.*

Chavez-Gonzalez, M. L., Contreras Esquivel, J. C., Prado Barragan, L. A., Rodriguez, R., Aguilera-Carbo, A. F. et al., (2012) Microbial and enzymatic hydrolysis of tannic acid: influence of substrate chemical quality. *Chemical Papers* 66 (3), 171-177.

Chhokar, V., Sangwan, M., Beniwal, V., Nehra, K. and Nehra, K. S. (2010) Effect of additives on the activity of tannase from *Aspergillus awamori* MTCC9299. *Appl. Biochem. Biotechnol.* 160 (8), 2256-65.

Chung, K. T., Wong, T. Y., Wei, C. I., Huang, Y. W. and Lin, Y. (1998) Tannins and human health: a review. *Crit. Rev. Food Sci. Nutr.* 38, 421-464.

Costa, A. M., Kadowaki, M. K., Minozzo, M. C., de Souza, G. M., Boer C. G., et al., (2012) Production, purification and characterization of tannase from *Aspergillus tamarii. Afri. J. Biotechnol.* 11 (2), 391-398.

Cruz-Hernúndez, M., Augur, C., Rodrĺguez, R., Contreras-Esquivel, J. and Aguilar, C. N. (2006) Evaluation of culture conditions for tannase production by *Aspergillus niger* GH1. *Food Technol. Biotechnol.* 44, 541-544.

Deschamps, A. M., Otuk, G. and Lebeault, J. M. (1983) Production of tannase and degradation of chestnut tannin by bacteria. *J. Ferment. Technol.* 61, 55-59.

Farias, G. M., Gorbea, C., Elkins, J. R. and Griffin, G. J. (1994) Purification, characterization, and substrate relationships of the tannase from *Cryphonectria parasitica. Physiol. Mol. Plant Pathol.* 44, 51-63.

Gaathon, A., Gross, Z. and Rozhanski, M. (1989) Propyl gallate: enzymatic synthesis in a reverse micelle system. *Enzyme Microb. Technol.* 11, 604-609.

García-Conesa, M. T., Ostergaard, P., Kauppinen, S. and Williamson, G. (2001) Hydrolysis of diethyl diferulates by a tannase from *Aspergillus oryzae*: breaking cross-links between plant cell wall polymers. *Carbohy. Poly.* 44, 319-324.

Giovanelli, G. (1989) Enzymatic treatment of malt polyphenols for stabilization. *Ind. Bevande.* 18, 497-502.

Gustavo, A. S. P., Selma, G. F., Leite, S. C. and Terzi, S. C. (2001) Selection of tannase-producing *Aspergillus niger* strains. *Braz. J. Microbiol.* 32, 24-26.

Hadi, T. A., Banerjee, R. and Bhattarcharyya, B. C. (1994) Optimization of tannase biosynthesis by a newly isolated *Rhizopus oryzae. Bioproc. Eng.* 11, 239-243.

Haslam, E. and Tanner, J. N. (1970) Spectrophotometric assay of tannase. *Phytochem.* 90, 2305-2309.

Hatamoto, O., Watarai, T., Kikuchi, M. and Sekhin, H. (1996) Cloning and sequencing of the gene encoding tannase and a structural study of the tannase subunit from *Aspergillus oryzae. Gene* 175, 215-221.

Huang, W., Ni, J. and Borthwick, A. G. L. (2005) Biosynthesis of valonia tannin hydrolase and hydrolysis of valonia tannin to ellagic acid by *Aspergillus* SHL 6. *Proc. Biochem.* 40, 1245-1249.

Iibuchi, S., Minoda, Y., and Yamad, K. (1968) Studies on tannin acyl hydrolase of microorganisms. Part III. Purification of the enzyme and some properties of it. *Agric. Biol. Chem.* 32, 803-809

Iibuchi, S., Minoda, Y. and Yamada, K. (1972) Hydrolyzing pathway, substrate specificity and inhibition of tannin acyl hydrolase of *Asp. oryzae* No. 7. *Agric. Biol. Chem.* 36, 1553-1562.

Iwamoto K., Tsuruta, H., Nishitaini, Y. and Osawa, R. (2008) Identification and cloning of a gene encoding tannase (tannin acylhydrolase) from *Lactobacillus plantarum* ATCC 14917 (T). *Syst. Appl. Microbiol.* 31,

269–277.

Jun, C. S. , Yoo, M. J. and Lee, W. Y. (2007) Ester derivatives from tannase–treated prunioside A and their anti-inflammatory activities, *Bull. Korean Chem. Soc.* 28, 73–76.

Kar, B. and Banerjee, R. (2000) Biosynthesis of tannin acyl hydrolase from tannin rich forest residue under different fermentation conditions. *J. Ind. Microbiol. Biotechnol.* 25, 29–38.

Kar, B. , Banerjee, R. and Bhattacharya, B. C. (1999) Microbial production of gallic acid by modified solid–state fermentation. *J. Ind. Microbiol. Biotechnol.* 23, 173–177.

Kar, B. , Banerjee, R. and Bhattacharyya, B. C. (2003) Effect of additives on the behavioural properties of tannin acyl hydrolase. *Proc. Biochem.* 38, 1285–1293.

Kasieczka–Burnecka, M. , Kuc, K. , Kalinowska, H. , Knap, M. and Turkiewicz, M. (2007) Purification and characterization of two cold-adapted extracellular tannin acyl hydrolases from an Antarctic strain *Verticillium* sp. P9. *Appl. Microbiol. Biotechnol.* 77, 77–89.

Kumar, R. A. , Gunasekaran, P. and Lakshmanan, M. (1999) Biodegradation of tannic acid by *Citrobacter freundii* isolated from a tannery effluent. *J. Basic Microbiol.* 39, 161–168.

Kumar, R. , Sharma, J. and Singh, R. (2007) Production of tannase from *Aspergillus ruber* under solid state fermentation using jamun (*Syzygium cumini*) leaves. *Microbiol Res.* 162, 384–390.

Laurence, D. , Du–Thumm, Szeles, L. H. , Sullivan, R. J. , Masters, J. G. and Robinson, R. S. (2005) Chewable antiplaque confectionery dental composition. US20050008582 dated 01/13/2005.

Lekha, P. K. and Lonsane, B. K. (1994) Comparative titres, location and properties of tannin acyl hydrolase produced by *Aspergillus niger* PKL 104 in solid-state, liquid surface and submerged fermentations. *Proc. Biochem.* 29, 497–503.

Lekha, P. K. and Lonsane, B. K. (1997) Production and application of tannin acyl hydrolase. State of the art. *Adv. Appl. Microbiol.* 44, 215–260.

Lewis, J. A. and Starkey, R. L. (1969) Decomposition of plant tannins by some soil microorganisms. *Soil Sci.* 107, 23S–241.

Lu M. J. and Chen, C. (2007) Enzymatic tannase treatment of green tea increases *in vitro* inhibitory activity against N–nitrosation of dimethylamine. *Proc. Biochem.* 42, 1285–1290.

Mahapatra, S. and Banerjee, D. (2009) Extracellular tannase production by endophytic *Hyalopus*. *J. Gen. Appl. Microbiol.* 55, 255–259.

Mahapatra, K. , Nanda, R. K. , Bag, S. S. , Banerjee, R. , Pandey, A. and Szakacs, G. (2005) Purification, characterization and some studies on secondary structure of tannase from *Aspergillus awamori* nakazawa. *Proc. Biochem.* 40, 3251–3254.

Marco, M. G. , Rodríguez, L. V. , Ramos, E. L. , Renovato, J. , Cruz-Hernάndez, M. A. et al. , (2009) A novel tannase from the xerophilic fungus *Aspergillus niger* GH1. *J. Microbiol. Biotechnol.* 19, 987–996.

Mata-Gómez, M. A. , Rodríguez, L. V. , Ramos, E. L. , Renovata, J. , CruzHernandez, M. A. , et al. , (2009) A novel tannase from the xerophilic fungus *Aspergillus niger* GH1. *J. Microbiol. Biotechnol.* 19 (9), 987–996.

Mondal, K. C. , Banerjee, R. and Pati, B. R. (2000) Tannase production by *Bacillus licheniformis*. *Biotechnol. Lett.* 22, 767–769.

Mondal, K. C., Banerjee, D., Banerjee, R. and Pati, B. R. (2001) Production and characterization of tannase from *Bacillus cereus* KBR9. *J. Gen. Appl. Microbiol.* 47, 263-267.

Murugan, K., Saravanababu, S. and Arunachalam, M. (2007) Screening of tannin acyl hydrolase (E. C. 3. 1. 1. 20) producing tannery effluent fungal isolates using simple agar plate and SmF process. *Biores. Technol.* 98, 946-949.

Nishitani, Y., Sasaki, E., Fujisawa, T. and Osawa, R. (2004) Genotypic analyses of *Lactobacilli* with a range of tannase activities isolated from human feces and fermented foods. *System. Appl. Microbiol.* 27, 109-117.

Noguchi, N., Ohashi, T. and Shiratori, T. (2007) Association of tannase-producing *Staphylococcus lugdunensis* with colon cancer and characterization of a novel tannase gene. *J. Gastroenterol.* 42 (5), 346-351.

Osawa, R., Kuroiso, K., Goto, S. and Shimzu, A. (2000) Isolation of tannin-degrading lactobacilli from humans and fermented foods. *Appl. Environ. Microbiol.* 66, 3093-3097.

Palanisami, S., Kannan, K. and Lakshmanan, U. (2011) Tannase activity from the marine bacterium *Phormidium valderium* BDU 140441. *J. Appl. Phycol.* Published online Nov. 2011. DOI: 10. 1007/sl0811-011-9738-4.

Parthasarathy, N. and Bose, S. M. (1976) Glycoprotein nature of tannase in *Aspergillus niger*. *Acta Biochim. Polon.* 23, 293-298.

Piater, L. A. (1999) *Aspergillus* tannase: purification, properties and application. M. Sc. thesis, University of the Free State, Blomfontein, South Africa.

Pourrat, H., Regerat, F., Pourrat, A. and Jean, D. (1982) Production of tannnase (tannin acyl hydrolase E. C. 3. 1. 1. 20) by a strain of *Aspergillus niger*. *Biotechnol. Lett.* 4, 583-588.

Pourrat, H., Regerat, F., Morvan, P. and Pourrat, A. (1987) Microbiological production of gallic acid from *Rhus coriaria* L. *Biotechnol. Lett.* 9, 731-734.

Rajakumar, G. S. and Nandy, S. C. (1983) Isolation, purification, and some properties of *Penicillium chrysogenium* tannase. *Appl. Environ. Microbiol.* 46, 525-527.

Ramirez-Coronel, M. A., Viniegra-Gonzalez, G., Darvill, A. and Augur, C. (2003) A novel tannase from *Aspergillus niger* with β-glucosidase activity. *Microbiol.* 149, 2941-2946.

Rana, N. and Bhat, T. (2005) Effect of fermentation system on the production and properties of tannase of *Aspergillus niger* van Tieghem MTCC 2425. *J. Gen. Appl. Microbiol.* 51, 203-212.

Reddy, S. B. and Rathod, V. (2012) Gallic acid production and tannase activity of *Pencillium purpurogenum* stoll employing agrobased waste through solid state fermentation: influence of pH and temperature. *Asian. J. Biochem. Pharm. Res.* 1 (2), 58-62.

Renovato, J., Gutiérrez-Sánchez, G., Rodríguez-Durán, L. V., Bergman, C., Rodríguez, R. and Aguilar, C. N. (2011) Differential properties of *Aspergillus niger* tannase produced under solid-state and submerged fermentations. *Appl. Biochem. Biotechnol.* 165, 382-385.

Ribereau-Gayon, M. (1973) *In les tannins des vegetaux*, Metche, M. and Girardin, M. (eds), France: Dunod, pp. 262-287.

Rout, S. and Banerjee, R. (2006) Production of tannase under MSSF and its application in fruit juice debittering. *Indian J. Biotechnol.* 5, 346-350.

Sabu, A., Pandey, A., Daud, M. J. and Szakacs, G. (2005) Tamarind seed powder and palm kernel cake: two novel agro residues for production of tannase under solid state fermentation by *Aspergillus niger* ATCC16620.

Biores. Technol. 96, 1223-1228.

Sabu, A., Augur, C., Swati, G. and Pandey, A. (2006) Tannase production by *Lactobacillus* sp. ASR-S1 under solid-state fermentation. *Proc. Biochem.* 41, 575-580.

Scalbert, A. (1991) Antimicrobial properties of tannins. *Phytochem.* 30, 3875-3883.

Seth, M. and Chand, S. (2000) Biosynthesis of tannase and hydrolysis of tannins to gallic acid by *Aspergillus awamori*, optimization of process parameters. *Proc. Biochem.* 36, 39-44.

Sharma, S. and Gupta, M. N. (2003) Synthesis of antioxidant propyl gallate using tannase from *Aspergillus niger* van Teighem in nonaqueous media. *Bioorg. Med. Chem. Lett.* 13 (3), 395-397.

Sharma, D. C. and Satyanarayana, T. (2006) A marked enhancement in the production of a highly alkaline and thermostable pectinase by *Bacillus pumilus* dcsr1 in submerged fermentation by using statistical methods, *Biores. Technol.* 97, 727-733.

Sharma, S., Agarwal, L. and Saxena, R. K. (2007) Statistical optimization for tannase production from *Aspergillus niger* under submerged fermentation. *Indian J. Microbiol.* 47, 132-138.

Soorej, M. Basheer (2008) Lipase production by marine fungus *Aspergillus awamori* Nagazawa. Ph. D. thesis, Cochin University of Science and Technology, India.

Tao, S., Peng, L., Beihui, L., Deming, L. and Zuohu, L. (1997) Solid state fermentation of rice chaff for fibrinolytic enzyme production by *Fusarium oxysporum*. *Biotechnol. Lett.* 19, 465-467.

Tejirian, A. and Xu, F. (2011) Inhibition of enzymatic cellulolysis by phenolic compounds, *Enzyme. Microb. Technol.* 48 (3), 239-247.

Tropp, B. E. and Freifelder, D. (2007) *Molecular Biology: Genes to Proteins*, pp. 1000. ISBN-13: 978-0-76-370916-7.

U. S. Patent No. 6482450. Goodsall Christopher William; Jones Timothy Graham, Mitei Joseph Kipsiele, Parry Andrew David, Safford Richard and Thiru Ambalavanar (2002). Cold water infusing leaf tea United States, Lipton, division of Conopco, Inc. (Englewood Cliffe, NJ), *http://www.freepatentsonline.com/*6482450.*html*.

Van de Lagemaat, J. and Pyle, D. L. (2001) Solid-state fermentation and bioremediation: development of a continuous process for the production of fungal tannases. *Chem. Eng. J.* 84, 115-123.

Van de Lagemaat, J. and Pyle, D. L. (2006) 'Tannase'. In: A. Pandey, C., Webb, C. R. Soccol, and C. Larroche (eds), *Enzyme Technology*, Springer, New York, pp. 380-397.

Yamada, D. and Tanaka, T. (1972) Wine making using tannase in fermentation process. Patent. Ger. Offen. 2.224 100 (cl. 12g). Japanese Patent Application no. 7, 133, 151.

Yamada, H., Adachi, O., Watanabe, M. and Sato, N. (1968) Studies on fungal tannase. Part I. Formation, purification and catalytic properties of tannase of *Aspergillus flavus*. *Agric. Biol. Chem.* 32, 1070-1078.

Zhu, Y., Smith, J., Knol, W. and Bol, J. (1994) A novel solid state fermentation system using polyurethane foam as inert carrier. *Biotechnol. Lett.* 16, 643-648.

第 12 章　有生物活性的咪唑相关二肽的合成和降解

Shoji Yamada, Kagoshima University, Japan

DOI：10.1533/9781908818355.3.295

摘要：一系列与咪唑相关的组分，像肌肽、鹅肌肽、高肌肽、蛇肉肽（丙酰胺甲基组氨酸）和 N-α-乙酰基组氨酸，广泛分布在陆生与水生脊椎动物的骨骼肌、神经系统以及其他组织中。由于其特殊的结构，1 个多世纪以来，人们对这些二肽和它们的类似物一直有着极大的兴趣。在本章综述了有关这一神秘组分的最新信息，集中介绍 6 种脊椎动物中的酶，它们催化这类二肽的生物合成和降解。而且，讨论了对于咪唑相关二肽酶的生物催化应用。

关键词：乙酰基组氨酸，鹅肌肽酶，鹅肌肽，肌肽酶，肌肽，咪唑相关二肽

12.1　引言

一种咪唑相关的二肽-肌肽，于 1900 年在 Liebig's 肉提取物中作为一种特殊结构二肽首先分离出来（Gulewitsch and Admiradzibi，1900）。到 20 世纪 60 年代，肌肽的 3 种结构类似物，包括鹅肌肽、高肌肽、蛇肌肽相继在脊椎动物的大脑和骨骼肌中被鉴定出来（Crush，1970）。在这些所有的二肽结构中都含有一个非 α 氨基酸和一个咪唑环。尽管人们因为这一系列二肽的不寻常的结构而对其感兴趣了 100 多年，但是它们确定的生物学功能仍然不清楚。除了这些二肽以外，其他咪唑相关的类似物，N-α-乙酰基组氨酸在变温脊椎动物，包括辐鳍鱼、两栖动物和爬行动物的脑和晶状体中以非常高的浓度普遍存在。已经有很多针对这些咪唑相关组分（主要是肌肽）生物学功能的研究。可能的功能有 H^+ 缓冲液，二价离子调节器，神经递质，非酶的自由基清除剂，抗氧化剂，血糖调节剂，水分子泵，防止晶状体白内障的保护因子。

已经有报道称 6 种脊椎动物的酶催化咪唑相关二肽的生物合成（3 种）与降解（3 种），即：肌肽合酶（EC 6.3.2.11），肌肽 N-甲基转移酶（EC 2.1.1.22）和组氨酸 N 乙

酰基转移酶（EC 2. 3. 1. 33）；降解酶有：β-丙氨酸-组氨酸 二肽酶（EC 3. 4. 13. 20），胞质特异性的二肽酶（EC 3. 4. 13. 18），Xaa-甲基-组氨酸二肽酶（EC 3. 4. 13. 5）（表 12. 1）。与合成酶不同，降解酶，通常被叫做肌肽酶，或鹅肌肽酶，已经被很好地研究了。特别是，许多医学研究现在开始关注于通过对β-丙氨酸-组氨酸二肽酶的易感基因位点的基因编码研究酶与糖尿病性肾病之间的遗传相关性（Janssen et al.，2005）。

表 12. 1　催化咪唑相关肽生物合成和降解的 6 种脊椎动物酶

在生物化学和分子生物学国际命名学会（NC-IUBMB）

Accepted names (NC-IUBMB)	Other names	EC numbers	Gene symbols	Sources: Organisms and tissues	Reactions[a]	MW of subunits (native forms)
Biosynthesis						
Carnosine synthase	Carnosine synthetase, carnosine-anserine synthetase	6.3.2.11	*CARNS1*	Mammal, bird/ skeletal muscle, brain, olfactory bulb	β-Ala + His → Car GABA + His → HCar β-Ala + Nπ-Methyl-His → Ans	100 kDa (tetramer)[b]
Carnosine N-methyltransferase		2.1.1.22	Unsequenced	Mammal, bird/ skeletal muscle, kidney, liver	Car + SAM → Ans + SAH His + SAM → Nπ-Methyl-His + SAH	? (native: 85 kDa)[c]
Histidine N-acetyltransferase	Acetylhistidine synthetase, histidine acetyltransferase	2.3.1.33	Unsequenced	Ray-finned fish, amphibia, reptile/ brain, lens	His + Acetyl-CoA → NAH + CoA	39 kDa (monomer)[c]
Degradation						
Cytosol nonspecific dipeptidase	Carnosine dipeptidase II, prolinase	3.4.13.18	*CNDP2*	Animals/all tissues	Dipeptides → Amino acids	53 kDa (dimer)[d]
β-Ala-His dipeptidase	Carnosine dipeptidase I, serum carnosinase	3.4.13.20	*CNDP1*	Vertebrates/brain, serum, skeletal muscle	Car → β-Ala + His HCar → GABA + His Ans → β-Ala + Nπ-Methyl-His	57 kDa (dimer)[e]
Xaa-Methyl-His dipeptidase	Anserinase, acetylhistidine deacetylase	3.4.13.5	*ANSN*	Ray-finned fish, amphibia, reptile/ brain, ocular fluid, skeletal muscle	Ans → β-Ala + Nπ-Methyl-His NAH → His + Acetate	55 kDa (dimer)[f]

注：a　Ans：鹅肌肽；β-Ala：β-丙氨酸；Car：肌肽；GABA：γ-氨基丁酸；HCar：高肌肽；NAH：Nα-乙酰组氨酸；SAM：S-腺苷蛋氨酸；SAH：S-腺苷高半胱氨酸.

b　Drozak et al.，2010.

c　Raghavan et al.，1992.

d　Yamada et al.，1995.

e　Teufel et al.，2003.

f　Yamada et al.，2005.

虽然对肌肽特殊的功能并不清楚，但毫无疑问的是酶和其天然底物（肌肽）都与疾病有关。因此研究酶的属性和分子结构对脊椎动物体内的咪唑相关二肽的生物活性的理解是十分重要的。

本章回顾了有关咪唑相关二肽的最新研究，集中关注脊椎动物酶对这些神奇组分的新陈代谢。同时也讨论了作用于咪唑相关二肽酶的可能的生物催化应用。

12.2 咪唑相关二肽的生物活性

12.2.1 结构

结构特殊的咪唑相关二肽，像肌肽（β-丙氨酰组氨酸）、鹅肌肽（β-丙氨酰-$N\pi$ 甲基-组氨酸）、蛇肉肽（β-并酰胺-Nτ-甲基-组氨酸）和高肌肽（γ-氨基丁酸-组氨酸），在 N 端含有非 α-氨基酸（β-丙氨酸或 γ-氨基丁酸）和在 C 末端一个咪唑相关的氨基酸（组氨酸，$N\pi$-甲基组氨酸或 $N\tau$-甲基组氨酸），这些二肽广泛存在于脊椎动物的骨骼肌和中枢神经系统中（图 12.1）。与 N 乙酰天门冬氨酸（Baslow and Yamada，1997；Baslow，2003）、N-乙酰谷氨酸（Caldovic and Tuchman，2003）和 N-乙酰甲硫氨酸（Smith et al.，2011）一样，$N\alpha$-乙酰基组氨酸是脊椎动物中天然存在的 N-乙酰化的氨基酸之一。

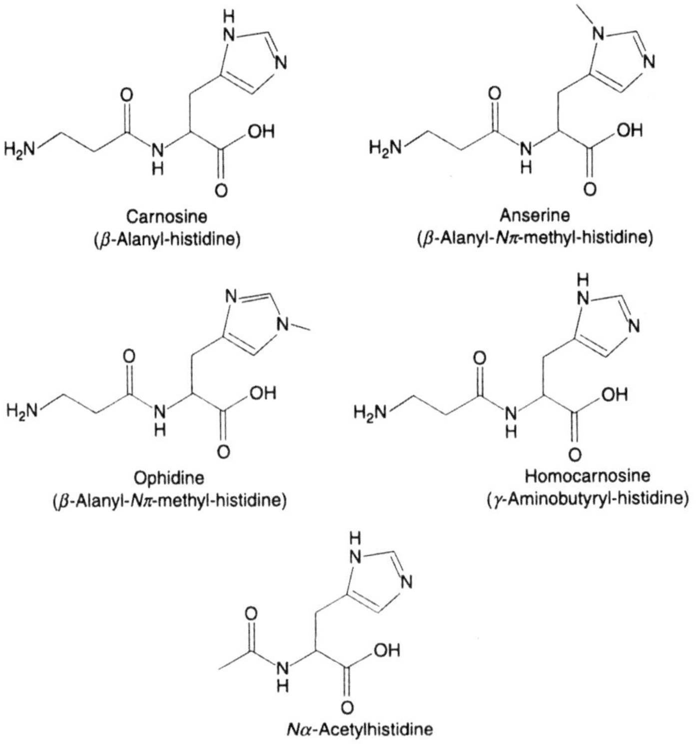

图 12.1 咪唑相关肽的分子结构

12.2.2 分布

Crush（1970），Chan 和 Decker（1994）总结了二肽（Nα-乙酰基组氨酸除外）在脊椎动物（和一些无脊椎动物）骨骼肌中的广泛分布。而且，Bonfantiet 等（1999）总结了这些二肽（蛇肉肽和 Nα-乙酰基组氨酸除外）在脊椎动物神经系统中的分布。

骨骼肌

肌肽、鹅肌肽、蛇肉肽和 Nα-乙酰基组氨酸存在于脊椎动物的骨骼肌中作为主要的非蛋白质类含氮成分（Crush，1970；Yamada et al.，2009）。没有报道证明高肌肽作为一种主要的成分存在于动物肌肉中，上述二肽在骨骼肌中的优势是物种依赖性的。一般来说，骨骼肌中咪唑相关的二肽成分在哺乳动物、鸟类和鱼类中的含量较高，这些动物都有持续紧张或全速运动的能力。

陆生动物含有大量的肌肽和鹅肌肽。例如，马的肌肉（中臀肌）中只含有肌肽（107.8 mmol/L）作为主要的成分，骆驼肌肉（中臀肌）含有肌肽（29.7 mmol/L）和鹅肌肽（37.5 mmol/L）（Dunnett and Harris，1977）。在牛、猪、Baikal 海豹肌肉中，肌肽含量要高于鹅肌肽（Crush，1970）。鹅肌肽主要存在于袋鼠、黑羚羊、长颈鹿、鼠兔、老鼠、大鼠、兔子、狗和猫的肌肉中。另一方面，水生哺乳动物，像鲸鱼、海豚含有大量的蛇肉肽，这些在陆生哺乳动物体内并没有发现。鸟纲中的动物既含有肌肽，又含有鹅肌肽，后一种组分占据主要的含量。事实上，鹅肌肽是以雁属的鹅来命名的，二肽也是从这种动物中分离出来的。就鸡肉来说，胸肌含有 4.1 mmol/L 的鹅肌肽，1.2 mmol/L 的肌肽。在爬行动物中，只有那些属于蛇亚目的动物组织中含有大量的蛇肉肽。蛇肉肽在水生哺乳动物和蛇体内有限的分布是十分神秘的。在龟鳖目骨骼肌中主要的咪唑相关的二肽是肌肽，而在蜥蜴亚目和鳄鱼中是鹅肌肽（Crush，1970）。

在鱼类中，Van Waarde（1988）已经将咪唑相关的组分包括组氨酸在骨骼肌中的分布划分出了以下 5 种模式：①迁徙远洋鱼类（金枪鱼科）含有非常高含量（>50 mmol/L）的组氨酸和鹅肌肽。②在鳗鲡科（鳗鱼）和鲟科（鲟鱼）肌肽是主要的咪唑相关的二肽组分。③鹅肌肽是鲑鱼科（鲑鱼）和鳕鱼科（鳕鱼）和板鳃亚纲的咪唑池的主要组分。④在鲱科和鲤科，只存在有高浓度的组氨酸，并且⑤在鲽科和鲈科中存在有很低浓度（<1 mmol/L）的咪唑相关组分。

并且，最近研究显示，Nα-乙酰基组氨酸是一些淡水鱼像罗非鱼（*Oreochromis niloticus niloticus*）、暹罗斗鱼（*Betta splendens*）和玻璃猫鱼（*Kryptopterus bicirrhis*）骨骼肌另一种主要成分（Yamada et al.，1992，2009）。一些关于无脊椎动物的研究显示，在软体动物和甲壳类只有很少量的肌肽，牡蛎中有少量的鹅肌肽（Lukton and Olcott，1958；Cameron，1989）。

神经系统

除了蛇肉肽，人们发现咪唑相关的二肽在脊椎动物的神经系统中十分丰富。在哺乳动物的神经系统中，肌肽和高肌肽都存在，但是鹅肌肽和 *Nα*-乙酰基组氨酸不存在（Baslow，1965，Bonfanti et al.，1999），一个例外就是在大鼠的脑中发现了 *Nα*-乙酰基组氨酸（O'Dowd et al.，1990）。肌肽在嗅球中含量高度丰富，但是在大脑中发现的鹅二肽的含量较低（Margolis，1974）。除了肌肽和高肌肽之外，鹅肌肽也存在于鸟类的神经系统中，像大脑、视网膜和嗅球（Tsunoo et al.，1963；Kanazawa and Sano，1967；Margolis and Grillo，1984）。在变温动物，包括辐鳍鱼，两栖类和爬行动物中，*Nα*-乙酰基组氨酸以很高的浓度（2~10 mmol/L）广泛分布在大脑中（Baslow，1965）。有趣的是，这些混合物在更为原始的鱼类的大脑中并没有被发现，像无颚鱼和软骨鱼。然而，爬行动物的脑中富含 *Nα*-乙酰基组氨酸；被认为从爬行动物进化而来的鸟类，在组织中并不含有这种乙酰化的氨基酸。虹鳟（*Oncorhynchus mykiss*）的内耳听毛细胞层被鉴定出含有高浓度的 *Nα*-乙酰基组氨酸（Drescher and Drescher，1991）。

其他组织

少量的肌肽存在于牛晶状体中（Krause，1936），而且在大鼠和蟾蜍的胃中也能检测出来（Wood，1957）。肌肽已经在鸡血细胞（2 510 nmol/g of cell）和鸡血浆（27 nmol/mL），大鼠血浆，兔血浆中检测出来（Seely and Marshall，1981）。而且，也有报道称在几种哺乳动物的富含多组胺的组织中存在肌肽（Flancbaum et al.，1990）。在鲣中，肌肽以低浓度（<0.3 mmol/L）存在于肠、肾脏、卵巢中。同时，相对较高浓度的鹅肌肽在眼（1.3 mmol/L）和幽门盲囊（1.1 mmol/L）中鉴定出来（Abe，1983）。*Nα*-乙酰基组氨酸被发现以非常高的浓度（>10 mmol/L）广泛存在于淡水鱼和咸水鱼的晶状体中，偶尔存在于它们的心脏中（Hanson，1966）。

12.2.3 可能的生理作用

肌肽、鹅肌肽和高肌肽

咪唑相关的二肽已经被假设有以下生物学功能：一种 H^+ 缓冲液（Abe and Okuma，1991；Sewell et al.，1992）；一种抗氧化剂（Baran，2000；Boldyrevet al.，2004；Aydin et al.，2010）；一种非酶自由基清除剂（Kohen et al.，1988；Guiotto et al.，2005）；一种神经递质（Rochel and Margolis，1982）；一种血糖调节剂（Nagai et al.，2003；Sauerhofer et al.，2007）。

肌肉的缓冲作用能力是一种防止肌肉中 ATP 厌氧分解引起的酸中毒（H^+ 积累）非常重要的因素。由于这些二肽的咪唑环的 pK_a 值接近生理的 pH，肌肽和鹅肌肽或许具有缓冲能力（Smith，1938）。Severin 等（1953）等已经报道了肌肽在青蛙的肌肉中扮演着胞内 pH 缓冲液的作用。与人相比，在马、狗中较高的肌肉缓冲能力似乎主要是因为这些物种肌肉中含有较高浓度的咪唑相关二肽的原因（Harris et al.，1990）。在 6 种鱼（鲣鱼、鲭鱼、鳟鱼、鲤鱼、鳗鱼、比目鱼），3 种哺乳动物（鲸、牛、猪）和一种鸟（鸡）的肌肉中，鲣鱼的白肌 pH 值在 6.5~7.5 显示出了更高的缓冲能力（Okuma and Abe，1992）。

肌肽和鹅肌肽已经被认为是一种抗氧化剂和非酶自由基清除剂。生理浓度下，这些多肽可以抑制生物膜上脂质组分的抗坏血酸依赖性的过氧化反应（Boldyrev et al.，1988）。肌肽也与一些低分子量的来源于不饱和脂肪酸的醛类反应，如对生物系统有毒性的反-2-己烯醛和 4-hydroxy-trans-2-nonenal（Zhou and Decker，1999）。

糖化是还原糖和蛋白质氨基酸的氨基首先反应生成希夫碱，然后生成 Amadori 产物，最终生成有毒的高级糖基化终产物（高度交联结构）的过程。肌肽的结构拥有靠近咪唑的氨基和羧基，这组成了蛋白质首选的糖基化位点。研究表明当修饰基团例如葡萄糖、果糖、甘油醛、丙二醛和乙醛反应时，肌肽可以保护蛋白质免受损伤（交联）（Hipkiss et al.，1997；Hipkiss and Chana，1998）。并且，肌肽已经作为一种抗衰老因子，通过与活性氧和糖化基团作用与羰基化合物反应，影响在衰老过程中蛋白质的后合成，并产生蛋白-羰基-肌肽复合物（Hipkiss and Brownson，2000）。因此，肌肽与蛋白羰基的反应也许能够抑制其与正常蛋白的进一步交联。

肌肽已被认为是嗅觉通路中的一种神经递质。最基本的化学受体神经元被发现富含肌肽（Ferriero and Marogolis，1975）。在钙存在下，通过高浓度 K^+或藜芦定碱的去极化，嗅球突触小体释放肌肽（Rochel and Margolis，1982）。一种脑片培养技术表明肌肽在嗅觉神经元和嗅球神经元之间作为一种兴奋的神经效应器（Kanaki et al.，1997）。

Nα-乙酰基组氨酸

在变温脊椎动物中 *Nα*-乙酰基组氨酸的准确作用了解很少。目前，在鱼晶状体中 *Nα*-乙酰基组氨酸已经推测具有两种生理学作用：一种水分子泵（Baslow，1998）和一种防止晶状体白内障的保护因子（Breck et al.，2005）。

在变温脊椎动物中，眼的晶状体和大脑中有高浓度的 *Nα*-乙酰基组氨酸。这种组分是由组氨酸和乙酰辅酶 A 通过在晶状体中的组氨酸 N-乙酰基转移酶合成（Baslow，1967）。另一方面，晶状体缺少降解酶即 Xaa-甲基组氨酸二肽酶（在 12.4.2 节中讨论）。Xaa-甲基组氨酸二肽酶存在于眼睛晶状体周围的液体中。因此，在眼睛中是彻底的区分合成酶与降解酶，在晶状体中合成 *Nα*-乙酰基组氨酸，并且在眼液的液体中水解成醋酸和组氨酸。Baslow（1998）提出 *Nα*-乙酰基组氨酸可以作为新陈代谢可回收梯度驱动的“分子水泵”来维持水的稳态以此保证晶状体胞外渗透压平衡。水的流入或新陈代谢水的形成可以作为钙依赖通道打开的开关，可以释放强的离子化的亲水性 *Nα*-乙酰基组氨酸分子及其交联的水，从而降低它的浓度梯度。通过计算，每一个 *Nα*-乙酰基组氨酸分子的蓝金水合复合物有33个偶极水分子传递进入细胞外眼流体。释放的 *N*α-乙酰基组氨酸的停留时间非常短暂，通过降解酶的作用迅速的脱乙酰化。释放的组氨酸从细胞外眼流体被转运到晶状体，在这里，通过合成酶的作用又被乙酰化。

挪威的研究团队提出了在鱼的晶状体中 *N*α-乙酰基组氨酸的其他生理学作用的假设（Breck et al.，2005）。在欧洲大西洋鲑鱼养殖业中，晶状体白内障已经报道了20多年，现在仍然继续造成硅鱼产量损失（Menzies et al.，2002）。与一些关于饲养的鲑鱼中白内障严重的事件同时发生的事情是，在饲养鲑鱼过程中疏忽利用哺乳动物的血粉来作为饲料中蛋白质来源（Wall，1998）。哺乳动物的血粉中组氨酸含量比较丰富，由于会引起疯牛病，欧洲早在20世纪90年代已经禁止将其作为饲料原料。Breck等（2003）揭示了通过饮食补充血粉或半胱氨酸组氨酸可以减少大西洋鲑严重白内障的发生率。因此，他们清楚地指明了通过将大西洋鲑经历parr-smolt转换后晶状体中 *N*α-乙酰基组氨酸水平降低与白内障的发展有关（Breck et al.，2005）。他们同时展示了在parr-smolt转换过程中，晶状体中 *N*α-乙酰基组氨酸合成速率将上调。因此，尽管 *N*α-乙酰基组氨酸防止白内障的确切机制仍然不明确，很明显它与维持鱼晶状体透明度有关。

Yamada等（1992）揭示了在罗非鱼骨骼肌中存在着异常大量的Nα-乙酰基组氨酸，浓度超过了那些大多数的自由氨基酸。随后，这种乙酰氨基酸被报道在其他淡水鱼种的骨骼肌中也有有限分布（Yamada et al.，2009）。通过饥饿（8周）的方法，观察到罗非鱼骨骼肌中 *N*α-乙酰基组氨酸水平快速减少的模式（Yamada et al.，1994）。另外，Abe和Ohmama（1987）报道了在日本鳗（*Anguilla japonica*）的肌肉中肌肽的水平和虹鳟鱼的鹅肌肽含量几乎与长期的饥饿（鳗鱼200 d，虹鳟62 d）无关。因此作者总结为鳗鱼和虹鳟鱼咪唑二肽也许是新陈代谢惰性的。一种好的生物策略也许是保持这些二肽在骨骼肌中较高的含量来维持 H^+ 缓冲能力。因为 *N*α-乙酰基组氨酸在罗非鱼饥饿过程中能够很容易地被骨骼肌利用，不像鳗鱼的肌肽和虹鳟鱼中的鹅肌肽，可以推测，*N*α-乙酰基组氨酸和这些二肽在鱼的肌肉中的生理学作用是不同的。*N*α-乙酰基组氨酸或许起着独特的作用，例如，在一些种类的鱼中，将 *N*α-乙酰基组氨酸作为骨骼肌主要的咪唑组分，这时 *N*α-乙酰基组氨酸作为一种组氨酸（一种必需氨基酸）的紧急储存来抵抗长期的食物缺乏（Yamada et al.，1994）。

12.3 咪唑相关二肽的生物合成

像在表12.1中展示的，3种催化咪唑相关二肽生物合成的脊椎动物酶，目前在国际生物化学与分子生物学联盟委员会注册为肌肽合酶（EC 6.3.2.11），肌肽N-甲基转移酶（EC 2.1.1.22），和组氨酸N乙酰基转移酶（EC 2.3.1.33）。肌肽合酶与鹅肌肽，高肌肽和肌肽的形成有关（图12.2）。肌肽N甲基转移酶利用S-腺苷-甲硫氨酸催化肌肽的甲基化形成鹅肌肽（图12.3）。*N*α-乙酰基组氨酸是由组氨酸和乙酰基辅酶A通过组氨酸N-

乙酰基转移酶合成的（图 12.4）。

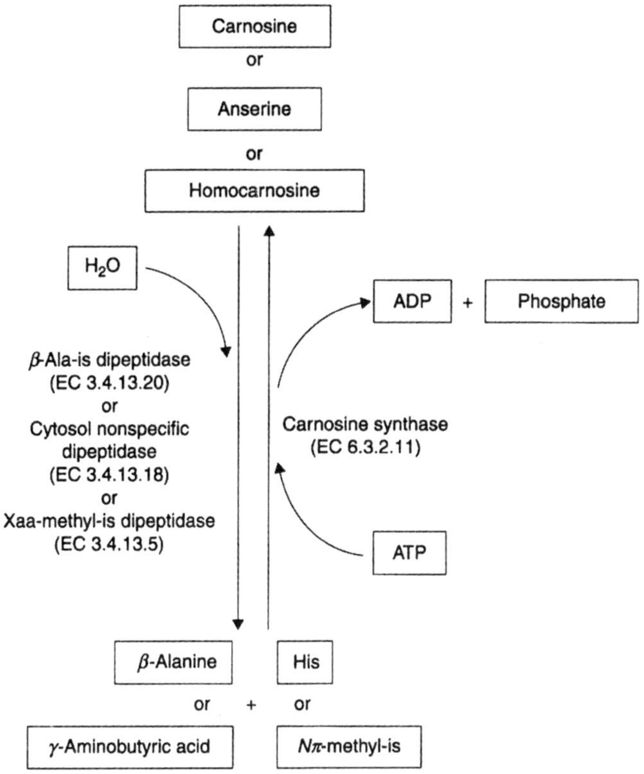

图 12.2　肌肽、丝氨酸、高肌氨酸的生物合成和降解

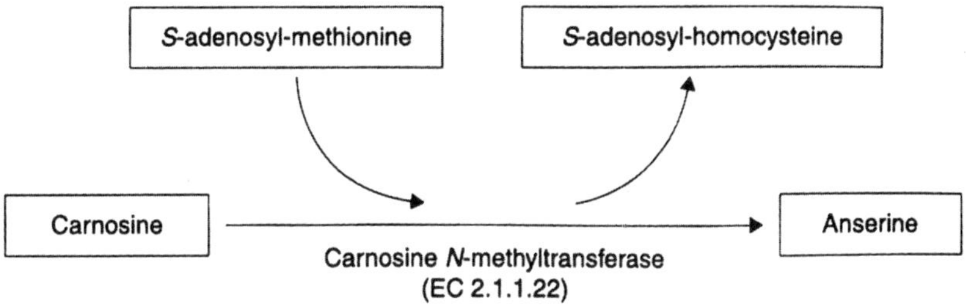

图 12.3　肌肽的 N−甲基化形成鹅肌肽

12.3.1　肌肽合酶

直到最近，关于催化肌肽、鹅肌肽和高肌肽生物合成的酶和途径的确切信息仍然是未知的。在 1959 年，两个研究团队（Kalyankar and Meister，1959；Winnick and Winnick，1959a）尝试从鸡胸肌肉中纯化肌肽合酶。随后，该酶从鼠的大脑（Skaper et al.，1973）和小鼠的嗅球（Horinishi et al.，1978）中纯化出来。并且通过细胞培养系统研究了肌肽

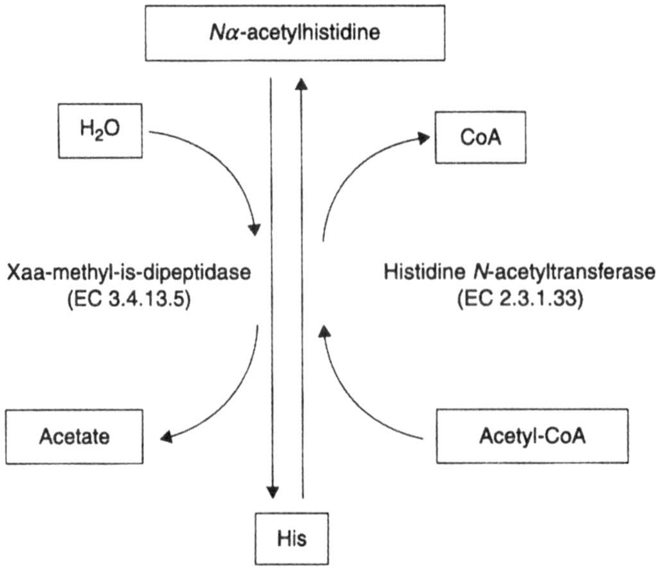

图 12.4 *N*α-乙酰组氨酸的生物合成及降解

的生物合成（Harding and Margolis，1976；Bauer et al.，1979，1982；Margolis et al.，1985；Schulz et al.，1989；Hoffmann et al.，1996）。肌肽是由 β-丙氨酸和 L-组氨酸通过 ATP 依赖性肌肽合酶（EC 6.3.2.11）的作用下合成的（图 12.2）。

L-组氨酸+β-丙氨酸+ATP → 肌肽+ADP + 磷酸盐

尽管肌肽合酶一开始被认为会产生 AMP 和焦磷酸盐（Kalyanker and Meister，1959；Stenesh and Winnick，1960），但是最近的分子鉴定结果（Drozak et al.，2010）证明生成的是 ADP。

由于其组织来源的酶的合成活性非常弱、不稳定，以及活性测定的困难，从动物组织中分离纯化这种酶的研究工作并没有取得很好的进展。早先的工作证明了鸟的肌肉要比哺乳动物肌肉中含有更高水平的肌肽合酶活性（Winnick and Winnick，1959a）。这种酶是从一种可溶的肌浆组分提取出来的，是一种粗酶并且是不稳定的形式（Stenesh and Winnick，1960）。Harding 和 Margolis（1976）报道了在小鼠的基本嗅球通路中发现的酶要比其他身体组织和大脑区域中的酶活性高，表明肌肽在嗅球神经传递中起作用。在 1981 年 Wood 和 Johnson（1981）报道了这种酶的一种有效纯化方法，就是利用 Cibacron Blue 亲和层析，会制备出电泳纯的样品。他们认为从鸡胸肌肉中纯化出的天然的酶是 119 kDa 的同源二聚体。在他们的研究中，揭示了从肌肉中纯化的酶的氨基酸组成，但当时并没有序列信息。

在 2010 年，Drozak 等（2010）报道了基因 *ATPGD*1（ATP-grasp domain-containing protein 1），主要在骨骼肌和人的大脑中表达，编码肌肽合酶。肌肽合酶属于 ATP 结合家族的连接酶。小鼠和人的 *ATPGD*1 在 HEK293T 细胞中表达。现在清楚的是 *ATPGD*1（最新提供

的基因代号 *CARNS*1）编码的酶在哺乳动物中合成肌肽。与 $N\pi$ 甲基-L-组氨酸相比，L-组氨酸是一种更好的底物，表明了这种酶在合成肌肽上要好于合成鹅肌肽。并且 β-丙氨酸与 γ-氨基丁酸相比是更好的底物，说明了这种酶能够更好地合成肌肽而不是高肌肽。此外，肌肽合酶可以与 $N\tau$-甲基-L 组氨酸反应形成蛇肉肽。因此通过 *CRNS*1 编码的单酶不仅能够催化肌肽的形成，还可以催化动物中鹅肌肽、高肌肽和蛇肉肽的形成。在他们的研究中，核苷酸数据库搜索表明在其他哺乳动物和鸡中，也存在着与 *CARNS*1 相近的基因序列。人和鸡的 *CARNS*1 序列的编码区分别可以翻译 950 个和 930 个氨基酸（将近 100 kDa 多肽）。小鼠和鸡来源的酶是同源四聚体（Drozak et al.，2010）。

相对地，在鱼类，例如基因组序列已经完全清楚的斑马鱼（*Danio rerio*）和发光的绿色河鲀（*Tetraodon nigroviridis*）中，并没有发现 *CARNS*1 同源基因。有意思的是，Tsubone 等（2007）报道称，在富含大量肌肽的日本鳗的骨骼肌中发现了一种可以催化咪唑相关二肽合成的新奇的酶。这种酶在合成反应中不需要 ATP，这与 *CARNS*1 编码的肌肽合酶不同。另外，日本鳗中纯化出来的酶也表现出了对这些二肽的水解活性，表明相同的酶既能催化咪唑多肽的合成，也能催化咪唑多肽的降解。之前已经预测，这种分子质量大约 275 kDa 的天然鳗鱼酶是一种由 43 kDa 亚基组成的六聚体或七聚体，并且根据末端氨基酸序列来判断是一种结合珠蛋白类似蛋白。尽管目前鳗鱼肌肉中肌肽合酶是否存在并不清楚，这种新的酶或许参与肌肉中肌肽的合成。因此，有必要进一步开展对包括鳗鱼在内的鱼类中肌肽合酶的鉴定和这种新奇双功能酶的研究。

12.3.2 肌肽 N-甲基转移酶

骨骼肌中鹅肌肽的生物合成通过两种途径：①$N\pi$-甲基-L-组氨酸与 β 丙氨酸的缩合；②肌肽的直接 N 甲基化。

（1）$N\pi$-甲基-L-组氨酸+β 丙氨酸+ATP → 鹅肌肽+ADP+磷酸

（2）肌肽+S-腺苷-L-甲硫氨酸 → 鹅肌肽+S-腺苷-L-高半胱氨酸

前一个反应是通过肌肽合酶进行催化，像之前在 12.3.1 章节中描述的一样。另一种酶，肌肽 N 甲基转移酶（EC 2.1.1.22）可以催化第二个反应（图 12.3）。然而，关于这种酶性质的相关信息十分有限。Winnick 和 Winnick（1959b）首先展示了 S-腺苷-L-甲硫氨酸可以作为鹅肌肽合成中甲基供体，并且更少量的作为 $N\pi$ 甲基-L-组氨酸的甲基供体。从鸡胸肌肉可溶性部分得到的肌肽 N 甲基转移酶纯化了 7.5 倍（McManus，1962）。对鸟和哺乳动物不同组织中的活性进行了研究，例如从鸡、豚鼠、老鼠、猫和兔子的骨骼肌、肝脏、肾脏、心脏、大脑、脾脏和肺中。我们知道这种酶的特异性针对肌肽的作为 S-腺苷-L-甲硫氨酸来源甲基的受体，在鱼中并没有这种酶的信息。这种酶的氨基酸序列至今也不清楚。

12.3.3 组氨酸 N 乙酰基转移酶

组氨酸 N 乙酰基转移酶催化乙酰基团从乙酰辅酶 A 转移到 L-组氨酸的 α 氨基上（图 12.4）。

$$\text{L-组氨酸+乙酰辅酶 A} \rightarrow N\alpha\text{-乙酰基-L-组氨酸 + 辅酶 A}$$

这种酶首先在鳉鱼（*Fundulus heteroclitus heteroclitus*）脑的冻干粉样品中被检测出来（Baslow，1996）。利用二维纸层析的方法检测在 $N\alpha$-乙酰基-L 组氨酸中^{14}C-L-组氨酸的量作为酶活测定方法。研究发现，$N\alpha$-乙酰基-组氨酸的合成是由乙酰辅酶 A 作为乙酰基的供体。Baslow 等（1969）利用体内实验研究还发现了在金鱼的晶状体和大脑中，这种组分是由 L-组氨酸合成的。尼罗罗非鱼利用等度反相高效液相色谱法检测反应产物，发现高活性的酶只在罗非鱼的大脑和晶状体中存在，而且，利用蓝琼脂糖凝胶柱亲和色谱层析法有效地从这种鱼的大脑中获得了高纯度的酶（4 166 倍）（Yamada et al.，1995）。这种酶有 pH 7.0~9.5 广泛的最适 pH，并且不需要任何的二价金属离子。当 Hg^{2+}、Ni^{2+} 和 Cu^{2+} 的浓度在 1 mmol/L 时，酶的活性被强烈抑制。由 Ni^{2+} 引起的抑制可以通过加入过量的 EDTA 进行彻底解除。这种酶显示出了严格的底物特异性，利用 L-组氨酸（和它的甲基衍生物）作为乙酰基受体并且利用乙酰辅酶 A 作为乙酰基的供体（表 12.2）。相对于 L-组氨酸的米氏常数（K_m）（0.45 mmol/L），其对以乙酰辅酶 A 为底物的米氏常数值（0.027 mmol/L）这是非常低。推测这种酶是分子量为 39 kDa，这在 SDS-PAGE 上非常直观的就可以看出是由单亚基组成的酶，因为通过凝胶过滤判断天然酶的分子量接近 39 kDa。

表 12.2　来源自尼罗罗非鱼（*Oreochromis nilotics niloticus*）的组氨酸 N-乙酰组氨酸的底物特异性

Substrate (1 mmol/L)	Relative activity/%
As an acetyl donor[b]	
Acetyl-coenzyme A	100
Acetylcholine	0
Sodium acetate	0
As an acetyl acceptor[c]	
L-Histidine	100
D-Histidine	0
L-Aspartic acid	0
L-Glutamic acid	0
L-Cysteine	0
$N\pi$-Methyl-L-histidine	23
$N\tau$-Methyl-L-histidine	71
Carmosine	0
Anserine	0
Hommocarnosine	0

a　Data from Yamada et al.,（1995）.

b　Using L-histidine as a substrate for an acetyl acceptor, the $N\alpha$-acetylhistidine synthesized was measured.

c　Using acetyl-coenzyme A as a substrate for an acetyl donor, the $N\alpha$-acetyl derivative synthesized was measured.

除了组氨酸 N-乙酰基转移酶之外，在脊椎动物中还有两种“N-乙酰氨基酸合成酶”：

氨基酸 N-乙酰基转移酶（EC 2. 3. 1. 1，也称作 N 乙酰谷氨酸合成酶）和天冬氨酸 N-乙酰基转移酶（EC 2. 3. 1. 17）。氨基酸-N-乙酰基转移酶是一种特异催化 L-谷氨酸和乙酰辅酶 A 形成 N-乙酰基谷氨酸的线粒体酶。这种酶并不会乙酰化 L-组氨酸（Sonoda and Tatibana，1983）。编码这种酶的基因，*NAGS*，主要在人的肝脏和小肠内表达（Morizono et al.，2004）。催化 N-乙酰天冬氨酸生物合成的天冬氨酸 N-乙酰基转移酶分布在哺乳动物的大脑中，哺乳动物大脑中含有大量的这种组分，而且这种酶的特异性将 L-天冬氨酸和乙酰辅酶 A 作为底物（Goldstein，1959）。最近报道称，*NAT8L* 是编码天冬氨酸 N-乙酰基转移酶的基因（Ariyannur et al.，2010；Wiame et al.，2010）。尽管到目前编码组氨酸 N-乙酰基转移酶的基因没有被鉴定出来，但是组氨酸 N-乙酰基转移酶与两种 N-乙酰基氨基酸合成酶很明显是不同的。这种酶仅仅分布在包括辐鳍鱼、两栖动物、爬行动物在内的变温哺乳动物有限的组织中（脑和晶状体），在这些动物中发现 *N*α-乙酰基-L-组氨酸广泛存在（图 12.5）。因此，在脊椎动物中组氨酸-N 乙酰基转移酶显示出了非常有趣的进化史。

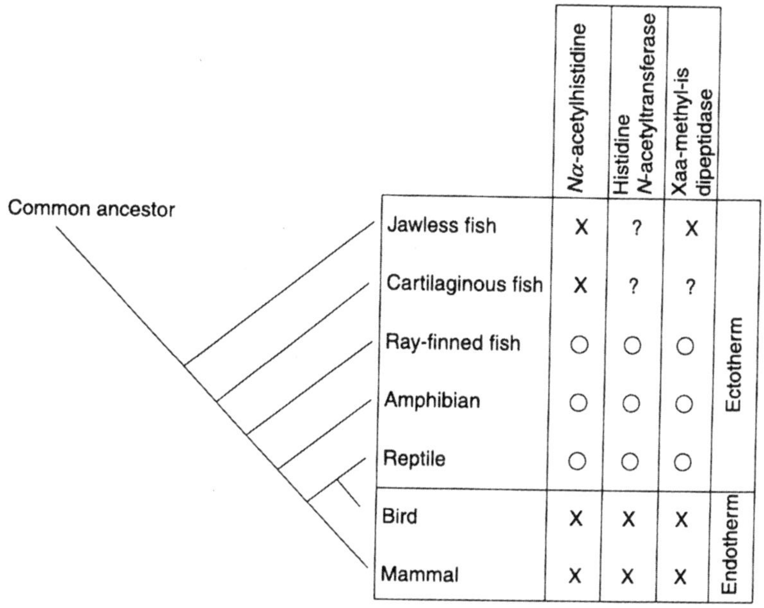

图 12.5　在脊椎动物大脑中催化 *N*α-乙酰组氨酸与生物合成和降解 *N*α-乙酰组氨酸的两种酶的系统进化分布

12.4　咪唑相关多肽的降解

3 种脊椎动物酶参与咪唑相关多肽的降解：β-丙氨酸-组氨酸二肽酶（EC 3.4.13.20），胞液非特异性二肽酶（EC 3.4.13.18）和 Xaa-甲基-组氨酸二肽酶（EC 3.4.13.5）（表 12.1）。与催化生物合成的酶相反，降解酶已经被相对很好地研究过了。

所有这些酶都能催化肌肽、高肌肽和/或鹅肌肽的水解，如下所示（图 12.2）：

肌肽+H_2O →β-丙氨酸+L-组氨酸

鹅肌肽+H_2O →β-丙氨酸+$N\pi$-甲基-L-组氨酸

高肌肽+H_2O → γ-氨基丁酸+L-组氨酸

关于这些酶对蛇肉肽的降解知道的很少，因为蛇肉肽真正的化合物是无法商业获得的。Xaa-甲基-组氨酸二肽酶也能够催化 Nα-乙酰基组氨酸的脱乙酰化，如下所示：

$N\alpha$-乙酰基组氨酸+H_2O→醋酸+L 组氨酸

很长的一段时间内，关于酶的分类一直困扰着人们，直到所有的这些酶从分子水平上被鉴定出来（Teufel et al.，2003；Yamada et al.，2005）。这揭示所有的这 3 种酶都属于 MEROPS 肽酶数据库中金属肽酶 H 家族的 M20 家族 A 亚族（M20A）（http：//merops. sanger. ac. uk/）（Rawlings et al.，2012）。在数据库中，β-丙氨酸-组氨酸二肽酶、胞液非特异性二肽酶和 Xaa-甲基-组氨酸二肽酶分别被分类为“肌肽二肽酶 Ⅰ”（MEROPS ID：M20. 006），“肌肽二肽酶Ⅱ”（MEROPS ID：M20. 005），和“Xaa-甲基-组氨酸二肽酶”（MEROPS ID：M20. 013）。另外两种肌肽切酶，“肽酶 V”（MEROPS ID：M20. 004）和“Xaa-组氨酸二肽酶”（MEROPS ID：M20. 007）来源于细菌。肽酶 V（EC 号还没有收录在 IUBMB 推荐中）属于 M20A 亚家族，而 Xaa-组氨酸二肽酶（之前归类为 EC 3. 4. 13. 3，但是在 2011 年删除了）属于不同的亚族，M20C。这些细菌酶在这一章节中不做讨论。（最近时间内，这 3 种脊椎动物酶的亚族在 MEROPS 数据库中从 MZOA 转到 MZOF）。

12. 4. 1 肌肽酶

由脊椎动物来源的叫做“肌肽酶”的酶有很长的分类史。Hanson 和 Smith（1949）首先描述了来源于猪肾脏的肌肽降解酶，锰可以强烈地激活这种酶。Rosenberg（1960a，1960b）从相同的组织中纯化了这种酶，并把它叫做“肌肽酶”。随后，各个研究者们研究了来源于几种物种不同组织的该酶，研究证据证明了两种不同种类的肌肽降解酶的存在（Perry et al.，1986；Lenney，1976，1990b；Lenney et al.，1977，1982，1985；Margolis et al.，1979，1983；Kunze et al.，1986；Jackson et al.，1991）。在这两种酶中，一种酶被命名为多个名字，像“锰依赖肌肽酶”（Margolis et al.，1979）、“人类组织肌肽酶”（Lenney et al.，1985）、“β-丙氨酸-精氨酸-水解酶”（Kunze et al.，1986）和“人类胞质非特异性二肽酶”（Lenney，1990a），实际上它不是肌肽特异性的酶，而是有着广泛的底物特异性，显示出了与其他已知的“胞液非特异性的二肽酶”（EC 3. 4. 13. 18）相同的特性。胞液非特异性二肽酶也被称为“脯肽酶”，“甘氨酰-亮氨酸二肽酶”，“甘氨酰-甘氨酸二肽酶”和“肌肽二肽酶Ⅱ”。其他的肌肽降解酶被描述成“血清肌肽酶”（Murphey et al.，1973）、“锰非依赖肌肽酶”（Margolis et al.，1979），“肌肽酶”（Kunze et al.，1986）和“人血清肌肽酶”（Jackson et al.，1991）是一种咪唑相关二肽特异性酶，有着非常严格的底物特异性。这种酶被IUBMB在1992年新注册为“β-丙氨酸-组氨酸二肽酶”(EC 3.4.13.20)。β-丙氨酸-组氨酸二肽酶也被叫做“肌肽二肽酶Ⅰ”。胞液非特异性二肽酶和β-丙氨酸-组氨酸二肽酶最终因为编码它们的人的基因，分别是*CNDP*2和*CNDP*1的发现而区分开来（Teufel et al.，2003)。尽管这两种“肌肽酶”结构相似，但是性质上有着很大不同。

胞液非特异性二肽酶

像名字中指示的，胞液非特异性二肽酶有着广谱特异性，而且既可以催化“普通”二肽，像L-脯氨酸-L-亮氨酸、L-丙氨酸-L-亮氨酸、L-亮氨酸-L-亮氨酸、L-丙氨酸-L-丙氨酸、甘氨酸-L-Ley、L-丙氨酸-L-组氨酸、L-脯氨酸-L-丙氨酸、甘氨酸-L-组氨酸和L-组氨酸-甘氨酸的水解，也能催化“特别”二肽肌肽的水解。以下组分并不能被这种酶水解：鹅肌肽、高肌肽、$N\alpha$-乙酰基-L-组氨酸、$N\alpha$-乙酰基-L-甲硫氨酸、甘氨酸-L-组氨酸-L-赖氨酸和甘氨酸-甘氨酸-甘氨酸（Lenney et al.，1985，Lenney，1990a)。

二硫苏糖醇或巯基乙醇可以稳定和激活这种酶，但冻干干燥可以破坏该酶（Lenney，1990a)。Mn^{2+}、Co^{2+}和Cd^{2+}也可以稳定酶活。因此，硫醇相关的组分（p-对氯高汞苯甲酸、N-乙马来酰亚胺和2-碘乙酰胺）和螯合剂（EDTA和1，10-邻二氮杂菲）可以抑制这种酶。与其他两种随后讨论到的降解酶不同，胞液非特异性二肽酶可以强烈、特异性地被苯丁抑制剂进行抑制，IC_{50}的值为4.2 nmol/L时。苯丁抑制剂是一种非常有效的亮氨酰氨肽酶和氨肽酶B的竞争抑制剂（Peppers and Lenney，1988)。这种抑制剂是一种将胞液非特异性二肽酶从其他两种降解酶中区别出来的简单有效的工具。

Teufel等（2003）从人类基因组科学公司人类大脑cDNA文库中鉴定出了属于金属蛋白酶家族的两个不同的克隆：一个是*CNDP*2编码胞液非特异性二肽酶；另一个是*CNDP*1编码β-丙氨酸-组氨酸二肽酶（在下一节中进行讨论)。纯化后重组人*CNDP*2蛋白在SDS-PAGE中显示的是54 kDa，通过MALDI-TOF确定的分子量是52.8 kDa，这与通过氨基酸序列进行推测的值一致（52.7 kDa)。通过排阻色谱判断，人*CNDP*2蛋白显然形成了同源二聚体。晶体结构研究（Unno et al.，2008）已经展示了老鼠胞液非特异性二肽酶的亚基，还有细菌肽酶V和Xaa-组氨酸二肽酶（Rowsell et al.，1997；Jozic et al.，2002；Chang et al.，2010)，组成含有二价金属离子像Zn^{2+}和Mn^{2+}活性位点的催化区。

胞液非特异性二肽酶广泛分布在真核细胞中（Bauer，2004）并且在几乎所有测试过的人组织中都广泛发现有活性的酶（Butterworth and Priestman，1982)。像两种鱼的*CNDP*2 mRNAs（罗非鱼（Yamada et al.，2005)，日本鳗（Oku et al.，2012））和人类（Teufel et al.，2003）在检测过的组织中都有表达，因此，*CNDP*2编码的胞液非特异性二肽酶很可能作为一种管家酶与细胞内的二肽类底物的分解代谢有关。人们认为在富含脯氨酸的胶原蛋白降解过程中，许多脯氨酸-Xaa类型的二肽都是由胞液非特异性二肽酶进行水解的（Bauer，2004)。因此可以推断胞液非特异性二肽酶并不是一种在生理条件下特异性地作用于肌肽的特异性酶。

β-丙氨酸-组氨酸二肽酶

在很长一段时间里，人们把β-丙氨酸-组氨酸二肽酶称作“血清肌肽酶”，直到 Perry 等（1986）首先报道了人血清对肌肽和鹅肌肽降解的观察。不像胞液非特异性二肽酶那样有着广泛的底物特异性，β-丙氨酸-组氨酸二肽酶是一种分泌性的酶，而且显示出只对一些咪唑相关的二肽，像肌肽、鹅肌肽、高肌肽、L-丙氨酸-L-组氨酸和甘氨酸-L-组氨酸等的狭窄的特异性（Teufel et al.，2003）。Jackson 等（1991）研究了有活性β-丙氨酸-组氨酸二肽酶在哺乳动物血清中的分布。6 种高等的灵长类动物包括人类、黑猩猩、长臂猿、大猩猩、猩猩和矮黑猩猩中发现有活性的酶。在 12 种非灵长类哺乳动物包括，狗、马、小牛、猪、鼠、家鼠、兔子、狒狒、豚鼠、犰狳和中国地鼠的血清中没有检测到有活性的酶。黄金鼠的血清中确实有酶活。尽管有报道称，人类 *CNDP*1 编码的β-丙氨酸-组氨酸二肽酶在大脑和肝脏中表达（Teufel et al.，2003），但是人血清中该酶的起源仍不清楚。

目前，有活性的β-丙氨酸-组氨酸二肽酶还未在非哺乳动物种类中检测到，然而，Partmann（1976）报道了在含有大量肌肽的欧洲鳗鱼（*Anguilla anguilla*）的骨骼肌中发现一种肌肽裂解酶的酶活。最近，已经有了关于β-丙氨酸-组氨酸二肽酶和来自日本鳗骨骼肌中的 Xaa-甲基-组氨酸二肽酶的纯化和分子鉴定的报道（Oku et al.，2012）。这是仅有的一篇报道关于从非哺乳动物中分离 *CNDP*1 编码的二肽酶的研究。表 12.3 显示了来自日本鳗的这两种酶的底物特异性。数据清楚地显示了鳗鱼β-丙氨酸-组氨酸二肽酶的底物特异性非常窄，然而鳗鱼 Xaa-甲基-组氨酸二肽酶的底物特异性非常广泛。BLAST 公共数据搜索库显示，在鸟、爬行动物、两栖动物和辐鳍鱼以及哺乳动物中有 *CNDP*1 的同源蛋白存在（Oku et al.，2012）。这表明β-丙氨酸-组氨酸二肽酶广泛分布在脊椎动物中，然而只有灵长类（和黄金鼠）的血清中有很高的活性。

表 12.3　来源于日本鳗（*Anguilla japonica*）[a] 的两种肌肉二肽酶的底物特异性

Substrate	Relative activity/%[b]	
	β-Ala-His dlpeptidase	Xaa-methyl-His dipeptidase
L-Ala-L-His	100	100
Carnosine	7	9
Anserine	78	33
Homocarnosine	4	14
Gly-L-Leu	2	352
Gly-D-Leu	0	15

续表

Substrate	Relative activity/%[b]	
	β-Ala-His dlpeptidase	Xaa-methyl-His dipeptidase
Gly-L-His	6	91
Gly-Gly	24	306
L-Cys-Gly	0	1
L-His-Gly	3	7
L-Pro-Gly	2	180
L-Leu-Gly	0	69
L-Ala-L-Pro	0	0
L-His-L-Leu	1	1
*N*α-Chloroacetyl-L-Leu	9	501
*N*α-Acetyl-L-Leu	1	53
*N*α-Acetyl-L-Met	1	80
*N*α-Acetyl-L-Glu	0	8
*N*α-Acetyl-L-Asp	0	3
*N*α-Acetyl-L-His	2	43
L-Ala-*p*-nitroanilide	0	0
L-Leu-*ρ*-nitroanilide	1	1
L-Leu-β-naphthylamide	0	0
Benzyloxycarbonyl-Gly-L-Leu	0	9

a　Data from Oku et al., (2012).

b　The purified enzymes were incubated at 37℃ for 1 h with 2 mmol/L substrata in 100 mmol/L N-ethylmorpholine-HCl buffer, pH 8.0, containing 0.1 mmol/L $CoSO_4$ and 1mg/mL bovine serum albumin.

人类 *CNDP*1 编码区翻译成 508 个氨基酸，包括含有 29 个氨基酸的信号肽，其中有亮氨酸重复和 3 个 N-糖基化位点（Teufel et al., 2003）。另外，日本鳗 *CNDP*1 编码 492 个氨基酸，包括丝氨酸重复和 1 个 N-糖基化位点的 11 个氨基酸组成的信号肽（Oku et al., 2012），成熟的来源于人类和日本鳗的酶都形成同源二聚体。

有报道称糖尿病肾病的敏感性与人 *CNDP*1 的多态性密切相关（Janssen et al., 2005）之后，过去的几年间许多研究已经专注于β-丙氨酸-组氨酸二肽酶与人类糖尿病肾病的遗传相关性（Vionnet et al., 2006; Iyengar et al., 2007; Wanic et al., 2008; McDonough et al., 2009; Kilis-Pstrusinska et al., 2009; Riedl et al., 2010; Ahluwalia et al., 2011; Chakkera et al., 2011; Peters et al., 2011）。*CNDP*1 的多态性影响β-丙氨酸-组氨酸二肽

酶分泌（Janssen et al.，2005）。也就是说，这种酶的低量分泌与人类 *CNDP*1 信号肽中亮氨酸重复（5-Leu）的数量少有关。研究小组提出，糖尿病病人 5-Leu 类型的 *CNDP*1 纯合子可以避免糖尿病肾病的发生，因为肌肽作为 β-丙氨酸-组氨酸二肽酶生理学的底物，扮演着一种抵抗高糖水平对肾细胞不利影响保护因子的角色。并且研究发现，除了 *CNDP*1 的亮氨酸重复多态性之外，酶的 N-糖基化对酶的正常分泌和活性也是至关重要（Riedl et al.，2010）。

12.4.2 “鹅肌肽酶”或“乙酰基组氨酸脱乙酰酶”

在变温脊椎动物中，Xaa-甲基-组氨酸二肽酶有两种可能的生理学底物：一种是鹅肌肽（在骨骼肌中），另一种是 $N\alpha$-乙酰基组氨酸（在脑和眼中）。这种酶曾经被 IUB 分别在 1961 年叫做“鹅肌肽酶”EC 3.4.3.4 和在 1972 年叫做“乙酰基组氨酸脱乙酰酶”EC 3.5.1.34。然而，依据 Lenney 等（1978）的研究发现鲣鱼乙酰基组氨酸脱乙酰酶和鳕鱼鹅肌肽酶两者是相似的，这“两种”酶在 1981 年被 IUB 判断为相同的，现在一起并入了 Xaa-甲基-组氨酸二肽酶（EC 3.4.13.5）。

Xaa-甲基-组氨酸二肽酶

像“骨骼肌”章节中提到的，鳕科与鲑鱼科的鱼的骨骼肌中含有鹅肌肽作为主要的咪唑相关组分。Jones（1995）发现在幼鳕（*Gadus callarias*）的骨骼肌中的鹅肌肽被一种内源性的酶水解，按照粗略的化学计量之比释放出等量 β-丙氨酸和 $N\pi$-甲基-组氨酸。这是首次报道证明之前叫做“鹅肌肽酶”的 Xaa-甲基-组氨酸二肽酶的存在。随后，这种幼鳕酶被证明是一种金属激活的二肽酶，有着很低的底物特异性（Jones，1956）。在另外一个独立的研究中，Baslow（1963）报道了鲔鱼脑中 $N\alpha$-乙酰基组氨酸在死后降解成组氨酸（$N\alpha$-乙酰基组氨酸在那时被叫做“neurosine”，因为它的结构还没有被鉴定出来）。然后从鲣鱼的脑中部分纯化“乙酰基组氨酸脱乙酰酶”的工作被同一个研究小组报道（Baslow and Lenney，1967）。这是利用色谱层析技术纯化 Xaa-甲基-组氨酸二肽酶的首次试验。从那以后，幼鳕肌肉和金枪鱼脑中的这种酶被高度纯化，并且对其底物特异性和不同组分对酶活的影响，分子质量和色谱层析行为进行了对比研究（Lenney et al.，1978）。研究结果说明“鹅肌肽酶”和“乙酰基组氨酸脱乙酰酶”是同一种酶。在 2005 年，第一个关于罗非鱼的相关酶分子鉴定完成，这也表明了 Xaa-甲基-组氨酸二肽酶和胞液非特异性二肽酶和 β-丙氨酸-组氨酸二肽酶是 M20A 亚族的一员（Yamada et al.，2005）。而且，Xaa-甲基-组氨酸二肽酶与 β-丙氨酸-组氨酸二肽酶从日本鳗骨骼肌中被鉴定出来（Oku et al.，2012）。

如表 12.3 所示，Xaa-甲基-组氨酸二肽酶对二肽和 $N\alpha$-乙酰基氨基酸有很广泛的底物特异性（Oku et al.，2012）。该二肽酶可以水解常规的二肽，如 Ala-L-His、Gly-L-Leu、Gly-Gly 与 L-Pro-Gly，同时也可以水解非常规的二肽，如鹅肌肽、肌肽和同源鹅肌肽。

该酶还具有很强的针对 *N*α-乙酰基氨基酸的脱乙酰基活性，如 *N*α-乙酰-L-His、*N*α-乙酰-L-Met、*N*α-乙酰-L-Leu、*N*α-氯乙酰-L-Leu。

Xaa-甲基-组氨酸二肽酶是一种亚基分子量约为 55 kDa 的同源二聚体（Yamada et al.，1993，2005）。7 个推测的活性位点氨基酸残基（两个活性位点残基和 5 个金属结合位点残基）在 Xaa-甲基-组氨酸二肽酶（Yamada, et al.，2005）和 *β*-丙氨酸-组氨酸二肽酶和胞液非特异性二肽酶（Teufel et al.，2003）中是完全保守的。像 *β*-丙氨酸-组氨酸二肽酶一样，Xaa-甲基-组氨酸二肽酶是一种分泌型的糖蛋白。鱼眼内液包括玻璃体和眼房水含有很强的酶活（Baslow，1967）。然而，找不到关于 Xaa-甲基-组氨酸二肽酶在鱼脑中亚细胞定位的信息。有人认为，成熟的酶或许存在于细胞外空间，有效地降解它的生理学底物 Nα-乙酰基组氨酸。

有意思的是，Xaa-甲基-组氨酸二肽酶与 *N*α-乙酰基组氨酸，只广泛分布在变温脊椎动物（辐鳍鱼，两栖类和爬行动物）的脑中（图 12.5）。恒温脊椎动物（哺乳动物和鸟）没有基因编码这种酶（Yamada et al.，2005）。除了无颌鱼类（未知的软骨鱼类），该基因在变温脊椎动物中是保守（Oku et al.，2011）。并且这种酶广泛分布在变温脊椎动物的眼液中。Xaa-甲基-组氨酸二肽酶在变温脊椎动物骨骼肌中的存在是物种特异性的。大西洋鳕（*Gadus morhua*）（Jones，1955）、黑线鳕（*Melanogrammus aeglefinus*）（Perez and Jones，1962）、太平洋鳕（*Gadus macrocephalus*）（Suzuki et al.，1988）、阿拉斯加鳕（*Theragra chalcogramma*）（Suzuki et al.，1988）和日本鳗（Oku et al.，2012）的骨骼肌中都显示出了很强的 Xaa-甲基-组氨酸二肽酶的酶活性，而鲣（*Katsuwonus pelamis*）（Lenney et al.，1978）、大西洋蓝鳍金枪鱼（*Thunnus thynnus*）（Suzuki et al.，1988）、远东拟沙丁鱼（*Sardinops sagax*）（Suzuki et al.，1988）、石鲽（*Kareius bicoloratus*）（Suzuki et al.，1988）、香鱼（*Plecoglossus altivelis altivelis*）（Suzuki et al.，1988）、鲤（*Cyprinus carpio carpio*）（Suzuki et al.，1988）、虹鳟（Suzuki et al.，1988；Yamada et al.，1991）和罗非鱼（Yamada et al.，1992）骨骼肌中没有显示出活性。

像上面提到的，Xaa-甲基-组氨酸二肽酶和组氨酸 N-乙酰基转移酶的分布模式与 *N*α-乙酰基组氨酸是完全一致的（图 12.5）。尽管 *N*α-乙酰基组氨酸在变温脊椎动物中枢神经系统中的生物学作用还不清楚，但是很明显地因为某种生理学目的，这种生理学底物在中枢神经系统中会普遍地被 Xaa-甲基-组氨酸二肽酶脱乙酰化。从这种酶在变温脊椎动物的中枢神经系统中的生物学特征的观点来看，“乙酰基组氨酸脱乙酰酶”或许是这种酶的一个更合适的名字。

12.4.3 相关降解酶的系统发育史

所有 3 种降解酶，包括胞液非特异性二肽酶、*β*-丙氨酸-组氨酸二肽酶和 Xaa-甲基-组氨酸二肽酶都属于同一个进化亚族 M20A，但是在分子系统发育树上是彼此分开的（图 12.6）。这 3 种降解酶的系统发育史在之前脊椎动物进化中被认为是如下模式（Yamada et al., 2005; Oku et al., 2011)。由于胞液非特异性二肽酶广泛分布在真核生物的现代物种中，可以推测在脊椎动物中，这种基因是原始基因的基因重复产生β-丙氨酸-组氨酸二肽酶和Xaa-甲基-组氨酸二肽酶。在尾索动物亚门（被囊类动物）和脊椎动物门（有头骨的动物）分开之后，胞液非特异性二肽酶的基因重复导致了额外的不受选择压力（第一次重复）的复制。

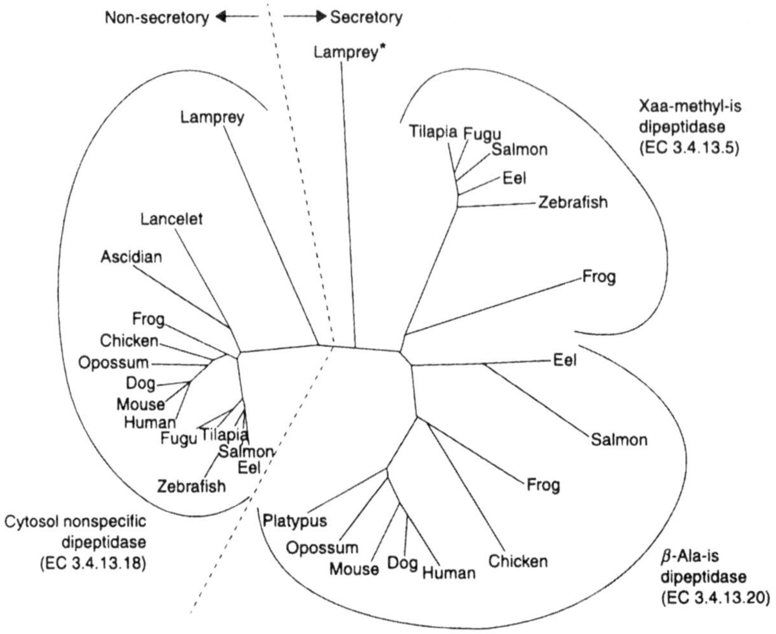

图 12.6　催化降解咪唑多肽的三种脊椎动物酶的系统发生树

注：七鳃鳗基因用星号标记是一个原始分泌，既不是Xaa-甲基-His二肽酶，也不是β-Ala-His二肽酶（在12.4.3中讨论）

这种新复制的基因成为原始的分泌类型，被一些无颚鱼的现代种类遗传，像远东溪七鳃鳗（*Lethenteron reissneri*）（在图12.6中标星号的基因）和七鳃鳗（*Lethenteron camtschaticum*）。分泌酶基因的第二次复制事件发生在颚口类（颌类脊椎动物）这一支。在辐鳍鱼（辐鳍鱼）和肉鳍鱼（肉鳍鱼和四腿的脊椎动物）分开之前，这两种基因或许已经各自进化成了Xaa-甲基-组氨酸二肽酶基因和β-丙氨酸-组氨酸二肽酶基因。然而β-丙氨酸-组氨酸二肽酶基因从辐鳍鱼到哺乳动物都有广泛分布，Xaa-甲基-组氨酸二肽酶基因在现代的温血鸟类或哺乳动物种类中并不保守。Xaa-甲基-组氨酸二肽酶（或许与组氨酸N-乙酰基转移酶一起）的消失似乎发生在从冷血到温血条件的进化过程中。

12.5 作用于咪唑相关二肽酶可能的生物催化应用

当今，肌肽和鹅肌肽作为一种可以促进健康的功能性食品成分出现在市场上。目前这些商业化的二肽都是从鸡肉、牛肉和金枪鱼肉中提取纯化。但是这些二肽在回收处理过程中的降解和损失是不可避免的。最近，为了制备大量的生物材料，将微生物细胞和酶应用在工业生产中的“白色生物技术”有了很大发展。可以预测这种技术或许也可以应用到生产咪唑相关的二肽中。

在肌肽合酶的分子结构被鉴定出来之前，Inaba 等（2010）报道了在酵母细胞中通过 β-丙氨酸-组氨酸二肽酶水解的反相反应将 L-组氨酸和 β-丙氨酸合成肌肽。它们构建了全细胞生物催化系统，用 α-凝集素作为锚蛋白在酵母细胞表面展示了 β-丙氨酸-组氨酸二肽酶，以 β-丙氨酸和 L-组氨酸作为底物合成肌肽。该研究表明，酵母细胞表面展示 β-丙氨酸-组氨酸二肽酶可以简单方便地用全细胞生物催化来合成肌肽。在章节 12.3.1 提到的，迄今为止，已经揭示了 *CARNS*1 作为编码肌肽合酶的基因。因此，关于肌肽合成酶的进一步的生物技术研究可能会推动其在生物催化方面的应用。

12.6 总结

由于咪唑相关二肽几乎专一的分布在颌类脊椎动物中，这些组分似乎是“有颌类特异性”的代谢产物。尽管有众多的研究报道了这些代谢产物的存在以及可能的功能和新陈代谢机制，但是每一个组分确定的生物学功能并不清楚。目前和未来对作用于咪唑相关二肽的酶的分子鉴定将会极大地推动对这些生物活性二肽功能的认识和这些酶生物催化的应用。

致谢

感谢 Steven M. Plakas 教授（FDA，Dauphin，Island AL，USA）的审稿。也感谢我的妻子 Hiromi Yamada，对我的帮助和鼓励。

参考文献

Abe，H.（1983）Distribution of free L-histidine and its related compounds marine fishes. *Nippon Suisan Gakkaishi* 49，1683-1687.

Abe，H. and Ohmama，S.（1987）Effect of starvation and sea-water acclimation the concentration of free L-histidine and related dipeptides in the muscle of eel，rainbow trout and Japanese dace. *Comp. Biochem. Physiol. B* 88 507-88 511.

Abe，H. and Okuma，E.（1991）Effect of temperature on the buffering capacities of histidine-related compounds and fish skeletal muscle. *Nippon Suisan Gakkaishi* 57, 2101–2107.

Ahluwalia, T. S., Lindholm, E. and Groop, L. C. (2011) Common variants in CNDP1 and CNDP2, and risk of nephropathy in type 2 diabetes. *Diabetologia* 54, 2295–2302.

Ariyannur, P. S., Moffett, J. R., Manickam, P., Pattabiraman, N., Arun, P., et al., (2010) Methamphetamine–induced neuronal protein NAT8L is the NAA biosynthetic enzyme: implications for specialized acetyl coenzyme A metabolism in the CNS. *Brain Res.* 1335, 1–13.

Aydin, A. F, Kusku–Kiraz, Z., Dogru-Abbasoglu, S., Gulluoglu, M., Uysal, M. and Kocak–Toker, N. (2010) Effect of carnosine against thioacetamide–induced liver cirrhosis in rat. *Peptides* 31, 67–71.

Baran, E. J. (2000) Metal complexes of carnosine. *Biochemistry (Mosc)* 65, 789–797.

Baslow, M. H. (1963) The enzymatic degradation of neurosine as an index of fish quality. *Am. Zool* 3, 536 (Abstract No. 266).

Baslow, M. H. (1965) Neurosine, its identification with N–acetyl–L–histidine and distribution in aquatic vertebrates. *Zoologica* 50, 63–66.

Baslow, M. H. (1966) N–Acetyl–L–histidine synthetase activity from the brain the killifish. *Brain Res.* 3, 210–213.

Baslow, M. H. (1967) N–Acetyl–L–histidine metabolism in the fish eye: evidence for ocular fluid–lens L–histidine recycling. *Expl. Eye Res.* 6, 336–342.

Baslow, M. H. (1998) Function of the N–acetyl–L–histidine system in the vertebrate eye. Evidence in support of a role as a molecular water pump. *J. Mol. Neurosci* 10, 193–208.

Baslow, M. H. (2003) N–Acetylaspartate in the vertebrate brain: metabolism and function. *Neurochem. Res.* 28, 941–953.

Baslow, M. H. and Lenney, J. F. (1967) α–N–Acetyl–L–histidine amidohydrolase activity from the brain of the skipjack tuna *Katsuwonus pelamis*. *Can. J. Biochem.* 45, 337–340.

Baslow, M. H., Turlapaty, P. and Lenney, J. F. (1969) N–Acetylhistidine metabolism in the brain, heart and lens of the goldfish, *Carassius auratus*, in vivo: evidence of rapid turnover and a possible intermediate. *Life Sci.* 8, 535–541.

Baslow, M. H. and Yamada, S. (1997) Identification of N–acetylaspartate in the lens of the vertebrate eye: a new model for the investigation of the function of N–acetylated amino acids in vertebrates. *Expl. Eye Res.* 64, 283–286.

Bauer, K. (2004) 'Cytosol nonspecific dipeptidase'. In *Handbook of Proteolytic Enzymes*, *2nd Ed.*, Barrett, A. J., Rawlings, N. D. and Woessner, J. F. (eds), Elsevier, London.

Bauer, K., Hallermayer, K., Salnikow, J., Kleinkauf, H. and Hamprecht, B. (1982) Biosynthesis of carnosine and related peptides by glial cells in primary culture. *J. Biol. Chem.* 257, 3593–3597.

Bauer, K., Salnikow, J., de Vitry, F., Tixier–Vidal, A. and Kleinkauf, H. (1979) Characterization and biosynthesis of ω–aminoacyl amino acids from rat brain and the C–6 glioma cell line. *J. Biol. Chem.* 254, 6402–6407.

Boldyrev, A., Bulygina, E., Leinsoo, T., Petrushanko, I., Tsubone, S. and Abe, H (2004) Protection of neuronal cells against reactive oxygen species by carnosine and related compounds. *Comp. Biochem. Physiol. B* 137, 81–88.

Boldyrev, A. A. , Dupin, A. M. , Pindel, E. V. and Severin, S. E. (1988) Antioxidative properties of histidine-containing dipeptides from skeletal muscles of vertebrates. *Comp. Biochem. Physiol. B* 89, 245-250.

Bonfanti, L. , Peretto, P. , De Marchis, S. and Fasolo, A. (1999) Carnosine-related dipeptides in the mammalian brain. *Prog. Neurobiol.* 59, 333-353.

Breck, O. , Bjerkas, E. , Campbell, P. , Arnesen, P. , Haldorsen, P. and Waagbo, R. (2003) Cataract preventative role of mammalian blood meal, histidine, iron and zinc in diets for Atlantic salmon (*Salmo salar* L.) of different strains. *Aqua. Nutri.* 9, 341-350.

Breck, O. , Bjerkas, E. , Sanderson, J. , Waagbo, R. and Campbell, P. (2005) Dietary histidine affects lens protein turnover and synthesis of N-acetylhistidine in Atlantic salmon (*Salmo salar* L.) undergoing parr-smolt transformation. *Aqua. Nutri.* 11, 321-332.

Butterworth, J. and Priestman, D. (1982) Fluorimetric assay for prolinase and partial characterisation in cultured skin fibroblasts. *Clin. Chim. Acta* 122, 51-60.

Caldovic, L. and Tuchman, M. (2003) N-Acetylglutamate and its changing role through evolution. *Biochem. J.* 372, 279-290.

Cameron, J. N. (1989) Intracellular buffering by dipeptides at high and low temperature in the blue crab *Callinectes sapidus*. *J. Exp. Biol.* 143, 543-548.

Chakkera, H. A. , Hanson, R. L. , Kobes, S. , Millis, M. P. , Robert, G. N. B. , et al. , (2011) Association of variants in the carnosine peptidase 1 gene (CNDP1) with diabetic nephropathy in American Indians. *Mol. Genet. Metab.* 103, 185-190.

Chan, K. M. and Decker, E. A. (1994) Endogenous skeletal muscle antioxidants. *Crit. Rev. Food Sci. Nutr.* 34, 403-426.

Chang, C. Y. , Hsieh, Y. C. , Wang, T. Y. , Chen, Y. C. , Wang, Y. K. , et al. , (2010) Crystal structure and mutational analysis of aminoacylhistidine dipeptidase from Vibrio alginolyticus reveal a new architecture of M20 metallopeptidases. *J. Biol. Chem.* 285, 39500-39510.

Crush, K. G. (1970) Carnosine and related substances in animal tissues. *Comp. Biochem. Physiol.* 34, 3-30.

Drescher, M. J. and Drescher, D. G. (1991) N-Acetylhistidine, glutamate, and β-alanine are concentrated in a receptor cell layer of the trout inner ear. *J. Neurochem.* 56, 658-664.

Drozak, J. , Veiga-da-Cunha, M. , Vertommen, D. , Stroobant, V. and Van Schaftingen, E. (2010) Molecular identification of carnosine synthase as ATP-grasp domain-containing protein 1 (ATPGD1) . *J. Biol. Chem.* 285, 9346-9356.

Dunnett, M. and Harris, R. C. (1997) High-performance liquid chromatographic determination of imidazole dipeptides, histidine, 1-methylhistidine and 3-methylhistidine in equine and camel muscle and individual muscle fibres. *J. Chromatogr. B* 688, 47-55.

Ferriero, D. and Marogolis, F. L. (1975) Denervation in the primary olfactory pathway of mice. II. Effects on carnosine and other amine compounds. *Brain Res.* 94, 75-86.

Flancbaum, L. , Fitzpatrick, J. C. , Brotman, D. N. , Marcoux, A. M. , Kasziba, E. and Fishes H. (1990) The presence and significance of carnosine in histamine-containing tissues of several mammalian species. *Agents Actions* 31, 190-196.

Goldstein, F. B. (1959) Biosynthesis of N-acetyl-L-aspartic acid. *J. Biol. Chem.* 234, 2702-2706.

Guiotto, A., Calderan, A., Ruzza, P., Osler, A., Rubini, C., et al., (2005) Synthesis and evaluation of neuroprotective α, β-unsaturated aldehyde scavenger histidyl-containing analogues of carnosine. *J. Med. Chem.* 48, 6156-6161.

Gulewitsch, W. and Admiradzibi, S. (1900) Uber das carnosine, eine neue organische Base des Fleischextraktes [On the carnosine, a new organic base of the flesh extracts]. *Ber. Dtsch. Chem. Ges.* 33, 1902-1903.

Hanson, A. (1966) Differences in *N*-acetylhistidine level in fresh water and seawater fish. [In French.] *C. R. Seances Soc. Biol. Fil.* 160, 265-268.

Hanson, H. T. and Smith, E. L. (1949) Carnosinase: an enzyme of swine kidney. *J. Biol. Chem.* 179, 789-801.

Harding, J. and Margolis, F. L. (1976) Denervation in the primary olfactory pathway of mice. III. Effect on enzymes of carnosine metabolism. *Brain Res.* 110, 351-360.

Harris, R. C., Marlin, D. J., Dunnett, M., Snow, D. H. and Hultman, E. (1990) Muscle buffering capacity and dipeptide content in the thoroughbred horse, greyhound dog and man. *Comp. Biochem. Physiol. A* 97, 249-251.

Hipkiss, A. R. and Brownson, C. (2000) A possible new role for the anti-ageing peptide carnosine. *Cell. Mol. Life Sci.* 57, 747-753.

Hipkiss, A. R. and Chana, H. (1998) Carnosine protects proteins against methylglyoxal-mediated modifications. *Biochem. Biophys. Res. Commun* 248, 28-32.

Hipkiss, A. R., Preston, J. E., Himswoth, D. T. M., Worthington, V. C. and Abbot, N. J. (1997) Protective effects of carnosine against malondialdehyde-induced toxicity towards cultured rat brain endothelial cells. *Neurosci. Lett.* 238, 135-138.

Hoffmann, A. M., Bakardjiev, A. and Bauer, K. (1996) Carnosine-synthesis in cultures of rat glial cells is restricted to oligodendrocytes and carnosine uptake to astrocytes. *Neurosci. Lett.* 215, 29-32.

Horinishi, H., Grillo, M. and Margolis, F. L. (1978) Purification and characterization of carnosine synthetase from mouse olfactory bulbs. *J. Neurochem.* 31, 909-919.

Inaba, C., Higuchi, S., Morisaka, H., Kuroda, K. and Ueda, M. (2010) Synthesis of functional dipeptide carnosine from nonprotected amino acids using carnosinase-displaying yeast cells. *Appl. Microbiol. Biotechnol.* 86, 1895-1902.

Iyengar, S. K., Freedman, B. I. and Sedor, J. R. (2007) Mining the genome for susceptibility to diabetic nephropathy: the role of large-scale studies and consortia. *Semin. Nephrol.* 27, 208-222.

Jackson, M. C., Kucera, C. M. and Lenney, J. F. (1991) Purification and properties of human serum carnosinase. *Clin. Chim. Acta* 196, 193-205.

Janssen, B., Hohenadel, D., Brinkkoetter, P., Peters, V., Rind, N., et al., (2005) Carnosine as a protective factor in diabetic nephropathy-association with a leucine repeat of the carnosinase gene CNDP1. *Diabetes* 54, 2320-2327.

Jones, N. R. (1955) The free amino acids of fish; 1-methylhistidine and β-alanine liberation by skeletal muscle anserinase of codling (*Gadus callarias*). *Biochem. J.* 60, 81-87.

Jones, N. R. (1956) 'Anserinase and other dipeptidase activity in skeletal muscle of codling (*Gadus callarias*)'. Abstract of the 352nd Meeting of the Biochemical Society. *Proceedings of the Biochemical Society*, 20p.

Jozic, D., Bourenkow, G., Bartunik, H., Scholze, H., Dive, V., et al., (2002) Crystal structure of the dinuclear zinc aminopeptidase PepV from *Lactobacillus delbrueckii* unravels its preference for dipeptides. *Structure* 10, 1097–1106.

Kalyankar, G. D. and Meister, A. (1959) Enzymatic synthesis of carnosine and related β–alanyl and γ–aminobutyryl peptides. *J. Biol. Chem.* 234, 3210–3218.

Kanaki, K., Kawashima, S., Kashiwayanagi, M. and Kurihara, K. (1997) Carnosine–induced inward currents in rat olfactory bulb neurons in cultured slices. *Neurosci. Lett.* 231, 167–170.

Kanazawa, A. and Sano, I. (1967) A method of determination of homocarnosine and its distribution in mammalian tissues. *J. Neurochem.* 14, 211–214.

Kilis–Pstrusinska, K., Zwolinska, D., Grzeszczak, W., Szczepanska, M. and Zachwieja, K. (2009) CNDP1 gene polymorphism and chronic kidney disease in the course of primary glomerulopathy. Results of the family–based–study. Pediatr. *Nephrol.* 24, 1883–1883.

Kohen, R., Yamamoto, Y., Cundy, K. C. and Ames, B. N. (1988) Antioxidant activity of carnosine, homocarnosine, and anserine present in muscle and brain. *Proc. Natl. Acad. Sci. U. S. A.* 85, 3175–3179.

Krause, A. C. (1936) Carnosine of the ocular tissues. *Arch. Ophthal.* 16, 986–989.

Kunze, N., Kleinkauf, H. and Bauer, K. (1986) Characterization of two carnosine–degrading enzymes from rat brain. Partial purification and characterization of a carnosinase and a β–alanyl–arginine hydrolase. *Eur. J. Biochem.* 160, 605–613.

Lenney, J. F. (1976) Specificity and distribution of mammalian carnosinase. *Biochim. Biophys. Acta* 429, 214–219.

Lenney, J. F. (1990a) Human cytosolic carnosinase: evidence of identity with prolinase, a non–specific dipeptidase. *Biol. Chem. Hoppe Seyler* 371, 167–171.

Lenney, J. F. (1990b) Separation and characterization of two carnosine–splitting cytosolic dipeptidases from hog kidney (carnosinase and non–specific dipeptidase). *Biol. Chem. Hoppe Seyler* 371, 433–440.

Lenney, J. F., Baslow, M. H. and Sugiyama, G. H. (1978) Similarity of tuna *N*–acetylhistidine deacetylase and cod fish anserinase. Comp. Biochem. *Physiol. B* 61, 253–258.

Lenney, J. F., George, R. P., Weiss, A. M., Kucera, C. M., Chan, P. W. and Rinzler, G. S. (1982) Human serum carnosinase: characterization, distinction from cellular carnosinase, and activation by cadmium. *Clin. Chim. Acta* 123, 221–231.

Lenney, J. F., Kan, S. C., Siu, K. and Sugiyama, G. H. (1977) Homocarnosinase: a hog kidney dipeptidase with a broader specificity than carnosinase. *Arch. Biochem. Biophys.* 184, 257–266.

Lenney, J. F., Peppers, S. C., Kucera–Orallo, C. M. and George, R. P. (1985) Characterization of human tissue carnosinase. *Biochem. J.* 228, 653–660.

Lukton, A. and Olcott, H. S. (1958) Content of free imidazole compounds in the muscle tissue of aquatic animals. *Food Res. Int.* 23, 611–618.

Margolis, F. L. (1974) Carnosine in the primary olfactory pathway. *Science* 184, 909–911.

Margolis, F. L. and Grillo, M. (1984) Carnosine, homocarnosine and anserine in vertebrate retinas. *Neurochem. Int.* 6, 207–209.

Margolis, F. L., Grillo, M., Brown, C. E., Williams, T. H., Pitcher, R. G. and Elgar, G. J. (1979) Enzymatic and immunological evidence for two forms of carnosinase in the mouse. *Biochim. Biophys. Acta* 570, 311-323.

Margolis, F. L., Grillo, M., Grannot-Reisfeld, N. and Farbman, A. I. (1983) Purification, characterization and immunocytochemical localization of mouse kidney carnosinase. *Biochim. Biophys. Acta* 744, 237-248.

Margolis, F. L., Grillo, M., Kawano, T. and Farbman, A. I. (1985) Carnosine synthesis in olfactory tissue during ontogeny: effect of exogenous β-alanine. *J. Neurochem.* 44, 1459-1464.

McDonough, C. W., Hicks, P. J., Lu, L. Y., Langefeld, C. D., Freedman B. I. and Bowden, D. W. (2009) The influence of carnosinase gene polymorphisms on diabetic nephropathy risk in African-Americans. *Hum. Genet.* 126, 265-275.

McManus, R. (1962) Enzymatic synthesis of anserine in skeletal muscle by *N*-methylation of carnosine. *J. Biol. Chem.* 237, 1207-1211.

Menzies, F. D., Crockford, T., Breck, O. and Midtlyng, P. J. (2002) Estimation of direct costs associated with cataracts in farmed Atlantic salmon (*Salmo salar*). *Bull. Eur. Assoc. Fish Pathol.* 22, 27-32.

Morizono, H., Caldovic, L., Shi, D. and Tuchman, M. (2004) Mammalian *N*-acetylglutamate synthase. *Mol. Genet. Metab.* 81 (Supplement), 4-11.

Murphey, W. H., Lindmark, D. G., Patchen, L. I., Housler, M. E., Harrod, E. K. and Mosovich, L. (1973) Serum carnosinase deficiency concomitant with mental retardation. *Pediatr. Res.* 7, 601-606.

Nagai, K., Niijima, A., Yamano, T., Otani, H., Okumra, N., et al., (2003) Possible role of L-carnosine in the regulation of blood glucose through controlling autonomic nerves. *Exp. Biol. Med.* (*Maywood*) 228, 1138-1145.

O'Dowd, J. J., Cairns, M. T., Trainor, M., Robins, D. J. and Miller, D. J. (1990) Analysis of carnosine, homocarnosine, and other histidyl derivatives in rat brain. *J. Neurochem.* 55, 446-452.

Oku, T., Ando, S., Hayakawa, T., Baba, K., Nishi, R., et al., (2011) Purification and identification of a novel primitive secretory enzyme catalyzing the hydrolysis of imidazole-related dipeptides in the jawless vertebrate *Lethenteron reissneri*. *Peptides* 32, 648-655.

Oku, T., Ando, S., Tsai, H.-C., Yamashita, Y., Ueno, H., et al., (2012) Purification and identification of two carnosine-cleaving enzymes, carnosine dipeptidase I and Xaa-methyl-His dipeptidase, from Japanese eel (*Anguilla japonica*). *Biochimie* 94, 1281-1290.

Okuma, E. and Abe, H. (1992) Major buffering constituents in animal muscle. *Comp. Biochem. Physiol. A* 102, 37-41.

Partmann, W. (1976) Eine Carnosine spaltende Enzymaktivitat im Skelettmuskel des Aales [A carnosine splitting enzyme activity in skeletal muscles of eels]. *Arch. Fisch Wiss.* 27, 55-62.

Peppers, S. C. and Lenney, J. F. (1988) Bestatin inhibition of human tissue carnosinase, a non-specific cytosolic dipeptidase. *Biol. Chem. Hoppe Seyler* 369, 1281-1286.

Perez, B. S. and Jones, N. R. (1962) Effects of tetracycline antibiotics on the products of anserinase action in chill stored haddock (*Gadus aeglefitius*) muscle. *J. Food Sci.* 27, 69-72.

Perry, T. L., Hansen, S. and Love, D. L. (1968) Serum-carnosinase deficiency in carnosinaemia. *Lancet* 1, 1229-1230.

Peters, V., Jansen, E. E. W., Jakobs, C., Riedl, E., Janssen, B., et al., (2011) Anserine inhibits carnosine degradation but in human serum carnosinase (CN1) is not correlated with histidine dipeptide concentration. *Clin. Chim. Acta* 412, 263-267.

Raghavan, M., Lindberg, U. and Schutt, C. (1992) The use of alternative substrates in the characterization of actin-methylating and carnosine-methylating enzymes *Eur. J. Biochem.* 210, 311-318.

Rawlings N. D., Barrett, A. J. and Bateman, A. (2012) MEROPS: the database of proteolytic enzymes, their substrates and inhibitors. *Nucleic Acids Res.* 40, D343-350.

Riedl, E., Koeppel, H., Pfistet, F., Peters, V., Sauerhoefer, S., et al., (2010) *N*-Glycosylation of carnosinase influences protein secretion and enzyme activity implications for hyperglycemia. *Diabetes* 59, 1984-1990.

Rochel, S. and Margolis, F. L. (1982) Camosine release from olfactory bulb synaptosomes is calcium-dependent and depolarization-stimulated. *J. Neurochem.* 38, 1505-1514.

Rosenberg, A. (1960a) The activation of carnosinase by divalent metal ions. *Biochim. Biophys. Acta* 45, 297-316.

Rosenberg, A. (1960b) Purification and some properties of carnosinase of swine kidney. *Arch. Biochem. Biophys.* 88, 83-93.

Rowsell, S., Pauptit, R. A., Tucker, A. D., Melton, R. G., Blow, D. M. and Brick, P. (1997) Crystal structure of carboxypeptidase G2, a bacterial enzyme with applications in cancer therapy. *Structure* 5, 337-347.

Sauerhofer, S., Yuan, G., Braun, G. S., Deinzer, M., Neumaier, M., et al., (2007) L-Camosine, a substrate of carnosinase-1, influences glucose metabolism. Diabetes 56, 2425-2432.

Schulz, M., Hamprecht, B., Kleinkauf, H. and Bauer, K. (1989) Regulation by dibutyryl cyclic AMP of carnosine synthesis in astroglia-rich primary cultures kept in serum-free medium. *J. Neurochem.* 52, 229-234.

Seely, J. E. and Marshall, F. D. (1981) Carnosine levels in blood. *Experientia* 37, 1256-1257.

Severin, S. E., Kirzon, M. V. and Kaftanova, T. M. (1953) Effect of carnosine and anserine on action of isolated frog muscles. [In Russian.] *Dokl. Akad. Nauk. SSSR* 91, 691-694.

Sewell, D. A., Harris, R. C., Marlin, D. J. and Dunnett, M. (1992) Estimation of the carnosine content of different fibre types in the middle gluteal muscle of the thoroughbred horse. *J. Physiol.* 455, 447-453.

Skaper, S. D., Das, S. and Marshall, F. D. (1973) Some properties of a homocarnosine-carnosine synthetase isolated from rat brain. *J. Neurochem.* 21, 1429-1445.

Smith, E. C. (1938) The buffering of muscle in rigor; protein, phosphate and carnosine. *J. Physiol.* 92, 336-343.

Smith, T., Ghandour, M. S. and Wood, P. L. (2011) Detection of N-acetyl methionine in human and murine brain and neuronal and glial derived cell lines. *J. Neurochem.* 118, 187-194.

Sonoda, T. and Tatibana, M. (1983) Purification of *N*-acetyl-L-glutamate synthetase from rat liver mitochondria and substrate and activator specificity of the enzyme. *J. Biol. Chem.* 258, 9839-9844.

Stenesh, J. J. and Winnick, T. (1960) Carnosine-anserine synthetase of muscle. 4. Partial purification of the enzyme and further studies of β-alanyl peptide synthesis. *Biochem. J.* 77, 575-581.

Suzuki, T., Hirano, T. and Suyama, M. (1988) Distribution of anserinase in organs of several fish. *Nippon Suisan Gakkaishi* 54, 541.

Teufel, M., Saudek, V., Ledig, J. P., Bernhardt, A., Boularand, S., et al., (2003) Sequence identification and characterization of human carnosinase and a closely related non-specific dipeptidase. *J. Biol. Chem.* 278,

6521-6531.

Tsubone, S., Yoshikawa, N., Okada, S. and Abe, H. (2007) Purification characterization of a novel imidazole dipeptide synthase from the muscle of the Japanese eel *Anguilla japonica*. *Comp. Biochem. Physiol. B* 146, 560-567.

Tsunoo, S., Horisaka, K., Kawasumi, M., Aso, K. and Tokue, S. (1963) Concerning the free amino acids in vhicken brain. The isolation and identification of histidine and anserine. [In German.] *J. Biochem.* 54, 355-362.

Unno, H., Yamashita, T., Ujita, S., Okumura, N., Otani, H., et al., (2008) Structural basis for substrate recognition and hydrolysis by mouse carnosinase CN2. *J. Biol. Chem.* 283, 27289-27299.

Van Waarde, A. V. (1988) Biochemistry of non-protein nitrogenous compounds in fish including the use of amino acids for anaerobic energy production. *Comp. Biochem. Physiol. B* 91, 207-228.

Vionnet, N., Marre, M., Groop, P. H., Tarnow, L., Lemainque, A., et al., (2006) The canosinase gene (CNDP1) is not a protection factor of diabetic nephropathy in type 1 diabetics. *Diabetes Metab.* 32, S50-S51.

Wall, A. E. (1998) Cataracts in farmed Atlantic salmon (*Salmo salar*) in Ireland, Norway and Scotland from 1995 to 1997. *Vet. Rec.* 142, 626-631.

Wanic, K., Placha, G., Dunn, J., Smiles, A., Warram, J. H. and Krolewski, A. S. (2008) Exclusion of polymorphisms in carnosinase genes (CNDP1 and CNDP2) as a cause of diabetic nephropathy in type 1 diabetes-results of large case-control and follow-up studies. *Diabetes* 57, 2547-2551.

Wiame, E., Tyteca, D., Pierrot, N., Collard, F., Amyere, M., et al., (2010) Molecular identification of aspartate *N*-acetyltransferase and its mutation in hypoacetylaspartia. *Biochem. J.* 425, 127-136.

Winnick, R. E. and Winnick, T. (1959a) Carnosine-anserine synthetase of muscle. I. Preparation and properties of soluble enzyme from chick muscle. *Biochim. Biophys. Acta* 31, 47-55.

Winnick, T. and Winnick, R. E. (1959b) Pathways and the physiological site of anserine formation. *Nature* 183, 1466-1468.

Wood, M. R. and Johnson, P. (1981) Purification of carnosine synthetase from avian muscle by affinity chromatography and determination of its subunit structure. *Biochim. Biophys. Acta* 662, 138-144.

Wood, T. (1957) Carnosine and carnosinase in rat tissue. *Nature* 180, 39-40.

Yamada, S., Kawashima, K., Baba, K., Oku, T. and Ando, S. (2009) Occurrence of a novel acetylated amino acid, *N*α-acetylhistidine, in skeletal musde of freshwater fish and other ectothermic vertebrates. *Comp. Biochem. Physiol. B* 152, 282-286.

Yamada, S., Tanaka, Y. and Ando, S. (2005) Purification and sequence identification of anserinase. *FEBS. J.* 272, 6001-6013.

Yamada, S., Tanaka, Y. and Furuichi, M. (1995) Partial purification and characterization of histidine acetyltransferase in brain of Nile tilapia (*Oreochromis niloticus*). *Biochim. Biophys. Acta* 1245, 239-247.

Yamada, S., Tanaka, Y., Sameshima, M. and Furuichi, M. (1991) *N*α-Acetylhistine metabolism in fish. 2. Distribution of *N*α-acetylhistidine-deacetylating enzyme in tissues of rainbow trout. *Nippon Suisan Gakkaishi* 57, 1601-1601.

Yamada, S., Tanaka, Y., Sameshima, M. and Furuichi, M. (1992) Occurrence of *N*α-acetylhistidine in the muscle and deacetylation by several tissues of Nile tilapia (*Oreochromis niloticus*). *Comp. Biochem. Physiol.*

B 103, 579–583.

Yamada, S., Tanaka, Y., Sameshima, M. and Furuichi, M. (1993) Properties of *N*α-acetylhistidine deacetylase in brain of rainbow trout *Oncorhynchus mykiss*. *Comp. Biochem. Physiol. B* 106, 309–315.

Yamada, S., Tanaka, Y., Sameshima, M. and Furuichi, M. (1994) Effects of starvation and feeding on tissue *N*α-acetylhistidine levels in Nile tilapia *Oreochromis niloticus*. *Comp. Biochem. Physiol. A* 109, 277–283.

Zhou, S. and Decker, E. A. (1999) Ability of carnosine and other skeletal muscle components to quench unsaturated aldehydic lipid oxidation products. *J. Agric. Food Chem.* 47, 51–55.

第 13 章 食草性海洋无脊椎动物的多糖降解酶

Takao Ojima, Hokkaido University, Japan

DOI：10.1533/9781908818355.3.333

摘要：食草性海洋软体动物，鲍、海兔和小蜗牛，拥有多种多糖降解酶，能够有效地降解海藻多糖，如褐藻胶、甘露聚糖、海带多糖和纤维素。这些酶在软体动物体内发挥着重要的作用，能够降解海藻叶，获得寡聚糖和单糖作为碳水化合物养分。另一方面，这些海藻多糖和寡糖可以发挥多种对人体有益的功能，例如，减轻肠壁细菌紊乱，缓和血压升高，抗癌和抗高血压功能。此外，这些海藻糖被认为是生物能源的潜在材料，而且不会像陆地作物在食物和能源之间产生剧烈冲突。在此背景下，能降解海藻多糖的酶被认为是从海藻中获得高值化碳水化合物产物的主要生物催化剂。与陆生生物相比，来自食草性海洋软体动物的酶有众多优势，因为软体动物来源的酶已经在海藻食物的选择压力下进行了筛选。因此，对这些酶的研究，对解决我们怎样用一个合理的方法降解海藻以生产高附加值的寡糖和单糖这一问题上提供有用的信息。本章中，我们将介绍食草性软体动物的多糖降解酶的一般特性，尤其集中在褐藻胶裂解酶、甘露聚糖酶、海带多糖酶和纤维素酶。

关键词：软体动物，海带，多糖降解酶，褐藻胶裂解酶，甘露聚糖酶，海带多糖酶，纤维素酶

13.1 引言

海藻在各种海洋生物群体中发挥着重要的生态作用，为生物提供一个舒适的栖息环境和常规食物（Mai et al.，1995；Takami et al.，1998；Johnston et al.，2005）。食草性海洋软体动物，像鲍、海兔和小蜗牛是海藻群体中普遍的栖息者，它们摄入海藻叶子，在消化腺中用合适的酶，如褐藻胶裂解酶、甘露聚糖酶、纤维素酶和海带多糖酶，降解海带多糖、褐藻胶盐、甘露聚糖、纤维素和海带多糖（Boyen et al.，1990；Heyraud et al.，1996；Suzuki et al.，2003；Shimizu et al.，2003；Andreotti et al.，2005；Suzuki et al.，2006；Ootsuka et al. , 2006; Nishida et al. , 2007; Hata et al. , 2009; Nikapitiya et al. , 2009; Kumagai and Ojima, 2009, 2010; Rahman et al. , 2010, 2011, 2012; Wang et al. , 2012; Zahura et al. , 2010, 2011, 2012)。这些软体动物直接吸收或者通过肠道微生物发酵获得海藻多糖的降解产物，作为碳水化合物养分吸收（Erasmus et al. , 1997; Sawabe et al. , 2003)。

另一方面，这些海藻多糖和其降解产物已表现出很多对人体有用的功能（Dhawan and Kaur, 2007; Moreira and Filho, 2008)。例如，海藻多糖可以作为膳食纤维减轻肠道内细菌紊乱，降低血液胆固醇水平。从海藻多糖酶解产生的寡糖具有抗癌和降高血压的作用，抑制 IgE 产物，改善肠道微生物种群。

这些发现已经推动了对具有降解海藻多糖能力酶的研究，除了它们作为生物活性物质的原料，目前，海藻多糖作为生物能源的原材料也引起人们的关注，因为大部分海藻还没被利用，并且海藻作为生物能量来源不会像陆地作物一样，与食物产生冲突（Takeda et al. , 2011; Wargacki et al. , 2012)。这个观点也使我们意识到能降解海藻多糖的酶是有用的生物催化剂。

在这些背景下，作者的团队已经开始研究食草性海洋软体动物的多糖降解酶。这些动物的消化液包含很多种酶类，它们能够有效地降解海藻细胞壁和细胞内多糖。在食草性腹足类动物的消化液中突出的酶是褐藻胶裂解酶、甘露聚糖酶、海带多糖酶和纤维素酶，分别作用于褐藻胶、甘露聚糖、海带多糖和纤维素。这些酶的研究将提供酶促反应的基本信息，如反应的最适条件，降解模式，一级结构等。我们希望借助这些研究，为基于酶降解的方法从海藻中获得寡糖和单糖方面提供有用信息。

本章主要描述了食草性海洋无脊椎动物多糖降解酶的一般特性，例如，褐藻胶裂解酶、甘露聚糖酶、海带多糖酶和纤维素酶，同时也介绍了这些酶降解产生的寡聚糖和单糖。

13.2 褐藻胶裂解酶

13.2.1 褐藻胶和褐藻胶裂解酶的基本信息

褐藻胶是一种酸性杂多糖，包括β-D-甘露糖醛酸（M）和α-L-古洛糖醛酸（G），按 M 或者 G 同聚物排列，或者 MG 杂多糖排列（Haug et al. , 1967; Gacesa, 1988, 1992; Wong et al. , 2000)。褐藻胶是存在于褐藻细胞壁和胞内基质中的结构物质，并且是某些细菌生物膜成分。因为褐藻胶溶液表现出的高黏稠度以及与二价金属盐形成弹性凝胶，它作为黏度剂和一种凝胶成分应用于很多领域，像食品和饮料，纸和印刷术以及制药工业。最近，酶降解的褐藻胶被发现有多种生物活性，比如，提高较高植物根的生长（Tomoda et al. , 1994; Sutherland, 1995; Xu et al. , 2003)，提高 *Bifidobacterium* sp. 的生长比例

(Akiyama et al., 1992)，提高 *Penicillium chrysogenum* 的盘尼西林（青霉素）的产量（Ariyo et al., 1998)，促进巨噬细胞的增殖和人类角化细胞及内皮细胞的生长（Iwamoto et al., 2003, 2005; Kawada et al., 1999; Yamamoto et al., 2007)，抑制 IgE（Yoshida et al., 2004)，降低血压（Tsuchida et al., 2001; Chaki et al., 2002)。此外，藻酸盐寡糖表现出抗氧化作用（Trommer et al., 2005; Bylund et al., 2006; Jeon et al., 2003; Konig et al., 1992)，抗凝集作用（Khodagholi et al., 2008; Rezaii and Khodagholi, 2009)，抗炎症作用（Mo et al., 2003)，病原菌感染的保护作用（An et al., 2009)，保护内质网和线粒体调控的氧化作用导致的细胞凋亡（Tusi et al., 2011)，改善糖基化最终产物抑制作用（Sattarahmady et al., 2007)。低分子量的褐藻胶（平均分子量 15 000）通过抗氧化作用可以有效地预防心血管和脑血管疾病（Xue et al., 1998; Liu et al., 2002)。这些发现表明褐藻胶寡糖在生物医学领域是非常有前景的活性物质。除了以上提到的应用，褐藻胶也被当做是生物能源潜在的碳水化合物来源（Takeda et al., 2011; Wargacki et al., 2012)。由于褐藻胶是海洋中含量最大的碳水化合物之一，并且它在生物上的应用不会像陆地作物那样与食物冲突，来源于褐藻胶的能源将会是一种可能的生物能源。

为了从褐藻生产功能性寡糖和发酵糖，褐藻胶裂解酶的使用，如褐藻胶裂解酶（EC 4.2.2.3 与 EC 4.2.2.11)，将会是一个主要技术。由于酸性条件会导致褐藻胶凝胶化，这会减弱褐藻胶糖链的水解。因此，褐藻胶的酸水解不容易进行。我们可以通过使用褐藻胶裂解酶解决这个问题，因为褐藻胶裂解酶可以在 pH 值 6~10 的范围内降解褐藻胶，此条件下褐藻胶是非凝胶化状态。

褐藻胶裂解酶通过 β-消除反应降解褐藻胶产生寡糖，在新形成的非还原性末端形成不饱和糖醛酸，即 4-deoxy-L-erythro-hex-4-enopyranosyluronate（Gacesa 1992; Wong 2000)。这种酶分布于食草性海洋软体动物（Nakada and Sweeney, 1967; Nishizawa et al., 1968; Elyakova and Favarov, 1974; Muramatsu et al., 1977, 1996; Boyen et al., 1990; Heyraud et al., 1996; Shimizu et al., 2003; Suzuki et al., 2006; Hata et al., 2009; Rahman et al., 2010, 2011, 2012; Wang et al., 2012)、海藻（Madgwick et al., 1973; Watanabe and Nishizawa, 1982)、海洋和土壤细菌（Murata et al., 1993; Wong et al., 2000; Sawabe et al., 1997; Sugimura et al., 2000; Zhang et al., 2004）以及 *Chlorella* 病毒（Suda et al., 1999; Ogura et al., 2009)。基于对底物的偏好性，褐藻胶裂解酶被分为甘露聚糖裂解酶（EC 4.2.2.3）和古洛糖醛酸裂解酶（EC 4.2.2.11)，分别作用于甘露聚糖（poly（M））和古洛糖醛酸（ploy（G)。除了 poly（M）和 poly（G）特异的裂解酶，还有既能降解 poly（M）又能降解 poly（G）的酶被发现（Sawabe et al., 1997; Iwamoto et al., 2001)。目前被研究的软体动物酶被认为是 poly（M）特异的酶（EC 4.2.2.1)。褐藻胶裂解酶通常是内切酶，然而，在 *Sphingomonas* sp. 和鲍中，少数外切褐藻胶裂解酶也被发现（Hashimoto et al., 2000; Suzuki et al., 2006)。根据一级结构疏水区分析，褐藻胶裂解酶被归类于多糖裂解酶家族（PL）-5，-6，-7，-14，-15，-17 和-18（http://www.cazy.org/)。软体动物褐藻

胶裂解酶是 PL 14 的主要成员。

13.2.2 软体动物褐藻胶裂解酶催化特性

已被研究的软体动物褐藻胶裂解酶的催化特性见表 13.1。鲍酶的研究来源是，*Haliotis rufescens*（Boyen et al.，1990），*Haliotis corrugate*（Boyen et al.，1990），*Haliotis tuberculata*（Heyraud et al.，1996），*Haliotis discus hannai*（Shimizu et al.，2003；Suzuki et al.，2006），*Haliotis iris*（Hata et al.，2009）。被研究其他软体动物来源的酶包括海螺 *Turbo cornutus*（Muranatsu et al.，1977，1996），小海洋蜗牛 *Littorina* sp.（Elyacova and Favarov，1974），*Omphalius rusticus*（Hata et al.，2009），*Littorina brevicula*（Hata et al.，2009；Rahman et al.，2012；Wang et al.，2012），海兔 *Dolabella auricular*（Nishizawa et al.，1968）和 *Aplysia kurodai*（Rahman et al.，2010，2011）。除了鲍酶 HdAlex 是一种具有降解褐藻胶外切酶活性的寡糖裂解酶，其他的软体动物酶都被鉴定为内切型 poly（M）特异的酶（EC 4.2.2.3）（Suzuki et al.，2006）。这些软体动物酶的分子量在 28~35 kDa，并且这些酶的主要降解产物是三糖和二糖，同时还有少量的单糖（4-脱氧-5-酮-糖醛酸）。另外，各种酶之间在最适 pH 值和温度稳定性方面差别较大。比如，来自 *H. discus hannai* 和 *O. rusticus* 酶的最适 pH 值为 8.0~8.5，然而 *A. kurodai* 和 *L. brevicula* 酶的最适 pH 值为 6.0~7.5（Hata et al.，2009；Rahman et al.，2010，2012；Wang et al.，2012）。*Haliotis* 和 *Omphalius* 的酶在 pH 值低于 6 高于 9 时，酶是失活的，然而 *Aplysia* 和 *Littorina* 的酶在一个较广的 pH 值范围 3~11 是稳定的。*Haliotis* 和 *Omphalius* 酶的最适温度为 35~45℃。*Aplysia* 和 *Littorina* 酶的最适温度为 40~55℃。引起 *Haliotis HdAly* 和 *Omphalius OrAly* 酶 20 min 孵育失去一半活性的温度是 43℃ 和 41℃，而对于 *Aplysia AkAly*30 和 *Littorina LbAly*28 酶，该温度分别是 48℃ 和 52℃。*Haliotis HdAly* 酶在 40℃ 和 45℃ 的一级失活比例常数分别是 0.06 min 和 1.80 min，这些数值是 *Littorina LbAly*35 的酶的 20 倍和 45 倍，即分别是 0.003 min 和 0.04 min（Wang et al.，2012）。软体动物间褐藻胶裂解酶的稳定性特点也许与产酶动物的栖息环境不同有关。比如，*Littorina* 栖息在潮汐带，其生活温度在夏天升高到 35℃ 以上，*Aplysia* sp. 和 *Littorina* sp. 的消化液 pH 值在 4~6。然而 *Haliotis* 的栖息温度通常低于 15℃，消化液的 pH 值在 6~7。这些动物栖息环境的不同也许会引起对褐藻胶降解酶的分子适应。在较高的温度和较广的 pH 值范围内，*Aplysia AkAly*30 和 *Littorina LbAly*28 酶比 *Haliotis* 酶具有更高的稳定性，这一特性也体现在大肠杆菌表达系统产生的重组酶中（Rahman et al.，2011，2012）。

表 13.1 一些软体动物褐藻胶裂解酶属性对比

生物（酶）	分子量/kDa	最适条件		稳定条件		参考文献
		pH 值	温度/℃	pH 值[a]	温度/℃[b]	
Haliotis discus hannai（*HdAly*）	28	8.0~8.5	45	6.0~9.0（15 min）	43（20 min）	Shimizu et al.，2003

续表

生物 （酶）	分子量 /kDa	最适条件		稳定条件		参考文献
		pH 值	温度/℃	pH 值[a]	温度/℃[b]	
Haliotis discus hannai (*HdAlex*)	32	7.1	42	ND	ND	Suzuki et al., 2006
Haliotis iris (*HiAly*)	34	8.0~8.5	35	6.0~9.0 (15 min)	38 (20 min)	Hata et al., 2009
Aplysia kurodai (*AkAly28*)	28	6.7	40	ND[c]	38 (20 min)	Rahman et al., 2010
Aplysia kurodai (*AkAly33*)	33	6.7	40	ND[c]	39 (20 min)	Rahman et al., 2010
Aplysia kurodai (*AkAly30*)	30	6.0	55	4.5~9.0 (30 min)	48 (20 min)	Rahman et al., 2011
Littorina sp. (*Alginate lyase VI*)	40	5.6	ND	4.0~8.0 (180 min)	42 (60 min)	Elyakova 和 Favarov 1974
Littorina brevicula (*LbAly28*)	28	7.0	53	5.0~9.0 (30 min)	52 (20 min)	Rahman et al., 2012
Littorina brevicula (*LbAly35*)	35	7.5	50	3.0~11.0 (15 min)	50 (20 min)	Hata et al., 2009
Omphalius rusticus (*OrAly*)	34	8.0~8.5	45	6.0~9.0 (15 min)	41 (20 min)	Hata et al., 2009

注：

a　pH 值范围是在 30℃经过 15 min，30 min 或 180 min 孵育后仍保留 80%原始活性的范围；

b　温度是在 pH 值 7.0 经过 20 min 或 60 min 孵育造成一半失活的温度；

c　仍未确定.

Aplysia 和 *Littorina* 的重组酶产量比 *Haliotis* 的高（Shimizu et al., 2003；Rahman et al., 2011，2012），这也许是因为 *Aplysia* 和 *Littorina* 的酶的结构稳定性比 *Haliotis* 的高。因此，*Aplysia* 和 *Littorina* 来源的酶在软体动物褐藻胶裂解酶的蛋白质工程研究中是更合适的模型。

13.2.3　软体动物褐藻胶裂解酶的一级结构

来自下面这些软体动物褐藻胶裂解酶的一级结构已经确定：*T. cornutus*（Muramatsu et al., 1996），*H. discus hannai*（Shimizu et al., 2003；Suzuki et al., 2006），*A. kurodai*

（Rahman et al.，2011），*L. brevicula*（Rahman et al.，2012；Wang et al.，2012）（图 13.1）。通过一级结构疏水基团分析，所有这些酶归类于 PL-14。软体动物褐藻胶裂解酶之间具有 44%~53%的氨基酸同源性，并且与 *Chlorella* 病毒酶 vAL-1 有将近 20%的同源性，*Chlorella* 病毒酶 vAL-1 实际上不是褐藻胶裂解酶，但可以通过 β 消除的方法降解 *Chlorella* 细胞壁多糖。序列比对结果揭示了几个保守区域（如图 13.1 提到）。这些区域对应的是 β 折叠的 A3-A6 和 loop 环 L1 区，它们组成了 Val-1 的活性口袋。最近通过 X-射线衍射解析了 Val-1 的 3-D 结构（Ogura et al.，2009）。Lys^{197}、Arg^{221}、Tyr^{233} 和 Tyr^{235} 是在 Val-1 中被确定的对催化活性至关重要的残基，它们的侧链都伸入到活性口袋中。而他们

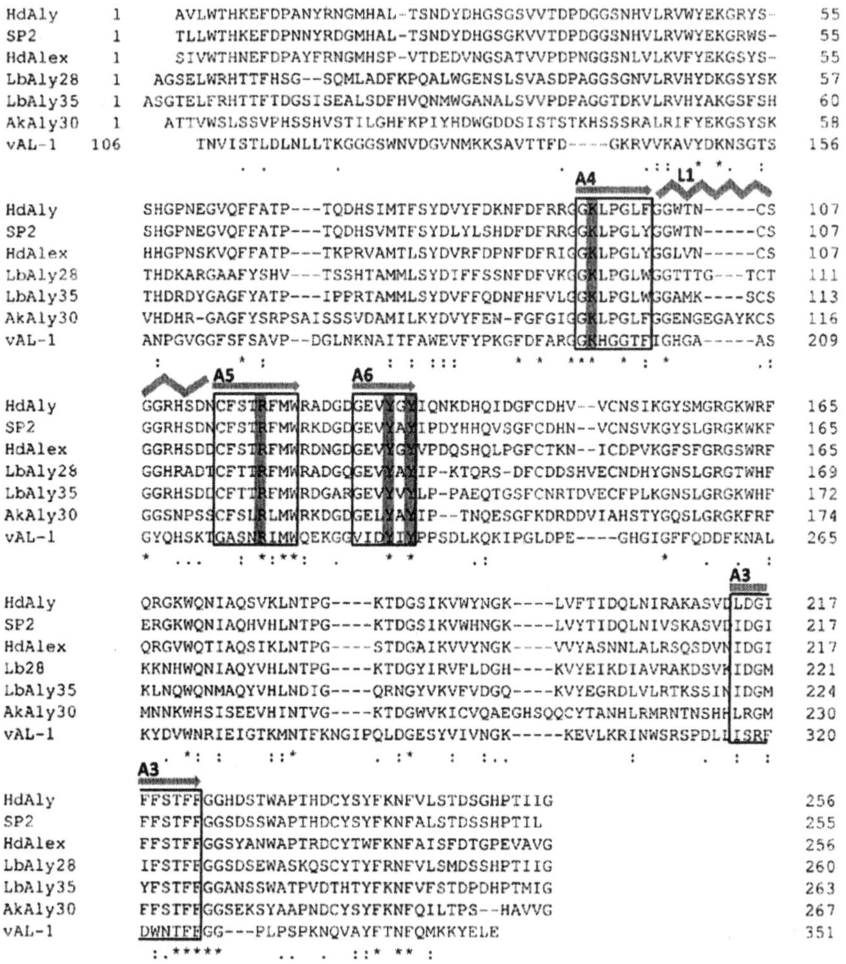

图 13.1　*LbAly*28 和其他 PL-14 酶的氨基酸序列对比

注释：这是 *Haliotis* HdAly（内切酶）和 HdAlex（外切酶）、*Turbo* SP2、*Littorina* LbAly28 和 LbAly35、Aplysia AkAly、Chlorella 病毒 vAL-1 的氨基酸片段对比。序列中同源性，高度保守性，保守残基由星号（＊），冒号（:），点（.）代表。在 PL-14 酶家族中保守氨基酸残基主要定位在 vAL-1 维空间结构 β-折叠 A3-A6（框）和 L-1 环中。紧密连接在催化反应区的氨基酸残基由阴影表明 .

对应的在 *Aplysia* 中的氨基酸（Lys^{99}、Arg^{128}、Tyr^{140}和 Tyr^{142}）在这些在软体动物酶中是高度保守。已经通过定点突变技术对这些残基在软体动物褐藻胶裂解酶（*Aplysia AkAly*30 和 *Littorina LbAly*28）中的重要性进行了验证（Rahman et al.，2011，2012），然而另两个在 Val-1 中重要催化残基（His^{213}和 Ser^{219}）在软体动物酶中是不保守的。即，在 *LbAly*28 和 *AkAly*30 中，His^{213}分别被 Arg^{115}和 Asn^{120}取代，而在 *LbAly*28 和 *LbAly*35 的两个 *Littorina* 酶中，Ser^{219}被 Thr^{121}取代。两个 Cys 对 C106-C115 和 C145-C150，推测在 turban-shell SP2 中形成二硫键，它们在 *Littorina* 和 *Haliotis* 的酶中也是保守的，而在 *Aplysia* 酶中不是保守的。在 SP2 中一个推测的糖基化残基 Asn^{105}，在 *Haliotis* 酶中是保守的，而在 *Aplysia* 和 *Littorina* 的酶中不是保守的。尽管有些变化出现在软体动物酶的保守区域中，但软体动物褐藻胶裂解酶的一级结构的整体特点似乎是非常保守的。

软体动物褐藻胶裂解酶的三级结构还未有报道。然而，基于 *Chlorella* 病毒 vAL-1 的结构数据基础上，一个预测的 *Aplysia AkAly*30 的结构被报道（Rahman et al.，2011）。*AkAly*30 的结构就类似于 *Chlorella* 病毒的 vAL-1 的 β-凝胶卷状结构。这个结构也被发现出现在 *Sphingomonas* sp. 的 PL-7 褐藻胶裂解酶中（Yamasaki et al.，2005）。PL14 和 PL17 的三级结构的相似性预示着 β-凝胶卷状结构是在长期趋同进化过程中形成的比较合理的结构。在 *AkAly*30 预测的三级结构中，重要的催化残基的排列与 Val-1 中的是一样的（Rahman et al.，2011）。

13.2.4 软体动物褐藻胶裂解酶的应用

褐藻胶裂解酶被用于褐藻的原生质体的制备（Butler et al.，1989；Boyen et al.，1990）。最近，使用重组鲍褐藻胶裂解酶显示它制备原生质的收获量比来自鲍肝胰腺的天然酶多（Yamamoto et al.，2008；Inoue et al.，2008，2011）。这可能跟天然酶中存在能够破坏原生质体的其他酶的污染有关，如蛋白酶、脂肪酶、核酸酶。除了原生质体制备，褐藻胶裂解酶也可以用在藻酸盐和它的寡糖的精细结构确定方面（Heyraud et al.，1996；Wang et al.，2000）。由于软体动物褐藻胶裂解酶不像细菌酶可以培养获得，因此，对其实际应用而言，重组酶的异源表达是必须。如上面提到的，*Aplysia* 和 *Littorina* 一些软体动物酶是适合在大肠杆菌表达系统中的表达。将来重组表达软体动物褐藻胶裂解酶在功能海藻酸盐寡糖制备和发酵糖中将会是最合适的选择。

13.3 甘露聚糖酶

13.1.1 甘露糖和甘露聚糖酶的基本信息

甘露聚糖是自然界第二大量的半纤维素，以结构多糖和贮存多糖的形式存在于高等植物和海藻中。甘露聚糖包括 4 种类型，即线性甘露聚糖，半乳甘露聚糖，葡甘露聚糖和半乳葡甘露聚糖（Moreira and Filho，2008）。甘露聚糖的高级排列结构因来源不同而不同。例如，来自绿藻 *Codium fragile* 的甘露聚糖是 β-1，4 连接的 D-甘露糖的线性同聚物，而来自魔芋根的葡甘露聚糖是 D-甘露糖和 D-葡萄糖 β-1，4 连接的同聚物。像刺槐豆胶和塔拉胶这些半乳甘露聚糖的主链是被 α-1，6 连接的 D-半乳糖支链修饰，（Moreira and Filho，2008）。另一方面，半乳葡甘露聚糖的主链是 D-甘露糖和 D-葡萄糖 β-1，4 混合连接物，并拥有 α-1，6 连接的 D-半乳糖支链。线性甘露聚糖是木材中的主要结构多糖，同时在象牙棕榈的果实和咖啡豆这样的植物种子（Aspinall，1959）以及 *Phytelephas macrocarpa* 之类的棕榈科胚乳（Petkowica et al.，2001）中以贮存多糖形式存在。线性甘露聚糖被发现在红藻 *Porphyra umbilicalis* 和绿藻 *Codium fragile* 中作为结构多糖的形式存在（Moreira and Filho，2008）。葡甘露聚糖也在一些植物的块茎、根和结节处以贮存多糖形式存在（Meier and Reid，1982）。半乳甘露聚糖在豆科家族的种子中以胚乳多糖存在。半乳葡甘露聚糖在裸子植物的木质结构中。这些甘露聚糖和它们的寡糖对人类健康有多种有益影响。例如，甘露聚糖作为食用纤维能缓解肠道细菌紊乱以及来自甘露聚糖的寡糖作为益生元改善肠道细菌生存环境。食品和制药工业领域的研究者正在关注甘露聚糖和它们的寡糖作为有潜力生物活性物质的开发（Moreira and Filho，2008；Dhawan and Kaur，2007）。

内切-β-1，4-甘露聚糖酶是可以应用从甘露聚糖中生产甘露寡糖的生物催化剂（EC 3.2.1.78，通常称为 β-甘露聚糖酶）。在各种工业过程中，β-甘露聚糖酶已被证明是一种有用的酶，如果汁的澄清，降低咖啡提取物的黏度，提高家禽饲料的可消化性，漂白纸浆以及在其他的工业进程中通过甘露聚糖的降解提高产品的质量（Dhawan and Kaur，2007）。而且，这个酶对从红藻中获得细胞物质和原生质体是有用的，这是藻类细胞工程中重要的材料（Percival and McDowell，1967；Polne-Fuller and Gibor，1984；Gall et al.，1993；Araki et al.，1994；Dai et al.，2004；Dipakkore et al.，2005；Ootsuka et al.，2006）。因此，β-甘露聚糖酶在各种领域的实际应用似乎依然持续增长。

到目前为止，β-甘露聚糖酶已经从细菌（Araki，1983；Akino et al.，1988；Talbot and Sygusch，1990；Braithwaite et al.，1995；Nakajima and Matsuura，1997；Li et al.，2000，2006；Politz et al.，2000；Kansoh and Nagieb，2004）、真菌（Johnson and Ross，1990；Stålbrand et al.，1993；Kurakake and Komaki 2001；Puchart et al.，2004；Naganagouda et al.，2009）、高等植物（Shimahara et al.，1975；Marraccini et al.，2001）和软体动物（Yamaura and Matsumoto，1993；Yamaura et al.，1996；Xu et al.，2002a，b；Ootsuka et al.，2006；Zzhura et al.，2010）中分离到。这些酶的最适温度和 pH 值分别是 40～90℃，pH 值 3～7.5，分子量是 30～80 kDa。因此，β-甘露聚糖酶性质根据酶的来源有很大不同。

在无脊椎动物中，食草性海洋软体动物是 β-甘露聚糖酶的主要生产者，这些软体动物食用植物组织，并通过消化液消化食物中的甘露聚糖。到目前为止，软体动物 β-甘露聚糖酶已经从以下这些动物中分离得到，即陆地腹足动物 *Pomacea insularus*（Yamaura and

Matsumoto，1993），淡水腹足动物 *Bimphalaria glabrata*（Vergote et al.，2005），海洋腹足动物 *Littorina brevicula*（Yamaura et al.，1996），*Haliotis discus hannai*（Ootsuka et al.，2006），*Aplysia kurodai*（Zahura et al.，2010，2011）以及海洋双壳类 *Mytilus edulis*（Xu et al.，2002a，b）。另一个被研究的甘露聚糖降解酶，β-甘露糖苷酶，也是来源于下面动物，即 *Turbo cornutus*（Muramatsu，1966），*Aplysia fasciata*（Andreotti et al.，2005）和 *A. kurodai*（Zahura et al.，2012）。来自 *A. kurodai* 的β-甘露糖苷酶能够完全降解甘露聚糖产生 D-甘露糖；然而，它对于聚合物甘露聚糖的活性不高。因此，*A. kurodai* 被认为利用两个甘露聚糖降解酶，也就是，β-甘露聚糖酶和β-甘露糖苷酶可以完全降解海藻甘露聚糖（Zahura et al.，2012）。但仅确定了软体动物β-甘露糖苷酶在 *A. kurodai* 酶中来源于 *A. kurodai* 的 *AkMnsd* 的一级序列（Zahura et al.，2012）。

13.3.2 软体动物β-甘露聚糖酶的酶学特点

从现有文献中得到的软体动物β-甘露聚糖酶的酶学特点总结在表 13.2 中。*Aplysia* 的酶 *AkMan* 的最适温度是 55℃，孵育 20 min，酶 *AkMan* 失去一半活性的温度是 52℃。这些温度是可以与被报道的来自 *M. edulis* 的β-甘露聚糖酶相比较的（Xu et al.，2002a，b）。其他软体动物的β-甘露聚糖酶比 *AkMan* 的最适温度低，稳定性差。例如，*Haliotis* 的酶 *HdMan* 最适温度是 45℃，孵育 30 min 使酶一半失活的温度是 40℃（Ootsuka et al.，2006）。酶 *AkMan* 的最适温度比酶 *HdMan* 高，也许与 *A. kurodai* 比 *H. discus* 的栖息环境温度高有关。也就是说 *A. kurodai* 在夏天的栖息温度增高到 25~30℃，*H. discus* 的栖息环境温度是 10~15℃，相差 10~15℃。*AkMan* 的最适 pH 值在 4.0~7.5，并且在 40℃酸性范围 pH 值孵育 20 min，它的活性并没有明显的降低。这些结果说明 *AkMan* 在酸性 pH 条件下是相当稳定的。在酸性 pH 值范围内 *AkMan* 的高稳定性，使得该酶能在 pH 值低于 6 的 *A. kurodai* 的消化液中有效地发挥活性。*Pomacea*，*Littorina* 和 *Mytilus* 的酶的最适 pH 值也是在酸性范围内，分别是 5.5、6.5 和 5.2，并且这些动物的消化液的 pH 值在酸性范围内。另外一个方面，*HdMan* 最适 pH 值为 7.5，其消化液的 pH 值也在 6.5~7.0 的范围。因此，软体动物酶的最适 pH 值是与这些动物消化液的 pH 值相对应。Fe^{3+} 和 Ag^{+} 使 *Aplysia* 的酶失活到原来活性的 70%和 50%，而 Co^{2+}、Fe^{2+} 和 Cu^{2+} 使 *Haliotis*、*Pomacea* 和 *Littorina* 的酶失活到 40%~50%，Ag^{+} 使这些酶完全失活（Ootsuka et al.，2006；Zahura et al.，2010；Yamaura and Matsumoto，1993；Yamaura et al.，1996）。因此，与其他软体动物的酶相比，*Aplysia* 酶对金属离子有较强的耐受性。0.6 mol/L 的 NaCl，也就是在海水中的浓度，它对 *Aplysia* 的酶活性没有实际影响，虽然这个酶来源于海洋腹足动物。

表 13.2　一些软体动物的甘露聚糖酶属性的对比

生物（酶）	分子量/kDa	最适条件		稳定条件		参考文献
		pH	温度/℃	pH 值[b]	温度/℃[c]	
A. kurodai（*AkMan*）	40	6.0	55	4~7.5	52（20 min）	Zahura et al., 2010
H. discus hanni（*HdMan*）	39	7.5	45	ND[a]	40（30 min）	Ootsuka et al., 2006
M. edulis（*MeMan5*）	39	5.2	50-55	4~9.0	50（20 min）	Xu et al., 2002a
L. brevicula	42	6.5	50	4~9.0	50（10 min）	Yamaura et al., 1996
P. insularus	44	5.5	50	5~10.5	45（10 min）	Yamaura and Matsumoto et al., 1996

注：a　仍未确定；

b　pH 值范围是在 30℃经过 30 min 孵育后仍保留 80 %原始活性的范围；

c　温度是在 pH 值 7.0 经过 10~30 min 孵育造成一半失活的温度.

Aplysia 和 *Haliotis* 的β-甘露聚糖酶 *AkMan* 和 *HdMan* 能分别降解从 *C. fragile* 来源的海藻多糖，然后是刺槐豆胶，但是不能降解塔拉胶和古尔胶（Ootsuka et al., 2006；Zahura et al., 2010, 2011）。对不同甘露聚糖底物的不同酶活性是由于甘露聚糖底物的结构不同。也就是说，*C. fragile* 的甘露聚糖是线性结构，而来自塔拉胶和古尔胶的半乳甘露聚糖有β-1，6 连接的半乳糖侧链。半乳甘露聚糖的半乳糖侧链能影响 *AkMan* 和甘露聚糖主链的结合，从而阻碍了甘露糖的降解。*C. fragile* 的β-1，4-甘露糖的主要降解产物是甘露三糖和甘露二糖，而来自塔拉胶的降解产物是四糖的寡糖。这个寡糖似乎是有半乳糖侧链的甘露三糖。通过来自 *Mytilus*, *Trichoderma reesei* 和 *Aspergillus niger* 的β-甘露聚糖酶，对象牙棕榈果实甘露聚糖有相似的降解产物（Xu et al., 2002a, Stålbrand et al., 1993；Ademark et al., 2001）。

葡甘露聚糖，是一种线性杂多糖，包含β-1，4 连接的 D-甘露糖和 D-葡糖基，也相对容易被 *AkMan* 降解。这说明酶 *AkMan* 能较好地作用于线性结构的甘露聚糖，产生包含葡萄糖和甘露糖的杂多糖。*AkMan* 能够降解比四糖大的寡聚甘露聚糖，并且从甘露聚糖五糖中产生三糖和二糖，从六糖中产生三糖。这些结果说明 *AkMan* 和底物结合位点的大小与甘露聚糖四糖-六糖的大小是相似的，并且可以裂解甘露聚糖六糖单元中心的甘露糖苷键。

13.3.3 软体动物甘露聚糖酶的一级结构

软体动物甘露聚糖酶的一级结构在以下这些酶中已被确定，来自 *M. edulis* 的酶 *MeMan*5A（Xu et al.，2002a，b），来自 *H. discus hannai* 的酶 *HdMan*（Ootsuka et al.，2006），来自 *Biomphalaria glabrata* 的酶 *BgMan*5A（Vergote et al.，2005）和来自 *A. kurodai* 的酶 *AkMan*（Zahura et al.，2011）。β-甘露聚糖酶已被归类于多糖水解酶家族 5（GHF5）和 26（GHF26）。GHF5 包括细菌和真核生物的甘露聚糖酶，而 GHF26 包括细菌甘露聚糖酶，还有一些来自厌氧真菌的独特甘露聚糖酶。在有些情况下，一个生物体包括来自不同家族的甘露聚糖酶，如 *Caldocellulosiruptor saccaharolyticus* 和 *Bacillus* sp.，都有 GHF5 和 GHF26 甘露聚糖酶（Gibbs et al.，1996；Sygusch et al.，1998；Hatada et al.，2005）。到目前为止鉴定的软体动物酶归类于 GHF5。软体动物 β-甘露聚糖酶的氨基酸序列比对见图 13.2。*AkMan* 与 *HdMan* 有 53.4%的相似性，与 *MeMan*5A 有 47.1%的相似性，与 *BgMan*5A 有 36.9%的相似性。参与 GHF5 酶催化活性的氨基酸残基在软体动物酶中是 Glu^{162}（推定的催化性的酸碱对）和 Glu^{293}（推定的催化性的亲核残基）（酶 *AkMan* 的残基数）中是保守的。位于 GHF5 酶催化位点的 5 个保守的氨基酸残基在软体动物的酶中也是完全保守的。在酶 *AkMan* 中是 Gly^{20}、Arg^{60}、His^{262}、Tyr^{264} 和 Trp^{322}。GHF5 包含大量来自原核和真核生物的 β-甘露聚糖酶。在序列相似性的基础上，真核生物的 β-甘露聚糖酶进一步归类于亚家族 7，而原核生物的 β-甘露聚糖酶归类于亚家族 8（Hilge et al.，1998）并且软体动物 β-甘露聚糖酶归于亚家族 10。在亚家族 10 和 7、8 之间的序列相似性是低于 15%（Larsson et al.，2006）。

来自 GHF5 和 GHF26 的 β-甘露聚糖酶的三维结构具有相似的 8（α/β）折叠，而这是糖苷水解酶 A 催化单元的主要特点（LeNours et al.，2005）。最近，来自 *Mytilus* 的 β-甘露聚糖酶 *MeMan*5A 的三维结构被发现也有 8（α/β）折叠结构（Larsson et al.，2006）。其他软体动物酶的三维结构还没有被解析，然而，*Aplysia* 酶 *AkMan* 的结构预测和 *MeMan*5A 的相似，有相同的 Glu^{162}（推定的催化酸碱对）和 Glu^{293}（推定的催化性的亲核残基）的空间构象。

13.3.4 甘露聚糖酶的应用

通常，在自然界的微生物和高等生物中，β-甘露聚糖酶的产量不高。因此，已经尝试了高效的重组酶表达。到目前为止，多种不同的异源表达系统已经被用于生产重组甘露聚糖酶（Stålbrand et al.，1993；Mendoza et al.，1995；Tang et al.，2001；Mingardon et al.，2005）。软体动物甘露聚糖酶的情况是，蓝色贻贝的酶 *MeMan*5A 已经通过毕赤酵母表达系统获得重组酶（Xu et al.，2002b），并且 *Aplysia* 的酶 *AkMan* 已经通过大肠杆菌表达系统得到重组酶（Zahura et al.，2011）。重组甘露聚糖酶的产量取决于表达系统的类型。事实

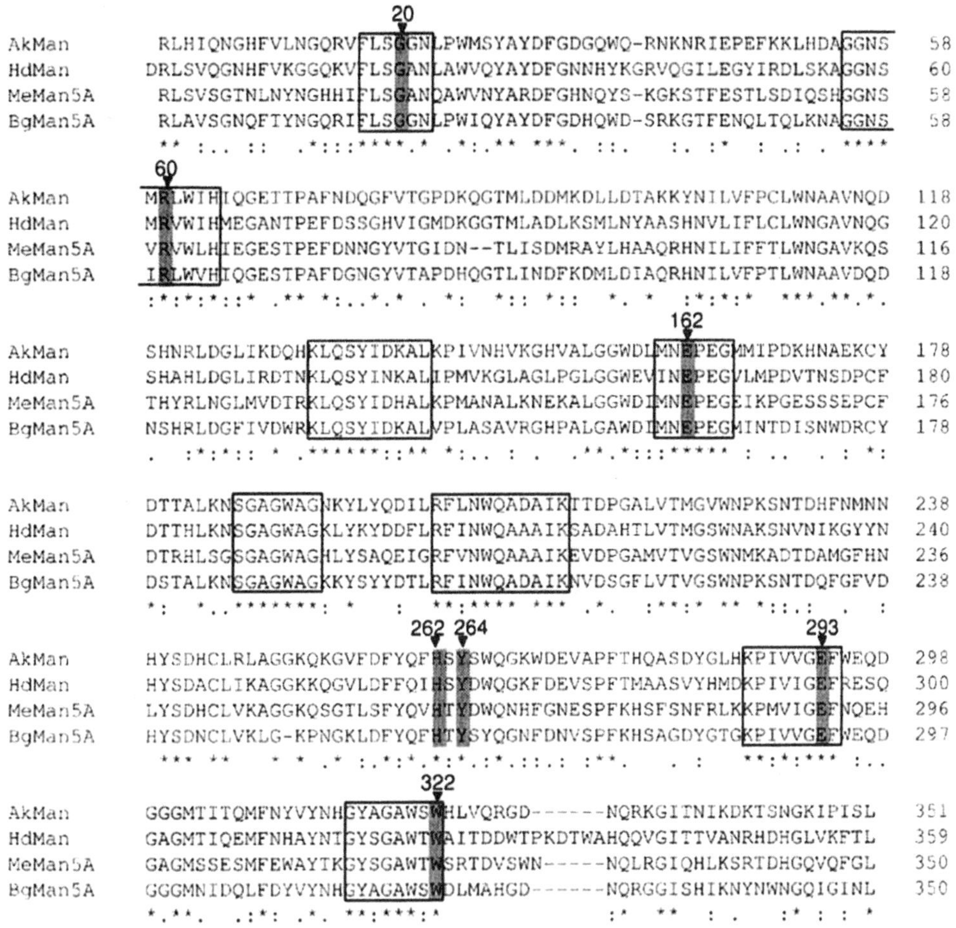

图 13.2 *AkMan* 和 GHF5 的软体动物的甘露聚糖酶氨基酸序列对比

注：*AkMan* 的氨基酸序列和 *HdMan*（DDBJ 编码，AB222081），*MeMan5*A（DDBJ 编码，AJ271365）和 *BgMan5*A（DDBJ 编码，AY678121）序列对比。序列中同源性，高度保守性，保守残基由星号（＊），冒号（:），点（.）代表。区域显示框中序列有高度同源性（高于 75%）。催化区的残基和 GHF5 酶的活性位点保守性在阴影区域标记出来 .

上，酶 *MeMan5*A 通过毕赤酵母表达系统的表达量是相当高的，然而，酶 *AkMan* 的大肠杆菌表达系统的表达量是适中的。

甘露聚糖酶应用于很多工业过程以提高产品质量，并且应用于生产食品添加剂和制药材料的寡聚甘露聚糖的生产（Kobayashi et al.，1987）。甘露聚糖酶可以应用在红藻的细胞工程中，因为在红藻细胞壁中包含 β-甘露糖。这个酶可以去除红藻的细胞壁而产生原生质体（Percival and McDowell，1967；Polne-Fuller and Gibor，1984；Gall et al.，1993；Dai et al.，2004；Dipakkore et al.，2005；Ootsuka et al.，2006）。甘露聚糖酶在各领域的应用是持续增加的（Moreira and Filho 2008）。

13.4 昆布多糖酶

13.4.1 昆布多糖和昆布多糖酶的基本信息

昆布多糖，是褐藻和微藻贮存的β-葡聚糖，包括β-1，3连接葡萄糖主链和β-1，6连接葡萄糖侧链（Maeda and Nisizawa，1968；Pang et al.，2005）。地衣多糖和蘑菇多糖分别是来自苔藓和蘑菇的β-葡聚糖，但是这些β-葡聚糖与昆布多糖的不同体现在主链β-1，6连接侧链的次数和主链中β-1，4键的出现。昆布多糖分布在多种褐藻和微藻中（Black et al.，1951；Smelcerovic et al.，2008）。昆布多糖高级结构的不同取决于它们的来源。例如，硅藻属昆布多糖的主链包含较少的β-1，6连接的侧链（Størseth et al.，2005）。然而 *Laminaria digitata* 的昆布多糖的主链包含适中次数的β-1，6连接侧链，β-1，6连接葡萄糖和β-1，3连接葡萄糖的比例接近1∶7（Hrmova and Fincher，1993）。*Eisenia bicyclis* 昆布多糖的侧链有相当多的β-1，6链接，并且在主链中也有β-1，6链接（Maeda and Nisizawa，1968；Usui et al.，1979）。昆布多糖和它的寡糖对人体显示多种生物活性。例如，来自 *E. bicyclis* 的昆布多糖能够刺激免疫反应并缓解糖尿病（Pang et al.，2005），而酶降解产生的昆布寡糖能够提高单核细胞的分泌物TNF-α（Miyanishi et al.，2003）。这些活性似乎来源于昆布多糖的特殊的高级结构有关，比如β-1，6链接的侧链结构。然而，昆布多糖的结构和活性之间的关系是不确定的。为了揭示昆布多糖的结构特点，通过昆布多糖的限制性降解产生的寡糖来分析局部或单元结构是必要的。为了达到这一目的，人们期望能够获能够降解昆布多糖产生特殊寡糖的酶。

β-葡聚糖降解酶，如β-1，3-葡聚糖酶，分布在各种生物体中，如细菌、酵母菌、真菌、植物、昆虫和无脊椎动物（Sova et al.，1970a，b；Privalova and Elyakova 1978；Erfle et al.，1988；Tangarone et al.，1989；Hrmova and Fincher，1993；Mrsa et al.，1993；Takami et al.，1998；Miyanishi et al.，2003；Kovalchuk et al.，2006；Kumagai et al.，2008；Kumagai and Ojima，2009，2010）。β-1，3-葡聚糖酶被归类于以下3种类型。

（1）内切型-1，3；1，4-β-葡聚糖酶（EC 3.2.1.6），既能裂解β-1，3；1，4-葡聚糖的β-1，3链接，又能断开β-1，3链接的还原末端附近的β-1，4链接。

（2）地衣多糖酶（EC 3.2.1.73）仅能断开β-1，3；1，4-葡聚糖中β-1，3连接的还原末端附近的β-1，4链接。

（3）葡聚糖内切型-1，3-β-D-葡聚糖酶（EC 3.2.1.39）对β-1，3连接是特殊的，并且需要至少两个邻近的β-1，3链接才能降解。

软体动物β-1，3-葡聚糖酶已经从以下这些生物中分离得到，即蛤蜊 *Spisula sachalinensis*（Sova et al.，1970b），扇贝 *Chlamis albidus* 和 *Patinopecten*（*Mizuhopecten*）*yessoensis*

(Privalova and Elyakova, 1978; Kovalchuk et al., 2006; Kumagai et al., 2008), 鲍 *Haliotis tuberculata* (Lépagnol-Descamps, et al., 1998), *Haliotis discus hannai* (Kumagai and Ojima, 2009), 以及海兔 *Aplysia kurodai* (Kumagai and Ojima, 2010)。这些酶被认为是第一类酶 (EC 3.2.1.6) 并被称为"昆布多糖酶", 因为它们的天然底物是昆布多糖, 并且 β-1, 3-葡聚糖酶具有在 β-1, 3-葡聚糖和拥有羟基的接收底物之间转糖基活性 (Bezukladnikov and Elyakova, 1990; Biarnés et al., 2006; Neustroev et al., 2006)。转糖基活性是保留式内切型寡聚糖水解酶的一个典型特点。昆布多糖酶的转糖基作用已经应用于制备人为设计的特殊寡糖, 这些寡糖包括昆布多糖寡聚糖, 各种糖类和非糖产物如酒精 (Kumagai et al., 2008)。这说明 β-1, 3-葡聚糖酶是制备非天然糖类化合物和 β-1, 3-寡糖的有用工具。

13.4.2 软体动物昆布多糖酶的酶学特点

最近, 作者的团队从鲍、海兔和扇贝这些软体动物中分离得到的昆布多糖酶的基本特性总结见表 13.3。昆布多糖酶通常被认为是内切酶。然而, 但是一种外切酶 *AkLam* 33 也从 *A. kurodai* 中分离到 (Kumagai and Ojima, 2010)。软体动物昆布多糖酶的分子量是在 36 ~37 kDa。最适温度和 pH 值分别是 40~50℃ 和 6。腹足动物的昆布多糖酶与扇贝的酶相比有相对较高的热稳定性。

表 13.3 一些软体动物的昆布多糖酶属性的对比

生物（酶）	分子量/kDa	最适条件		稳定条件	活性/(U/mg)	作用方式	参考文献
		pH 值	温度/℃	温度/℃[a]			
H. disucus hannai (*HdLam*33)	37	6.0	50	50 (20 min)	68	内型	Kumagai and Ojima, 2009
A. kurodai (*AkLam*33)	36	5.7	50	45 (20 min)	38	外型	Kumagai and Ojima, 2010
A. kurodai (*AkLam*36)	36	6.0	40	42 (20 min)	172	内型	Kumagai and Ojima, 2010
P. yessoensis (*PyLam*38)	37	6.0	45	35 (20 min)	162	内型	Kumagai et al., 2008

a 温度是在 pH 值 7.0 经过 10~30 min 孵育造成一半失活的温度.

软体动物昆布多糖酶的底物偏好与动物的食物习惯有关, 即动物食用包含什么种类昆布多糖的藻类。例如, 来自鲍和扇贝的酶 *HdLam*33 和 *PyLam*38 对来自 *L. digitata* 的昆布多糖显示高活性，而对来自 *E. bicyclis* 的昆布多糖没有活性，酶 *AkLam*33 降解来自 *L. digitata* 和 *E. bicyclis* 的昆布多糖。鲍的日常食物是属于 *Laminaria* sp. 的褐藻，而扇贝的是如硅藻类的微藻。这些食物包含带有一些 β-1，6 链接侧链的昆布多糖。然而海兔则食用很多在主链和侧链都有很多 1，6 链接的褐藻，包括 *Eisenia* sp. 和 *Sargassum* sp. 的昆布多糖。软体动物昆布多糖酶降解 *L. digitata* 的昆布多糖产生昆布二糖和葡萄糖作为终产物，也产生 6-O-葡糖苷-昆布三糖。6-O-葡糖苷-甘露三糖是从 β-1，6-链接葡萄糖侧链产生的，昆布二糖很难被软体动物酶降解。然而，它可以在昆布三糖存在下通过二糖和三糖的转糖基作用被降解（Kumagai and Ojima，2009，2010）。这说明软体动物昆布多糖酶的转糖基作用能够提高海藻昆布糖的糖化作用。然而，转糖基作用在消化液中是否是一个重要的生理事件还是不确定的，因为转糖基作用需要相对高的昆布二糖浓度，并且通过昆布糖酶产生的昆布二糖将会很容易被消化液中的 β-葡糖苷酶降解。

13.4.3 软体动物昆布多糖酶的一级结构

下面这些软体动物中的昆布多糖酶的一级结构已被测定，包括蛤蜊、扇贝、鲍和海兔（Kozhemyako et al.，2004；Kovalchuk et al.，2006，2009；Kumagai et al.，2008；Kumagi and Ojima，2009，2010）。这些软体动物昆布多糖酶被归类于多糖水解酶家族 16（GHF16）。这些软体动物昆布多糖酶的一级结构总结见表 13.3。组成 GHF16 酶催化中心的两个谷氨酸残基在这些酶中是保守的，两个 Cys 残基被报道在扇贝酶中形成二硫键（Kovalchuk et al.，2006），它们在软体动物酶中也是保守的。来源于扇贝和蛤蜊的酶的氨基酸相似性非常高，达到 90%，然而鲍和蚌类之间氨基酸残基相似性为 56%～61%。因此，酶的一级结构在腹足动物和蚌类之间差别是非常大的。关于腹足动物昆布多糖酶的信息与蚌类酶相比还是很少的。

13.4.4 昆布多糖酶的应用

目前，来自真菌和蜗牛的 β-1，3-葡聚糖酶已经商业化。然而，这些酶对工业应用来说似乎是太贵。因此，期望得到较便宜的酶以便工业应用。最近发现，来自扇贝中消化腺的液体表现出相对高效的昆布多糖酶活性（Kumagai et al.，2008）。扇贝消化腺在日本被认为是传统的渔业废料，这些液体似乎可以作为较便宜的昆布多糖酶的来源。因此，昆布多糖酶已经从扇贝消化液中分离到，并且它的基本特性也被研究（Kumagai et al.，2008）。纯化的扇贝昆布多糖酶显示更高的转糖基活性，并能利用昆布四糖和各种各样的受体底物像单糖衍生物、寡糖、糖已醇、酒精和带有羟基的氨基酸产生异型寡糖（Kumagai et al.，2008）。这些表明扇贝昆布多糖酶可以应用于生产作为新奇糖衍生物的多种非天然异型寡糖。

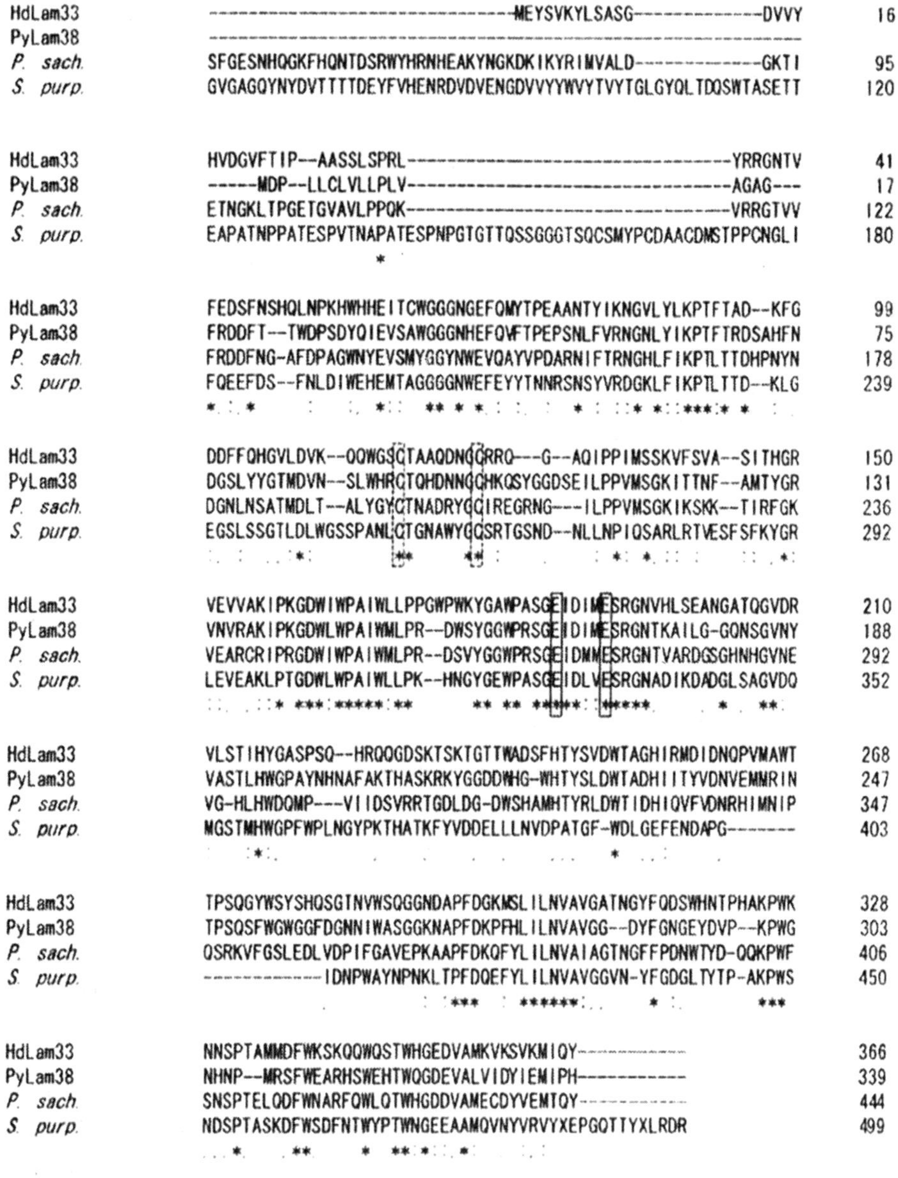

图 13.3 无脊椎动物昆布多糖酶的氨基酸序列比对

注：*Haliotis HdLam33*、*Patinopecten PyLam38*、*Pseudocardium* 酶（*P. sach.*）和 *Strongylocentrotus purpuratus* 酶（*S. purp.*）的氨基酸序列罗列对比；同源残基的位置，高度保守底物，保守底物和间隙由星号（*），冒号（:），点（.）和横杠（-）代表；GHF16 酶，即 E186 和 E191 中作为亲核试剂或者质子供体在催化区域中的残基被框标记出来；海洋无脊椎动物酶，即 C117 和 C125 保守半胱氨酸残基由折线标记出 .

13.5 纤维素酶

13.5.1 纤维素和纤维素酶的基本信息

纤维素，是植物细胞壁、细菌生物膜和海鞘纲动物外被的结构成分，是有β-1，4链接的D-葡糖线性多糖。纤维素酶能够水解纤维素的β-1，4糖苷键产生纤维寡糖如纤维二糖和三糖（Tomme et al.，1995）。这个酶粗略地分为两种类型，即内切型酶（内型-1，4-β-葡聚糖酶，EC 3.2.1.4）和外切型酶（纤维二糖水解酶，EC 3.2.1.91）。纤维素完全降解成葡萄糖需要另一种类纤维寡糖降解酶，1，4-β-葡糖苷酶（EC 3.2.1.21）。高效的通过纤维素酶降解纤维素产生的葡萄糖已成为世界上一个社会性的重要事件，因为纤维素来源的葡萄糖能够用于生物乙醇的生产。纤维素降解酶已经吸引了很多研究者，他们尝试应用纤维素降解酶在纤维素材料的糖化作用和生物能源制备方面（Lin and Tanaka，2006）。到目前为止，纤维素酶已经从以下生物中分离到，细菌和真菌（Tomme et al.，1995）、植物（Brummell et al.，1994）、霉菌（Blume and Ennis，1991）、动物肠道微生物（Moriya et al.，1998）以及食草性软体动物如节肢动物（Tokuda et al.，1997；Watanabe et al.，1998；Xue et al.，1999）、线虫类（Smant et al.，1998）、软体动物（Yokoe and Yasumasu，1964；Marshall and Grand 1976；Anzai et al.，1984；Xu et al.，2000；Suzuki et al.，2003；Wang et al.，2003；Sakamoto et al.，2008）和棘皮动物（Nishida et al.，2007）。

无脊椎动物的纤维素酶曾经被认为是它们消化腺中的共生微生物或腐败食物产生的（Cleveland，1924；Martin and Martin，1978）。然而，最近基因组学研究揭示白蚁、龙虾、线虫、蚌类、鲍、甲壳虫、甲壳纲动物和棘皮动物的酶都是这些动物自身产生的（Byrne et al.，1999；Yan et al.，1998；Linton et al.，2006；参见前面的引文）。通过疏水基团分析，来自植物寄生线虫和甲壳虫的纤维素酶归类于GHF5，来自蚌类的纤维素酶归类于GHF45，大多数纤维素酶，像来自白蚁、龙虾、蟑螂、鲍和海胆的酶归类于GHF9。基因组学和表达序列标签分析表明GHF9的纤维素酶基因广泛的分布于多细胞动物（Davison and Blaxter，2005；Suzuki et al.，2003；Nishida et al.，2007），并且动物纤维素酶基因的很多内含子都是处在相同位置（Lo et al.，2003；Davison and Blaxter，2005）。这些研究充分的说明GHF9动物纤维素酶基因是从多细胞动物祖先中遗传来的，并且很可能通过微生物间基因的水平转移获得了纤维素酶基因。在自然界中，产纤维素酶的生物体被认为在植物的碳循环中发挥着重要的作用。包括软体动物的食草性无脊椎动物，居住在海岸、微咸水、沼泽和红树林中，他们能够利用纤维素酶消化来自植物废物的纤维素材料（Kristensen，2008；Lee 2008；Cannicci et al.，2008；Sakamoto et al.，2009）。像这样的能产生纤维素酶的无脊椎动物和可分解纤维素的微生物已经变成海洋生态系统中主要的纤维素降解者。

13.5.2 来自海洋无脊椎动物纤维素酶的催化特点

最近，作者团队研究了两个无脊椎动物内切型$-\beta-1$，4-葡聚糖酶，HdEG66和SnEG54的生物化学特点，这两个酶分别来自太平洋鲍（*Haliotis discus hannai*）和日本紫海胆（*Strongylocentrotus nudus*），因为这两个动物是海藻富集区主要的栖息者（Suzuki et al.，2003；Nishida et al.，2007）。这些纤维素酶作为代表性酶，它们的基本特点总结如下。酶HdEG66的最适温度是38℃、pH值6.3；而酶SnEG54的最适温度是35℃、pH值6.5，因此，它们之间是相似的。当然在两者之间也发现微小的不同。例如，酶SnEG54在10~30℃的低温范围内比酶HdEG66表现更高的活性。在pH值低于5.5或者高于7.5时，酶HdEG66的活性是降低的，而酶SnEG54在这样的pH值范围内保持相对高水平的活性。酶SnEG54被认为在宽的pH值范围内比酶HdEG66更稳定。另一方面，酶SnEG54比酶HdEG66更不稳定，也就是说，在pH值7.0处理30 min活性降低到50%时，酶SnEG54和酶HdEG66的处理温度分别是37℃和42℃。

鲍酶HdEG66和海胆酶SnEG54能够很好地降解非晶形纤维（磷酸纤维素），主要产物是纤维三糖和纤维二糖（Suzuki et al.，2003；Nishida et al.，2007）。另一方面，酶HdEG66和酶SnEG54在降解纤维寡糖方面显示微小的不同。即酶SnEG54能快速地降解纤维六糖和五糖产生纤维四糖、三糖和二糖，能缓慢的降解四糖产生三糖、二糖和葡萄糖；然而，但是几乎不能降解纤维三糖和二糖。另一方面，酶HdEG66能快速的降解六糖和四糖产生三糖和二糖，也能降解三糖产生二糖和葡萄糖。酶HdEG66和酶SnEG54对于纤维寡糖降解性质的不同，也许与这些酶的结构组成或者底物结合部位的大小不同有关。

酶HdEG66和酶SnEG54都属于GHF9，而一些蚌类的纤维素酶归类于GHF5和GHF45。尽管软体动物纤维素酶的详细结构还未被确定，一级结构差别也许导致纤维素酶降解活性的不同，如底物偏好和最小结合底物。

13.5.3 海洋无脊椎动物纤维素酶的一级序列

作者团队用cDNA和基因片段确定了鲍HdEG66和海胆SnEG54的一级结构（Suzuki et al.，2003；Nishida et al.，2007）。酶HdEG66包含579个残基，计算分子量为63 197 Da，而酶SnEG54包含427个残基，计算分子量是47 448 Da。与其他GHF9家族的纤维素酶相比，酶HdEG66 C末端的453个残基和酶SnEG54的全长被看做是GHF9类型的催化结构域（图13.4）。这些区域与以下生物的GHF9纤维素酶的相应区域有高于40%的氨基酸相似性，即白蚁（Watanabe et al.，1998），龙虾（Byrne et al.，1999）和*Thermomonospora fusca*（Jung et al.，1993）（褐色高温单胞菌）。GHF9纤维素酶的重要催化残基（Tomme

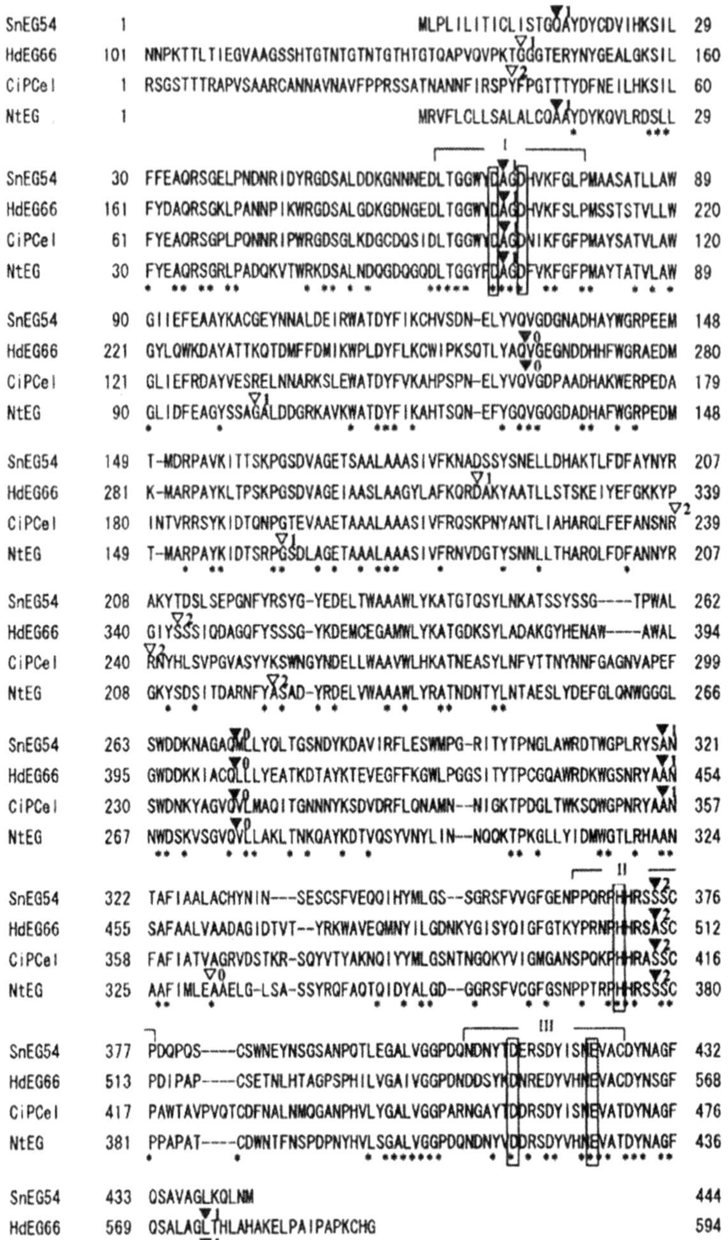

图 13.4　无脊椎动物纤维素酶 GHF9 家族的氨基酸序列比对

注：*Strongylocentrotus nudus* 纤维素酶 SnEG54、*Haliotis discus hannai* 纤维素酶 HdEG66、*Ciona intestinalis* 纤维素酶 CiPCel 和 *Nasutitermes takasagoensis* 纤维素酶 NtEG 氨基酸序列对比；同源残基的位置由星号表示；GHF9 酶中作为亲核试剂或者质子供体在催化区域中的残基被框标记出来；片段上方的实心箭头标记至少在两种片段中相同位置的内含子；空心箭头表示每个片段中独特的内含子；在每个箭头的右侧 3 个位置用 0、1、3 所标记 .

et al., 1991; Sakon et al., 1997; Khademi et al., 2002)，即 His^{506}、Asp^{200}、Asp^{203}、Asp^{550}和 Glu^{559}在 HdEG66 序列中是高度保守的。

鲍酶 HdEG66 N 末端的 126 个残基被看做是一个多糖结合域（CBM），因为它与纤维单胞菌（*Cellulomonas fimi*）的纤维素酶 CenA 的 CBM（Wong et al., 1986）有 27%的序列相似性。酶 CenA 的 CBM 属于 CBM 家族Ⅱ，目前，CBM 家族Ⅱ是 5 个最重要家族中最大的一个，即家族Ⅰ-Ⅳ和Ⅵ（Tomme et al., 1995）。在 CBM 家族Ⅱ中高度保守的 4 个 Trp 残基和两个 Cys 残基在酶 HdEG66 推测的 CBM 中是相对保守的。也就是说，4 个 Trp 中的 3 个在残基即在位置 24、43 和 79 对应的氨基酸残基被发现，而剩下的一个在 57 号位置被 Asp 取代。两个 Cys 残基在酶 HdEG66 中不是保守的。然而，在 33、58 和 90 号位置发现了 3 个 Cys 残基。除了推定的 CBM，一个富含 Thr 和 Gly 的连接区域也发现存在于 N 末端延伸区域和催化结构域。这些特点说明酶 HdEG66 的 N 末端区域包含与 CBM 家族Ⅱ及其连接域相对应。酶 HdEG66 是动物纤维素酶中第一个含有 N 末端区域 CBM 家族Ⅱ的 GHF9 家族水解酶。

通过对酶 HdEG66 和酶 SnEG54 基因片段的核酸序列分析（DDBJ 编号，AB092979，AB2826787 和 AB282679），揭示了几个内含子的位置。通过对动物纤维素酶的氨基酸序列的内含子位置比较，酶 SnEG54 基因片段的 5 个内含子位置被发现存在于其他纤维素酶基因的相同位置（图 13.4）。例如，与 Ala^{70}、Glu^{272}-Met^{273}和 Ser^{374}相对应的 3 个间隔子的位置，在白蚁、海鞘和鲍纤维素酶基因中有相似的位置。其他两个 SnEG1DNA 内含子的位置与另一个物种的基因至少有一个相同位置。就像 Lo 等（2003）指出的，动物纤维素酶基因公共位置处内含子的出现，也许说明了无脊椎动物从一个共同祖先中继承了纤维素酶基因，很可能在百万年前通过微生物基因间的水平转移获得了纤维素酶基因。一个广泛的纤维素酶基因的研究将会为追溯动物纤维素酶起源提供更多可靠的信息。

13.5.4 软体动物纤维素酶的应用

真菌纤维素酶似乎是目前唯一一个应用于工业目的的纤维素酶。因此，生产大量便宜的纤维素酶对低成本生产生物乙醇是相当重要的。到目前为止，很多种生物的纤维素酶基因已经获得，并且重组纤维素酶和携带纤维素酶基因的微生物已经得到。在这种情况下，软体动物纤维素酶在工业中的应用似乎并不重要。然而，作者想要指出扇贝纤维素酶是一个特例。因此，像之前描述的，扇贝消化腺在扇贝加工过程中被大量丢弃，而纤维素酶可以从废物中制备，就像昆布多糖酶。作者的团队证实扇贝纤维素酶的产量是相当高的（从 1 000 扇贝中得到将近 10 000 U）。除了纤维素酶和昆布多糖酶，内脏中也存在一定量的甘露聚糖酶和淀粉酶。我们很可能从扇贝废物中生产酶混合物应用于实际生产。

13.6 结论

本章介绍了能够有效降解海藻多糖的软体动物多糖降解酶的基本特性。到目前为止，已研究了很多细菌和真菌的多糖降解酶，然而，这些微生物和真菌酶的分离鉴定是不容易的。这很可能是因为在粗酶制品中，存在有相似结构和分子量的很多同工酶。因此，很多研究集中在酶基因的克隆和重组酶的生产。与微生物酶相比，软体动物酶的分离鉴定是相对简单的。从工业应用的角度看，细菌和真菌酶是很有前景的。然而，在纯化和产量方面重组酶比天然酶更有优势。对于软体动物酶，重组酶的生产研究也是很有前景的。当然，扇贝天然酶似乎是一个例外，因为这些酶可以从内脏废物中制备。而在日本，这些内脏在扇贝加工过程中被大量丢弃。对来自消化海藻生物的多糖降解酶的进一步研究，将会提高海藻多糖作为生物功能材料和生物能源新材料的广泛应用。

参考文献

Ademark, P., de Vries, R. P., Hagglund, P., Stalbrand, H. and Visser, J. (2001) Cloning and characterization of *Aspergillus ttiger* genes encoding an α-galactosidase and a β-mannosidasc involved in galactomannan degradation *Eur. J. Biochem.* 268, 2982-2990.

Akino, T., Nakamura, N. and Horikoshi, K. (1988) Characterization of three β-mannanases of an alkalophilic *Bacillus* sp. *Agric. Biol. Chem.* 52, 773-779.

Akiyama, H., Endo, T., Nakakita, R., Murata, K., Yonemoto, Y. and Okayama, K. (1992) Effect of depolymerized alginates on the growth of Bifidobacteria, *Biosci. Biotechnol. Biochem.* 56, 355-356.

An, Q. D., Zhang, G. L., Wu, H. T., Zhang, Z. C., Zheng, G. S., et al., (2009) Alginate-deriving oligosaccharide production by alginase from newly isolated *Flavobacterium* sp. LXA and its potential application in protection against pathogens. *J. Appl. Microbiol.* 106, 161-170.

Andreotti, G., Giordano, A., Tramice, A., Mollo, E. and Trincone, A. (2005) Purification and characterization of a β-mannosidase from the marine anaspidean *Aplysia fasciata*. *J. Biotechnol.* 119, 26-35.

Anzai, H., Nisizawa, K. and Matsuda, K. (1984) Purification and characterization of a cellulase from *Dolabella auricularia*. *J. Biochem.* 96, 1381-1390.

Araki, T. (1983) Purification and characterization of an endo-β-mannanase from *Aeromonas* sp. F-25. *J. Fac. Agr. Kyushu Univ.* 27, 89-98.

Araki, T., Hayakawa, M., Tamaru, Y., Yoshimatsu, K. and Morishita, T. (1994) Isolation and regeneration of haploid protoplasts from *Bangia atro-purpurea* (Rhodophyta) with marine bacterial enzymes. *J. Phycol.* 30, 1040-1046.

Ariyo, B., Tamerler, C., Bucke, C. and Keshavarz, T. (1998) Enhanced penicillin production by oligosaccharides from batch culture of *Penicillium chrysogenum* in stirred-tank reactors. *FEMS Microbiol. Lett.* 166, 165-170.

Aspinall, GO. (1959) Structural chemistry of the hemicelluloses. *Adv. Carbohydr. Chem.* 14, 429-468.

Bezukladnikov, P. W. and Elyakova, L. A. (1990) Transglycosylation and multiple attack of endo- (1-3) -β-D-glucanase L-IV from *Spisula sachalinensis*: a new approach to the evaluation of the degree of multiple attack on polysaccharides. *Carbohydro. Res.* 203, 119-127.

Biarnés, X., Nieto, J., Planas, A. and Rovira, C. (2006) Substrate distortion in the michaelis complex of *Bacillus* 1, 3-1, 4-β-glucanase. *J. Biol. Chem.* 281, 1432-1441.

Black, W. A. P., Cornhill, W. J., Dewar, E. T. and Woodward, F. N. (1951) Manufacture of algal chemicals. III. Laboratory-scale isolation of laminarin from brown marine algae. *J. Appl. Chem.* 1, 505-517.

Blume, J. E. and Ennis, H. L. (1991) A *Dictyostelium discoideum* cellulase is a member of a spore germination-specific gene family. *J. Biol. Chem.* 266, 15432-15437.

Boyen, C., Kloareg, B., Polne-Fuller, M. and Gibor, A. (1990) Preparation of alginate lyases from marine molluscs for protoplast isolation in brown algae. *Phycol.* 29, 173-181.

Braithwaite, K. L., Black, G. W., Hazlewood, G. P., Ali, B. R. and Gilbert, H. J. (1995) A non-modular endo-beta-1, 4-mannanase from *Pseudomonas fluorescens subspecies cellulosa*. *J. Biochem.* 305, 1005-1010.

Brummell, D. A., Lashbrook, C. C. and Bennett, A. B. (1994) Plant endo-1, 4-β-D-glucanases. *ACS Symp. Ser.* 566, 100-129.

Butler, D. M., Ostgaard, K., Boyen, C., Evans, L. V., Jensen, A., Kloareg, B. (1989) Isolation conditions for high yields of protoplasts from *Laminaria saccharina* and *L. digitata* (Phaeophyceae). *J. Exp. Bot.* 40, 1237-1246.

Bylund, J., Burgess, L. A., Cescutti, P., Ernst, R. K. and Speert, D. P. (2006) Exopolysaccharides from *Burkholderia cenocepacia* inhibit neutrophil chemotaxis and scavenge reactive oxygen species. *J. Biol. Chem.* 281, 2526-2532.

Byrne, K. A., Lehnert, S. A., Johnson, S. E. and Moore, S. S. (1999) Isolation of a cDNA encoding a putative cellulase in the red claw crayfish *Cherax quadricarinatus*. *Gene* 239, 317-324.

Cannicci, S., Burrows, D., Fratini, S., Smith III, T. J., Offernberg, J. and Dahdouh-Guebas, J. (2008) Faunal impact on vegetation structure and ecosystem function in mangrove forests: a review. *Aquatic Botany* 89, 186-200.

Chaki, T., Nishimoto, S., Hiura, N., Satou, R., Tou, Y. and Kakinuma, S. (2002) Effect of a powdered drink containing sodium alginate oligosaccharide on blood pressure in mild hypertensive and high normal blood pressure subjects. [In Japanese.] *J. Nutr. Food.* 5, 41-54.

Cleveland, L. R. (1924) The physiological and symbiotic relationships between the intestinal protozoa of termites and their host, with special reference to *Reticulitermes flavipes* (Kollar). *Biol. Bull. Mar. Biol. Lab.* 46, 117-227.

Dai, J., Yang, Z., Liu, W., Bao, Z., Han, B., et al., (2004) Seedling production using enzymatically isolated thallus cells and its application in *Porphyra* cultivation. *Hydrobiologia* 512, 127-131.

Davison, A. and Blaxter, M. (2005) Ancient origin of glycosyl hydrolase family 9 cellulase genes. *Mol. Bio. Evol.* 22, 1273-1284.

Dhawan, S. and Kaur, J. (2007) Microbial mannanases: an overview of production and applications. *Grit. Rev. Biotechnol.* 27, 197-216.

Dipakkore, S., Reddy, C. R. K. and Jha, B. (2005) Production and seeding of protoplasts of *Porphyra okhaensis*

(Bangiales, Rhodophyta) in laboratory culture. *J. Appl. Phycology* 17, 331–337.

Elyakova, L. A. and Favarov, V. V. (1974) Isolation and certain properties of alginate lyase VI from the mollusk *Littorina sp. Biochim. Biophys. Acta* 358, 341–354.

Erasmus, J. H., Cook, P. A. and Coyne, V. E. (1997) The role of bacteria in the digestion of seaweed by the abalone *Haliotis midae. Aquaculture* 155, 377–386.

Erfle, J. D., Teather, R. M., Wood, P. J. and Irvin, J. E. (1988) Purification and properties of a 1, 3–1, 4–β–D–glucanase (lichenase, 1, 3–1, 4–β–D–glucan 4–glucanohydrolase, EC 3. 2. 1. 73) from *Bacteroides suc-ctnogenes* cloned in *Escherichia colt*, *Btochem. J.* 255, 831–441.

Gaecva. P. (1992) Ritcyttuttfe degradation of alginates. *Int. J. Birtchem.* 24, 545–552.

Gaeesa, P. (1988) Alginates, *Cathohydr. Polyat.* 8, 161–182.

Gall. E A., Chiang, Y–M. and Kloareg, B. (1993) Isolatifm amd regeneration of promplasts from *Potpbyra detu-afa* and *Porphyra entpata. Eur. J. Phycol.* 28, 277–245.

Gibbs, M. D., Elinder, A, U., Reeves, R. A. and Befgquist. P. L (1996) Sequencing, cloning and expression of the β–1, 4–mannanase gene manA from the extremely thermophilic anaerobic bacterium *Caldicelluhwrupter* Rt8B. 4. *FFMS Microbiol*, *lett.* 141, 37–43.

Hashimom, W., Miyake, O., Momma, K., Kawai, S. and Murata. K. (2000) Molecular identification of oli-goalginate lyase of *Sphingomonas* sp. stram A1 as one of the enzymes required for complete depolymerization of alginate. *J. Bcatenol.* 182, 4572–4577.

Hata, M., Kumagai, Y., Rahman, M. M., Chiba, S., Tanaka, H., et al, (2009) Comparative study on gener-al properties of alginate lyases from some marine gastropod mollusks. *Fish. Sci.* 75, 755–763.

Hatada, Y., Takeda, N., Hirasawa, K., Otha, Y., Usami, R., et al, (2005) Sequence of the gene for a high–alkaline mannanase from an alkaliphilic *Bacillus* sp strain JAMB–750, its expression in B. *subtilis* and charac-terization of the recombinant enzyme. *Extremophiles* 9, 497–500.

Haug, A., Larsen, B., Smidsrod, O. (1967) Studies on the sequence of uranic acid residues in alginic acid. *Acta Chem. Scand.* 21, 691–704.

Heyraud, A., Colin–Morel, P., Girond, S., Richard, C. and Kloareg, B. (1996). HPLC analysis of saturat-ed or unsaturated oligoguluronates and oiigomannuronates. Application to the determination of the action pattern of *Haliotis tuberculata* alginate lyase. *Carbohydr. Res.* 291, 115–126.

Hilge, M., Gloor, S. M., Rypniewski, W., Sauer, O., Heightman, T. D., et al., (1998) High–resolurion n-ative and complex structure of thermostable β–tmannanase from *Thermomonospora fusca*–substrate specificity in glycosyl hydrolase family 5. *Structure* 6, 1433–1444.

Hrmova, M. and Fincher, G. B, (1993) Purification and properties of three (1–>3) –β–D–glucanase isoenzymes from young leaves of barley (*Hordeum vulgare*), *Btochem. J.* 289, 453–461.

Inoue, A., Kagaya, M. and Ojima, T. (2008) Preparation of protoplasts from Laminaria japonica using native and recombinant abalone alginate lyase. *J. Appl Phycol* 20, 633–640.

Inoue, A., Mashino, C, Kodama, T. and Ojima, T. (2011) Protoplast preparation from Laminaria japonica with recombinant alginate lyase and cellulase. *Mar. Biotechnol.* 13, 256–263.

Iwamoto, M., Kurachi, M., Nakashima, T., Kim, D., Yamaguchi, K., et al., (2005) Structure–activity rela-tionship of alginate oligosaccharides in the induction of cytokine production from RAW264. 7 cells. *FEBS Lett*

579, 4423-4429.

Iwamoto, Y., Araki, R., Iriyama, K., Oda, T., Fukuda, Ft, et al., (2001) Purification and characterization of bifunctional alginate lyase from *Alteromonas* sp. strain no. 272 and its action on saturated oligomeric substrates. *Biosci. Biotechnol. Biochem.* 65, 133-142.

Iwamoto, Y., Xu, X., Tamura, T., Oda, T. and Muramatsu, T. (2003) Enzymatically depolymerized alginate oligomers that cause cytotoxic cytokine production in human mononuclear cells. *Biosci. Biotechnol. Biochem.* 67, 258-263.

Jeon, T. I., Hwang, S. G., Park, N. G., Jung, Y. R., Shin, S. I., et al., (2003) Antioxidative effect of alginate on chronic carbon tetrachloride induced hepatic injury in rats. *Toxicology* 187, 67-73.

Johnson, K. G. and Ross, N. W. (1990) Enzymatic properties of β-mannanase from Polyporus versicolor. *Enzyme Microb. Techttol.* 12, 960-964.

Johnston, D., Moltschaniwskyj, N. and Wells, J. (2005) Development of the radula and digestive system of juvenile blacklip abalone (*Haliotis rubra*): potential factors responsible for variable weaning success on artificial diets. *Aquaculture* 250, 341-355.

Jung, E. D., Lao, G., Irwin, D., Barr, B. K., Benjamin, A. and Wilson, D. B. (1993) DNA sequences and expression in *Streptomyces lividans* of an exoglucanase gene and an en-doglucanase gene from *Thermomottospora fusca*. *Appl. Environ. Microbiol.* 59, 3032-3043.

Kansoh, A. L. and Nagieb, Z. A. (2004) Xylanase and mannanase enzymes from *Streptomyces galbus* NR and their use in biobleaching of soft wood kraft pulp. *Anton. Van. Leeuwonhoek.* 85, 103-114.

Kawada, A., Hiura, N., Tajima, S. and Takahara, H. (1999) Alginate oligosaccharides stimulate VEGF-mediated growth and migration of human endothelial cells. *Arch. Dermatol. Res.* 291, 542-547.

Khademi, S., Guarino, L. A., Watanabe, H., Tokuda, G. and Meyer, E. F. (2002) Structure of an endoglucanase from termite, *Nasutitermes takasagoensis*. *Acta Crystallogr. D Biol. Crystallogr.* 58, 653-659.

Khodagholi, F., Eftekharzadeh, B. and Yazdanparast, R. (2008) A new artificial chaperone for protein refolding: sequential use of detergent and alginate. *Protein J.* 27, 123-129.

Kobayashi, Y., Echigen, R., Mada, M. and Mutai, M. (1987) 'Effects of hydrolyzates of konjac mannan and soybean oligosaccharides on intestinal flora in man and rats'. In: Mitsuoka, T. (ed.), *Intestinal Flora and Food Factors*. Gakkai Shuppan Centre, Tokyo, 79-97.

Konig, B., Friedl, P., Pedersen, S. S. and Konig, W. (1992) Alginate-its role in neutrophil responses and signal transduction towards mucoid *Pseudomonas aeruginosa bacteria*. *Int. Arch. Allergy Immunol.* 99, 98-106.

Kovalchuk, S. N., Bakunina, I. Y., Burtseva, Y. V., Emelyanenko, V. I., Kim, N. Y., et al., (2009) An endo-(1-3)-β-D-glucanase from the scallop *Chlamys albidus*: catalytic properties, cDNA cloning and secondary-structure characterization. *Carbohydr. Res.* 344, 191-197.

Kovalchuk, S. N., Sundukova, E. V., Kusaykin, M. I., Guzev, K. V., Anastiuk, S. D., et al., (2006) Purification, cDNA cloning and homology modeling of endo-1, 3-β-D-glucanase from scallop Mizuhopecten yessoensis. *Comp. Biochem. Physiol. B* 143, 473-485.

Kozhemyako, V. B., Rebrikov, D. V., Lukyanov, S. A., Bogdanova, E. A., Marin, A., ct al., (2004) Molecular cloning and characterization of an endo-1, 3-β-D-glucanase from the mollusk *Spisula sachalinensis*. *Comp. Biochem. Physiol. B* 137, 169-178.

Kristensen, E. (2008) Mangrove crabs as ecosystem engineers; with emphasis on sediment processes. *J. Sea Res.* 59, 30-43.

Kumagai, Y., Inoue, A., Tanaka, H. and Ojima, T. (2008) Preparation of β-1, 3-glucanase from scallop mid-gut gland drips and its use for production of novel heterooligosaccharides. *Fish. Sci.* 74, 1127-1136.

Kumagai, Y. and Ojima, T. (2009) Enzymatic properties and the primary structure of a β-1, 3-glucanases from the digestive fluid of the Pacific abalone *Haliotis discus hattnai. Comp. Biochem. Physiol. B.* 154, 113-120.

Kumagai, Y. and Ojima, T. (2010) Isolation and characterization of two types of beta-1, 3-glucanases from the common sea hare *Aplysia kurodai. Comp. Biochem. Physiol. B.* 155, 138-144.

Kurakake, M. and Komaki, T. (2001) Production of beta-mannanase and beta-mannosidase from *Aspergillus awamori* K4 and their properties. *Curr. Microbiol.* 42, 377-380.

Larsson, A. M., Anderson, L., Xu, B., Munoz, I. G., Uson, I., et al., (2006) Three-dimensional crystal structure and enzymatic characterization of β-mannanase Man5A from blue mussel *Mytilus edulis. J. Mol. Biol.* 357, 1500-1510.

Lee, S. Y. (2008) Mangrove macrobenthos: Assemblages, services, and linkages. *J. Sea Res.* 59, 16-29.

LeNours, J., Anderson, L., Stoll, D., Stalbrand, H., Lo Leggio, L. (2005) The structure and characterization of a modular endo-β-1, 4-mannanase from *Cellulomonas fimi. Biochemistry.* 44, 12700-12708.

Lépagnol-Descamps, V., Richard, C., Lahaye, M., Potin, P., Yvin, J. C. and Kloareg, B. (1998) Purification and determination of the action pattern of Haliotis tubeculata laminarinase. *Carbohydr. Res.* 310, 283-289.

Li, W., Dong, Z. and Cui, F. (2000) Purification and characterization of an endo-beta-1, 4-mannanase from *Bacillus subtilis* BM9602. [In Chinese.] *Wei Sheng Wu XueBao.* 40, 420-424.

Li, Y. N., Meng, K., Wang, Y. R. and Yao, B. (2006) A beta-mannanase from *Bacillus subtilis* B36: purification, properties, sequencing, gene cloning and expression in *Escherichia coli. Z Naturforsch C.* 61, 840-846.

Lin, Y. and Tanaka, S. (2006) Ethanol fermentation from biomass resources: current state and prospects. *Appl. Microbiol. Biotechnol.* 69, 627-642.

Linton, S., Greenaway, P. and Towle, D. W. (2006) Endogenous production of endo-β-1, 4-glucanase by decapod crustaceans. *J. Comp. Physiol. B.* 176, 339-348.

Liu, Y., Jiang, X. L., Liao, W. and Guan, H. S. (2002) Analysis of oligoguluronic acids with NMR, electrospray ionization-mass spectrometry and high-performance anion-exchange chromatography. *J. Chromatogr. A.* 968, 71-78.

Lo, N., Watanabe, H. and Sugimura, M. (2003) Evidence for the presence of a cellulase gene in the last common ancestor of bilaterian animals. *Proc. R. Soc. Lond. B.* 270, S69-S72.

Madgwick, J., Haug, A. and Larsen, B. (1973) Alginate lyase in the brown alga *Laminaria digitata* (Huds.) Lamour. *Acta. Chem. Scand.* 27, 711-712.

Maeda, M. and Nisizawa, K. (1968) Fine structure of laminaran of *Eisettia bicyclis. J. Biochem.* 63, 199-206.

Mai, K., Mercer, J. P. and Donlon, J. (1995) Comparative studies on the nutrition of two species of abalone, Haliotis tuberculata L, and *Haliotis discus hannailno.* III. Response of abalone to various levels of dietary lipid. *Aquaculture* 134, 65-80.

Marraccini, R, Rogers, W. J., Allard, C., Andre, M. L., Caillet, V., et al., (2001) Molecular and biochemical characterization of endo-beta-mannanases from germinating coffee (*Coffea arabica*) grains. *Planta* 213, 296-308.

Marshall, J. J. and Grand, R. J. A. (1976) Characterization of a β-1, 4-glucan hydrolase from the snail, Helix pomatia. *Comp. Biochem. Physiol. B.* 53, 231-237.

Martin, M. M. and Martin, J. S. (1978) Cellulose digestion in the midgut of the fungus-growing termite *Macrotermes natalensis*: the role of acquired digestive enzymes. *Science* 199, 1453-1455.

Meier, H. and Reid, J. S. G. (1982) 'Reserve polysaccharides other than starch in higher plants'. In: Loewus, F. A. and Tanner, W. (eds), *Encyclopedia of Plant*, Springer, Berlin, 13*A*, 418-471.

Mendoza, N. S., Arai, M., Sugimoto, K., Udea, M., Kawaguchi, T. and Jonson, L. M. (1995) Cloning and sequencing of β-mannanase gene from *Bacillus subtilis* NM-39. Biochim. *Biophys. Acta* 1243, 552-554.

Mingardon, F., Perret, S., Belaich, A., Tardif, C., Belaich, J. P. and Fierobe, H. P. (2005) Heterologus production, assembly, and secretion of a minicellulosome by *Clostridium acetobutylicum* ATCC 824. *Appl. Environ. Microbiol.* 71, 1215-1222.

Miyanishi, N., Iwamoto, Y., Watanabe, E. and Oda, T. (2003) Induction of TNF-a production from human peripheral blood monocytes with β-1, 3-glucan oligomer prepared from laminarin with /3-1, 3-glucanase from *Bacillus clausii* NM-1. J. *Biosci. Bioeng.* 95, 192-195.

Mo, S. J., Son, E. W., Rhee, D. K. and Pyo, K. (2003) Modulation of TNF-α-induced ICAM-1 expression, NO and H202 production by alginate, allicin and ascorbic acid in human endothelial cells. *Arch. Pharm. Res.* 26, 244-251.

Moreira, L. R. and Filho, E. X. (2008) An overview of mannan structure and mannan-degrading enzyme systems. *Appl. Microbiol. Biotechnol.* 79, 165-178.

Moriya, S., Ohkuma, M. and Kudo, T. (1998) Phylogenetic position of symbiotic protest *Dinemympha exilis* in the hindgut of the termite *Reticulitermes speratus* inferred from the protein phylogeny of elongation factor 1 alpha. *Gene* 210, 221-227.

Mrsa, V., Klebl, F. and Tanner, W. (1993) Purification and characterization of *the Saccharomyces cerevisiae* BGL2 gene product, a cell wall endo-β-1, 3-glucanase. *J. Bacteriol.* 175, 2102-2106.

Muramatsu, T., Hirose, S. and Katayose, M. (1977) Isolation and properties of alginate lyase from the mid-gut gland of wreath shell *Turbo cornutus. Agric. Biol. Chem.* 41, 1939-1946.

Muramatsu, T., Komori, K., Sakurai, N., Yamada, K., Awasaki, Y., et al., (1996) Primary structure of mannuronate lyases SP1 and SP2 from *Turbo cornutus* and involvement of the hydrophobic C-terminal residues in the protein stability. *J. Protein Chem.* 15, 709-719.

Muramatsu, T. (1966) Purification of α-D-mannosidase and β-D-mannosidase from marine gastropods. Arch. *Biochem. Biophys.* 115, 427-429.

Murata, K., Inose, T., Hisano, T., Abe, S., Yonemoto, Y., et al., (1993) Bacterial alginate lyase: enzymology, genetics and application. *J. Ferment. Bioeng.* 76, 427-437.

Naganagouda, K., Salimath, P. V. and Mulimani, V. H. (2009) Purification and characterization of endo-beta-1, 4 mannanase from *Aspergillus niger* for application in food processing industry. *J. Microbiol. Biotechnol.* 19, 1184-1190.

Nakada, H. I. and Sweeny, P. C. (1967) Alginic acid degradation by eliminases from abalone hepatopancreas. *J. Biol. Chem.* 242, 845-851.

Nakajima, N. and Matsuura, Y. (1997) Purification and characterization of kojack glucomannan degrading enzyme from anaerobic human intestinal bacterium, *Clostridium butyricum-Clostridium beijerinckii group. Biosci. Biotechnol. Biochem.* 61, 1739-1742.

Neustroev, K. N., Golubev, A. M., Sinnott, M. L., Borriss, R., Krah, M., et al., (2006) Transferase and hydrolytic activities of the laminarinase from *Rhodothermus marinus* and its M133A, M133C, and M133W mutants. *Glycoconj. J.* 23, 501-511.

Nikapitiya, C., Oh, C., Whang, I., Kim, C. G., Lee, Y. H., et al., (2009) Molecular characterization, gene expression analysis and biochemical properties of α-amylase from the disk abalone, *Haliotis discus discus. Comp. Biochem. Physiol. B.* 152, 271-281.

Nishida, Y., Suzuki, K., Kumagai, Y., Tanaka, H., Inoue, A. and Ojima, T. (2007) Isolation and primary structure of a cellulase from the Japanese sea urchin *Strongylocentrotus ttudus. Biochimie* 89, 1002-1010.

Nishizawa, K., Fujibayashi, S. and Kashiwabara, Y. (1968) Alginate lyases in the hepatopancreas of a marine mollusk, *Dolabella auricular Solander. J. Biochem.* (*Tokyo*) 64, 25-37.

Ogura, K., Yamasaki, M., Yamada, T., Mikami, B., Hashimoto, W. and Murata, K. (2009) Crystal structure of family 14 polysaccharide lyase with pH - dependent modes of action. *J. Biol. Chem.* 284, 35572-35579.

Ootsuka, S., Saga, N., Suzuki, K. I., Inoue, A. and Ojima, T. (2006) Isolation and cloning of an endo-β-1, 4-mannanase from pacific abalone *Haliotis discus hannai. J. Biotechnol.* 125, 269-280.

Pang, Z., Otaka, K., Maoka, T., Hidaka, K., Ishijima, S., et al., (2005) Structure of β-glucan oligomer from laminarin and its effect on human monocytes to inhibit the proliferation of U937 cells. *Biosci. Biotechnol. Biochem.* 69, 553-558.

Percival, E. and McDowell, R. H. (1967) 'Other Neutral polysaccharides, food reserve and structural food reserve'. In: Percival, E. and McDowell, R. H. (eds), *Chemistry and Enzymology of Marine Algal Polysaccharides. Academic Press, London*, pp. 73-96.

Petkowicz, C. L. O., Reicher, F., Chanzy, F. I., Taravel, F. R. and Vuong, R. (2001) Linear mannan in the endosperm of *Schizolobium amazonicum. Carbohydr. Polym.* 44, 107-112.

Politz, O., Krah, M., Thomsen, K. K. and Borris, R. R. (2000) A highly thermostable endo- (1, 4) -beta-mannanase from the marine bacterium *Rhodothermus marinus. Appl. Microbiol. Biotechnol.* 53, 715-721.

Polne-Fuller, M. and Gibor, A. (1984) Developmental studies in *Porphyra*, Ⅰ. Blade differentiation in *Porphyra perforata* as expressed by morphology, enzymatic digestion, and protoplast regeneration. *J. Phycol.* 20, 609-616.

Privalova, N. M. and Elyakova, L. A. (1978) Purification and some properties of endo-β- (1-3) -glucanase from the marine bivalve *Chlamys abbidus. Comp. Biochem. Physiol. B* 60, 225-228.

Puchart, V., Vrsanska, M., Svoboda, P., Pohl, J., Ogel, Z. B. and Biely, P. (2004) purification and characterization of two forms of endo-beta-1, 4-mannanase from a thermotolerant fungus, *Aspergillus fumigatus* IMI 385708 (formerly *Thermomyces lanugittosus* IMI 158749). *Biochem. Biophys. Acta.* 1674, 239-250.

Rahman, M. M., Inoue, A., Tanaka, H. and Ojima, T. (2011) cDNA cloning of an alginate lyase from a marine gastropod *Aplysia kurodat* and assessment of catalytically important residues of this enzyme. *Biochimie* 93, 1720-1730.

Rahman, M. M., Inoue, A., Tanaka, H. and Ojima, T. (2010) Isolation and characterization of two alginate lyase isozymes, AkAly28 and AkAly33, from the common sea hare *Aplysia kurodai. Comp. Biochem. Physiol. B* 157, 317-325.

Rahman, M. M., Ling W., Inoue, A. and Ojima, T. (2012) cDNA cloning and bacterial expression of a PL-14 alginate lyase from a herbivorous marine snail *Littorina hrevicula. Carbohydr. Res.* 360, 69-77.

Rezaii, N. and Khodagholi, E (2009) Evaluation of chaperone-like activity of alginate: Microcapsule and water-soluble forms. *Protein J.* 28, 124-130.

Sakamoto, K. and Toyohara, H. A. (2009) Comparative study of cellulase and hemicellulose activities of brackish water clam Corbicula japonica with those of other marine *Veneroida bivalves. J. Exp. Biol.* 212, 2812-2818.

Sakamoto, K., Uji, S., Kurokawa, T. and Toyohara, H. (2008) Immunohistochemical, in situ hybridization and biochemical studies on endogeneous cellulase of *Corbicula japonica. Comp. Biochem. Physiol. B* 150, 216-221.

Sakon, J., Irwin, D., Wilson, D. B. and Karplus, P. A. (1997) Structure and mechanism of endo/exocellulase E4 from *Thermomonospora fusca. Nat. Struct. Biol.* 4, 810-818.

Sattarahmady, N., Khodagholi, E, Moosavi-Movahedi, A. A., Heli, H. and Hakimelahi, G. H. (2007) Alginate as an antiglycating agent for human serum albumin. *Int. J. Biol. Macromol.* 41, 180-184.

Sawabe, T., Ohtsuka, M. and Ezura, Y. (1997) Novel alginate lyases from marine bacterium *Alteromonas sp.* strain H-4. *Carbohydr. Res.* 28, 69-76.

Sawabe, T., Setoguchi, N., Inoue, S., Tanaka, R., Ootsubo, M., et al., (2003) Acetic acid production of *Vibrio halioticoli* from alginate: a possible role for establishment of abalone-*Vibrio halioticoli* association. *Aquaculture* 219, 671-679.

Shimahara, H., Suzuki, H., Sugiyama, N. and Nisizawa, K. (1975) Partial purification of beta-mannanase from the konjac tubers and their substrate specificity in relation to the structure of konjac glucomannan. *Agric. Biol. Chem.* 39, 301-312.

Shimizu, E., Ojima, T. and Nishita, K. (2003) cDNA cloning of an alginate lyase from abalone, *Haliotis discus hannai. Carbohydr. Res.* 338, 2841-2852.

Smant, G., Stokkermans, J. P. W. G., Yan, Y., De Boer, J. M., Baum, T. J., et al., (1998) Endogenous cellulases in animals: isolation of β-1, 4-endoglucanase genes from two species of plant-parasitic cyst nematodes. *Proc. Natl. Acad ScL USA* 95, 4906-4911.

Smelcerovic, A., Knezevic-Jugovic, Z. and Petronijevic, Z. (2008) Microbial polysaccharides and their derivatives as current and prospective pharmaceuticals. *Curr. Pharm. Des.* 14, 316-395.

Sova, V. V., Elyakova, L. A. and Vaskovsky, V. E. (1970a) The distribution of laminarinases in marine invertebrates. *Comp. Biochem. Physiol. B* 32, 459-464.

Sova, V. V., Elyakova, L. A. and Vaskovsky, V. E. (1970b) Purification and some properties of β-1, 3-glucan glucanohydrolase from the crystalline style of bivalvia, *Spisula sachalineitsis. Biochim. Biophys. Acta* 212, 111-115.

Stalbrand, S., Siika-aho, M., Tenkanen, M. and Viikari, L. (1993) Purification and characterization of two β-

mannanases from *Trichoderma reesei*. *J. Biotechnol.* 29, 229-242.

Storseth, T. R., Hansen, K., Reitan, K. I. and Skjermo, J. (2005) Structural characterization of β-D- (1, 3) -glucans from different growth phases of the marine diatoms *Chaetoceros miilleri* and *Thalassiosira weissflogii*. *Carbohydr. Res.* 340, 1159-1164.

Suda, K., Tanji, Y., Hori, K. and Unno, H. (1999) Evidence for a novel *Chlorella* virus-encoded alginate lyase. *FEMS Microbiol. Lett.* 180, 45-53.

Sugimura, I. I., Sawabe, T. and Ezura, Y. (2000) Cloning and sequence analysis of *Vibrio halioticoli* genes encoding three types of polyguluronate lyase. *Mar. Biotechnol.* 2, 65-73.

Sutherland, I. W. (1995) Polysaccharide lyases. *FEMS Microbiol. Rev.* 16, 323-347.

Suzuki, H., Suzuki, K., Inoue, A. and Ojima, T. (2006) A novel oligoalginate lyase from abalone, *Haliotis discus hannai*, that releases disaccharide from alginate polymer in an exolytic manner. *Carbohydr. Res.* 341, 1809-1819.

Suzuki, K., Ojima, T. and Nishita, K. (2003) Purification and cDNA cloning of a cellulase from abalone *Haliotis discus hannai*. *Eur. J. Biochem.* 270, 771-778.

Sygusch, J., Ethier, N. and Talbot, G. (1998) Gene cloning, DNA sequencing, and expression of thermostable β-mannanase from *Bacillus stearothermophilus*. *Appl. Environ. Microbial.* 64, 4428-4432.

Takami, H., Kawamura, T. and Yamashita, Y. (1998) Development of polysaccharide degradation activity in postlarval abalone *Haliotis discus hannai*. *J. Shellfish Res.* 17, 723-727.

Takeda, H., Yoneyama, E, Kawai, S., Hashimoto, W. and Murata, K. (2011) Bioethanol production from marine biomass alginate by metabolically engineered bacteria. *Energy Environ. Sci.* 4, 2575-2581.

Tangarone, B., Royer, J. C., Nakas, J, P. (1989) Purification and characterization of an endo- (1, 3) -β-glucanase from *Trichoderma longibrachiatum*. *Appl. Environ. Microbiol.* 55, 177-184.

Talbot, G. and Sygusch, J. (1990) Purification and characterization of thermostable beta-mannanase and alpha-galactosidase from *Bacillus stearothermophilus*. *Appl. Environ. Microbiol.* 56, 3505-3510.

Tang, C. M., Waterman, L. D., Smith, M. H. and Thurston, C. E (2001) The cel4 gene of *Agaricus bisporus* encodes a β-mannanase. *Appl. Environ. Microbiol.* 67, 2298-2303.

Tokuda G., Watanabe, H., Matsumoto, T. and Noda, H. (1997) Cellulose digestion in the wood-eating higher termite, *Nasutitermes takasagoensis* (Shiraki): distribution of cellulases and properties of endo-β-1, 4-glucanase. *Zool. Sci.* 14, 83-93.

Tonime, P., Chauvaux, S., Beguin, P., Millet, J., Aubert, J. P. and Claeyssens, M. (1991) Identification of a hist idyl residue in the active center of endoglucanase D from *Clostridium thermocellum*. *J. Biol. Chem.* 266, 10313-10318.

Tomnic, P., Warren, R. A. J. and Gilkes, N. R. (1995) Cellulose hydrolysis by bacteria and fungi. *Adv. Microbiol. Physiol.* 37, 1-81.

Tomoda, Y., Umemura, K. and Adachi, T. (1994) Promotion of barley root elongation under hypoxic conditions by alginate lyase-lysate. *Biosci. Biotechnol. Biochem.* 58, 202-203.

Trommer, H. and Neubert, R. H. (2005) The examination of polysaccharides as potential antioxidative compounds for topical administration using a lipid model system. *Int. J. Pharm.* 298, 153-163.

Tsuchida, T., Chaki, T. and Ogawa, H. (2001) Effect of sodium alginate oligosaccharide on human blood pressure. [In Japanese.] *Med. Cons. New-Remed.* 38, 555-560.

Tusi, S. K., Khalaj, L., Ashabi, G., Kiaei, M. and Khodagholi, F. (2011) Alginate oligosaccharide protects against endoplasmic reticulum-and mitochondrial-mediated apoptotic cell death and oxidative stress. *Biomaterials* 32, 5438-5458.

Usui, T., Toriyama, T. and Mizuno, T. (1979) Structural investigation of laminaran of *Eisenia bicyclis*. *Agric. Biol. Chem.* 43, 603-611.

Vergote, D., Bouchut, A., Sautiere, P. E., Roger, E., Galinier, R. and Rognon, A. (2005) Characterization of proteins differentially present in the plasma of *Biomphalaria glabrata* susceptible or resistant to *Echinostoma caproni*. *Int. J. Parasitol.* 35, 215-224.

Wang, J., Ding, M., Li, Y., Chen, Q., Xu, G. and Zhao, F. (2003) Isolation of a multi-functional endogeneous cellulase gene from mollusc, *Ampullaria crossean*. *Acta Biochim. Biophys. Sin.* 35, 941-946.

Wang, L., Rahman, M. M., Inoue, A. and Ojima. T. (2012) Heat-stability and primary structure of the major alginate lyase isozyme LbAly35 from *Littorina brevicula*. *Fish. Sci.* 78, 889-896.

Wargacki, A. J., Leonard, E., Win, M. N., Regitsky, D. D., Santos, C. N., et al., (2012) An engineered microbial platform for direct biofuel production from brown macroalgae. *Science* 335, 308-313.

Watanabe, H., Noda, H., Tokuda, G. and Lo, N. (1998) A cellulase gene of termite origin. *Nature* 394, 330-331.

Watanabe, T. and Nishizawa, K. (1982) Enzymatic studies on alginate lyase from Undaria pinnatifida in relation to texture-softening prevention by ash-treatment (Haiboshi). *Bull. Jap. Soc. Sci. Fish.* 48, 243-249.

Wong, T. Y., Preston, L. A. and Schiller, N. L. (2000) Alginate lyase: review of major sources and enzyme characteristics, structure-function analysis, biological roles, and applications. *Annu. Rev. Microbiol.* 54, 289-340.

Wong, W. K., Gerhard, B., Guo, Z. M., Kilburn, D. G., Warren, A. J. and Miller, R. C. Jr. (1986) Characterization and structure of an endoglucanase gene cenA of *Cellulomonas fimi*. *Gene* 44, 315-324.

Xu, B., Hagglund, P., Stalbrand, H. and Janson, J. C. (2002a) Endo-β-1, 4. mannanases from blue mussel, *Mytilus edulis*: purification, characterization and mode of action. *J. Biotechnol.* 92, 267-277.

Xu, B., Heilman, U., Ersson, B., Janson, J. C. (2000) Purification, characterization and amino-acid sequence analysis of a thermostable, low molecular mass endo-β-1, 4-glucanase from blue mussel, *Mytilus edulis*. *Eur J Biochem.* 267, 4970-4977.

Xu, B., Sellos, D. and Janson, J. C. (2002b) Cloning and expression in *Pichia pastoris* of a blue mussel (*Mytilus edulis*) β-mannanase gene. *Eur. J. Biochem.* 269, 1753-1760.

Xu, X., Iwamoto, Y., Kitamura, Y., Oda, T. and Muramatsu, T. (2003) Root growth-promoting activity of unsaturated oligomericuronates from alginate on carrot and rice plants. *Biosci. Biotechnol. Biochem.* 67, 2022-2025.

Xue, C., Yu, G., Hirata, T., Terao, J. and Lin, H. (1998) Antioxidative activities of several marine polysaccharides evaluated in a phosphatidylcholineliposomal suspension and organic solvents. *Biosci. Biotechnol. Biochem.* 62, 206-209.

Xue, X. M., Anderson, A. J., Richardson, N. A., Anderson, A. J., Xue, G. P. and Mather, P. B. (1999) Characterisation of cellulase activity in the digestive system of the redclaw crayfish (*Cherax quadricarinatus*).

Aquaculture 180, 373–386.

Yamamoto, S., Sahara, T., Sato, D., Kawasaki, K., Ohgiya, S., et al., (2008) Catalytically important amino-acid residues of abalone alginate lyase HdAly assessed by site-directed mutagenesis. *Enzyme Microb. Tech.* 43, 396–402.

Yamamoto, Y., Kurachi, M., Yamaguchi, K. and Oda, T. (2007) Stimulation of multiple cytokine production in mice by alginate oligosaccharides following intraperitoneal administration. *Carbohydr. Res.* 342, 1133–1137.

Yamasaki, M., Ogura, K., Hashimoto, W., Mikami, B. and Murata, K. (2005) A structural basis for depolymerization of alginate by polysaccharide lyase family-7. *J. Mol. Biol.* 352, 11–21.

Yamaura, I. and Matsumoto, T. (1993) Purification and some properties of endo-1, 4-beta-D-mannanase from a mud snail, *Pomacea insularus* (de Ordigny). *Biosci. Biotech. Biochem.* 57, 1316–1319.

Yamaura, I., Nozaki, Y., Matsumoto, T. and Kato, T. (1996) Purification and some properties of an endo-1, 4-beta-D-mannanase from a marine mollusc *Littorirui brevicula. Biosci. Biotech. Biochem.* 60, 674–676.

Yan, Y., Smant, G., Stokkermans, J., Qin, L., Helder, J., et al., (1998) Genomic organization of four β-1, 4-endoglucanase genes in plant-parasitic cyst nematodes and its evolutionary implications. *Gene* 220, 61–70.

Yokoe, Y. and Yasumasu, I. (1964) The distribution of cellulase in invertebrates. *Comp. Biochem. Physiol.* 13, 323–338.

Yoshida, T., Hirano, A., Wada, H., Takahashi, K. and Hattori, M. (2004) Alginic acid oligosaccharide suppresses Th2 development and IgE production by inducing IL-12 production. *Int. Arch. Allergy Immunol.* 133, 239–247.

Zahura, U. A., Rahman, M. M., Inoue, A. and Ojima, T. (2012) Characterization of a β-D-mannosidase from a marine gastropod, *Aplysia kurodai. Comp. Biochem. Physiol. B* 162, 24–33.

Zahura, U. A., Rahman, M. M., Inoue, A., Tanaka, H. and Ojima, T. (2010) An endo-β-1, 4 mannanase, AkMan, from the common sea hare *Aplysia kurodai. Comp. Biochem. Physiol. B* 157, 137–143.

Zahura, U. A., Rahman, M. M., Inoue, A., Tanaka, H. and Ojima, T. (2011) cDNA cloning and bacterial expression of an endo-b-1, 4-mannanase, AkMan, from *Aplysia kurodai. Comp. Biochem. Physiol. B* 159, 227–235.

Zhang, Z., Yu, G., Guan, H., Zhao, X., Du, Y. and Jiang, X. (2004) Preparation and structural elucidation of alginate oligosaccharides degraded by alginate lyase from *Vibrio* sp. 510. *Carbhydr. Res.* 339, 1475–1481.

第 14 章　海洋解烃微生物

Valeria Cafaro, University of Naples Federico Ⅱ, Italy, Viviana Izzo, University of Salerno, Italy, Eugenio Notomista, University of Naples Federico Ⅱ, Italy and Alberto Di Donato, University of Naples Federico Ⅱ, Italy

DOI：10. 1533/9781908818355. 3. 373

摘要：在全世界范围内，已从被污染海水中分离出了几种专门降解烃类的海洋细菌。其中一些细菌可以专一性地利用烃类作为生长基质，被称作“专性解烃菌”（OHCB）。海洋解烃细菌不仅是有效的生物修复工具，而且是单、双加氧酶，氧化酶、脱氢酶和在区域或者立体选择性生物合成中具有活性酶类的优良资源库。此外，为适应生长在富含烃类的环境中，海洋解烃微生物通常合成一些奇特的化合物，例如，具有工业开发前景的有高乳化活性生物去污剂和存储的大分子物质。本章详细描述了其特征和解烃细菌这个小家族最有趣的产品。

关键词：烃类，加氧酶，氧化酶，生物表面活性剂，储存脂质

14.1　引言

烃是最广泛的污染物。自然和人工资源都会导致芳香和脂肪族烃弥散在几乎每种环境中，从温带区城市土地再到极地冰川。

同时专一的微生物代谢不断地产生简单的烃类，如甲烷、乙烷、丙烷、丁烷等及复杂高分子烃类如异十八烃、植烷等（Ladygina et al., 2006；Taylor et al., 2000）。植物也产生大量不同的烃类（Fuentes et al., 2000）。高分子量异戊二烯烃类主要存在于昆虫表皮中（Golebiowski et al., 2011），一些多环芳香烃类产生于森林大火（Vergnoux et al., 2011）。

最重要人工来源的烃类包括发动机和焚烧炉内的燃烧，工业废料的直接排放，很明显也包括化石燃油的泄露。原油和所有来自于原油蒸馏得到的燃料是释放到环境中的主要烃类来源。漏油不仅仅发生在石油平台事故中，如碰撞和搁浅，同时发生在常规的日常操作中，包括在陆地和海底石油井的挖掘，石油装载、运输和卸货过程。

原油及燃料包含饱和、不饱和及芳香烃类（Henry，1998；King，1988）。这些片段是数以百计化合物的复杂的混合物。例如，芳香片段，这个组分占有总烃类物质的10%～40%多，包含（多聚）烷基苯，PAHs，（多聚）烷基PAHs，环烷等混合物。

大多数烃类是相当稳定并且会在环境中富集。此外，它们是富集能源和碳的一个主要来源。因此，有相当数量的微生物进化成为专门以其作为基质而生长并不意外（Rosenberg 1992；Van Hamme et al.，2003和其他文献）。这个能力是广泛分布的并且可以几乎在每种微生物中发现，包括细菌、真菌、嗜温菌和极端环境菌。有趣的是，一些微生物可以利用不止一种烃类作为碳源和能源，而一些物种则是专一性的，这样的话就可以专门利用烃类。这些令人惊讶的存在于整个海洋细菌中的特征将在14.2中展开讨论。

化学稳定性和烃类多样性促使微生物发展出多种多样有趣的独特酶活性，这涉及烃类的活化和片段化形成更小分子，使其可以进入中心能源和生物合成代谢过程。通常，在有氧代谢中，烃类活化是通过增加一到两个氧原子来介导，它提供了进行后续反应的活性分子。例如，甲烷和链烷烃的代谢开始于终端单加氧反应，这提供了一个初级醇，它可以继续氧化成羧酸（Rojo，2009；van Beilen and Funhoff，2007）（图14.1a，b）。

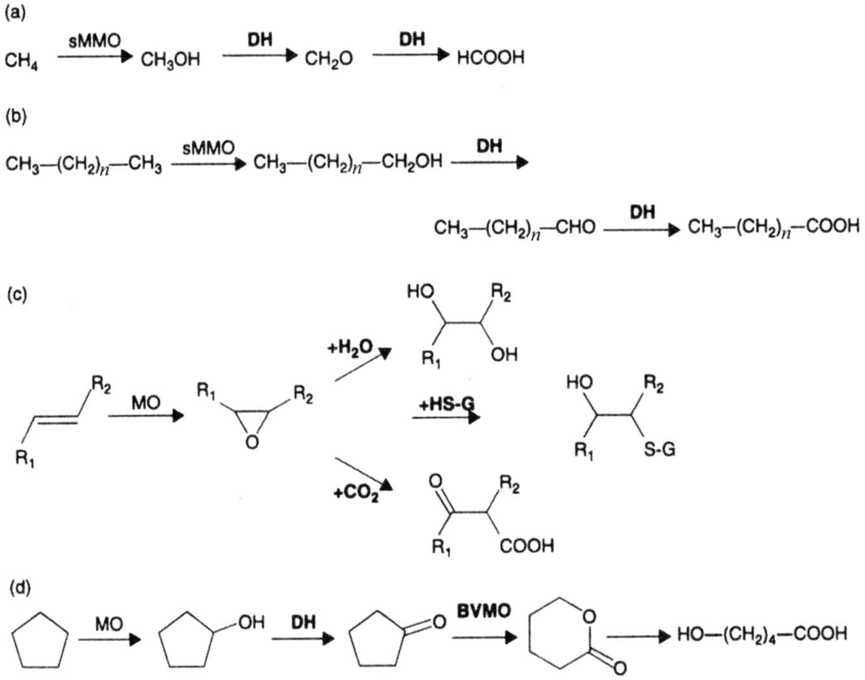

图14.1 脂肪族烃的通常降解途径

注：（a）甲烷；（b）链烷烃；（c）烯烃；（d）环戊烷；sMMO，可溶性甲烷单氧酶；MO，单氧酶；DH，脱氢酶；BVMO，氢过氧化物单氧酶；HS-G，谷胱甘肽.

烯烃，例如，乙烯和苯乙烯被氧化成环氧，它可以通过一些不同的途径被进一步代谢

(Mooney et al., 2006; van Beilen and Funhoff, 2007; van Hylckama Vlieg and Janssen, 2001)(图 14.1c)。环烯烃被羟化成二醇，氧化成醌继而氧化成酯最后被水解（图 14.1d）。这种分支的烯烃尤其芳香烃的代谢更加复杂（Seo et al., 2009）。图 14.2 展示了一些已知的相对简单的单环芳香烃甲苯降解的不同通路（Jindrova et al., 2002）。最初的氧化发生在甲基基团，因此使苯甲酸或在环上产生甲基邻苯二酚或者甲基对苯二酚。邻苯二酚中两个羟基可以通过两个连续羟化反应或者一个双氧化反应被添加。同时邻苯二酚芳香环也可以通过两个不同方式开环——外二醇和内二醇分裂（图 14.3）——每个开始于一个不同的“更低的通路”它提供了降解中心的中间体。未来将会发现更多的可能性。

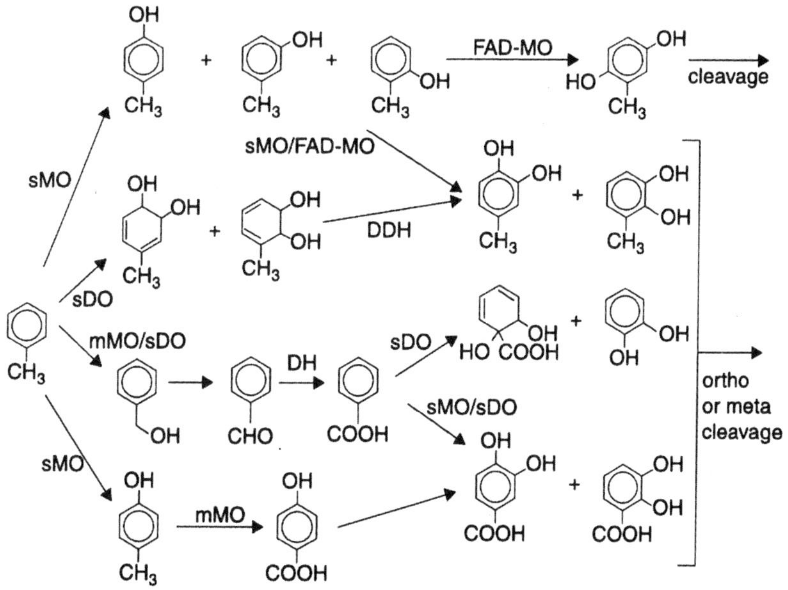

图 14.2　可能的甲苯降解上游通路

sMO：可溶性单氧酶；mMO：膜单氧酶；FAD-MO：FAD 依赖性单氧酶；sDO：可溶性双加氧酶；DDH：二氢二醇脱氢酶；DH：脱氢酶

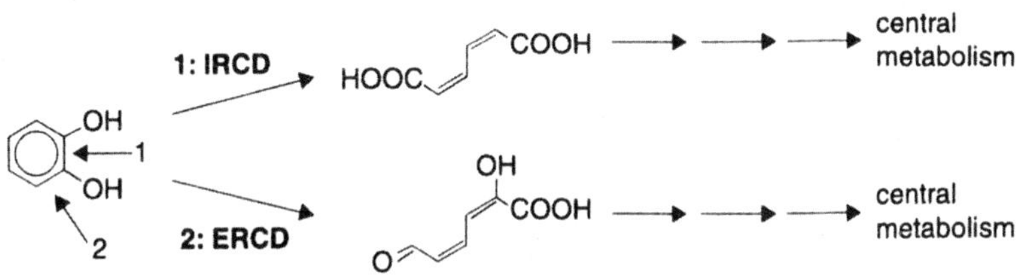

图 14.3　可能的针对邻二苯酚降解通路

ERCD：外二醇环解双加氧酶（元裂解）；IRCD：内二醇解环双加氧酶（邻位裂解）.

PAHs 的降解类似于单环，即使通常环是被顺序的羟基化与裂解（Seo et al., 2009），就像在图 14.4 展示的甲基萘一样。氮和氧的代谢包含起始的芳香杂环化合物的侧向或者有一定角度的双氧化（图 14.5a）（Seo et al.，2009），然而，喹啉和异喹啉是通常在吡啶环上单羟基化（图 14.5b）。含硫的杂环芳香化合物的代谢例如二苯并噻吩起始于环上（图 14.5a）或者硫分子的氧化（Seo et al.，2009）。

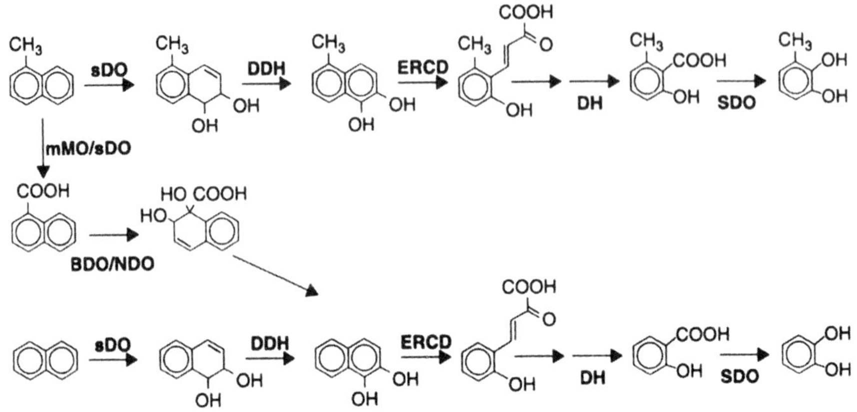

图 14.4　PAHs 降解途径

sMO：可溶性单氧酶；mMO：膜单氧酶；FAD-MO：FAD 依赖性单氧酶；sDO：可溶性双加氧酶；DDH：二氢二醇脱氢酶；DH：脱氢酶；ERCD：外二醇环解双加氧酶；SDO：水杨酸双加氧酶；BDO/NDO：苯甲酸，萘双加氧酶.

所有涉及这些复杂通路的单加和双加氧酶都是非常有趣的酶，它们具有独特的化学性质，同时可以作为生物合成的潜在工具。一个值得关注的例子是细菌中可溶性甲烷氧化酶，这种细菌生长在依赖于甲烷作为唯一碳源和能源的培养基中（Tinberg and Lippard，2011）。这些可溶性胞质多组分酶包括了还原酶组分，它氧化了 NADPH 为 $(\alpha\beta\gamma)_2B_2$结构的氧化酶复合物提供了两个电子。

这个 B 亚基作为一个调控组分控制电子的流向和位于 α 亚基上活性位点的暴露程度。这个活性位点包含两个非血红素离子（Ⅲ）连接的羟基负离子。通过还原酶提供的电子还原铁（Ⅲ）-OH-铁（Ⅲ）到铁（Ⅱ）-OH（H）-铁（Ⅱ），它会转而与氧气反应产生高价态的二价铁聚合物，它们具有足够强的氧化能力可以将氧原子插入到甲烷的 C-H 键中。当然整个反应机理的细节仍然在讨论之中；然而这些研究将为开发合成类似物作为化学工业中的生物催化剂提供可能（Do LH，2011）。

甲烷单氧化酶属于细菌多组分单氧化酶（BMMs），它包括了几个酶的家族，它们都有一个相似机制，但具有不同的底物特异性（Notomista et al.，2003），例如环氧烯烃，苯甲苯二甲苯和 PHs。BTX-MOs 和 PHs 是芳香环特异的区域选择性单氧酶，这是一个利用常规催化剂很难重现的反应，反应通常会产生复杂的混合物产物，包括聚合副产物，从而降低产率并使纯化变得艰难。因此，一些研究尝试去开发这些酶来发展生物合成过程就不足

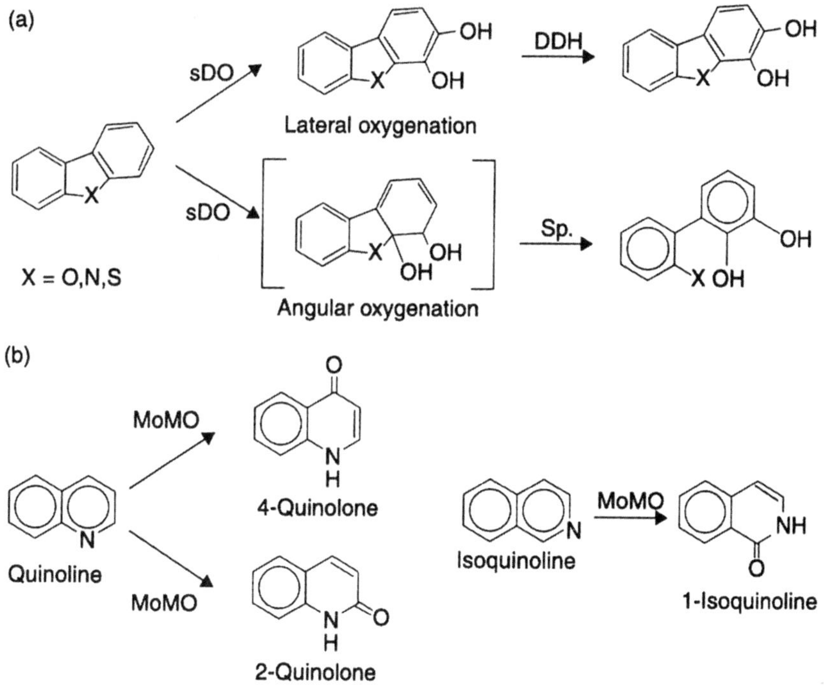

图 14.5 芳香杂环组分（a）和喹啉（b）的降解途径

sMO：可溶性单氧酶；DDH：二氢二醇脱氢酶；Sp：自发转化；MoMO：钼依赖性单氧酶.

为奇了（Nolan and O'Connor，2008；Notomista et al.，2011）。

膜单氧酶是与非铁血红素家族无关的，而专门作用于脂肪族的单加氧（AlkBlike 链烷单加氧酶），或单和多环烃的芳族甲基（XylM 状单加氧酶）（Austin et al.，2003；van Beilen et al.，2007）。XylM 类单加氧酶产生的苄醇和苯甲醛通过脱氢酶转化成苯甲酸，萘甲酸等（图 14.2 和图 14.4）。但他们结合到膜上这一事实阻碍了对这些单氧酶的特征及其开发研究。

在喹啉和异喹啉中，杂环氮原子使吡啶环不活跃；从而初始羟基化通常是由特殊的使用含有钼作为辅因子（molybdopteryn）氧化还原反应中心的单氧酶来完成（Bonin et al.，2004；Hille，2005）。这些 molybdopteryn 依赖性加氧酶（MoMOs）可以催化其他难以完成的反应，如黄嘌呤的单加氧和一氧化碳到二氧化碳的氧化（Hille，2005）。

与此相反，酚羟基提高了芳香环的反应性；的确，一些苯酚羟化酶使用 FAD 或 FMN 而非金属去激活双氧（Ballou，et al.，2005）。一些 FAD/ FMN-PHS 羟化位点与预先存在的羟基处于对位，从而制备对苯二酚而非邻二苯酚，产物是非铁血红素依赖的 PHS。

另一个不相关家族的 MOs 是 Baeyer-Villiger MOs，它能催化酮与酯的转换（Leisch et al.，2011）。它们参与环烷类和一些环烷烃的代谢（图 14.1）（Seo et al.，2009）。由于酮的单加氧反应是化学工业一个非常有用的反应，BVMOs 被认为是开发生物催化工艺非常有价值的工具（Leisch et al.，2011）。

可溶性羟基化双加氧酶在几个方面类似于 BMM。它们是复杂的多组分酶，包括氧化 NADH 并且为羟化酶复合物提供电子的一个氧化还原酶复合物（Ferraro et al.，2005）。这个羟化酶含有 Fe（Ⅲ）作为辅因子，因而一个单一的铁离子结合到活性位点。因此，两个电子被依次传送，双氧被激活成一种活性状态，它转移两个原子到芳香环从而产生二氢二醇（图 14.2）。sDO 能够催化羟基化芳香甲基从而模仿膜 MOs。像 BTX-MOS 和 PHs，sDO 是用于生物合成的两个单独的非常有趣酶的工具，既可以单独使用以制备二氢二醇，这是化学合成有用的中间分子，又可以与脱氢酶配合，以产生二羟基芳族分子（Boyd and Bugg，2006；Nolan and O'Connor et al.，2008）。

外环和内环裂解 DOs（ERCD 和 IRCD）是由一个或两个亚单位构成的简单的可溶性酶（Vaillancourt et al.，2006）。两者都使用铁作为辅因子，邻苯二酚结合铁离子作为二合配位体。这两种类型的 DOs 之间的主要区别是 ERCDs 含有的 Fe（Ⅱ）离子和邻二苯酚结合作为单价阴离子，而 IRCDs 含有 Fe（Ⅲ）和邻二苯酚结合为二价阴离子（Vaillancourt et al.，2006）。这些差异决定了不同的区域选择性。但迄今为止，其生物合成的潜力还未得到发挥。

除了 RCDs 和 MoMOs 外，上述所有加氧酶均需要电子源——通常是 NAD（P）H。这使得它们在工业生产过程中的使用更复杂，从而更促使我们去寻找有趣和富有想象力的解决方案。在某些情况下，全细胞表达重组酶被用作催化剂（Notomista et al.，2011）。在其他情况下，加氧酶已经耦合到第二个酶当中，使其氧化像甲酸或亚磷酸盐的廉价底物，重新生成的还原型辅酶（Leisch et al.，2011）。一些 BVMOs 已用“回收酶”成功制备了融合蛋白（Leisch et al.，2011）。

解烃微生物是一些非常有趣的其他相关产品的来源，如具有高乳化活性和聚合存储物质的生物去污剂，这与其适合于生长在富烃环境有关（Rosenberg 1992；Van Hamme et al.，2003）。

海洋解烃菌是酶和符合工业利益的非酶促生物分子极为丰富的来源。接下来的章节将详细描述解烃菌的特征和其最有趣的产物。

14.2 海洋解烃微生物

在过去的几十年里海洋微生物降解碳氢化合物的研究已经广泛开展。石油降解细菌已经从世界的不同地方（Head et al.，2006）被分离出来，说明这些细菌在各种环境中广泛存在。解烃微生物通常在海洋环境中是低丰度的。在污染的区域如受石油碳氢化合物的污染可能刺激这些细菌的生长，并引起其微生物群体在结构上和成分上发生一些变化。

对生物降解污染物关键微生物的确认是了解、评价和开发原位生物修复策略的关键。此外，由于来源于解烃微生物酶在生物技术领域的不断应用，现已在描述菌种群体，确定可能的降解者，阐明可能发挥作用的酶蛋白方面做出了努力。

与利用广泛有机底物的陆地烃类化合物降解者相比，海洋拥有相对专一的烃类化合物降解者。它们组成了海洋烃解微生物（Yakimov et al.，2007）。其中大多数微生物都具有窄的底物范围，对烃类化合物（烷烃、多环芳香烃）和少数有机酸来说几乎是专一的，因此又称为“专职烃解微生物”（OHCB）。它们主要的细菌种属是 *Alcanivorax*（Yakimov et al.，1998），*Cycloclasticus*（Dyksterhouse et al.，1995）、*Marinobacter*（Gauthier et al.，1992）、*Neptunomonas*（Hedlund et al.，1999）、*Oceanobacker*（Teramoto et al.，2009）、*Oleiphilus*（Golyshin et al.，2002）、*Oleispira*（Yakimov et al.，2003）和 *Thalassolituus*（Yakimov et al.，2004）。

OHCB 可以根据其降解脂肪族或者芳香族烃类化合物划分为两个类群。*Alcanivorax*、*Marinobacter*、*Oleiphilus*、*Oleispira*、*Oceanobacter* 和 *Thalassolituus* spp. 能够利用多种分支或者直链的饱和烃，而 *Neptunomonas* 和 *Cycloclasticus* spp. 进化出可以使用多环芳香烃。但是许多非专职烃解微生物菌属、*Vibrio*、*Pseudoalteromonas*、*Marinomonas* 和 *Halomonas* 已被作为海洋微生物分离出来，它们能够降解萘、菲和䓛（Melcher et al.，2002）。

Alcanivorax 和 *Cycloclasticus* 包含两个专职烃解微生物，如 *A. borkumensis*，*A. jadensis*（Fernandez-Martinez et al.，2003；Golyshin et al.，2002），*A. dieselolei*（Liu and Shao，2005），*C. pugetii*（Dyksterhouse et al.，1995）和 *C. oligotrophus*（Robertson et al.，1998）以及有很少限制的底物类型和多种营养来源的菌种如 *A. venustensis*（Chung and King 2001）和 *C. spirillensus*（Fernandez-Martinez et al.，2003）。

Oleiphilus messinensis（Golyshin et al.，2002），*Oleispira antarctica*（Yakimov et al.，2003），*Thalassolituus oleivorans*（Yakimov et al.，2004）和 *Neptunomonas naphthovorans*（Hedlund et al.，1999）是它们群体中最具代表性的 OHCB 种群。

一些研究表明海洋中石油涌入的地方会引起 OHCB 密度很快增长到占所有微生物种群的 90%。*Alcanivorax* 和 *Cycloclasticus* 显示是烃类化合物污染后最快和最强被选择的菌属，分别成为降解脂肪族和芳香族烃类化合物的关键微生物（Golyshin et al.，2005；Harayama et al.，1999，2004；Head et al.，2006）。

例如，针对 *Alcanivorax* spp. 研究表明，起初在海水中检测不到，但在污染海水中仅用 1~2 个周的时间，就占原核细胞的 70%~90%（Harayama et al.，1999；Kasai et al.，2002a；Syutsubo et al.，2001）。这说明 *Alcanivorax* spp. 的快速生长在于与其他烃类化合物降解细菌相比，它能够更有效地利用支链烷烃，使之更具选择优势（Hara et al.，2003）。在污染区域，经过起初种群数量的快速增长之后，短短几周就会降到很低的数量。这种现象与饱和烃类化合物的消耗有关，这提示了 *Alcanivorax* spp. 降解这些烃类化合物的特点（Head et al.，2006）。

对 *Cycloclasticus* spp. 来说也是一个类似的情况，它在油污染的海水中被发现（Head，et al.，2006；Kasai et al.，2002b；Mckew et al.，2007；Niepceron et al.，2010），并且在降

解芳香族烃类化合物中起到重要作用。值得注意的是，通过在微观和宏观石油降解研究中发现 *Alcanivorax* 和 *Cycloclasticus* spp. 都是在石油降解的早期达到高的丰度（Cappello et al.，2007；Roling et al.，2002）。

14.3 来源于解烃微生物的生物活性化合物

自从发现海洋解烃微生物具有能够有效地降解烃类的能力，并对缓解石油泄露危害的使用潜能进行了评估，现已经成为人们日益需要的以生物技术为基础的催化生物活性化合物生产的酶（Yakimov et al.，2007）（14.3.2），也作为具有新奇酶活性的来源（14.4）。

14.3.1 生物表面活性剂

对于源于微生物生物表面活性剂的兴趣在过去一段时期内得到稳定增长。相比于化学表面活性剂，生物表面活性剂有一定的优势，例如低毒性，环境相容性好和生物可降解性（Lima et al.，2011；Zajic et al.，1977）。生物表面活性剂的商业用途已经涵盖（Banat et al.，2000；Desai and Banat，1997）化妆品、食品、健康护理、造纸工艺、煤、陶瓷和金属工业等方面。还被发现在治疗和生物药学方面也有重要的用途：有抗菌、抗真菌、抗病毒等性质，同时也对其他微生物有抗吸附性（Cameotra and Makkar，2004；Rodrigues et al.，2006；Singh and Cameotra，2004）。当然，最有前景的应用是处理油类，清洁油泄漏污染恢复原油和原位生物修复处理烃类，重金属和其他污染物（Mulligan，2005；Slizovskiy et al.，2011）。

为什么海洋解烃微生物在新型生物表面活性剂合成中是有如此价值的资源呢？在石油衍生的烃类化合物环境中生长的细菌已经进化出来了一些策略去消化吸收这些生物可利用度非常低的烃类化合物（McGenity et al.，2012；Wentzel et al.，2007）。细胞表面疏水性和合成生物表面活性剂（Berrtrand et al.，1993；Desai and Banat，1997；Satpute et al.，2010；Weiner，1997）对调节微生物和基质（Ron and Rosenberg，2001）结合和分离上起到很重要的作用。通过两方面可以改善微生物细胞对烃类物质的生物可利用性，一是通过增强水-油乳胶液的形成；二是通过增加细菌和水不溶性烃类的接触区域面积（Ron and Rosenberg，2001，2002）。这样增加了烃类的溶解性和被微生物利用的程度。

源于微生物生物表面活性剂是两性的胞外化合物，包括亲水基和疏水基，因此可以通过降低表面的张力使烃类物质吸收与乳化更加容易（Ron and Rosenberg，2001）。它们大体上分为低分子和高分子量的生物表面活性剂两个组群（Ron and Rosenberg，1999）。低分子量和高分子量的生物表面活性剂都包含脂类部分连接到不同分子上，形成糖脂、脂蛋白、酯类多聚物等（Ron and Rosenberg，2001，2002）。低分子量的分子能够有效地降低表面和界面的张力，而高分子量的多聚物与这表面紧紧连接在一起（Ron and Rosenberg，2002）。

值得提出的是，因为生产过程中的低产量和较高的复性及纯化费用，导致基于这些分子的产品的大规模生产还没有实现。因此，为降低生产成本而使用更经济的材料是人们渴望实现的（Makkar and Cameotra，2002；Mukherjee et al.，2006）。已经被报道出的不同的廉价的用于进行生物表面活性剂产品的生产材料包括植物油类及其废弃物；在此背景下，解烃类细菌将成为具有连接生物修复和生物技术的独特机遇。

低分子表面活性剂

低分子量的表面活性剂通常是糖脂，其中糖与一个长链的脂肪酸或者脂肽相连。糖脂表面活性剂如鼠李糖脂、藻糖脂、槐糖脂都是二糖与长链脂肪酸或羟基脂肪酸酰化形成的。

食盐菌（*Alcanivorax*）是已研究的正烷烃降解者，它是最具代表性的一种表面活性剂的生产者。最近，阐述了由 *Alcanivorax borkumensis* SK_2（Yakimov et al.，1998）产生的一类新的糖脂（Abraham et al.，1998）。这种菌以脂肪烃作为生长的主要碳源并产生阴离子葡萄糖脂表面活性剂。这种葡萄糖化合物由葡萄糖β-异头物与β-羟基癸酸四聚物相连。4个β-羟基脂肪酸通过酯键和葡萄糖苷键 C_1 相连在一起。Golyshin 等（2003）报道在指数增长后期阶段的细胞产生的葡萄糖脂有两种形式：一种是含甘氨酸的细胞结合前体，它能够增加细胞的疏水性，对水中悬浮的油滴具有亲和力；另一种是甘氨酸缺乏的形式，是有细胞释放到培养基中促进形成水油溶胶微粒从而增加油的生物可利用性。

Alcanivorax hongdengensis 是一种从马来西亚和新加坡海水表面分离出来的烷烃降解细菌，产生一种脂肽作为它的生物表面活性剂（Wu et al.，2009）。菌株 A-11-3 能够利用正烷烃的碳原子数量范围是 C_8-C_{36}。当 C_{16}作为唯一的碳源时，A-11-3 产生一种包含脂肽样化合物的表面活性剂。脂肽中主要的脂肪酸是 $C_{15:0}$（46.3%）和 $C_{17:0}$（40.2%），其他已发现的脂肪酸只占一小部分，包括 $C_{13:0}$（3.3%），$C_{17:1}$（6.5%）和 $C_{17:2}$（3.8%）。在氨基酸分析中检测出两个峰值，其中一个检测确认的峰值是酪氨酸；另一个没有被检测出来。

Schulz 等（1991）从黑尔法兰岛岛屿（北海）附近收集的正烷烃降解微生物样本中分离出了 3 种产生生物表面活性剂的海洋细菌。一株菌被确定为是 *Alcaligenes* sp.，MM 1 产生一种葡萄糖脂，另两株菌种属于 *Arthrobacter* sp.。通过一些特征证明是来自细胞外的高分子量乳化在纯培养的 *Arthrobacter* sp. SI 1 中被检测到，而海藻糖脂质主要是由 *Arthrobacter* sp. EK 1 产生的生物表面活性剂。化学分析和核磁共振测量显示主要组分是阴离子型的 2，3，4，2-海藻糖四酯，包含的脂肪酸长度范围从 8~14，琥珀酸部分与海藻糖 C_2 原子相连（Passeri et al.，1991）。

Pepi 等（2005）从南极洲罗斯海泰拉诺瓦湾站的冰下海水中分离出细菌 *Halomonas* ANT-3b 菌株，使用柴油燃料作为唯一碳源和能量来源。这种菌产生 18 kDa 的包含脂肪酸和混合糖（甘露糖，半乳糖，葡萄糖）糖脂乳化剂。

高分子细菌表面活性剂

高分子量细菌表面活性剂是由高分子量的聚合化合物组成，由大量不同菌属的细菌产生。组成成分主要有多糖、蛋白、脂多糖、脂蛋白以及这些生物聚合物的复杂混合物。高分子量的表面活性剂不能有效地减小表面张力，但是能有效地包裹油滴，防止其聚结。这些高分子量的生物表面活性剂显示有很高的底物特异性。例如，一些能够有效地乳化脂肪族和芳香族（环状烷烃）碳氢化合物的混合物，不能乳化纯的脂肪族和芳香族（环状烷烃）碳氢化合物。其他的只能乳化高分子量的纯的碳氢化合物。

研究最成熟的生物表面活性剂是由不同菌种 *Acinetobacter*（Rosenberg and Ron，1998）产生的乳化剂（Rosenberg and Ron，1998）。其中来自 *Acinetobacter calcoaceticus* RAG-1 的乳化是研究最透彻的。*A. calcoaceticus* RAG-1（ATCC 31012）是从地中海分离的革兰氏阴性石油降解海洋细菌，能够代谢多种碳源包括原油、长链烃、醇、脂肪酸以及甘油三酸酯（Reisfeld et al.，1972）。在生长阶段能够分泌一种阴离子脂杂多糖，这就是被人们所知的表面活性剂乳化剂（Rosenberg et al.，1979；Zuckerberg et al.，1979）。这种乳化剂是一种高分子杂多糖，包含 N-乙酰-D-半乳糖，N-乙酰半乳糖胺糖醛酸和未确定的 N-乙酰氨基糖的甘油三酸酯重复单元（Zhang et al.，1997）。脂肪酸通过共价 O-酯键与多糖相连（Belsky et al.，1979；Desai and Banat 1997；Zuckerberg et al.，1979）。尽管 emulsan 不能像小分子表面活性剂一样有效降低界面的张力，但是它能够紧密结合到油滴表面从而防止其结聚在一起。C_{10}-C_{18}的饱和或单不饱和脂肪酸通过 O-或 N—酰键与杂多糖相连构成，其占聚合物组成的 5%~23%（*w/w*）（Zhang et al.，1997）。亲水阴离子糖主链与疏水脂肪酸侧链连接在一起使聚合物具有两亲特征。

海洋细菌 *A. calcoaceticus* subsp. *anitratus* SM7 是从泰国宋卡环礁湖石油泄漏海水中分离出来的可以产生乳化作用的菌株。当生长在正十七烷作为唯一碳源和能量来源时，菌株 SM7 产生一种细胞外高分子乳化物（Phetrong et al.，2008）。来自于其他 *Acinetobacter* 菌种如 *A. calcoaceticus* RAG-1，BD4 和 BD413 生产的生物表面活性剂只能乳化脂肪族和芳香族碳氢化合物的混合物（Kaplan and Rosenherg 1982，Rosenberg et al.，1979），而菌种 SM7 产生的生物表面活性剂能够乳化纯的脂肪族和芳香族碳氢化合物（Phetrong et al.，2008）。而且这种生物表面活性剂具有 pH 值、温度和盐度的稳定性，使之非常适合在环境保护中应用。

一种极耐盐的菌种 *Marinobacter hydrocarbonoclasticus*，（Gauthier et al.，1992）生长在十二烷中时会合成细胞外高分子乳化剂，它能够黏附在烷烃表面，而且在吸收前不会被溶解掉。在十二烷中生长时，乳化作用与黏附作用拉近了碳氢化合物与细胞的距离。

菌株 *Halomonas* spp. 可以生产细胞外糖蛋白生物乳化剂（Calvo et al.，2002，Martinrz -Checa et al.，2002；Pepi et al.，2005）。两种海洋 *Halomonas* 菌株 TG39 和 TG67 由于其具有分泌对十六烷油高乳化活性的能力被筛选出来（Gutierrez et al.，2007）。对由 TG39 和

TG67 菌种产生的两种细胞外可水溶性乳化剂，被分别命名为 HE39 和 HE67，部分纯化和化学物理性质显示有糖醛酸组分的存在。有趣的是，这些生物聚合物因含有高密度的带电性，因而可能对有毒金属生物修复起作用，因为研究表明其他有相似成分的化合物具有此性质（Slizovskiy et al.，2011）。

14.3.2 OHCB 中脂质的储存

当有合适的碳源而氮和磷受到限制时，几乎所有的原核生物都会合成脂质储存物质。储存的脂质可作为饥饿时的能源和碳源（Manilla-Perez et al.，2010）。一些菌种合成的疏水聚合物如 PHB 或者其他类型的 PHAs，而 TAGs 和 Wes 的是不常见的（Wältermann and Steinbüchel, 2006）。

原油的泄漏污染形成了碳过度，而相关的可以利用的氮源是有限的一个环境条件，这促进海洋解烃细菌积聚储存化合物。*Alcanivorax*（Bredemeier et al.，2003；Kalscheuer et al.，2007），*Acinetobacter*（Wältermann and Steinbüchel，2006）和 *Marinobacter*（Rontani et al.，1999）种属被频繁报道可以产生 TAGs 和 Wes。

尽管大多数细菌种类的甘油三酯和蜡酯储存在细胞内，但在海洋细菌中脂质储存却不仅局限在细胞质中。一些作者已经描述了由 *Alcanivorax*，*Marinobacter* 和 *Acinetobacter*，菌株在十六烷存在的生长过程中产生的细胞外蜡酯（Bredemeier et al.，2003；DeWitt et al.，1982；Rontani et al.，1999）。这些化合物的重要性在可作为末端产物或前体来制备已被广泛认可的与生物技术相关的化学品。

OHCB 具有特异性代谢几乎所有碳氢化合物的能力，因此可以用来生产储存化合物，如像 PHAs，TAGs 和/或作为填充化学物质的 Wes，同时这些化合物的细胞外沉积也代表了纯化工艺的主要优势。脂质产品分泌至培养基中而不是在细胞内累积可以明显减少产品回收的成本。更重要的是，微生物生产的蜡酯有比其他生物资源更好的优点，就是蜡酯成分可以通过合理的选择起始物质和生长环境对其进行控制。

Alcanivorax jadensis T9 一个在德国北海岸发现的海洋革兰氏阴性细菌（Bruns and Berthe-Corti，1999），当在正烷烃作为碳源的环境中进行培养，其积累 TAG 和大量胞外的 WEs（Manilla-Perez et al.，2010）。这株菌被认为是利用生物技术进行胞外中性脂质生产最合适的候选者。

相似的特性在 *A. borkumensis* SK2 中被报道出来，这种菌同样在正烷烃的培养环境中生长，产生胞内和胞外的 TAGs 和 WEs（Kalscheuer et al.，2007）。尽管细菌储存化合物，例如 PHAs，TAGs，WEs 生物合成的酶基因编码在 *A. borkumensis* SK2 的基因组（Sabirova et al.，2008；Scheiker et al.，2006）中存在，但是似乎只有与碳源积累化合物的有关 TAGs 和 Wes 被合成（Kalscheuer et al.，2007）。不同碳源可以调控合成不同的储存产物。在丙酮酸或醋酸盐存在的培养条件下，*A. borkumensis* SK2 主要产生 TAGs，而在烷烃存在的培养条件下如十六烷，正十八烷等，则主要合成 WEs（Manilla-Perez et al.，2010）。特别是

TAG 的积累生产，据报道 *A. borkumensis* SK2 能够积累 TAG 达到细胞干重的 23%（Kalscheuer et al.，2007），是大规模生产 TAG 产品最好的候选者。

Marinobacter 和 *Acinetobacter* spp. 是人们所知的可以积累蜡酯为储存材料的。其中 *Marinobacter hydrocarbonoclasticus* SP17（Klein et al.，2008）和 *Acinetobacter* sp. M-1（Ishige et al.，2002）菌种在限制氮源条件下培养在正烷烃中能够积累大量的蜡酯。在限制氮源或磷的条件下 *Marinobacter hydrocarbonoclasticus* DSM 8798 已被报道培养在叶绿醇中能够产生类异戊二烯蜡酯储存化合物（Holtzapple and Schmidt-Dannert，2007）。

Rontani 等报道（1999）在法国地中海福斯湾不同地点收集的碳氢化合物污染的沿海沉积物和泡沫中分离出的 4 种菌，在有氧培养生长过程中，类异戊二烯蜡酯的形成也同时被发现。*Acinetobacter* sp. 菌种 PHY9，*Pseudomonas nautica*（IP85/617），*Marinobacter* sp. 菌种 CAB（DSMZ 11874），*Marinobacter hydrocarbonoclasticus*（ATCC 49840）培养在游离叶绿醇和 6，10，14-三甲基十五烷-2-酮中时合成大量胞外类异戊二烯蜡酯，培养产生的两种类异戊二烯化合物广泛分散于海洋沉积物中（Brassell et al.，1981；Volkman and Maxwell 1986）。

由于对脂质生物技术有较大的兴趣，已对与蜡酯和 TAGs 合成相关的一些酶进行了表征。低特异性的酶可以接受一个广泛的底物范围，这对体外新脂质的生物合成是非常有趣的。其中来源于 *Acinetobacter* sp. 菌种 ADP1 的蜡酯合成酶/乙酰辅酶（acyl-CoA）：甘油二酯酰基转移酶是研究最透彻的一种。这种双功能酶能够生物合成蜡酯和三酰甘油，证明其具有很低的底物特异性（Kalscheuer and Steinbuchel，2003），能够接受广泛的底物范围，如各种链长的饱和或不饱和脂肪醇，acyl-CoA 1，16-十六烷二醇，1-单棕榈甘油都可作为底物。对于 WS/DGAT 的酰基受体也包括烷基硫醇，例如 1-巯基十六烷，1，8-辛二硫醇和 1-S-单棕榈甘油辛硫醇，能够使不寻常的含硫蜡酯或硫代蜡酯的生物合成在体内体外都能进行。

14.4 对于工业可能有意义的烃降解微生物酶

14.4.1 烃降解微生物的鉴定

对于具有独特的生理学特征的海洋微生物的分类和研究可以成为探究生物技术酶的有力工具（Sanchez-Amat et al.，2010）。海洋生物催化所表现的不同寻常或者特异的生物进程与其生活环境有关，例如耐盐性、热稳定性、嗜压、冷适应性，这些特征可能是生物工程产业化所期待的（Trincone，2011）。

作为 OHCB 是相对近期被发现的，而且是一个具有新的生理生化特征，因此，它们被认为是一类至今未被研究，且在酶催化的精细化工产品和高附加值产品应用方面有巨大潜力的酶资源库（Yakimov et al.，2007）。

在此背景下，与这个现实有关值得我们重视的是，尽管不同降解石油烃的细菌已经被发现和表征，但大部分烃降解细菌都是未被发现的，主要的原因在于生长在海洋环境中的大量的微生物是不可培养的（Eilers et al.，2000）。然而，污染环境下非可培养微生物种群的表征和相关特定酶基因编码基因的获得，可以通过分子生物学的方法开展。

在公众数据库中测定的海洋与陆地原油降解细菌基因组的增长，以及关于微生物烃降解通路知识的积累，事实上使得我们可以相对容易地鉴定特定微生物的分子标志物。功能标记基因，编码关键特征通路酶，经常被用作特异微生物通路的标志物，去归属他们在特定环境中的功能和分离可以具有新应用的生物技术的酶。

就 PAH 的生物降解而言，最常见的标志基因（Habe and Omori，2003）是 ARHDs 催化的大亚基，该亚基负责其催化底物的特异性（Gibson and Parales，2000）。有一些独特的双氧酶基因也从海洋细菌中发现，这些细菌主要属于 *Cycloclasticus*（Geiselbrecht et al.，1998；Kasai et al.，2003），*Nocardioides*（Saito et al.，2000），或者 *Neptunomonas* 与 *Pseudoalteromonas*（Hedlund et al.，1999；Hedlund and Staley，2006）。

作为一个例子，从解环菌属（*Cycloclasticas* sp.）的 A5 中分离出了一整套系统表征过的 ARHD 编码基因，能够在取代萘、硫芴、菲和有或没有烷基取代的烷基芴的条件下生长。基因编码了硫铁双加氧酶，铁氧原蛋白和铁氧蛋白还原酶的 α（大亚基）和 β（小亚基）亚基，分别称为 *phnA*1，*phnA*2，*phnA*3，*phnA*4（Kasai et al.，2003）。大肠杆菌细胞拥有的 *phnA1A2A3A4* 基因能够转化菲、萘、甲基萘、氧芴和硫芴至羟化形式，这也证明了基于同源序列分析假设的功能。烷烃羟化酶的 *AlkB* 编码基因负责正烷烃降解的第一步，确实是烷烃降解菌最重要的标志基因（van Beilen and Funhoff，2007；Vomberg and Klinner，2000；Wang et al.，2010）。*alkB* 基因的生化和分子方面及其编码的酶已经有了相对较好的研究，这能够推动海洋环境中 *alkB* 基因作为分子工具的研究（Heiss-Blanquet et al.，2005；Kloos et al.，2006；Sei et al.，2003）。

14.4.2 烃降解海洋细菌生物催化氧化系统的应用

氧化系统不仅仅在烃降解微生物的鉴定方面有用途，它们的酶也是生物技术工业的一个主要热点。事实上，有机物的氧化或者羟基化反应对于多聚物和药物的合成具有重要价值。然而，选择具有氧化功能的有机物底物可能是有机合成中一个重要的问题，因为这些反应都需要强氧化剂，经常发生在几乎没有化学，局部和对映选择性的条件下（Notomista et al.，2011）。因此，发展生物转化已经成为近几年关注的重点，他们可以使用纯化后的酶或者整个细胞对工业上感兴趣的有机物分子的进行氧化。相对于已经建立的化学过程，这些方法是一个很有潜力的替代方法，可以在温和实验条件下以及没有有毒原料存在的情况下获得有活性有机化合物（Notomista et al.，2011）。值得注意的是，特别对于氧化作用，酶的存在可以避免使用含有卤素，或者通过使用氧气或干净可选择的氧化剂来满足环境上的要求（Hollmann et al.，2011）。

烃降解细菌在生物催化氧化碳氢化合物，如现成的石化产品（如：苯、二甲苯、辛烷）和可再生的植物油（如：柠檬烯、朱栾倍半萜）上具有重要意义，是最有价值的生物转化合成应用之一。对应的化学反应因为存在副反应，和由于产物产量低而导致的高产品回收成本而没有竞争力（Park，2007）。

由于其固有的立体性、区域选择性和高效性，氧化酶作为各种生物技术过程中潜在的生物催化剂已引起了关注。这些酶成功的商业应用可能通过采用新的方法，如在反应混合物中使用有机溶剂，固定化完整微生物或分离的酶制剂到各种支持介质上和基因工程技术（Sariaslani，1989）。完成这类反应的酶中最具代表性的种类是加氧/羟化酶、氧化物酶和漆酶（Di Gennaro et al.，2011）。这些酶完成的反应是通过转化或完全矿化的有机分子在维护全球碳循环中发挥了重要作用。

到目前为止，仅从海洋细菌中分离出来有限的应用于生物合成的氧化系统。仅有少许加氧酶和漆酶已经用于生物催化；这强烈提示应该鼓励未来海洋环境酶生物勘探活动。

在可能的生物催化底物中，最近许多同行已经致力于多环芳香族化合物的氧化，特别是萘。值得注意的是，在他们的分子结构中含有萘，使得其成为有生物活性的化学物质、药物和商业化学物质的候选。这些新颖的化合物可以用于进一步修饰，如：异戊烯化，含有一个或多个异戊烯基的天然产物已被证明具有抗菌、氧化、抗炎活性（Shindo et al.，2011）。通过海洋细菌 *Cycloclasticus* sp. 的 A5 的 PhnA1A2A3A4 中的芳香羟基双加氧酶、*Streptomyces* sp. 中的 CL190 或 SCO 7190 中的异戊烯转移酶 NphB、*Streptomyces Coelicolor* 中的二甲烯基丙基转移酶的联合反应，Shindo 和同事报道了 2-甲基萘、1-甲基氧化萘和 1-乙氧萘转化成 10 个异常的含异戊二烯基的萘-OLS。这些新颖的含异戊二烯基的萘-ols 都针对老鼠大脑匀浆模型显示出强大的抗氧化活性（Shindo et al.，2011）。

各种取代萘的生物转化也是由 Misawa 等（2011）完成的实验内容，使用的底物含有 1-甲氧基与 1-乙氧基-萘，甲基萘，二甲基萘和萘甲基羧酸酯。这些底物的生物转化利用重组大肠杆菌表达的 *Cycloclasticus* sp. 的 A5 的多环芳烃（PAH）-羟化双加氧酶。各种新颖的单羟基化衍生从这些取代萘中制备出来（Misawa et al.，2011）。

菲是被用于生物转化过程中的另一种多环芳香族化合物。海洋细菌中 *Nocardioides* sp. 的 KP7 的 phdABCD 基因簇，编码一个多组分酶—菲双加氧酶，是第一次在高拷贝数载体 pIJ6021 中被可诱导的硫链丝菌素启动子的下游 PtipA 操纵和定位，然后引入到革兰氏阳性、居于土壤中的、丝状细菌链霉菌属的 *Streptomyces lividans* 中（Chun et al.，2001）。重组美国 *S. lividans* 细胞将菲转换成环己二烯双醇形式，后来被鉴定为环-3，4-二羟基-3 4-二氢菲。虽然这个例子主要讨论重组链霉菌属菌株多环芳烃污染土壤的生物修复，但应该记得的是：环己二烯双醇被认为是不对称合成构建模块，这是非常有用的（Hudlicky et al.，1999）。因此，*Nocardioides* sp. KP7 的菲加双氧酶，在异源表达宿主上证明了其功能。基于其酶活性，它是未来生物催化发展的重要目标。

代谢多样性，以及今后生物技术的潜力，使这种酶已经在各种三环融合芳香族化合物如氟、氧芴，硫芴，咔唑，吖啶，菲啶中被测试。这些实验是利用大肠杆菌转化株细胞表达的来源于海洋细菌的 *Nocardioides* sp. 的 KP7 的菲双加氧酶（*phdABCD*）基因，转换所有这些三环芳香族化合物（Shindo et al. , 2001）。链霉菌属中携带菲双加氧酶基因的 *Streptomyces lividans* 转化株对各种三环融合芳香族化合物的生物转化做了评估。在转化中，这种放射菌类的能力类似于携带相应基因的大肠杆菌。利用含有这些重组细菌细胞转化的芳香族化合物的产物被纯化和鉴定。几个产物，包括从芴转换，4-羟基菊成，从菲啶转换来的顺-1 2-二羟基-1，2-二氢菲啶，顺-9，10-二羟基-9，10-二氢菲啶，10-二氢菲啶，这些都是新颖的化合物（Shindo et al. , 2001）。

正如在这一节前面强调的，是漆酶也是很重要的酶系统，可以用来做生物工程上有意义的底物的氧化。漆酶是蓝色含铜多功能氧化酶类，在环境和工业生物技术上有潜在应用功能。在 Fang 和同事的一项研究中，使用序列筛选策略从中国南海的海洋微生物基因组得到一个新的 1. 32 kb 的细菌漆酶基因（Fang et al. , 2011）。这个蛋白含有 439 个氨基酸（命名为 Lac15），包含 3 个保守的 Cu^{2+}结构域。但是与细菌所有的多功能氧化酶类有低于 40%的序列同源性。在大肠杆菌中重组表达 Lac15，对丁香醇连氮在 pH 6. 5~9 内有活性，最佳 pH 值是 7. 5，最高的活性温度是 45℃。Lac15 在 pH 值 5. 5~9 及温度 15~45℃时是稳定的。区别于真菌漆酶，Lac15 的活性在氯化物浓度低于 700 mmol/L 时会提高两倍，实际上在 1 000 mmol/L 氯化物中保持原水平。此外，Lac15 显示在碱性条件下脱色几个活性偶氮类工业染料的能力。依靠弱碱性活性的性能、高氯耐受性和染料脱色能力使新漆酶 Lac15 具有独特的工业应用（Fang et al. , 2011）。

参考文献

Abraham, W. R. , Meyer, H. and Yakimov, M. (1998) Novel glycine containing glucolipids from the alkane using bacterium Alcanivorax borkumensis. *Biochim Biophys Acta* 1393, 57-62.

Austin, R. N. , Buzzi, K. , Kim, E. , Zylstra, G. J. and Groves, J. T. (2003) Xylene monooxygenase, a membrane-spanning non-heme diiron enzyme that hydroxylates hydrocarbons via a substrate radical intermediate. *J Biol Ittorg Ghent* 8, 733-740.

Ballou, D. R, Entsch, B. and Cole, L. J. (2005) Dynamics involved in catalysis by single-component and two-component flavin-dependent aromatic hydroxylases. *Biochem Biophys Res Commun* 338, 590-598.

Banat, I. M. , Makkar, R. S. and Cameotra, S. S. (2000) Potential commercial applications of microbial surfactants. *Appl Microbiol Biotechnol* 53, 495-508.

Belsky, I. , Gutnick, D. L. and Rosenberg, E. (1979) Emulsifier of Arthrobacter RAG-1: determination of emulsifier-bound fatty acids. *FEBS Lett* 101, 175-178.

Bertrand, J. C. , Bonin, P. , Goutx, M. and Mille, G. (1993) Biosurfactant production by marine microorganisms: Potential application to fighting hydrocarbon marine pollution. *J Mar Biotechnol* 1, 125-129.

Bonin, L, Martins, B. M. , Purvanov, V. , Fetzner, S. , Huber, R. and Dobbek, H. (2004) Active site geometry and substrate recognition of the molybdenum hydroxylase quinoline 2-oxidoreductase. *Structure* 12, 1425–1435.

Boyd, D. R. and Bugg, T. D. (2006) Arene cis-dihydrodiol formation: from biology to application. *Org Biomol Ghent* 4, 181–192.

Brassell, S. C., Wardroper, A. M., Thomson, I. D., Maxwell, J. R. and EgJinton, G. (1981) Specific acyclic isoprenoids as biological markers of methanogenic bacteria in marine sediments. *Nature* 290, 693–696.

Bredemeier, R., Hulsch, R., Metzger, J. O. and Berthe-Corti, L. (2003) Submersed culture production of extracellular wax esters by the marine bacterium *Fundibacter jadensis*. *Mar Biotechnol (NY)* 5, 579–583.

Bruns, A. and Berthe-Corti, L. (1999) Fundibacter jadensis gen. nov., sp. nov., a new slightly halophilic bacterium, isolated from intertidal sediment. *Int J Syst Bacteriol* 49 *Pt* 2, 441–448.

Calvo, C., Martinez-Checa, F., Toledo, F. L., Porcel, J. and Quesada, E. (2002) Characteristics of bioemulsifiers synthesised in crude oil media by Halomonas eurihalina and their effectiveness in the isolation of bacteria able to grow in the presence of hydrocarbons. *Appl Microbiol Biotechnol* 60, 347–351.

Cameotra, S. S. and Makkar, R. S. (2004) Recent applications of biosurfactants as biological and immunological molecules. *Curr Opin Microbiol* 7, 262–266.

Cappello, S., Denaro, R., Genovese, M., Giuliano, L. and Yakimov, M. M. (2007) Predominant growth of Alcanivorax during experiments on 'oil spill bioremediation' in mesocosms. Microbiol Res 162, 185–190.

Chun, H. K., Ohnishi, Y., Misawa, N., Shindo, K., Hayashi, M., et al., (2001) Biotransformation of phenanthrene and 1-methoxynaphthalene with Streptomyces lividans cells expressing a marine bacterial phenanthrene dioxygenase gene cluster. *Biosci Biotechnol Biochem* 65, 1774–1781.

Chung, W. K. and King, G. M. (2001) Isolation, characterization, and polyaromatic hydrocarbon degradation potential of aerobic bacteria from marine macrofaunal burrow sediments and description of *Lutibacterium anuloederans* gen. nov., sp. nov., and *Cycloclasticus spirillensus* sp. *nov*. *Appl Environ Microbiol* 67, 5585–5592.

Desai, J. D. and Banat, I. M. (1997) Microbial production of surfactants and their commercial potential. *Microbiol Mol Biol Rev* 61, 47–64.

DeWitt, S., Ervin, J. L., Howes-Orchison, D., Dalietos, D., Neidleman, S. L. and Geigert, J. (1982) Saturated and unsaturated wax esters produced by *Acinetobacter* sp. HOl-N grown on C_{16}–C_{20} n-Alkanes. *J Am Oil Chem Soc* 59, 69–74.

Di Gennaro, P., Bargna, A. and Sello, G. (2011) Microbial enzymes for aromatic compound hydroxylation. *Appl Microbiol Biotechnol* 90, 1817–1827.

Do, L. H. and Lippard, S. J. (2011) Evolution of strategies to prepare synthetic mimics of carboxylate-bridged diiron protein active sites. *J Inorg Biochem* 105, 1774–1785.

Dyksterhouse, S. E., Gray, J. P., Herwig, R. P., Lara, J. C. and Staley, J. T. (1995) *Cycloclasticus pugetii* gen. nov., sp. nov., an aromatic hydrocarbon-degrading bacterium from marine sediments. *Int J Syst Bacteriol* 45, 116–123.

Eilers, H., Pernthalei; J., Glockner, F. O. and Amann, R. (2000) Culturability and *in situ* abundance of pelagic bacteria from the North Sea. *Appl Environ Microbiol* 66, 3044–3051.

Fang, Z., Li, T., Wang, Q., Zhang, X., Peng, H. (2011) A bacterial laccase from marine microbial metagenome exhibiting chloride tolerance and dye decolorization ability. *Appl Microbiol Biotechnol* 89, 1103–1110.

Fernandez-Martinez, J. , Pujalte, M. J. , Garcia-Martinez, J. , Mata, M. , Garay, E. and Rodriguez-Valeral, F. (2003) Description of *Alcanivorax venustensis* sp. nov. and reclassification of *Fundibacter jadensis* DSM 1 21 78T (Bruns and Berthe-Corti 1999) as *Alcanivorax jadensis* comb, nov. , members of the emended genus Alcanivorax. *Int J Syst Evol Microbiol* 53, 331-338.

Ferraro, D. J. , Gakhar, L. and Ramaswamy, S. (2005) Rieske business: structure-function of Rieske non-heme oxygenases. *Biochem Biophys Res Comtnun* 338, 175-190.

Fuentes, J. D. , Lerdau, M. , Atkinson, R. , Baldocchi, D. , Bottenheim, J. W. , et al. , (2000) Biogenic hydrocarbons in the atmospheric boundary layer: a review. *Bull Am Met Soc* 81, 1537-1576.

Gauthier, M. J. , Lafay, B. , Christen, R. , Fernandez, L. , Acquaviva, M. , et al. , (1992) *Marinobacter hydrocarbonoclasticus* gen. nov. , sp. nov. , a new, extremely halotolerant, hydrocarbon - degrading marine bacterium. *Int J Syst Bacteriol* 42, 568-576.

Geiselbrecht, A. D. , Hedlund, B. P. , Tichi, M. A. and Staley, J. T. (1998) Isolation of marine polycyclic aromatic hydrocarbon (PAH) -degrading Cycloclasticus strains from the Gulf of Mexico and comparison of their PAH degradation ability with that of puget sound Cycloclasticus strains. *Appl Environ Microbiol* 64, 4703-4710.

Gibson, D. T. and Parales, R. E. (2000) Aromatic hydrocarbon dioxygenases in environmental biotechnology. *Curr Opin Biotechnol* 11, 236-243.

Golebiowski, M. , Bogus, M. I. , Paszkiewicz, M. and Stepnowski, P. (2011) Cuticular lipids of insects as potential biofungicides: methods of lipid composition analysis. *Anal Bioanal Chem* 399, 3177-3191.

Golyshin, P. N. , Chernikova, T. N. , Abraham, W. R. , Lunsdorf, H. , Timmis, K. N. and Yakimov, M. M. (2002) *Oleiphilaceae fam.* nov. , to include Oleiphilus messinensis gen. nov. , sp. nov. , a novel marine bacterium that obligately utilizes hydrocarbons. *Int J Syst Evol Microbiol* 52, 901-911.

Golyshin, P. N. , Harayama, S. , Timmis, K. N. and Yakimov, M. M. (2005) Family II. Alcanivoraceae fam. nov. In *Bergey's Manual of Systematic Bacteriology. The Proteobacteria, part B, The Gammaproteobacteria.* Brenner, D. J. , Krieg, N. R. , Staley, J. T. and Garrity, G. M. (eds), 295-298. New York: Springer.

Golyshin, P. N. , Martins Dos Santos, V. A. , Kaiser, O. , Ferrer, M. , Sabirova, Y. S. , et al. , (2003) Genome sequence completed of Alcanivorax borkumensis, a hydrocarbon-degrading bacterium that plays a global role in oil removal from marine systems. *J Biotechnol* 106, 215-220.

Gutierrez, T. , Mulloy, B. , Black, K. and Green, D. H. (2007) Glycoprotein emulsifiers from two marine Halomonas species: chemical and physical characterization. *J Appl Microbiol* 103, 1716-1727.

Habe, H. and Omori, T. (2003) Genetics of polycyclic aromatic hydrocarbon metabolism in diverse aerobic bacteria. *Biosci Biotechnol Biochem* 67, 225-243.

Hara, A. , Syutsubo, K. and Harayama, S. (2003) Alcanivorax which prevails in oil-contaminated seawater exhibits broad substrate specificity for alkane degradation. *Environ Microbiol* 5, 746-753.

Harayama, S. , Kasai, Y. and Hara, A. (2004) Microbial communities in oil-contaminated seawater. *Curr Opitt Biotechnol* 15, 205-214.

Harayama, S. , Kishira, H. , Kasai, Y. and Shutsubo, K. (1999) Petroleum biodegradation in marine environments. *J Mol Microbiol Biotechnol* 1, 63-70.

Head, I. M. , Jones, D. M. and Roling, W. F. (2006) Marine microorganisms make a meal of oil. *Nat Rev Microbiol* 4, 173-182.

Hedlund, B. P., Geiselbrecht, A. D., Bair, T. J. and Staley, J. T. (1999) Polycyclic aromatic hydrocarbon degradation by a new marine bacterium, *Neptunomonas naphthovorans* gen. nov., sp. nov. *Appl Environ Microbiol* 65, 251-259.

Hedlund, B. P. and Staley, J. T. (2006) Isolation and characterization of Pseudoalteromonas strains with divergent polycyclic aromatic hydrocarbon catabolic properties. *Environ Microbiol* 8, 178-182.

Heiss-Blanquet, S., Benoit, Y., Marechaux, C. and Monot, F. (2005) Assessing the role of alkane hydroxylase genotypes in environmental samples by competitive PCR. *J Appl Microbiol* 99, 1392-1403.

Henry, J. A. (1998) Composition and toxicity of petroleum products and their additives. *Hum Exp Toxicol* 17, 111-123.

Hille, R. (2005) Molybdenum-containing hydroxylases. *Arch Biochem Biophys* 433, 107-116.

Hollmann, F., Arends, I. W. C. E., Buehler, K., Schallmey, A. and Biihler, B. (2011) Enzyme-mediated oxidations for the chemist. *Green Chem* 13, 226-265.

Holtzapple, E. and Schmidt-Dannert, C. (2007) Biosynthesis of isoprenoid wax ester in *Marinobacter hydrocarbonoclasticus* DSM 8798: identification and characterization of isoprenoid coenzyme A synthetase and wax ester synthases. *J Bacteriol* 189, 3804-3812.

Hudlicky, T., Gonzalez, D. and Gibson, D. T. (1999) Enzymatic dihydroxylation of aromatics in Enantioselective synthesis: expanding asymmetric methodology. *Aldrichim Acta* 32, 35-62.

Ishige, T., Tani, A., Takabe, K., Kawasaki, K., Sakai, Y. and Kato, N. (2002) Wax ester production from n-alkanes by Acinetobacter sp. strain M-1: ultrastructure of cellular inclusions and role of acyl coenzyme A reductase. *Appl Environ Microbiol* 68, 1192-1195.

Jindrova, E., Chocova, M., Demnerova, K. and Brenner, V. (2002) Bacterial aerobic degradation of benzene, toluene, ethylbenzene and xylene. *Folia Microbiol (Praha)* 47, 83-93.

Kalscheuer, R. and Steinbuchel, A. (2003) A novel bifunctional wax ester synthase/acyl-CoA: diacylglycerol acyltransferase mediates wax ester and triacylglycerol biosynthesis in Acinetobacter calcoaceticus ADP1. *J Biol Chem* 278, 8075-8082.

Kalscheuer, R., Stoveken, T., Malkus, U., Reichelt, R., Golyshin, P. N., et al., (2007) Analysis of storage lipid accumulation in *Alcanivorax borkumensis*: evidence for alternative triacylglycerol biosynthesis routes in bacteria. *J Bacteriol* 189, 918-928.

Kaplan, N. and Rosenberg, E. (1982) Exopolysaccharide distribution of and biocmulsifier production by *Acinetobacter calcoaceticus* BD4 and BD413. *Appl Environ Microbiol* 44, 1335-1341.

Kasai, Y., Kishira, H., Sasaki, T., Syutsubo, K., Watanabe, K. and Harayama, S. (2002a) Predominant growth of Alcanivorax strains in oil-contaminated and nutrient-supplemented sea water. *Environ Microbiol* 4, 141-147.

Kasai, Y., Kishira, H. and Harayama, S. (2002b) Bacteria belonging to the genus cycioclasticus play a primary role in the degradation of aromatic hydrocarbons released in a marine environment. *Appl Environ Microbiol* 68, 5625-5633.

Kasai, Y., Shindo, K., Harayama, S. and Misawa, N. (2003) Molecular characterization and substrate preference of a polycyclic aromatic hydrocarbon dioxygenase from Cycioclasticus sp. strain AS. *Appl Environ Microbiol* 69, 6688-6697.

King, R. W. (1988) Petroleum: its composition, analysis and processing. *Occup Med* 3, 409-430.

Klein, B., Grossi, V., Bouriat, P., Goulas, P. and Grimaud, R. (2008) Cytoplasmic wax ester accumulation during biofilm-driven substrate assimilation at the alkane-water interface by *Marinobacter hydrocarbonoclasticus* SP17. *Res Microbiol* 159, 137-144.

Kloos, K., Munch, J. C. and Schloter, M. (2006) A new method for the detection of alkane-monooxygenase homologous genes (alkB) in soils based on PCR-hybridization. *J Microbiol Methods* 66, 486-496.

Ladygina, N., Dedyukhina, E. G. and Vainshtein, M. B. (2006) A review on microbial synthesis of hydrocarbons. *Process Biochemistry* 41, 1001-1014.

Leisch, H., Morley, K. and Lau, P. C. (2011) Baeyer-Villiger monooxygenases: more than just green chemistry. *Chem Rev* 111, 4165-4222.

Lima, T. M., Procopio, L. C., Brandao, ED., Carvalho, A. M., Totola, M. R. and Borges, A. C. (2011) Biodegradability of bacterial surfactants. *Biodegradation* 22, 585-592.

Liu, C. and Shao, Z. (2005) *Alcanivorax dieselolei* sp. nov., a novel alkanedegrading bacterium isolated from sea water and deep-sea sediment, *Int J Syst Evol Microbiol* 55, 1181-1186.

Makkar, R. S. and Cameotra, S. S. (2002) An update on the use of unconventional substrates for biosurfactant production and their new applications. *Appl Microbiol Biotechnol* 58, 428-434.

Manilla-Perez, E., Reers, C., Baumgart, M., Hetzler, S., Reichelt, R., et al., (2010) Analysis of lipid export in hydrocarbonoclastic bacteria of the genus Alcanivorax: identification of lipid export-negative mutants of *Alcanivorax borkumensis* SK2 and *Alcanivorax jadensis T9*. *J Bacterial* 192, 643-656.

Martinez-Checa, E, Toledo, EL., Vilchez, R., Quesada, E. and Calvo, C. (2002) Yield production, chemical composition, and functional properties of emulsifier H28 synthesized by *Halomonas eurihalina* strain H-28 in media containing various hydrocarbons. *Appl Microbiol Biotechnol* 58, 358-363.

McGenity, T. J., Folwell, B. D., McKew, B. A. and Sanni, G. O. (2012) Marine crude-oil biodegradation: a central role for interspecies interactions. *Aquat Biosyst* 8, 10.

McKew, B. A., Coulon, F., Osborn, A. M., Timmis, K. N. and McGenity, T. J. (2007) Determining the identity and roles of oil-metabolizing marine bacteria from the Thames estuary, UK. *Environ Microbiol* 9, 165-176.

Melcher, R. J., Apitz, S. E. and Hemmingsen, B. B. (2002) Impact of irradiation and polycyclic aromatic hydrocarbon spiking on microbial populations in marine sediment for future aging and biodegradability studies. *Appl Environ Microbiol* 68, 2858-2868.

Misawa, N., Nodate, M., Otomatsu, T., Shimizu, K., Kaido, C., et al., (2011) Bioconversion of substituted naphthalenes and beta-eudesmol with the cytochrome P450 BM3 variant F87V. *Appl Microbiol Biotechnol* 90, 147-157.

Mooney, A., Ward, P. G. and O'Connor, K. E. (2006) Microbial degradation of styrene: biochemistry, molecular genetics, and perspectives for biotechnological applications. *Appl Microbiol Biotechnol* 72, 1-10.

Mukherjee, S., Das, P. and Sen, R. (2006) Towards commercial production of microbial surfactants. *Trends Biotechnol* 24, 509-515.

Mulligan, C. N. (2005) Environmental applications for biosurfactants. *Environ Pollut* 133, 183-198.

Niepceron, M., Portet-Koltalo, F., Merlin, C., Motelay-Massei, A., Barray, S. and Bodilis, J. (2010) Both Gycloclasticus spp. and Pseudomonas spp. as PAH-degrading bacteria in the Seine estuary (France). *FEMS*

Microbiol Ecol 71, 137–147.

Nolan, L. C. and O'Connor, K. E. (2008) Dioxygenase-and monooxygenase-catalysed synthesis of cis–dihydrodiols, catechols, epoxides and other oxygenated products. *Biotechnol Lett* 30, 1879–1891.

Notomista, E., Lahm, A., Di Donato, A. and Tramontano, A. (2003) Evolution of bacterial and archaeal multicomponent monooxygenases. *J Mol Evol* 56, 435–445.

Notomista, E., Scognamiglio, R., Troncone, L., Donadio, G., Pezzella, A., et al., (2011) Tuning the specificity of the recombinant multicomponent toluene o–xylene monooxygenase from Pseudomonas sp. strain OX1 for the biosynthesis of tyrosol from 2–phenylethanol. *Appl Environ Microbiol* 77, 5428–5437.

Park, J. B. (2007) Oxygenase-based whole-cell biocatalysis in organic synthesis. *J Microbiol Biotechnol* 17, 379–392.

Passeri, A., Lang, S., Wagner, F. and Wray, V. (1991) Marine biosurfactants, 11. Production and characterization of an anionic trehalose tetraester from the marine bacterium Arthrobacter sp. EK 1. Z *Naturforsch C* 46, 204–209.

Pepi, M., Cesaro, A., Liut, G. and Baldi, F. (2005) An antarctic psychrotrophic bacterium Halomonas sp. ANT–3b, growing on n–hexadecane, produces a new emulsyfying glycolipid. *FEMS Microbiol Ecol* 53, 157–166.

Phetrong, K., H–Kittikun, A. and Maneerat, S. (2008) Production and characterization of bioemulsifier from a marine bacterium, *Acinetobacter calcoaceticus* subsp. anitratus SM7. Songklanakarin. *J Sci Technol* 30, 297–305.

Reisfeld, A., Rosenberg, E. and Gutnick, D. (1972) Microbial degradation of I crude oil: factors affecting the dispersion in sea water by mixed and pure I-cultures. *Appl Microbiol* 24, 363–368.

Robertson, B. R., Button, D. K. and Koch, A. L. (1998) Determination of the biomasses of small bacteria at low concentrations in a mixture of species with forward light scatter measurements by flow cytometry. *Appl Environ Microbiol* 64, 3900–3909.

Rodrigues, L., Banat, I. M., Teixeira, J. and Oliveira, R. (2006) Biosurfactants: potential applications in medicine. *J Antimicrob Chemother* 57, 609–618.

Rojo, F. (2009) Degradation of alkanes by bacteria. *Environ Microbiol* 11, 2477–2490.

Roling, W. F., Milner, M. G., Jones, D. M., Lee, K., Daniel, F., et al. (2002) Robust hydrocarbon degradation and dynamics of bacterial communities during nutrient–enhanced oil spill bioremediation. *Appl Environ Microbiol* 68, 5537–5548.

Ron, E. Z. and Rosenberg, E. (2001) Natural roles of biosurfactants. *Environ Microbiol* 3, 229–236.

Ron, E. Z. and Rosenberg, E. (2002) Biosurfactants and oil bioremediation. *Curr Opin Biotechnol* 13, 249–252.

Rontani, J. F., Bonin, RC. and Volkman, J. K. (1999) Production of wax esters during aerobic growth of marine bacteria on isoprenoid compounds. *Appl Environ Microbiol* 65, 221–230.

Rosenberg, E. (1992) The hydrocarbon–oxidizing bacteria. In *The Prokaryotes. A Handbook on the Biology of Bacteria: Ecophysiology, Isolation, Identification, Applications*, Balows, A., Truper, H. G., Dworkin, M., Hardee, W. and Schleifer, K. H. (eds), 446–459.

Rosenberg, E., Perry, A., Gibson, D. T. and Gutnick, D. L. (1979) Emulsifier of Arthrobacter RAG–1: specificity of hydrocarbon substrate. *Appl Environ Microbiol* 37, 409–413.

Rosenberg, E. and Ron, E. Z. (1998) 'Surface active polymers of Acinetobacter'. In *Biopolymers from Renewable Sources*, Kaplan, D. (eds), 281-291.

Rosenberg, E. and Ron, E. Z. (1999) High-and low-molecular-mass microbial surfactants. *Appl Microbiol Biotechnol* 52, 154-162.

Rosenberg, E., Zuckerberg, A., Rubinovitz, C. and Gutnick, D. L. (1979) Emulsifier of Arthrobacter RAG-1: isolation and emulsifying properties. *Appl Environ Microbiol* 37, 402-408.

Sabirova, J. S., Chernikova, T. N., Timmis, K. N. and Golyshin, P. N. (2008) Niche-specificity factors of a marine oil-degrading bacterium *Alcanivorax borkumensis* SK2. *FEMS Microbiol Lett* 285, 89-96.

Saito, A., Iwabuchi, T. and Harayama, S. (2000) A novel phenanthrene dioxygenase from Nocardioides sp. Strain KP7: expression in *Escherichia coli*. *J Bacteriol* 182, 2134-2141.

Sanchez-Amat, A., Solano, F. and Lucas-Elio, P. (2010) Finding new enzymes from bacterial physiology: a successful approach illustrated by the detection of novel oxidases in *Marinomonas mediterranea*. *Mar Drugs* 8, 519-541.

Sariaslani, F. S. (1989) Microbial enzymes for oxidation of organic molecules. *Crit Rev Biotechnol* 9, 171-257.

Satpute, S. K., Banat, I. M., Dhakephalkar, P. K., Banpurkar, A. G. and Chopade, B. A. (2010) Biosurfactants, bioemulsifiers and exopolysaccharides from marine microorganisms. *Biotechnol Adv* 28, 436-450.

Schneiker, S., Martins dos Santos, V. A., Bartels, D., Bekel, T., Brecht, M., et al. (2006) Genome sequence of the ubiquitous hydrocarbon-degrading marine bacterium *Alcanivorax borkumettsis*. *Nat Biotechnol* 24, 997-1004.

Schulz, D., Passeri, A., Schmidt, M., Lang, S., Wagner, F., et al., (1991) Marine biosurfactants, I. Screening for biosurfactants among crude oil degrading marine microorganisms from the North Sea. Z *Naturforsch C* 46, 197-203.

Sei, K., Sugimoto, Y., Mori, K., Maki, H. and Kohno, T. (2003) Monitoring of alkane-degrading bacteria in a sea-water microcosm during crude oil degradation by polymerase chain reaction based on alkane-catabolic genes. *Environ Microbiol* 5, 517-522.

Seo, J. S., Keum, Y. S. and Li, Q. X. (2009) Bacterial degradation of aromatic compounds, *Int J Environ Res Public Health* 6, 278-309.

Shindo, K., Ohnishi, Y., Chun, H. K., Takahashi, H., Hayashi, M., et al., (2001) Oxygenation reactions of various tricyclic fused aromatic compounds using *Escherichia coli* and *Streptomyces lividans* transformants carrying several arene dioxygenase genes. *Biosci Biotechnol Biochem* 65, 2472-2481.

Shindo, K., Tachibana, A., Tanaka, A., Toba, S., Yuki, E., et al., (2011) Production of novel antioxidative prenyl naphthalen-ols by combinational bioconversion with dioxygenase PhnAlA2A3A4 and prenyltransferase NphB or SCO7190. *Biosci Biotechnol Biochem* 75, 505-510.

Singh, P. and Cameotra, S. S. (2004) Potential applications of microbial surfactants in biomedical sciences. *Trends Biotechnol* 22, 142-146.

Slizovskiy, I. B., Kelsey, J. W. and Hatzinger, P. B. (2011) Surfactant-facilitated remediation of metal-contaminated soils: efficacy and toxicological consequences to earthworms. *Environ Toxicol Chem* 30, 112-123.

Syutsubo, K., Kishira, H. and Harayama, S. (2001) Development of specific oligonucleotide probes for the identification and in situ detection of hydrocarbon-degrading Alcanivorax strains. *Environ Microbiol* 3, 371-379.

Taylor, S. W., Sherwood Lollar, B. and Wassenaar, L. I. (2000) Bacteriogenic ethane in near-surface aquifers: implications for leaking hydrocarbon well bores. *Environ Sci Technol* 34, 4727-4732.

Teramoto, M., Suzuki, M., Okazaki, F., Hatmanti, A. and Harayama, S. (2009) Oceanobacter-related bacteria are important for the degradation of petroleum aliphatic hydrocarbons in the tropical marine environment. *Microbiology* 155, 3362-3370.

Tinberg, C. E. and Lippard, S. J. (2011) Dioxygen activation in soluble methane monooxygenase. *Acc Chem Res* 44, 280-288.

Trincone, A. (2011) Marine biocatalysts: enzymatic features and applications. *Mar Drugs* 9, 478-499.

Vaillancourt, F. H., Bolin, J. T. and Eltis, L. D. (2006) The ins and outs of ring-cleaving dioxygenases. *Crit Rev Biochem Mol Biol* 41, 241-267.

van Beilen, J. B. and Funhoff, E. G. (2007) Alkane hydroxylases involved in microbial alkane degradation. *Appl Microbiol Biotechnol* 74, 13-21.

Van Hamme, J. D., Singh, A. and Ward, O. P. (2003) Recent advances in petroleum Hmicrobiology. *Microbiol Mol Biol Rev* 67, 503-549.

van Hylckama Vlieg, J. E. and Janssen, D. B. (2001) Formation and detoxification Kof reactive intermediates in the metabolism of chlorinated ethenes. *J Biotechnol* 85, 81-102.

Vergnoux, A., Malleret, L., Asia, L., Doumenq, P. and Theraulaz, F. (2011) Impact of forest fires on PAH level and distribution in soils. *Environ Res* 111, 193-198.

Volkman, J. K. and Maxwell, J. R. (1986) 'Acyclic isoprenoids as biological markers'. In *Biological Markers in the Sedimentary Record*, Johns, R. B. (ed) 1-46. Amsterdam: Elsevier.

Vomberg, A. and Klinner, U. (2000) Distribution of alkB genes within n-alkane-degrading bacteria. *J Appl Microbiol* 89, 339-348.

Wältermann, M. and Steinbiichel, A. (2006) 'Wax ester and triacylglycerol inclusions'. In *Inclusions in Prokaryotes*, Shively, J. M. (ed) 137-166. Heidelberg: Springer-Verlag.

Wang, W., Wang, L. and Shao, Z. (2010) Diversity and abundance of oil-degrading bacteria and alkane hydroxylase (alkB) genes in the subtropical seawater of Xiamen Island. *Microb Ecol* 60, 429-439.

Weiner, R. M. (1997) Biopolymers from marine prokaryotes. *Trends Biotechnol* 15, 390-394.

Wentzel, A., Ellingsen, T. E., Kotlar, H. K., Zotchev, S. B. and Throne-Hoist, M. (2007) Bacterial metabolism of long-chain n-alkanes. *Appl Microbiol Biotechnol* 76, 1209-1221.

Wu, Y., Lai, Q., Zhou, Z., Qiao, N., Liu, C. and Shao, Z. (2009) *Alcanivorax hongdengensis* sp. nov., an alkane-degrading bacterium isolated from surface seawater of the straits of Malacca and Singapore, producing a lipopeptide as its biosurfactant. *Int J Syst Evol Microbiol* 59, 1474-1479.

Yakimov, M. M., Giuliano, L., Denaro, R., Crisafi, E., Chernikova, T. N., et al., (2004) *Thalassolituus oleivorans* gen. nov., sp. nov., a novel marine bacterium that obligately utilizes hydrocarbons. *Int J Syst Evol Microbiol* 54, 141-148.

Yakimov, M. M., Giuliano, L., Gentile, G., Crisafi, E., Chernikova, T. N., et al., (2003) *Oleispira antarctica* gen. nov., sp. nov., a novel hydrocarbonoclastic marine bacterium isolated from Antarctic coastal sea water. *Int J Syst Evol Microbiol* 53, 779-785.

Yakimov, M. M., Golyshin, P. N., Lang, S., Moore, E. R., Abraham, W. R., et al., (1998) *Alcanivorax borkumensis* gen. nov., sp. nov., a new, hydrocarbon-degrading and surfactant-producing marine bacterium. *Int J Syst Bacteriol* 48 *Pt* 2, 339-348.

Yakimov, M. M., Timmis, K. N. and Golyshin, P. N. (2007) Obligate oil-degrading marine bacteria. *Curr Opin Biotechnol* 18, 257-266.

Zajic, J. E., Guignard, H. and Gerson, D. E (1977) Properties and biodegradation of a bioemulsifier from *Corynebacterium hydrocarboclastus*. *Biotechnol Bioeng* 19, 1303-1320.

Zhang, J., Gorkovenko, A., Gross, R. A., Allen, A. L. and Kaplan, D. (1997) Incorporation of 2-hydroxyl fatty acids by *Acinetobacter calcoaceticus* RAG-1 to tailor emulsan structure. *Int J Biol Macromol* 20, 9-21.

Zuckerberg, A., Diver, A., Peeri, Z., Gutnick, D. L. and Rosenberg, E. (1979) Emulsifier of Arthrobacter RAG-1: chemical and physical properties. *Appl Environ Microbiol* 37, 414-420.

第 15 章　海洋真菌来源的木质素水解酶：制备与应用

Lara Durães Sette, São Paulo State University, Brazil, and Rafaella Costa Bonugli Santos, University of Campinas, Brazil

DOI：10.1533/9781908818355.3.403

摘要：由于微生物产生的用于工业生产的酶可以参与多个过程，因此对其需求快速增长。木质素水解酶在全球碳循环中具有重要作用。当前最重要的应用是对已污染环境的修复，主要应用于微生物系统降解污染物或脱色。考虑到污染大多发生于高盐环境，而且海洋微生物能适应海洋生态系统，因此海洋真菌来源的酶可能具有潜在的应用价值。

关键词：海洋真菌，木质素水解酶，生物技术应用，生物降解，生物修复

15.1　引言

15.1.1　木质素水解酶

木质素水解酶属于氧化酶和过氧化物酶类。氧化酶是一类能够进行氧化还原反应，以氧分子(O_2)为电子受体的酶；在此反应中氧原子被还原形成 H_2O 或 H_2O_2。过氧化物酶是将过氧化氢的氧原子转化到合适的底物并氧化底物的一类酶。此类酶定义为木质素水解酶是因为其具有水解木质素的能力。“Lignin” 来源与拉丁文木材 “Lignum”（Wong，2009）。木质素是一种复杂的芳香族聚合物，是维管植物细胞壁特有的，化学和生物的方法很难降解（Ruiz-Dueñas and Martínez，2009）。木质素的降解依赖木质素水解酶。木质素水解酶或木质素降解酶参与不同的氧化还原反应，打开木质素芳香族结构单体间之间的化学键（Kuhad et al.，1997）。表 15.1 和图 15.1 总结了木质素降解的关键酶、酶的底物以及反应原理。在木质素生物降解的过程中，木质素首先和木质素水解酶相互作用，然后结构发生变化以达到最佳状态，然后导致自由基的形成，从而打断木质素中的各种化学键（Chen et al.，2011）。

表 15.1　木质素水解酶及主要反应

酶及缩写	辅助因子	底物	反应
木质素过氧化物酶，LiP	H_2O_2	藜芦醇	芳香环氧化为阳离子自由基
锰过氧化物酶，MnP	H_2O_2	锰，有机酸如螯合剂、硫醇、不饱和脂肪酸	Mn（Ⅱ）氧化为 Mn（Ⅲ）；螯合的 Mn（Ⅲ）氧化酚类化合物为苯氧自由基；氧化其他化合物
多功能过氧化物酶，VP	H_2O_2	锰，藜芦醇，与 LiP 和 MnP 类似的化合物	Mn（Ⅱ）氧化为 Mn（Ⅲ）；氧化酚类化合物和非酚类化合物以及染料
漆酶，Lac		酚类化合物，介质如 ABTS	氧化酚类化合物为苯氧自由基；其他介质的反应
乙二醛氧化酶，GLOX		乙二醛，甲基乙二醛	乙二醛氧化为乙二酸；产生 H_2O_2
芳基乙醇氧化酶，AAO		芳香醇（甲氧苯基，藜芦醇）	芳香醇氧化为醛；产生 H_2O_2
其他产生 H_2O_2的酶		多种有机化合物	O_2生成 H_2O_2

资料来源：根据 Hatakka 2001 修改。

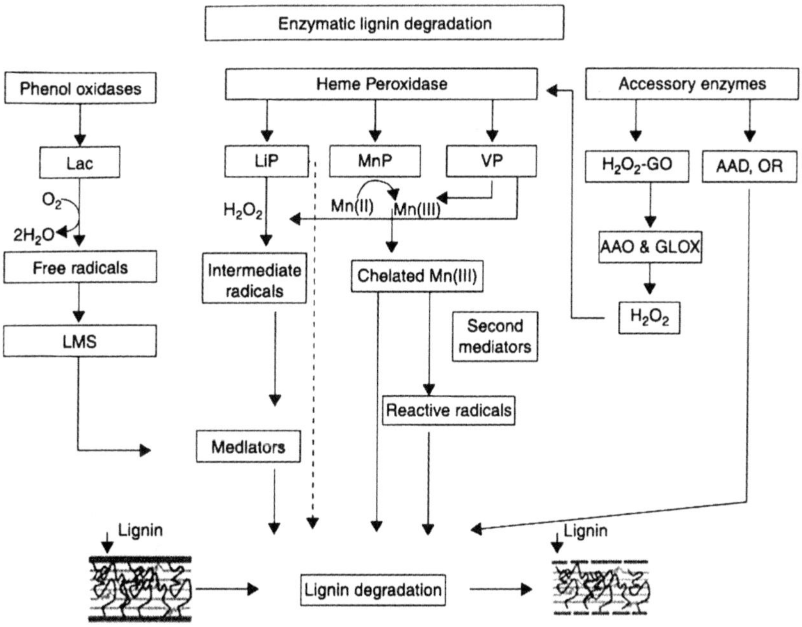

图 15.1　木质素水解原理

资料来源：根据 Dashtban et al.，2010 修改 .

最主要水解木质素的过氧化物酶包括木质素过氧化物酶（LiP）、锰过氧化物酶（MnP）和多功能过氧化物酶（VP）。过氧化物酶通常分泌到细胞外或存在于酶体中，广泛存在于生物体的各器官。大部分的过氧化物酶为血红素过氧化物酶，其活性位点有铁卟啉 IX（Hofrichter et al.，2010）。根据氨基酸序列、来源宿主和生理功能，非动物过氧化物酶超家族主要包括三类：Ⅰ类包括细胞内非动物过氧化物酶；Ⅱ类包括真菌分泌的过氧化物酶；Ⅲ类包括植物分泌的过氧化物酶（Hofrichter et al.，2010）。Ⅱ类过氧化物酶是由真菌分泌的血红素过氧化物酶，包括木质素修饰的过氧化物酶（LiP、MnP 和 VP）。LiP（EC 1.11.1.14；H_2O_2氧化还原酶）和 MnP（EC 1.11.1.13）是在 25 年前首先被发现和描述，其在木质素水解酶模型中已被深入研究（Tien and Kirk，1984）。LiPs 是含血红素的一类糖蛋白，催化依赖 H_2O_2的非酚基木质素复合物（二芳基丙烷）水解，包括β-O-4 非酚基木质素水解模型及多种酚基（如愈创木酚、4-羟基-3，5-二甲氧基苯甲醇）（Wong，2009）。LiPs 催化氧化底物包括多步电子转移，形成自由基中间产物，包括含酚基自由基和藜芦醇自由基的阳离子。与 MnP 不同，LiP 能彻底氧化非酚基芳香族底物，而且不需要其他辅基的参与（图 15.1），因为其具有高氧化还原能力。典型的 MnP（EC 1.11.1.13；Mn（Ⅱ）：H_2O_2氧化还原酶）比 LiPs 和 VPs 多 20~30 个氨基酸残基，在 C-末端 Mn^{2+}结合位点附近包含一个第五半胱氨酸形成的二硫键（Sundaramoorthy et al.，1994）。MnPs 是位于细胞外的糖蛋白，具有多种同工酶，这些同工酶中的辅基为含有铁卟啉 IX 的血红素（Asgher et al.，2008）。MnP 是专一的利用 Mn（Ⅱ）还原底物，其原理与 LiP 类似。MnPs 将酚基氧化为苯氧基，其机制是通过 H_2O_2将 Mn（Ⅱ）氧化为 Mn（Ⅲ），有机酸（在自然界中为草酸和苹果酸）螯合 Mn（Ⅲ），螯合后的 Mn（Ⅲ）将木质素的酚基氧化为苯甲基自由基，同时自身降解（Hofrichter 2002）。对于非酚基底物，在第二个介质存在的情况下，Mn（Ⅲ）的氧化涉及有反应活性的自由基的产生（Reddy et al.，2003）。

第三种水解木质素及具有高氧化还原潜能的Ⅱ类过氧化物酶是一种 LiP 类似的复合锰过氧化物酶，称为多功能过氧化物酶（VP，EC 1.11.1.16，活性黑 5:H_2O_2氧化还原酶；表 15.1），首先在植物根和树干的白腐菌 *Pleurotus eryngii*（Hofrichter et al.，2010）中被发现。与传统的 MnP 不同，典型的 VP 可以不依赖 Mn^{2+}氧化简单的胺类化合物和单体酚（Ruiz-Dueñas，Martínez，2009；Hofrichter et al.，2010），而且在氧化酚状态时比 LiP 更稳定（Hofrichter et al.，2010）。与 MnP 的机制类似，Mn（Ⅲ）从 VPs 上释放出来，并作为氧化剂氧化木质素酚基和未形成交联的苯氧基，血红素位于蛋白质内部，通过图 15.1 所展示的由 MnP 和 LiP 形成的通道与外界交流（Dashtban et al.，2010）。

除了 LiP、MnP 和 VP，过氧化物酶还包括超过 16 种酶（Hofrichter et al.，2010）。在这些酶中，最新发现的两种血红素过氧化物酶：芳香族过氧化物酶（APOs）和 DyP-type 过氧化物酶（DyPs），这两种酶均具有明显的酶活力。然而，关于这两种酶降解木质素的潜能研究还仍然处于初期阶段。Sugano（2009）报道，DyPs 只氧化非酚基底物，例如，二聚体木质素和蒽醌衍生物，尽管这些反应与木质素的生物降解相关，但是 APOs 和 DyPs

的具体功能仍不清楚（Hofrichter et al.，2010）。因此有关在真菌定殖过程中，利用天然木质素作为底物条件下酶的产生仍有待研究。

Hofrichter 与合作者（2010）在生存于木材和草本植物中的黑杨树菇 *Agrocybe aegerita*（也叫 *A. cylindracea*）中发现了另一种含有血红素-硫醇基的过氧化物酶，这种酶后来证明是一种环化酶，能够有效地把过氧化物的氧原子转移给多种底物。由于其具有通过还原过氧化氢使芳香族化合物环氧化和羟基化，接着通过一系列过氧化物酶转移氧（O_2）来氧化芳香族底物，这些酶现在大多数归类为 APOs。尽管现在没有光谱法检测到 APO 中间活性产物的存在，但是推测其催化机理可能是通过血红素过氧化物酶各个组分（例如 CiP，CfuCPO）和 P450 单氧酶的过氧化氢“分流”途径组成循环氧化底物。

DyP 是在染料脱色研究（Kim et al.，1995）中发现，DyPs 显示出对蒽醌染料比含氮染料更高的脱色活性以及对含酚基化合物（如 2，6-二甲氧基苯酚，愈创木酚和藜芦醇）不同的降解谱。这些属性揭示 DyP 可能是一种新型的过氧化物酶，与其他过氧化物酶不同（Dunford，1999；Sugano，2009）并具有独特的结构。DyP 特有的结构决定了 Dyp-类型过氧化物酶的催化机制。特别地，DyP 的 H_2O_2 结合位点（即酸碱催化剂）为 Asp，然而其他血红素过氧化物酶 H_2O_2 结合位点为 His（Sugano，2009）。

漆酶（EC 1.10.3.2，苯二酚：氧原子氧化还原酶）属于多铜离子酶家族，与过氧化物酶共同参与木质素的生物合成（Wong，2009）。真菌漆酶通常具有多种同工酶，包括各种单体酶和多亚基结构，这些酶具有相似的结构，包括 3 个按顺序排列的结构域组成的一个 β-筒状结构，这种结构与小的蓝铜蛋白有关（Wong，2009）。漆酶可以结合 4 个铜离子，分别称为 Cu T1（与还原性底物结合）以及三铜离子簇 T2/T3（电子从Ⅰ铜转移到Ⅱ型和Ⅲ型铜离子组成三环簇/ 还原三环还原性氧化产生水）（Gianfreda et al.，1999）。

漆酶催化氧化底物 4 个 $1e^-$ 并伴随着 2 个还原 $2e^-$，将氧分子还原形成水［$4RH + O_2 4R + 2H_2O$］（Wong，2009）。漆酶氧化底物时形成自由基。木质素的降解需要苯氧自由基去氧化 α-碳或者切断 α-碳和 β-碳之间的共价键。这种氧化反应导致了氧化中心的自由基，此反应可以被其他酶催化反应生成苯醌。苯醌可以和自由基聚合（Thurston，1994）。漆酶参与的催化反应可以通过插入其他中间物而扩展到非酚基底物的反应。这些中间物为低分子量可以被漆酶氧化的有机复合物。高活性的阳离子自由基氧化漆酶不能单独氧化的非酚基化合物，通常产生的中间物为 1-羟基苯并三唑（HOBT）、N-羟基邻苯二甲酰亚胺（NHPI）、2′-联氨-双-3-乙基苯并噻唑啉-6-磺酸（ABTS）及 3-羟基邻氨基苯甲酸。漆酶氧化 ABTS 的速度高于 HOBT（Wong，2009）。

木质素水解酶可以用于降解多种化合物及其他生物反应，在真菌中具有多种作用。漆酶在增加孢子抗性、根状体形成、子实体形成、色素合成及木质素降解中发挥重要作用（Claus，2004；Minussi et al.，2007）。LiPs 主要应用于木质素生物降解和抵抗真菌病原体（Piontek et al.，2001；Trejo-Hernandez et al.，2001）。除了降解木质素，MnPs 在真菌菌体直接相互作用中发挥了重要作用（Trejo-Hernandez et al.，2001）。Sugano（2009）合理地推测真菌通过 DyPs 的降解作用来保护真菌的菌株避免一些有毒物质的损伤，包括抗生素。APOs 在自然界中的作用仍有待研究；其多变性可使其参与非特异的氧化反应降解植物成分或细菌代谢产物，并转化木质素来源的化合物（Hofrichter et al.，2010）。

15.1.2 降解木质素的真菌

根据降解方式和自然条件下的降解作用，降解木质素的真菌主要有三类：褐腐菌、软腐菌和白腐菌（表 15.2）。白腐菌降解木质素的效率最高，产物为一种白色纤维，因此它们被认为是最好的木质素水解酶产生菌。不同的白腐菌具有不同的木质素水解酶系统，不同的白腐菌具有一种甚至多种木质素水解酶（Arora and Shamar，2010）。然而，通过 Tuomela 等（2000）的研究说明其他种类真菌来源的木质素水解酶研究很少，例如，存在于海洋生态系统中的子囊菌类——软腐菌，它可以在高湿度的情况下降解木质素，该菌主要存在水环境中，如海洋系统中。

表 15.2 降解木质素的微生物

微生物	门	木质素降解	所处环境	属
白腐菌	担子菌门（子囊菌门）	木质素矿化，选择或非选择性降解	主要在硬木	*Phanerochaete* *Phlebia* *Trametes*
褐腐菌	担子菌门	木质素修饰	主要在软木	*Poria Polyporus*
软腐菌	子囊菌门或变形菌门	限制性的木质素降解	水环境	*Chaetomium* *Paecilomyces* *Fusarium*

资料来源：Tuomela et al.，2000.

15.1.3 源于海洋真菌的木质素水解酶

地球表面超过 70%被海洋覆盖，因此海洋可以看做是自然资源的宝库。然而目前对海洋生物多样性，特别是对微生物种类仍了解较少（Menezes et al.，2010）。海洋真菌微生物领域的研究历史较短，目前根据生长和繁殖的方式将海洋真菌分为两类（Kohlmeyer and Kohlmeyer，1979），其中专性海洋菌只能生长于海洋或河流入海口，兼性海洋菌来自淡水或陆地环境，在海洋环境中也可以生存（可能产生孢子）。大多数来自海洋的真菌不是专性就是兼性海洋菌。因此，更多常见的海洋来源真菌的表述方式被采用（Kohlmeyer and Kohlmeyer，1979）。海洋环境中发现的具有代表性的有 Basidiomycota、Ascomycota、Zygomycota、anamorphic fungi 和 yeast（Surajit et al.，2006；da Silva et al.，2008；Menezes et al.，2010）。

Sutherland 与合作者（1982）是最早的研究者，他们通过 ^{14}C 示踪木质素底物的方法检测到海洋真菌降解木质素水解生成 CO_2 和水溶性产物。所有研究过的海洋真菌由于具有较强的水解硬木和软木中木质素的能力而被划为木质素水解菌。

多种专性海洋真菌能够产生木质素水解酶，其中以漆酶为主（Pointing et al.，1999）。近期，分布在红树、海草、树叶及海底沉积物中的真菌成为发现木质素水解酶的潜在资源（D'Souza et al.，2006，2009）。红树和水草含有 50%的木质素材料作为结构聚合物，这些植物是海岸海洋环境木质纤维素底物的主要来源。生长在此的专性和兼性海洋真菌能产生细胞壁水解酶（如木质素水解酶）分解底物，将产生可溶性有机物（DOC）和水溶性有机物颗粒（POC）分解（Raghukumar，2008）。此外，从海洋无脊椎动物中分离到的真菌会产生非常有意义的次级代谢产物，它们具有细胞毒性、抗菌、抗真菌、抗病毒及抗炎活性。但是从海洋生物（如海绵和腔肠动物）中分离得到的真菌产生的酶研究很少。多种大型动物是滤食性动物，生活在含有污染物的水中，富集来自浮游植物或其他悬浮杂质。因此，有理由相信这些大型生物体内的真菌能产生水解酶来转化这些营养物质（Wang，2006）。考虑到这些情况，Bonugli-Santos 与合作者（2010a）报道了高盐条件下生长的两种子囊菌和一种变形菌（来自巴西的一种刺细胞动物）具有水解木质素的能力。此外，在海洋高盐和无盐环境下来源于海绵的担子菌产生的 MnP 和漆酶具有显著的水解木质素活性（Bonugli-Santos et al.，2010b，2012）。有趣的是，在大多数水解木质素的海洋真菌的相关研究中，子囊菌比孢子菌和白腐菌更有优势。水生子囊菌的优势在于应用培养技术时其更易被发现。并且孢子的存在使其更适合水生生态系统，这有利于促进其漂浮和附着（Prasannarai and Sridhar，2001；Vijaykrishna et al.，2006）。

15.1.4 海洋真菌产生的木质素水解酶对工业和生物技术的重要性

木质素水解酶最重要的在生物工程领域应用的特点是其具有低的底物特异性，因为木质素具有多种结构，以及从合适的底物产生自由基的能力。因此，这些酶在不同工业生产需要的酶处理技术和生物技术的应用中具有重要作用，包括化学、燃料、食品、农业、造纸、纺织化妆品等（表 15.3）。

表 15.3　木质素分解酶的工业与生物技术应用

酶	工业和生物技术应用				
	食品工业	纸浆和造纸业	纺织品工业	生物修复	有机合成，药业，制药，化妆品，纳米技术应用
漆酶	食品和饮料含有的酚类 抗坏血酸测定 糖甜菜果胶凝胶化	木质素解聚物 木纸浆的脱木素 牛皮纸纸浆的漂白	纺织染料的降解和漂白	异生素的生物降解 多环芳香烃（PAHs）的降解	苯酚和类固醇的多聚产物，医疗物质 C-N 键的组成，复杂天然产物合成，个人卫生产品 生物传感器和生物载体

续表

酶	工业和生物技术应用				
	食品工业	纸浆和造纸业	纺织品工业	生物修复	有机合成，药业，制药，化妆品，纳米技术应用
LiP	天然芳香烃的来源， 香草醛的产物	牛皮纸纸浆磨流出物的脱色	纺织染料的降解	偶氮，杂环，反应和聚合染料的降解， 环境污染物的矿化， 异生素和农药的降解	
MnP	天然芳香香料的产物	牛皮纸纸浆的漂白	纺织染料的降解	PAH 的降解合成染料， 造纸用料 DDT、PCB、TNT 的漂白	

资料来源：Maciel et al.，2010.

陆生真菌产生的木质素水解酶在生物技术应用领域已有广泛研究。然而，近期分离的非陆生真菌产酶效率高于陆生真菌，这引起了对不同生理生态和分类真菌的广泛关注。Raghukumar（2008）的研究表明，在海洋环境中生物技术应用有两个问题值得考虑：①对生态系统的充分了解有助于寻找新的基因；②生物技术产品受到海洋环境中特殊环境的影响。

海洋环境中真菌产生的木质素水解酶在生物技术中具有重要作用，因其具有高稳定性、耐碱性环境、低水势以及高盐浓度、低温、低营养条件和静水压力。海水的盐度为 33~35，而淡水盐度约为 0.5。低水势是海水存在的问题。生活在海水中的生物需要保持细胞内的水势低于海水，以保证水的摄入，为此，大多数海洋真菌通过积累盐离子浓度梯度来实现。海水中高钠离子浓度也使海洋真菌具有适应海水生活的独特特性。钠离子，即使是低浓度钠离子将可以对大多数陆地及淡水环境中的细胞产生毒性（Raghukumar，2008）。

15.2 海洋真菌产生的木质素水解酶

15.2.1 真菌分离

海洋真菌主要从支持它们生长的介质中分离得到。海洋环境中木质素水解酶最重要的底物是红木，通过寻找长在树木上的海洋真菌（生长在死亡的树干、树叶、树枝或碎片的真菌）来发现。此外，Hyde 和 Jones（1988）证实红树真菌构成了海洋真菌的第二大生态系统，这些真菌在古老和新生的红树（分布于大西洋、太平洋和印度洋）中广泛存在。其他可以分离降解木质素真菌的介质包括海藻、海草、高盐沼泽及其他红树属植物（Jones，2011）。

子囊菌是红树上数量最多、分类最广的真菌。担子菌在陆地上占主导地位，但海洋中含量很少（Jones，2011）。Pointing 和 Hyde（2000）指出大多数腐生海洋真菌为软腐菌，以及少量白腐菌。然而，Verma 等（2010）报道了从印度红树沼泽的碎石和腐烂的木屑中分离的两株担子菌。由于担子菌特别是白腐菌被认为是产生木质素水解酶最大的生产者，因此担子菌的分离对于从海洋环境中分离木质素水解酶具有重要作用。这样，当提供新的底物时，可以发现多种不同种属的真菌（Jones，2011）。因此，担子菌可以从寄生于植物或者大型海洋动物的共生物中被分离出来。3 株可以水解木质素的担子菌 *Marasmiellus* sp. CBMAI 1602、*Peniophora* sp. CBMAI 1603 和 *Tinctoporellus* sp.，CBMAI 1601 在巴西圣保罗州北海岸的海绵 *Amphimedon viridis* 和 *Dragmacidon reticulatum* 中发现（Menezes et al.，2010）。

从根本上说，海洋微生物基因资源分离技术的应用是基于培养基中的底物。对于海洋中产木质素水解酶真菌的发现，文献中报道了几种培养基：低氮（LN）培养基（Raghukumar et al.，1994）、麦芽糖（MEA）培养基（Menezes et al.，2010）以及 Boyd 和 Kohlmeyer（B & K）培养基，一种专门针对海洋真菌的培养基（Kohlmeyer and Kohlmeyer，1979）。海洋真菌分离的共同特征是在培养基中加入盐离子，特别是海水。

大多数研究中应用的培养基中添加了染料 Poly-R 478（Raghukumar et al.，1994）。如果 Poly-R 培养基发生脱色，则证明微生物产生了降解木质素的酶。这是分离过程中的一种筛选方法。但是，如果分离的目的是选择性的分离担子菌或木质素水解真菌，那么杀菌剂或木质素水解酶底物的应用可以成为筛选的策略，正如 Gao 等（2011）在分离陆地白腐菌的中所述。

如上文所述，传统的水解木质素的真菌可以从培养基依赖的分离技术中得到。但是，科学家发现只有 1%~10%的微生物能用现有的培养基从环境中获得（Pace，1996）。培养基非依赖的以分子生物学为基础的方法涉及直接分离微生物群落 DNA（宏基因组学），此方法证明是研究培养基中不能分离的微生物代谢的潜在方法。在此意义上，大量多种不同的方法值得研究，以获得具有商业开发前景的及低成本技术，来促进工业用酶产品的制备。

15.2.2 海洋真菌的保存

在基因资源开发领域，正确的鉴定和保存对于研究酶和其生物技术应用是极为重要的。当前真菌分类学是基于多种方法，包括传统分析（大型生物和微生物形态学）以及分子生物学分析（DNA 序列和进化）、蛋白质组学分析（MALDI-TOF ICMS）和其他方法及多种方法联合。

为了成功保存筛选出的菌种，必须使用多种方法保存菌种以确保维持具有的生物技术应用潜能。OECD《生物资源中心最佳实践指南》（OECD，2007）数据显示，丝状真菌采用了表 15.4 描述的两种方法，其中之一是低温或冻干（长期保存）。

表 15.4 丝状真菌的推荐保存方法

孢子品系	非孢子品系
在-140℃以下冷冻保存（最佳的）	油或水下
在-80℃以下冷冻保存（可接受的）	冻干
冻干	冷冻保存
液体干燥	

资料来源：OECD，2007.

除保存方法外，还应开展定期的分析，包括纯度评估和细胞活力检测。就海洋真菌来说，我们团队获得的结果显示保存在 4℃水中（Castellani 法）是有效的。

15.2.3 筛选海洋真菌产生的木质素水解酶

定性分析是筛选木质素水解酶的有效工具，可以针对酶的分泌给出阳性或阴性指标。培养皿中加入染料是最适合定性验证木质素水解酶。多项研究表明木质素水解酶能使一些多聚染料脱色，其脱色的能力与木质素的降解能力是相关的（Chet et al.，1985）。真菌分泌的 LiP、MnP 和漆酶使多聚染料 Poly-R 478 脱色。由于这种蓝紫色染料易于观察，因此这种简单的检测可以给出明确的结果。

雷马素马斯亮蓝 R（RBBR）是另一种鉴定木质素水解真菌的染料。RBBR 是一种芳香族聚合物（PAH），可以作为制备多种高分子染料原料的聚蒽衍生物。Chroma 等（2002）证实 RBBR 染料脱色和 PAH 降解存在很好的联系。然而在某种意义上有些例外，有些热带海洋真菌产生的漆酶不能降解 Poly-R。此外，Da Silva 等（2008）评价了在固体培养基和液体培养基中 RBBR 的脱色能力。其中 13 种海洋丝状真菌能降解液体培养基中的染料，*Cladosporium cladosporioides* CBMAI 857 只能在菌丝体周围形成透明圈。Bonuli-Santos 等（2010a）报道了类似的结果，*M. racemosus* CBMAI847 产生大量 LiP、MnP 和漆酶，但是不能使用 RBBR 的固体培养基进行筛选。这说明 RBBR 染料筛选的方法存在缺陷。

通过酶底物进行定性实验的方法也得到了应用。Pointing 等（1999）报道，多种底物（包括 ABTS、对二氨基联苯、丁香醛连氮和愈创木酚）可以用在“Well test”的琼脂生长培养基中鉴定木质素水解酶。在“Well test”筛选时，几滴反应液加入后，孔内在 30 min 内出现紫色证明木质素水解酶产生了。然而，为了证明过氧化物酶活性，必须选择漆酶表现是阴性的或者比较弱的（Pointing et al.，1999）。

木质素水解酶底物也可以补充加入到培养基中。在固体培养基中加入愈创木酚后产生木质素水解酶的真菌周围和下面变成棕色（图 15.2），添加 ABTS 后漆酶使其变为绿色（Verma et al.，2010）。

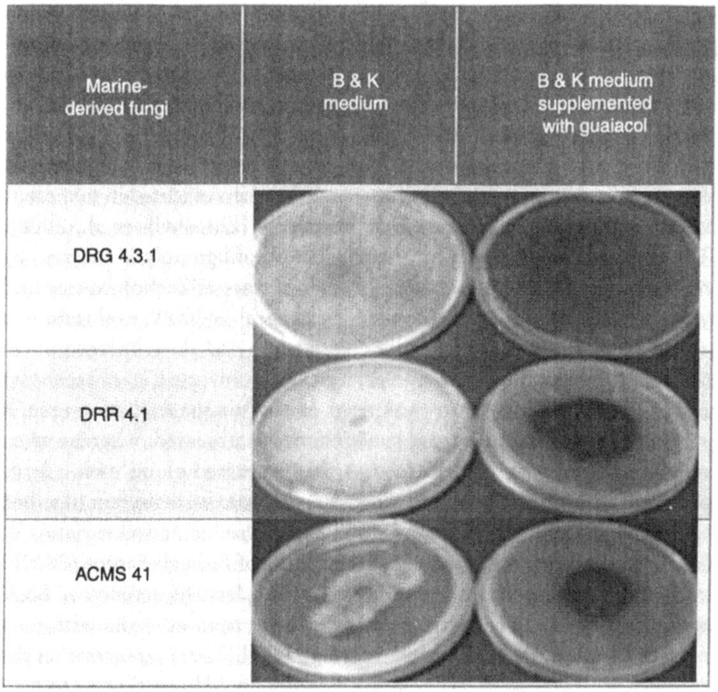

图 15.2　鉴定水解木质素的专性海洋菌

阳性结果：添加愈创木酚的 B & K 培养基中出现深棕色（图中深灰色）环（28℃培养 7 d）

15.2.4　海洋真菌产生木质素水解酶：制备和优化

pH、发酵周期、温度、氧气含量、转速、添加诱导物和碳/氮比等因素都可能对酶活性造成影响，这取决于真菌种类和生理条件。有研究表明，木质素水解酶（如漆酶、LiP 和 MnP）所需的条件不同。因此针对感兴趣的酶，设计和研究哪些因素是相关的，是非常重要的。木质素水解酶的液体和固体发酵都进行了大量的研究，但值得注意的是，这些酶产量不仅与真菌的生理生化性质有关，而且跟培养基的组成也相关（Baldrian and Gabriel，2003）。关于木质素水解酶的许多综述和工作可以帮助改善实验设计，主要是漆酶的制备（Levin et al.，2002；Elisashvili et al.，2008）。

碳源和氮源在木质素水解酶的制备中是至关重要的，它们对陆生白腐菌的影响已经进行了大量的研究（Elisashvili et al.，2008）。海洋真菌生产木质素最主要的培养基是用葡萄糖或麦芽糖作为碳源，酵母粉作为氮源。D'Souza-Ticlo 等（2009）研究了海洋担子菌

Cerrena unicolor MTCC 5159 生产漆酶时，合成培养基中成分的影响和相互作用。当柠檬酸盐作为唯一碳源时，漆酶的表达量低但是更稳定；然而当葡萄糖作为唯一碳源时，漆酶在较大温度范围内均有大量表达。当葡萄糖和柠檬酸盐作为共同碳源时，菌体量在增加，而酶产量有暂时降低的变化。Bonugli-Santos（2010b）对 3 种海洋担子菌的研究表明，B & K 培养基中表达的漆酶活性均低于麦芽糖作为唯一碳源的培养基。Arora 和 Gill（2001）证实麦芽糖培养基漆酶高表达可能是由于提取的麦芽糖中含有芳香族氨基酸酪氨酸和色氨酸。另一方面，从海洋刺细胞动物中分离的两株子囊菌和一株接合菌中在麦芽糖培养基培养中检测到了 MnP、LiP 和漆酶。然而当这些真菌在含有葡萄糖和麦麸的普通培养基生长时，检测不到 LiP 表达，但是 MnP 和漆酶的表达量显著增加（Bonugli-Santos et al.，2010a）。

关于 pH 值对木质素水解酶合成的影响几乎没有研究。针对这个问题的研究说明接种菌体前，pH 值通常在 4.5~6.0，但是在发酵过程中通常不控制 pH。Chen 等（2011）报道海洋真菌 *Pestalotiopsis* sp.，J63 表达漆酶最适 pH 值为 4.5~5。丝状真菌合成木质素水解酶的最佳温度通常为 25~30℃（Baldrian，2006）。

在通过增加诱导物来提高木质素水解酶的产量方面已经进行了大量的研究报道。这些化合物大部分是木质素或其他芳香族化合物（Farnet et al.，1999）。藜芦醇是一种芳香族化合物，在木质素合成和水解过程中发挥重要作用。此外，铜离子和酚的化合物（麦麸和愈创木酚）可以诱导多种真菌产生漆酶（Pointing and Hyde，2000；Elisashvili et al.，2008）。Bonugli-Santos（2010b）和 Passarini（2012）报道，当培养基中加入 $CuSO_4$ 和麦麸时，海洋来源的子囊菌和担子菌产生的漆酶活性增加。来自农作物和食品加工的农产品废弃物是大量存在的可再生资源，其可以通过微生物转化成其他不同附加值的产品（Elisashvili et al.，2008），这已经引起了广泛关注，因为来源于农产品废弃物（如麦麸、甘蔗渣、玉米棒及稻草）用于单一能源和碳源有很多优势，而且是环境友好的。Chen 等（2011）证实选择农产品废弃物和植物残渣作为生长基质具有增加专性海洋真菌 *Pestalotiopsis* sp.，J63 合成木质素水解酶的活性。此研究分别探究了水稻秸秆、麦麸、豆类、甘蔗渣、玉米秆及柚子皮的效果。水稻秸秆固体发酵体系产生的漆酶具有最高的活性，然而与其他复合物相比，未处理的甘蔗残渣在液体发酵中可以获得最高的酶产量。

盐离子浓度对于海洋真菌合成木质素水解酶的影响已有报道。Bonugli-Santos 等（2010a）研究了从海洋刺细胞动物中分离的真菌在含有 5%~23%（*w/v*）NaCl 的培养基中合成木质素水解酶的情况，结果显示从 12.5%~23%的 NaCl 浓度会产生高活性的 MnP 和漆酶。Chen 等（2011）报道，海洋真菌 *Pestalotiopsis* sp.，J63 在盐离子浓度为 30 g/L 的培养基中漆酶产量最高。D'Souza 等（2006）也报道了类似的结果，漆酶产量最高时是在 25×10^{-3}环境下。此外，还有担子菌合成漆酶时更喜欢使用 NaCl 存在的海水（Bonugli-Santos et al.，2010b；D'Souza-Ticlo et al.，2009）。

可以肯定的是培养基的成分对木质素水解酶的产生有影响，提高这些酶的产量最好的方法是运用实验设计并统计分析后得出结论。这种方法已经成功用于陆生真菌（Levin et al.，2008)，但是很少用于海洋真菌。以统计为基础的实验设计提供了一种高效的方法来确定酶产量最高的最佳培养条件，这同时也会有助于生产过程的优化。D'Souza-Ticlo 等(2009) 运用反应曲面法研究了培养基的成分与作用，以通过优化海洋真菌制备漆酶。基于统计学分析生物量和漆酶产量的关系获得的表面图表明了葡萄糖和 NH_4Cl 的相互关系。低浓度的 NH_4Cl 和高浓度的葡萄糖可以增加菌体产量。葡萄糖系数是阳性但是变化程度低表明，随着葡萄糖浓度的增加，菌体数量的增加将会达到一定的限度。葡萄糖和 NH_4Cl 的浓度具有相反的作用，因此漆酶的生产符合倒钟形曲线。因此，Bonugli-Santos 等(2010a) 通过中心组分设计，说明了专性海洋真菌 *Mucor racemosus* CBMAI 847 产生 MnP 的量可能与盐浓度有关。

基于水解底物导致颜色变化，木质素水解能力的检测可以通过分光光度法。其活性可以用每毫升比色单位或每毫升每分钟产物生成的微摩尔数表示。漆酶用不同的方法检测其活性，例如，愈创木酚、ABTS、丁香醛连氮（Bonugli-Santos et al.，2010b；Arora and Shamar，2010)。Tien 和 Kirk（1994）通过藜芦醇氧化的方法检测 LiP 活性，用酚红氧化的方法检测 MnP 活性（Kuwahara et al.，1984)。此外，MnP 可以根据 Mn（Ⅲ）-丙二酸复合物在 270 nm 有光吸收检测（Wariishi et al.，1992)。根据 Beer-Lambert 定律，酶活性可以通过公式（15.1）计算，单位为 $U \cdot L^{-1}$：

$$U \cdot L^{-1} = \Delta A \times V \times 10^6 / \varepsilon \times R \times t \tag{15.1}$$

其中：ΔA 代表起始和最终吸光度的差值；

V 代表反应体系（L)；

10^6 代表浓度（ε）从 mol/s 到 μmol/s；

ε 代表消光系数（$M^{-1} \cdot cm^{-1}$)；

R 代表培养基体积（L)；

t 代表反应时间（min)。

为了推动酶功能和进化的研究，含有或者潜在含有胞外木质素水解酶活性的基因研究极为重要。大量的与木质纤维素相关的陆生微生物基因组正在测序，这推动了在生物技术领域的研究（Martínez et al.，2009)。在此方面海洋真菌的研究还很少。然而，最近的研究报道了来自海洋的三组担子菌和一种子囊菌具有漆酶基因多样性，可能含有新的推测的漆酶基因（Bonugli-Santos et al.，2010b；Passarini，2012)。

15.3 海洋真菌木质素水解酶的生物技术应用

海洋真菌产木质素水解酶的能力推动了其在生物技术领域研究的发展，包括纸浆脱色、染色废水处理、糖蜜酒精（Raghukumar，2008）以及溢出的石油中 PAHs 的生物降解。

染料脱色和纺织工业废水是最主要的应用领域之一。废液中含有的少量染料能阻止光透入水中，抑制光合作用，阻止绿色植物生长，干扰了水体中空气的溶解（Enayatzamir et al., 2009)。此外，有颜色的不同工业废水，特别是纺织废水，可能含有诱变剂、致癌物和/或有毒物质（Chung et al., 1992）。降解和分解染料和纺织工业废水的研究从20世纪90年代中期已经开始，并在寻找替代物去处理这些污染物。Vyas和Molitoris（1995）是最早的研究者之一，他们报道了*Pleurotus ostreatus* 304产生的木质素水解酶参与了RBBR的脱色。

在过去的20年中，大量的研究表明木质素水解酶在不同种类的纺织染料和废水处理中具有潜力（Kaushik and Malik, 2009）。然而，工业废水中含有极端pH值和高浓度盐是陆生真菌面临的一个问题（Muthukumar and Selvakumar, 2004）。在此背景下，海洋真菌可以考虑用于高盐环境中的生物降解，包括处理纺织工业废水。

Raghukumar（2004）报道了产生木质素水解酶的11种海洋真菌可以用于工业染料脱色和木质素水解酶的制备，其中70%具有漆酶活性，82%具有水解纤维素活性。从印度珊瑚潟湖腐烂的海草中分离的一种白腐担子菌*Flavodon flavus*，可以分泌各种木质素水解酶（LiP，MnP和漆酶）在染料脱色过程中具有重要作用。Bonugli-Santos等（2012）研究发现了分离的3株海洋担子菌在含盐和无盐培养基中都能分泌LiP和MnP，其对降解RBBR的效率极高（100%）。因此，从真菌中提取酶用于染料脱色是一种高效的方法。这表明酶可以用于在真菌无法生长的环境中进行脱色。这些耐盐真菌和耐盐酶在多种生物技术中应用，在高盐和无盐环境中用于污染物的生物修复降解。然而，Jones（2011）报道尽管多种真菌具有使多种染料和污水脱色的能力，但是至今仍未得到商业化应用。

PAHs，特别是具有4个及以上环的化合物，其分解产物具有潜在危险。因为其具有有害的生物学效应，可能导致急性和慢性中毒，诱发突变和致癌。PAHs的真菌降解通过两条代谢途径：细胞外木质素水解复合物的非特异性水解和细胞色素P-450加氧酶的羟基化。这与哺乳动物的代谢途径类似，包括一系列的反应（Passarini, 2012; Capotorti et al., 2004）。PAHs可存在于石油中，在石油工业（如石油运输）中释放。这些复合物可能随着海洋意外溢油而流出，对环境、社会和经济造成损害。由于其耐受高盐环境，推动了海洋真菌降解在高盐环境中生物修复的应用。然而，海洋真菌用于此领域几乎未见报道。Passarini（2011）研究了8种已经证实具有使RBBR脱色，降解苯骈（a）芘能力的海洋真菌，其中*Aspergillus sclerotiorum* CBMAI 849分别在8 d和16 d时具有降解苯（99.7%）和苯骈（a）芘（76.6%）的能力，*Mucor racemosus* CBMAI 847也具有降解大量苯骈（a）芘（> 50.0%）的能力。近期研究表明，3种海洋担子菌（*Marasmiellus* sp. CBMAI 1602、*Peniophora* sp. 1603和*Tinctoporellus* sp. CBMAI 1601）在培养7 d、14 d和21 d后具有木质素水解酶（特别是漆酶）活性，具有分解PAHs芘及BaP的能力。*Marasmiellus* sp. CBMAI 1602是最有效的菌株，在最初的7 d内降解超过90%的最初的PAHs（Magrini, 2012）。

15.4 结论

考虑到海洋真菌分泌木质素水解酶和在高盐和无盐环境下分解海洋污染物的潜力，此方面研究必会得到发展，其涉及微生物酶的生产、代谢、毒理及生物修复。

致谢

感谢 FAPESP、CNPq 和 CAPES 的基金和奖学金在海洋真菌调查中的支持。同时感谢 Alvaro E. Migotto（Centro de Biologia Marinha-CEBIMAR）和 Roberto G. S. Berlinck（Universidade de São Paulo，USP）在海洋真菌取样时逻辑和技术上的支持。

参考文献

Arofa, D. S. and Gill, P. K. (2001) Comparison of two assay procedures for lignin peroxidase. *Enzyme Microb. Tech.* 28, 602–605.

Arora, D. S. and Sharma, R. K. (2010) Ligninolytic fungal laccases and their biotechnological applications. *Appl. Biochem Biotechnol.* 160, 1760–1788.

Asgher, M., Bhatti, H. N., Ashraf, M. and Legge, R. L. (2008) Recent developments in biodegradation of industrial pollutants by white rot fungi and their enzyme system. *Biodegradation* 19, 771–783.

Baldrian, P. and Gabriel, J. (2003) Lignocellulose degradation by *Pleurotus ostreatus* in the presence of cadmium. *FEMS Microbiol. Let.* 220, 235–240.

Baldrian, P. (2006) Fungal laccases—occurrence and properties. *FEMS Microbiol. Rev.* 30, 215–242;

Bonugli–Santos, R. C., Durrant, L. R., da Silva, M. and Sette, L. D. (2010a) Production of laccase, manganese peroxidase and lignin peroxidase by Brazilian marine–derived fungi. *Enzyme Microb. Tech.* 46, 32–37.

Bonugli–Santos, R. C., Durrant, L. R. and Sette L. R. (2010b) Laccase activity and putative laccase genes in marine–derived basidiomycetes. *Fungal Biol.* 114, 863–872.

Bonugli–Santos, R. C., Durrant, L. R. and Sette L. R. (2012) The production of ligninolytic enzymes by marine–derived basidiomycetes and their biotechnological potential in the biodegradation of recalcitrant pollutants and the treatment of textile effluents. *Water Air Soil Pollut.* 223, 2333–2345.

Capotorti, G., Digianvincenzo, P., Cesti, P., Bernardi, A. and Guglielmetti, G. (2004) Pyrene and benzo (a) pyrene metabolism by an *Aspergillus terreus* strain isolated from a polycylic aromatic hydrocarbons polluted soil. *Biodegradation* 15, 79–85.

Chen, M., Zeng, G., Jiang, M., Li, H., Liu, L., et al., (2011) Understanding lignin–degrading reactions of ligninolytic enzymes: binding affinity and interactional profile. *PLos one* 6 (9), 1–8.

Chet, I., Trojanowis, J. and Huettermann, A. (1985) Decolourization of the Poly B–411 and its correlation with lignin degradation by fungi. *Microbiol. Lett.* 29, 37–43.

Chroma, L., Mackova, M., Kucerova, P., der Wiesche, C. and Macek, T. (2002) Enzymes in plant metabolism of PCBs and PAHs. *Acta Biotechnol.* 22, 35–41.

Chung, K. T., Stevens, S. E.. Jr. and Cerniglia, C. R. (1992) The reduction of azo dyes by the intestinal microflora. *Crit. Rev. Microbiol.* 18, 175–190.

Claus, H. (2004) Laccases: structure, reactions, distribution. *Micron.* 35 (1–2), 93–96.

D'Souza, D. T., Tiwari, R., Sah, A. K. and Raghukumar, C. (2006) Enhanced production of laccase by a marine fungus during treatment of colored effluents and synthetic dyes. *Enzyme Microb. Tech.* 38, 504–511.

D'Souza–Ticlo, D., Garg, S. and Raghukumar, C. (2009) Effects and interactions of medium components on laccase from a marine–derived fungus using response surface methodology. *Mar. Drugs* 7, 672–688.

da Silva, M., Passarini, M. R. Z., Bonugli, R. C. and Sette, L. D. (2008) Cnidarian–derived filamentous fungi from Brazil: isolation, characterisation and RBBR decolourisation screening. *Environ. Technol.* 29, 1331–1339.

Dashtban, M., Schraft, H., Syed, T. A. and Qin, W. (2010) Fungal biodegradation and enzymatic modification of lignin. *Int. J. Biochem. Mol. Biol.* 1 (1), 36–50.

Dunford, H. B. (1999) *Heme Peroxidases*, Wiley–VCH, New York, 281–308.

Elisashvili, V., Kachlishvili, E. and Penninck, M. (2008) Effect of growth substrate, method of fermentation, and nitrogen source on lignocellulose–degrading enzymes production by white-rot basidiomycetes. *J. Ind. Microbiol. Biotechnol.* 35, 1531–1538.

Enayatzamir, K., Tabandeh, F., Yakhchali, B., Alikhani, H. A. and Couto, S. R. (2009) Assessment of the joint effect of laccase and cellobiose dehydrogenase on the decolouration of different synthetic dyes. J. Hazard Mater. 169, 176–181.

Farnet, A. M., Tagger, S. and Le Petit, J. (1999) Effects of copper and aromatic inducers on the laccases of the white rot fungus *Marasmius quercophilus. CR. Acad. Sci. Paris*, *Sciences de la Vie/Life Sciences* 322, 499–503.

Gao, H., Wang, Y., Zhang, W., Wang, W. and Mu, Z. (2011) Isolation, identification and application in lignin degradation of an ascomycete GHJ–4. *Afr. J. Biotechttol.* 10 (20), 4166–4174.

Gianfreda, L., Xu, F. and Bollag, J. M. (1999) Laccases: a useful group of oxidoreductive enzymes. Bioremediat. J. 3 (1), 1-25.

Hatakka, A. (2001) 'Biodegradation of lignin, Germany'. In: *Biopolymers*, Hofrichter, M. and Steinbüchel, A. (eds). Wiley-VCH, 129–180.

Hyde, K. D. and Jones, E. B. G. (1988) Marine mangrove fungi. *Mar. Ecol.* 9, 15–33.

Hofrichter, M. (2002) Review: lignin conversion by manganese peroxidase (MnP). *Enzyme Microb. Tech.* 30, 454–466.

Hofrichter, M., Ullrich, R., Pecyna, M. J., Liers, C. and Lundell, T. (2010) New and classic families of secreted fungal heme peroxidases. *Appl. Microbiol. Biotechnol.* 87, 871–897.

Jones, E. B. G. (2011) Fifty years of marine mycology. *Fungal Diversity* 50, 73–112.

Kaushik, P. and Malik, A. (2009) Fungal dye decolourization: recent advances and future potential. *Environ, bit.* 35, 127–141.

Kim, S. J., Ishikawa, K., Hirai, M. and Shoda, M. (1995) Characteristics of a newly isolated fungus, *Geotrichum candidum* Dec 1, which decolorizes various dyes. *J. Ferment. Bioeng.* 79, 601–607.

Kohlmeyer J. and Kohlmeyer E. (1979) *Marine Micology*: The Higher Fungi, Academic Press, New York.

Kuhad, R. C., Singh, A. and Eriksson, K. -E. L. (1997) 'Microorganisms and enzymes involved in the degradation of plant fiber cell walls, Germany'. In K. -E. L. Eriksson (ed.). *Advances in Biochemical Engineering Biotechnology*, 46–125.

Kuwahara, M., Glenn, J. K., Morgan, M. A. and Gold, M. H. (1984) Separation and characterization of two extracellular H_2O_2 dependent oxidases from ligninolytic cultures of *Phanerochaete chrysosporium. FEBS Lett.*

169, 247-250.

Levin, L., Forchiassin, F. and Ramos, A. M. (2002) Copper induction of ligninmodifying enzymes in the white-rot fungus *Trametes trogii*. *Mycologia* 94, 377-383.

Levin, L., Herrmann, C. and Papinutti, V. L. (2008) Optimization of lignocellulolytic enzyme production by the white-rot fungus *Trametes trogii* in solid-state fermentation using response surface methodology. *Biochem. Eng. J.* 39, 207-214.

Maciel, M. J. M., Silva, A. C. and Ribeiro, H. C. T. (2010) Industrial and biotechnological applications of ligninolytic enzymes of the basidiomycota: a review. *Electron. J. Biotechn.* 13 (6), 1-12.

Magrini, M. J. (2012) *Produção de enzimas ligninolíticas e degradação de HPAs por fungos basidiomicetos derivados de esponjas marinhas* [PAH degradation and lignolytic enzyme production by marine-sponge-derived basidiomycetes fungi]. Master dissertation in Genetics and Molecular Biology, University of Campinas, Brazil, 1-100.

Martínez, Á. T., Ruiz-Duenãs, F. J., Martínez, M. J., del Río, J. C. and Gutierrez, A. (2009) Enzymatic delignification of plant cell wall: from nature to mill. *Curr. Opin. Biotechn.* 20, 348-357.

Menezes, C. B. A., Bonugli-Santos, R. C, Miqueletto, P. B., Passarini, M. R. Z., Silva C. H. D., et al., (2010) Microbial diversity associated with algae, ascidians and sponges from the north coast of São Paulo state, Brazil. *Microbiol. Res.* 165, 66-482.

Minussi, R. C., Miranda, M. A., Silva, J. A., Ferreira, C. V., Aoyama, H., et al., (2007) Purification, characterization and application of laccase from *Trametes versicolor* for colour and phenolic removal of olive mill wastewater in the presence of 1-hidroxybenzotriazole. *Afric. J. Biotechn.* 6 (10), 1248-1254.

Muthukumar, M. and Selvakumar, N. (2004) Studies on the effect of inorganic salts on decolouration of acid dye effluents by ozonation. *Dyes Pigments* 62, 221-228.

OECD (2007) Best Practice Guidelines for Biological Resources Centres, Organisation for Economic Co-operation and Development (*http://wunv: oecd. org/dataoecd/7/13/38777417.pdf*).

Passarini, M. R. Z., Rodrigues, M. V. N., da Silva, M. and Sette, L. D. (2011) Marine-derived filamentous fungi and their potential application for polycyclic aromatic hydrocarbon bioremediation. *Mar. Pollut. Bull.* 62, 364-370.

Passarini, M. R. Z. (2012) *Caracterização da diversidade de fungos filamentosos associados a esponjas marinhas e avaliação da produção de lacase* [Diversity of filamentous fungi associated with marine sponges and evaluation of laccase production]. PhD thesis in genetics and molecular biology. University of Campinas, Brazil.

Pace, N. R. (1996) New perspective on the natural microbial world: molecular microbial ecology. *ASM News* 62, 463-470.

Piontek, K., Smith, A. T. and Blodig, W. (2001) Lignin peroxidase structure and function. *Biochem. Soc. Trans.* 29 (2), 11-116.

Pointing, S. B., Buswell, J. A., Jones, E. B. G. and Vrijmoed, L. L. P. (1999) Extracellular cellulolytic enzyme profiles of five lignicolous mangrove fungi. *Mycol. Res.* 103 (6), 696-700.

Pointing, S. B. and Hyde, K. D. (2000) Lignocellulose-degrading marine fungi. *Biofouling* 15, 221-229.

Prasannarai, K. and Sridhar, K. R. (2001) Diversity and abundance of higher marine fungi on woody substrates along the west coast of India. *Curr. Sci.* 81 (3), 304-311.

Raghukumar, C., Raghukumar, S., Chinnaraj, A., Chdranohan, D., D'Souza, T. M. and Reddy, C. A. (1994) Laccase and other lignoceluloses modifying enzymes of marine fungi isolated from the coast of India. *Bot. Mar.* 37, 515–523.

Raghukumar, C. (2004) 'Marine fungi and their enzymes for decolorization of colored effluents'. In: *Marine Microbiology: Facets & Opportunities*, Ramaiah, N. (ed.), 145–158. National Institute of Oceanagraphy, India.

Raghukumar, C. (2008) Marine fungal biotechnology: an ecological perspective. *Fungal Divers.* 31, 19–35.

Reddy, G. V. B., Sridhar, M. and Gold, M. H. (2003) Cleavage of nonphenolic β-1 diarylpropane lignin model dimers by manganese peroxidase from Phanerochaete chrysosporium. *Eur. J. Biochem.* 270, 284–292.

Ruiz-Dueñas, F. J. and Martínez, A. T. (2009) Microbial degradation of lignin: how a bulky recalcitrant polymer is efficiently recycled in nature and how we can take advantage of this. *Microbial Biotechn.* 2 (2), 164–177.

Sakayaroj, J., Preefanon, S., Supaphon, O., Jones, E. B. G. and Phongpaichit, S. (2010) Phylogenetic diversity of endophyte assemblages associated with tropical seagrass *Enhalus acoroides* from Thailand. *Fungal Divers.* 41, 1–19.

Sugano, Y. (2009) DyP-type peroxidases comprise a novel heme peroxidase family. *Cell. Mol. Life Sci.* 66, 1387–1403.

Sundaramoorthy, M., Kishi, K., Gold, M. H. and Poulos, T. L. (1994) The crystal structure of manganese peroxidase from *Phanerochaete chrysosporium* at 2.06-A resolution. *J. Biol. Chem.* 269, 32759–32767.

Surajit, D., Lyla, P. S. and Ajmal Khan, S. (2006) Marine microbial diversity and ecology: importance and future perspectives. *Curr. Sci.* 90 (10), 1325–1335.

Sutherland, J. B., Crawford, D. L. and Speedie, M. K. (1982) Decomposition of C-labeled maple and spruce lignin by marine fungi. *Mycologia* 74 (3), 511–513.

Tien M. and Kirk, T. K. (1984) Lignin-degrading enzyme from *Phanerochaete chrysosporium*: purification, characterization and catalytic properties of unique H_2O_2 requiring oxygenase. *Proc. Natl. Acad. Sci. USA* 81, 2280–2284.

Trejo-Hernandez, M. R., Lopez-Munguia, A. and Quintero Ramirez, R. (2001) Residual compost of *Agaricus bisporus* as a source of crude laccase for enzymic oxidation of phenolic compounds. *Process Biochem.* 36 (7), 635–639.

Thurston, C. F. (1994) The structure and function of fungal laccases. *Microbiology* 140 (1), 19–26.

Tuomela, M., Vikman, M., Hatakka, A. and Itavaara, M. (2000) Biodegradation of lignin in a compost environment: a review. *Bioresource Technol.* 72, 169–183.

Verma, A. K., Raghukumar, C., Verma, P., Shouche, Y. S. and Naik, S. G. (2010) Four marine-derived fungi for bioremediation of raw textile mill effluents. *Biodegradation* 21, 217–233.

Vijaykrishna, D., Jeewon, R. and Hyde, K. D. (2006) Molecular taxonomy, origins and evolution of freshwater ascomycetes. *Fungal Diversity* 23, 351–390.

Vyas, B. R. M. and Molitoris, H. P. (1995) Involvement of an extracellular H_2O_2-dependent ligninolytic activity of the white rot fungus *Pleurotus ostreatus* in the decolorization of Remazol Brilliant Blue R. *Appl. Environ. Microbiol.* 61 (11), 3919–3927.

Wang, G. (2006) Diversity and biotechnological potential of the sponge-associated microbial consortia. *J. Ind.*

Microbiol. Bioty. 33 (7), 545–551.

Wariishi, H., Valli, K. and Gold, M. H. (1992) Manganese (Ⅱ) oxidation by manganese peroxidase from the basidiomycete Phanerochaete chrysosporium. Kinetic mechanism and role of chelators. *J. Bio. Chem*. 267, 23688–23695.

Wong, D. W. S. (2009) Structure and action mechanism of ligninolytic enzymes. *Appl. Biochem. Biotechnol*. 157, 174–209.

第 16 章　海洋细菌中的多糖降解酶

Gurvan Michel and Mirjam Czjzek, University of Pierre and Marie Curie, Paris

DOI：10.1533/9781908818355.3.429

摘要：海洋细菌是褐藻、红藻、绿藻多糖专一性降解酶分离的珍贵资源。海洋酶的发现基本上包括基于特定活力的分离纯化而进行，由于工业生产偏向的引导，经常会导致一些稀有的海洋多糖降解酶缺乏。有趣的是，首批发现的具有活性的来源于细菌的海藻多聚糖酶揭示出他们同陆地来源的多糖降解酶具有非常远的结构相似性，通常属于新糖降解酶家族。在这一部分，我们回顾起源于海洋细菌的多糖降解酶的一般特性，尤其关注琼脂水解酶和卡拉胶酶，然而也包括褐藻胶裂解酶、岩藻聚糖酶和石莼聚糖酶。本章还涵盖了相关酶的应用潜力的展望和需要更系统的方法来解决现代生物科技需求。

关键词：琼脂水解酶，卡拉胶酶，岩藻糖酶，石莼聚糖裂解酶，褐藻胶裂解酶，底物特异性，结构功能

16.1　引言

海洋环境是当今被关注的最大的生物多样性来源，但在基因、蛋白质和酶的精确生物功能性归属研究方面进展最慢（Azam and Malfatti，2007）。这种矛盾使得海洋环境最吸引生物研究者（Arnaud-Haond et al.，2011）去发现新的功能和分子，但是精确信息的缺乏导致高通量筛选技术在探索大量酶和蛋白质方面极为低效。因此，海洋酶的发现基本上还是基于特定活性的筛选与纯化为基础（Michel et al.，2006；Wong et al.，2000；Fu and Kim，2010），并且后续还需要大量的劳动和相应的底物。这是在本章叙述的大多数藻类糖酶活性的情况。值得注意的是，首批发现的具有活性的来源于细菌的海藻多糖酶表现出与它们陆地来源的同源酶具有非常远的结构相似性，通常属于新糖类水解酶家族（即 GH82，GH86，GH96，GH107，GH117，GH118 等）。这个事实说明在通过“推测的保守蛋白”的技术去发现这些酶需要更多精细的研究。

近年来，伴随着海洋基因组学的快速发展，大量的生物信息学上的信息以及大尺度的

基因组学方法已经被应用于新的海洋酶的发现（Ekborg et al.，2006；Groisillier et al.，2010；Shin et al.，2010）。此外，新的海洋生物基因组（Bauer et al.，2006；Glöckner et al.，2003；Weiner et al.，2008）揭示出至今已经发现和表征的酶是整个降解和同化海洋有机碳源中非常有限的一部分。现如今，有研究将整个海藻多糖的降解作为一个整体去研究，从而可以在海洋环境中，从整个碳循环的角度去分析基于海洋微生物的海洋碳循环。其中最典型的一个例子是用宏基因组学技术研究在浮游植物暴发时营养（多糖）诱导的海洋微生物（Teeling et al.，2012）。

近年来在寻找有关潜在原料的研究中，海洋大型藻类得到大量的关注，大型藻类作为替代性可再生资源可以用于生物质燃料的生产和通过微生物发酵技术生产可商品化的化学物质，尤其是在远东，一些海藻的传统应用有着较长历史的国家中，正在积极地考虑在海洋中大规模种植大型海藻用于生产生物乙醇（Aizawa，2007；Gao and McKinley，1994；Jensen，1993）。海藻多糖由于含量丰富，已经成为一种最广泛的利用海洋生物质资源。的确，多糖是组成海洋环境中有机碳类化合物中含量最多的成分之一（Benner et al.，1992；Craigie，1990；Ogawa et al.，2001）。因此，如果能够开发有效的褐藻和红藻多糖生物转化技术，褐藻和红藻多糖就可以用于生物燃料或工业化学品的生产。海洋微生物及藻类相关细菌越来越多地被称作是“综合细胞工厂”，用于提供必要的催化机制来执行物质间的转换。此外，与陆地环境相比，海洋环境存在的海洋微生物具有更独特的遗传特征和生存习惯（Harmsen et al.，1997；Stach et al.，2003）。海洋环境中既有营养丰富的区域，也有极少量生物能够赖以生存的营养贫瘠的地方。海洋环境具有高度复杂性，主要包括高盐、高压、低温和特殊的光照，这使得海洋微生物产生的酶类和陆地微生物产生的同源性酶类之间存在显著的差异，从而导致近年来海洋微生物酶技术的快速发展以及高值化产品的生产（Trincone，2011）。但是，海洋大型藻类的细胞壁多糖（存储多糖）（Popper et al.，2011）具有独特性和巨大的多样性，这是与海洋中存在大量新颖的、特异性的酶类相适应的。在这些用以彻底消化海洋多糖的酶类中，迄今为止仅有少数酶已经被详细地进行生化表征。需要特别指出的是，某些酶类能够产生单糖或者能够转化某些特殊的藻类单糖，如3，6-脱水-L-半乳糖，以彻底消化吸收这些单糖，迄今为止，在代谢途径中缺少这些酶。

本章内容重点介绍能够降解主要海洋多糖化合物，如琼脂、卡拉胶和褐藻胶的酶类的研究现状，对其中的每一大类酶都分节重点介绍。另一节总结了各种其他海洋多糖降解酶，如褐藻胶裂解酶、岩藻聚糖和石莼聚糖酶，对这些酶，迄今为止仅有少量例子被发现并报道，尽管这不足以反映出这些酶的重要性或者用途，这些酶与两大类主要降解酶相比而言至少是同等重要的。最后一部分总结了有关生物催化下的海洋生物物质降解的整体研究状况。

16.2 琼脂降解多糖酶

琼脂是产琼脂红藻类细胞壁的主要成分，代表着一个复杂的多糖家族。这些复杂的多糖家族成员可以大致分为两大类，其中一类被命名为琼脂糖，为中性分子，也就是传统的多糖。另一大类统称为琼脂胶，包括所有其他修饰化的琼脂，其分子中含有较高程度的修饰化（5%~8%）。琼脂分子骨架中的规律性结构通常被不同的化学修饰成分所掩盖，比如硫酸脂键基团（S），甲基化基团（M）或者丙酮酸缩醛基团等（P）（Lahaye and Rochas，1991）。最常见的化学修饰是L-半乳糖单体（LA单体）分子内出现的3，6-缩水键，琼脂分子中主要的重复单元是琼脂二糖（G-LA，相当于DP2）（Knutsen et al.，1994）。因此，琼脂糖是中性线性多糖，由交替出现的3-O-β-D-吡喃型半乳糖和4-O-3，6-缩水-α-L-吡喃型半乳糖残基组成，分子中还有极少量的硫酸化成分（大约1%）。另一种较常见的化学修饰是发生在L-半乳糖残基的C6上的硫酸化作用，产生α-L-半乳糖-6-硫酸（L6S）。从紫菜中提取的琼脂中发现，这种L6S单体的含量丰富，是LA单体含量的两倍多（Anderson and Rees，1965）。这种琼脂通常被称为紫菜胶，其分子中主要的重复单位是G-L6S二糖单位，在这里被称为紫菜胶二糖。L6S在洋菜植物如江蓠属（*Gracillaria*）和石花菜属（*Gelidium* spp.）中提取的琼脂中经常被发现。β-D-吡喃型半乳糖单体的C6位置上经常发生甲基化，使得紫菜胶的结构变得更加复杂化（Correc et al.，2011；Lahaye et al.，1989）。要实现从琼脂到发酵糖的彻底转化，至少需要两种类型的酶来催化β-1，4和β-1，3糖苷键两种连接方式的断裂。

16.2.1 琼脂降解酶的出现

在20世纪初，首次从海水中分离得到琼胶降解细菌，但是直到70年代初期（Gran，1902），才首次报道了能够降解琼胶的酶（Day and Yaphe，1975；Groleau and Yaphe，1977）。这些β-琼脂酶能够断裂β-D-半乳糖和α-L-3，6-缩水半乳糖，产生系列新琼胶类寡糖。如今，催化这一反应的酶在各种不同糖苷水解酶（GH）家族中都有被发现，如GH16、GH50、GH86和GH118（see http：//www. cazy. org and Michel et al.，2006）。能够催化断裂α-1，3糖苷键的酶非常少见，至今为止，只有GH96家族中的两种α-琼脂酶被研究报道（Flament et al.，2007；Seok et al.，2012）。最新的研究还报道了几种外切方式的α-半乳糖苷酶。这些酶属于一个新的海洋来源GH酶家族（GH117家族），包含了各种能够参与琼胶代谢的末端途径中的酶和α-1，3-（3，6-缩水）-L-半乳糖苷酶AhgA（Ha et al.，2011；Hehemann et al.，2012b；Rebuffet et al.，2011）。

β-琼脂酶大多数属于GH16家族；从Blast检索中发现100多个基因被注释为β-琼脂酶，其中有25个基因已经被证实确认为β-琼脂酶（http：//www. cazy. org）。除此之外，GH16

是一种多特异性酶家族，大约包含有 2 400 个成员，还包括卡拉胶酶、昆布多糖酶和其他陆生生物多聚糖降解特异性酶（Michel et al.，2006）。属于其他 GH 家族的 β-琼胶酶比较少见，迄今为止发现大约有 100 种属于 GH50，40 种属于 GH86，和 GH118 家族中的 6 种序列（http：//www.cazy.org）。通常来说，这些家族包含几乎所有从各种海洋原核生物有机体中分离出的 β-琼胶酶。非常值得一提的是其中一些基因同样在人体肠道菌群中发现，分析可能是通过横向基因传递（HGT），而从食品相关海洋微生物中获得了这些特征（Hehemann et al.，2010）。GH16 家族的琼脂酶是研究最彻底、最深入的，有关酶催化的一些详细内容会在下部分讲述。对几种 GH50 家族中 β-琼脂酶的生化研究表明其中一些酶具有纯外切活性（Kim et al.，2010；Temuujin et al.，2012），能够从非还原末端断裂，产生新琼胶二糖。其他的 GH50 家族成员具有内切和外切活性，能够同时作用于新琼胶寡糖，如琼胶四糖、琼胶六糖和琼胶等（Fu et al.，2009；Ohta et al.，2005；Sugano et al.，1993a）。至今为止，只有少数 GH86 β-琼脂酶被报道（Belas，1989；Ekborg et al.，2006），而且具有内切和外切活性。最后，一种 GH118 β-琼脂酶的生化特性已被详细阐明（Ma et al.，2007），但是没有得到其晶体结构。最近的一篇有关琼胶降解以及降解所需要的酶的综述中，列举了大量已经鉴定的琼脂酶以及他们的原始出处（Chi et al.，2012）。

然而，由于天然存在的琼胶中并不是单纯由琼脂糖组成，因此需要更多的酶的参与才能实现琼胶这种生物碳资源的彻底降解和消化。在一种多糖降解海洋细菌中，通常存在大量琼胶相关的酶，这也反映了琼胶的降解需要多种酶的参与（Ekborg et al.，2006；Hehemann et al.，2010，2012a；Shin et al.，2010）。在这一点上，第一个紫菜胶酶已经在 *Z. galactanivorans* 中被鉴定，该酶能够特异性的断裂半乳糖与前面或者后面的 6-硫酸 L-半乳糖（L6S-G）之间的 β-1，4 糖苷键，属于 *Z. galactanivorans* 的基因组中已有的 9 种催化琼胶水解的 GH16 酶（Hehemann et al.，2012a），这些酶对于纯的琼胶是没有活性的。有趣的是，最近发现的一种由人肠道细菌（*Bacteroides plebeius*）产生的被定义为 β-紫菜胶酶的 GH86 酶，其酶和底物分子的复合物的晶体结构已经解析（PDB：4AW7），但是与此相关的文献仍未见发表。这些被最初自动定义为潜在的 β-琼脂酶，其更多的有关底物特异性仍然存在变化，这也恰恰说明了为什么大量的酶都属于同一个 GH 家族，以及为什么不同的 GH 家族中存在大量的基因都编码所谓的 β-琼脂酶。

16.2.2 琼脂降解酶的生物催化特性

对于不同的已分离的 β-琼胶酶的描述和表征，其水平是非常不同的，而且对不同 β-琼脂酶的琼胶水解活性的检测也受到定性检测方法的限制（比如琼脂糖凝胶结构的破坏，琼胶寡糖产物的检测），还有其他的琼脂酶只是在一级序列水平得到表征。然而，一些属于 GH16 家族的 β-琼脂糖酶，已经开展了深入的生化表征，包括酶分子或者酶与底物复合物 3D 晶体结构的测定。酶与特定底物的晶体结构的信息与生化数据的结合分析使得我们能够分析这些 GH16 酶家族成员对底物分子的要求，精确地阐明酶的特异性（图 16.1）。

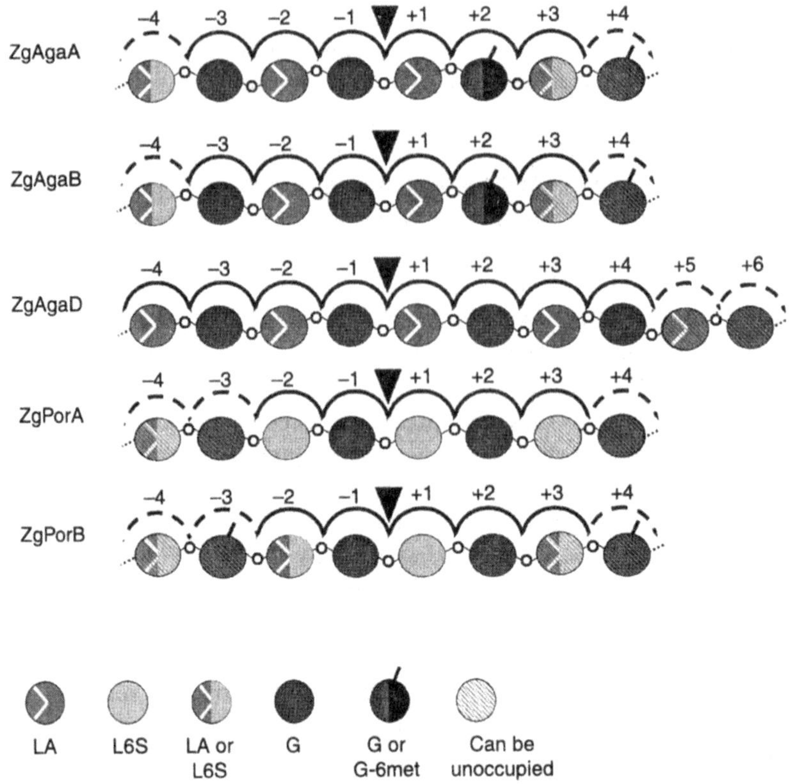

图 16.1 经过生化特性和 3D 晶体数据得到不同琼脂糖酶亚结合位点特异性分析

注：界定亚连接位点的粗线代表至少一个氨基酸在结合糖单位上有最近距离（小于 4 Å）；界定外部亚结合位点的中断线表示了当分析生化降解模式进行时，这些位置有着重要影响，但没有结构的证据被发现；刻线糖单位表明这些位置可能不被占满，但酶仍然具有水解活性；所有酶属于 GH16 家族并且从 *Zobellia galactanivorans* 中克隆出来 .

到现在为止，所有已报道的 GH16 β-琼胶酶都表现出一种内切活性，随机切断多糖链产生各种终产物的混合物，这些终产物主要包括新琼胶四糖和新琼胶六糖，以及少量的新琼胶二糖。天然的琼脂或琼胶的水解产物，成分更为复杂，因为多糖链的结构更为混杂，而且存在一些碳水化合物修饰性子结合位点，比如甲基化位点和硫酸化位点，图 16.1 可以看出琼脂分子中存在不同的亚结合位点，可以进行各种不同的修饰（Correc et al. , 2011；Hehemann et al. , 2012a；Jam et al. , 2005）。尽管这些研究通常采用的是可溶性底物，必须指出的是，与其他 β-琼脂酶相比，ZgAgaA 以琼脂糖凝胶为底物能够表现出明显的高活性（Jam et al. , 2005）。这主要归因于在酶分子表面除了活性位点的凹槽外还具有琼脂糖结合位点（Allouch et al. , 2004），而且在凝胶相更易于采用连续的作用模式完成反应。有趣的是，这是迄今为止发现的唯一一种已经鉴定的表面结合位点的 β-琼脂酶。而且，据报道，很多这种类型的 β-琼脂酶，不仅是属于 GH16 家族，而且都包含其他的模块，比如已经报道的琼脂糖结合模块（CBM，碳水化合物结合模块）（Ekborg et al.，2006；Henshaw et al.，2006；Jam et al.，2005）。这些额外模块能够促进酶对固体或者凝胶底物的催化效率（Gilbert et al.，2008），但是没有生化研究能够解释这与β-琼脂酶活性相关的问题。

CY24 是在 *Pseudoalteromonas* sp. 中发现的表征最详细的 GH118 β-琼脂酶，它也表现出内切活性（Ma et al.，2007），并且与 GH16 ZaAgaD（Hehemann et al.，2012a）作用非常类似。这些β-琼脂酶通常产生的是分子量较大的 DP8 或者 DP10 的琼脂寡糖，分子量最小的最终产物是新琼脂四糖。相反，大多被报道的 GH50 酶表现出外切酶活性，产生一种终产物-新琼脂二糖（Kim et al.，2010；Ohta et al.，2005；Temuujin et al.，2012）。有关这些 GH118 和 GH50 酶，均没有相关的结构数据，而且如上所述，唯一有关 GH86 的晶体结构却没有文献报道。尽管对琼脂酶已经开展了大量的研究，但是对这些酶家族成员的生化表征等仍然存在很多亟待破解的问题。比如，邻近的琼脂糖结合 CBMs 对酶的催化效率有什么影响？修饰后的和改性的琼脂能够被 GH50、GH86 和 GH118 酶所识别吗？需要什么酶一起反应以实现协同催化？随着日益增长的寻求新的可再生生物能源的兴趣，以及致力于对含有琼脂的藻类的彻底糖化的尝试，对上述最后一个问题已经开始着手解决。的确，在最近的一个研究中，一个所谓的“琼脂糖降解平台”已经被应用于彻底转化琼脂糖为单糖，供后续的发酵利用。这一平台像一个包含各种酶的“鸡尾酒”，含有 GH16 内切β-琼脂酶，能够产生新琼脂二糖的外切酶（GH50）和 AghA，以及外切 α-琼脂水解酶（Kim et al.，2012a）。尽管这项研究的结果很有希望，有关 3，6-缩水 L-半乳糖单元的转化的问题仍然有待解决，因为 3，6-缩水 L-半乳糖占琼脂糖多糖的 50%左右。

16.3 卡拉胶降解多糖酶

16.3.1 卡拉胶结构的复杂性

在衫藻苔目（如 *Chondrus crispus*，*Kappaphycus alvarezii*，*Gigartina skottsbergi*）红藻的细胞壁中，琼胶被另一类硫酸化半乳聚糖-卡拉胶所替代（Popper et al.，2011）。这一大类水胶体也是由线性链状半乳糖组成，分子中 α-1，3 键和 β-1，4 键交替出现，但是分子中只含有 D-半乳糖残基（Rees，1969）。Knutsen 与其同事引入一种卡拉胶的分类系统，主要依据是基于硫酸酯键的数量和位置，以及在 α-链接的残基（DA -单体）（Kuntsen et al.，1994）中 3，6-缩水键的出现。在工业应用上最有价值的 3 种卡拉胶是 κ-卡拉胶，能够形成离子-温度可逆凝胶的 ι-卡拉胶，还有一种不能成胶，但是黏度较高的高分子 λ-卡拉胶。其分子中主要的重复二糖单位由于具有 1，2 和 3 个硫酸脂键基团而有所不同，分别是：κ-卡拉胶（DA-G4S），ι-卡拉胶（DA2S-G4S）和 λ-卡拉胶（D2S6S-G2S）。卡拉胶的结构进一步被甲酯和丙酮酸化等修饰基团所修饰。像琼脂一样，卡拉胶通常存在一种杂合，在多糖链的主链上结合不同的角叉二糖单位（Campo et al.，2009；Van de Velde，2008；Van de Velde et al.，2002）。

16.3.2 卡拉胶酶的发现

直到现在，所有已知的卡拉胶酶都是专一性的作用于一种类型的卡拉胶，能够区分 κ-卡拉胶，ι-卡拉胶和 λ-卡拉胶，后来被分别称作卡帕酶、艾欧塔酶和拉姆达酶。与琼脂酶不同的是，所有已知的卡拉胶酶只能够断裂相应底物分子中的 β-1，4 糖苷键。继上一篇有关卡拉胶酶的综述（Michel et al.，2006）之后，又报道了几种新的卡拉胶降解微生物，大都属于 *Flavobacteria*（genera：*Maribacter*（Barbeyron et al.，2008a）*Mariniflexile*（Barbeyron et al.，2008b），*Tamlana*（Sun et al.，2010）和 *Gammaproteobacteria*（genera：*pseudoalteromonas*（Ohta and Hatada，2006；Zhou et al.，2008；Ma et al.，2010；Liu et al.，2011；Kobayashi et al.，2012），*Pseudomonas*（Khambhaty et al.，2007）和 *Microbulbifer*（Hatada et al.，2011）。从几种红藻中分离出一种少见的卡拉胶降解菌 *Actinobacteria*（Beltagy et al.，2012）。从这些细菌中分离纯化了几种野生型卡拉胶酶，对其酶学功能进行了测定，但是只有基因被克隆（Hatada et al.，2011；Kobayashi et al.，2012；Liu et al.，2011；Ohta and Hatada，2006）。然而，如今海洋微生物基因组数据的快速增加也同样为新型卡拉胶酶的发现提供了宝贵的资源（Glöckner and Joint，2010），我们会在这里展示我们最近得到的原始数据。这些新型卡拉胶酶的序列将在与已有的卡帕酶（Michel et al.，2001a）和艾欧塔酶（Michel et al.，2001b，2003；Rebuffet et al.，2010）的晶体结构对比的基础上进行解释。

卡帕-卡拉胶酶

从 *Pseudoalteromonas carrageenovora* 中发现的卡帕酶的基因是第一个被克隆（Barbeyron et al.，1994）和过量表达（Michel et al.，1999）的基因。与从 *Zobellia galactanivorans* 中发现的卡帕酶（41%序列一致性）一起，他们共同代表了 GH16 家族（Barbeyron et al.，1998）中的一个子家族。卡帕酶的催化机制是能够保留异头碳构象的机制（Potin et al.，1995）。从 *P. carrageenovora* 中发现的卡帕酶折叠成 β-三明治样空间构象，是典型的 GH16 家族酶的构象，表现出一种通道样活性中心，表明这种酶的催化方式是内切连续的模式（Michel et al.，2001a）。这种假设后来得到生化研究的进一步证实（Lemoine et al.，2009）。从此以后，只有两种其他卡帕-卡拉胶酶的基因被克隆，一个是从 *Pseudoalteomonas porphyrae* L11 中发现的（Liu et al.，2011），另一种是从 *Pseudoalteromonas tetraodonis*（Kobayashi et al.，2012）发现的。但是这两种卡帕-卡拉胶酶与从 *P. carrageenovora* 中发现的卡拉胶酶几乎是相同的（分别具有 92%和 98% 的相似性），在 C-端的区域都存在一个 BIG2 模块。但是，最近我们在一些生物的基因组中发现了新的卡帕-卡拉胶酶，包括一些海洋细菌，如浮霉菌门的

Rhodopirellula baltica（菌株 SH1 和 WH47）（Glöckner et al.，2003；Wecker et al.，2010），疣微菌门的 *Coraliomargarita akajimensis*（Mavromatis et al.，2010），产黄菌属的 *Cellulophaga lytica*（Pati et al.，2011），更令人惊奇的是在人肠道细菌中的 *Bacteroides ovatus* 菌株 CL02T12C04（Gevers et al.，2012）中也有发现。在与 Rudolf Amann 课题组的合作研究中，我们证实了 *R. baltica* 具有降解卡帕-卡拉胶（Dabin，2008）的能力。通过两两比对，这些酶与 *P. carrageenovora* 卡帕酶的序列相似性程度为 31%~47%。然而，这些酶都表现出 GH16 家族酶的催化特征序列 ExDx(x)E（图 16.2）。

更重要的是，在卡帕-卡拉胶酶对底物的识别中起关键作用的几个残基（Michel et al.，2001a）在序列上是高度保守的：Trp^{144}和 Arg^{260}（-1 位），His^{183}（+1 位）。而且，其他 3 个可能的结合残基也是非常保守的：Trp^{194}，Arg^{196}和 Asn^{269}。这种高度保守性进一步证实了这些残基确实参与底物卡帕-卡拉胶的结合这一假设（Michel et al.，2001a）。值得指出的是β5 和β6 链不是高度保守的，在其他的酶分子中明显较短（从 *R. baltica* 中发现的卡帕酶除外），β5 和β6 链主要参与组成 *P. carrageenovora* 卡帕酶的表面沟穴。因此，这些新的卡帕酶一般不会存在沟穴拓扑异构构象，一般都采用的是随机内切的降解方式。

艾欧塔-卡拉胶酶

第一个被克隆的艾欧塔酶的基因来源于 *Alteromonas fortis* 和 *Z. galactanivorans*，被定义为糖苷水解酶家族的 82 个成员（GH82）（Barbeyron et al.，2000）。整整一个世纪，唯一一个表达过的艾欧塔酶是来自 *A. fortis*（Michel et al.，2000）。与卡帕酶不同的是，该酶的作用方式是一种异头体构型转化机制（Barbeyron et al.，2000）。CgiA 的晶体结构表明 GH82 酶家族成员采用的是右手β-螺旋折叠（Michel et al.，2001b），与卡帕酶（Michel et al.，2001a）的β-凝胶卷折叠没有任何关系。与其他蛋白质中的β-螺旋折叠不同的是，CgiA 在 C-端位置存在两个其他的结构域（A 和 B）。结构域 A 表现为α/β折叠，这种α/β折叠方式在一些 DNA/RNA 结合蛋白质中曾有发现，结构域 B 基本由凸环组成，靠两个二硫键稳定构象。CgiA 和一分子 ι-角叉二糖及一分子角叉四糖的复合物的结构使我们能够确认参与底物结合的几个残基为：Asn^{123}、Arg^{125}、Lys^{163}、Lys^{394}、Arg^{243}、Arg^{303}、Gly^{423}、Gln^{424}（β-螺旋核心），Tyr^{341}、Arg^{321}和 Arg^{353}（结构域 A）（Michel et al.，2003）。当与底物结合时，结构域 A 的构象发生很大的改变，变成β-螺旋裂缝，形成一个能够容纳艾欧塔-四糖的通道，同时艾欧塔-二糖与 N-端的区域结合。这种构象的转变，从一种开放的转变成闭合通道的构象，能够很好地执行 CgiA 的高效催化性能。艾欧塔-卡拉胶与酶分子共同孵育后，连续电镜观察可以看到纤维变得细长，而不是纤维结构的破坏，这是与酶的作用方式直接有关的。因为研究发现单一多糖链是从纤维上逐渐被降解下来而不是整条纤维的随机降解（Michel et al.，2003）。定点突变研究表明，在 *A. fortis* 艾欧塔酶分子中，Glu^{245}发挥了催化性质子供体的作用，Asp^{247}作为广义碱，激活催化性水分子。其他 3 个残基与酶活性间接相关：Gln^{222}同时与催化性水分子和氯离子结合，在活性中心通过活化水

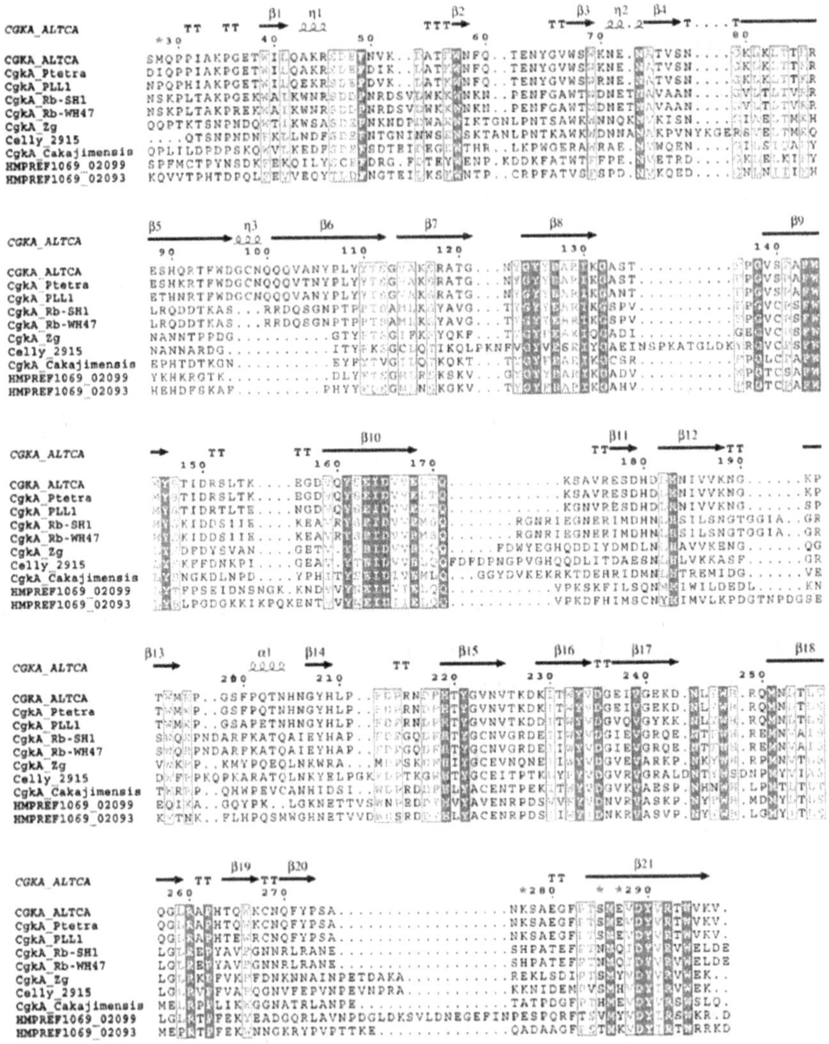

图 16.2 GH16 kappacarrageenases 的多序列比对

注：片段上方的二级结构对应这些从 *Pseudoalteromonas carrageenovora*（PDB：1DYP）卡帕酶的晶体结构中提取的．对应下面序列和来源物种的信息：CGKA_ALTCA，P43478（基因编码）、*Pseudoalteromonas carrageenovora*；CgkA_Ptetra、BAJ61957；*Pseudoalteromonas tetraodonis*；CgkA_PLL1、ADD92366；*Pseudoalteromonas* sp. LL1；CgkA_Rb-SH1、NP_865103、*Rhodopirellula baltica* SH1；CgkA_Rb-WH47、EGF25839、*Rhodopirellula baltica* WH47；CgkA_Zg、AAC27890、*Zobellia galactanivorans*；Celly_2915、YP_004263603、*Cellulophaga lytica* DSM 7489；CgkA_Cakajimensis、YP_003547637、*Coraliomargarita akajimensis* DSM 45221；HMPREF1069_02093、EIY64564、*Bacteroides ovatus* CL02T12C04；HMPREF1069_02099、EIY64570、*Bacteroides ovatus* CL02T12C04.

资料来源：此图用 ESPRIPT 准备（Gouet et al.，NAR，2003）

系统发挥作用；His[281]参与艾欧塔卡拉胶的结合，还可能参与 Glu[245]质子供体之间的质子传递作用；Glu[310]稳定了过渡态底物中间物构象（Rebuffet et al.，2010）。

Z. galactanivorans 基因组序列也表明这一海洋产黄菌属不仅拥有一个艾欧塔酶的基因，而是总共拥有 3 个酶的基因。基因 *cgiA*（Barbeyron et al.，2000）被重新命名为 *cgiA*1，另外两个被命名为 *CgiA*2 和 *CgiA*3。*CgiA*2 和 *CgiA*3 蛋白的序列远远短于 *CgiA*1 蛋白的长度。*CgiA*2 和 *CgiA*3 两种蛋白中都缺少结构域 A，*CgiA*3 蛋白中还缺少结构域 B。然而，*CgiA*1 和 *CgiA*3 都成功地在 *E. coli* 中得到高表达，而且均表现出活性（Rebuffet et al.，2010）。其他艾欧塔酶的基因后续在深海细菌中 *Microbulbifer thermotolerans* JAMB-A94 被克隆，并在 *Bacillus subtilis* 中高表达（Hatada et al.，2011）。该酶与从 *A. fortis* 和 *Z. galactanivorans* 中发现的艾欧塔酶 CgiA1 在进化关系上是较远的，分别只有 16%和 19%的相似性。艾欧塔酶家族成员的结构比对表明 *M. thermotolerans* 艾欧塔酶缺少结构域 A，在 β13 和 β14 链间插入了一段长达 67 个残基的序列（图 16.3）。Hatada 及其同事使 Glu[245]残基失活（*M. thermotolerans* 艾欧塔酶中的 Glu[351]），证实了 Glu[245]参与酶的催化机制。该酶没有保守的 Asp[247]残基，也没有发现酸碱催化机制（Hatada et al.，2011）。最后，我们最近研究的在几种海洋细菌的全基因组中发现了新的基因并推测为艾欧塔酶：在 *Cellulophage algicola* DSM14237（Abt et al.，2011）和 *Pseudoalteromonas atlantica* T6 C 中分别发现一个基因，在 *Cellulophage lytica* DSM7489（Pati et al.，2011）中发现两个基因，在 *Aquimarina agarilytica* ZC1（Lin et al.，2012）中发现 3 个基因，在一株未鉴定的命名为 S85（Oh et al.，2011）产黄菌属的菌株中发现了 3 个基因。这些艾欧塔酶中有 3 个酶具有保守的结构域 A 和 B（Celal_ 3965，Celly_ 2877 和 FbacS_ 07655），与从 *P. carrageenovora* 和 *Z. galactanivorans* 中发现的 CgiA1 进化为一支（图 16.4）。

其他的艾欧塔酶缺少结构域 A，GH82 家族成员的系统进化分析表明这些蛋白分化为至少两支：分支 B 主要包含从 *M. thermotolerans* 中发现的特征性艾欧塔酶，及从 *P. atlantica* T6 C 中发现的同源蛋白质 Patl_0879；分支 C 包含了其他的序列（FbacS_11850 除外），但是分支 C 相对不固定，将来很有可能被划分为不同的亚组（Gloster et al.，2008）。在 *A. fortis*（Michel et al.，2003）CgiA 已经鉴定出的主要残基中 Glu[245]（质子供体），His[281]（质子传递体），Gln[222]（结合 Cl 离子），Arg[243]（底物结合）在 GH82 家族中是严格保守的。Asp[247]的碱催化机制在所有分支 A 的艾欧塔酶中是保守的，在 CgiA3_Zg，AagaZ_16007 和 FbacS_11850 中也是保守的，但是在其余的序列中是缺失的。在一个糖苷水解酶家族中不保留相同催化残基的现象是非常少见的，但也不是没有先例。比如 GH97 家族中包含 inverting 及 retainingα-糖苷酶。在 inverting 糖苷酶中谷氨酸（Glu）发挥广义碱催化功能，而在其他催化成员中，不同位点的天冬氨酸（Asp）作为亲核催化基团。有关 GH82 家族是否代表着一种新的非寻常进化机制，已成为一个公开的问题。关键的问题是要确定 Asp[247]缺失艾欧塔酶的酸碱催化机制，和这些酶的作用方式是属于分支 A 艾欧塔酶的 inverting 机制，还是属于 GH97 家族中发现的 retaining 机制。最后，Asn[123]、Arg[125]、

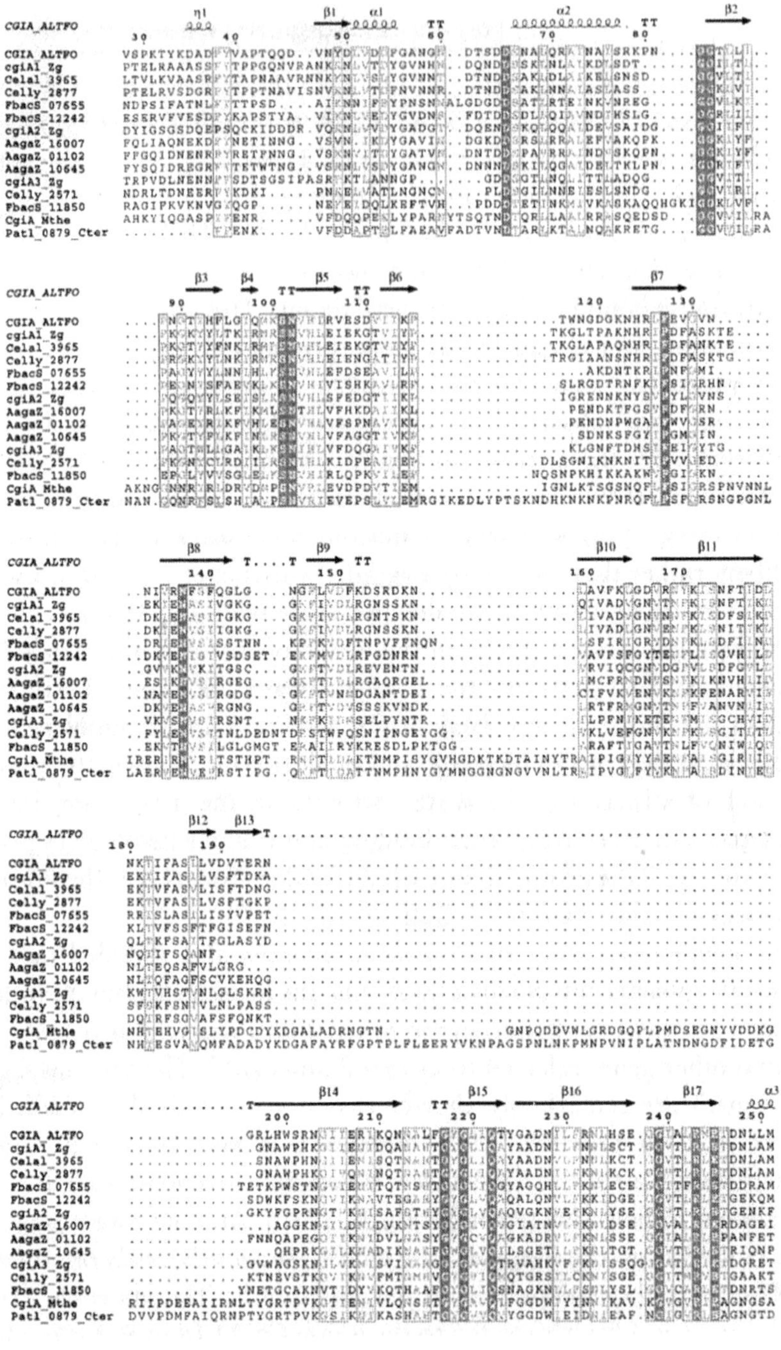

图 16.3　GH82 iotacarrageenase 的序列比对

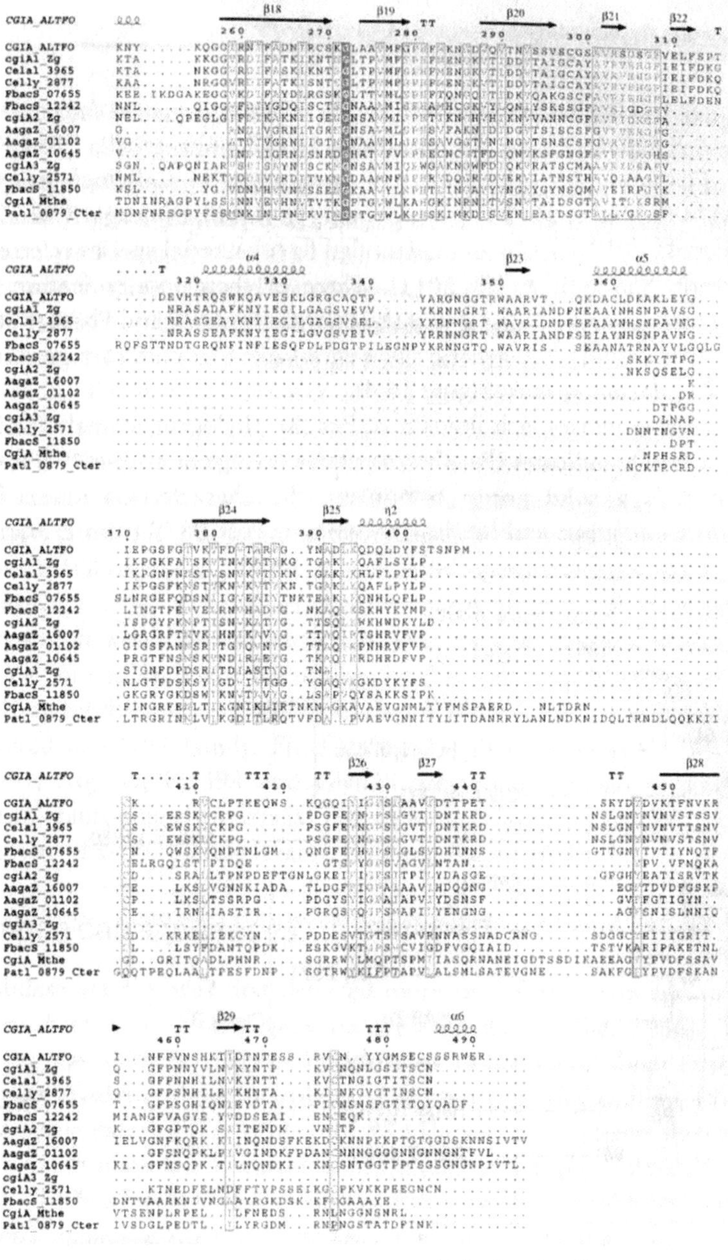

图 16.3　GH82 iotacarrageenase 的序列比对（续）

注：序列上方的二级结构是从对应的 *Alteromonas fortis*（PDB：1KTW）卡拉胶酶 CgiA 的晶体结构中提取的．标签对应下面序列和来源物种信息如下：CGIA_ALTFO，Q9F518，*Alteromona fortis*；cgi A1_Zg，YP_004738679，*Zobellia galactanivorans*；Celal_3965，YP_004166707，*Cellulophaga algicola* DSM 14237，Celly_2877，YP_004263565，*Cellilophaga lytica* DSM 7489；FbacS_07655，ZP_09497777，*Flavobacteriaceae* bacterium S85；FbacS_12242，ZP_09498608，*Flavobacteriaceae* bacterium S85，cgiA2_Zg，YP_004736593，*Zobellia galactanivorans*，AagaZ_16007，ZP_10842596，*Aquimarina agarilytica* ZC1；cgiA3_Zg，YP_004736413，*Zobellia galactanivorans*；Celly_2571，YP_004263259，*Cellulophaga lytica* DSM 7489，FbacS_11850，ZP_09498608，*Flavobacteriaceae bacterium* S85；CgiA_Mthe，BAJ40863，*Microbulbifer thermotolerans*；Patl_0879，YP_660459，*Pseudoalteromonas atlantica* T6c.

资料来源：此图用 ESPRIPT 准备（Gouet et al.，NAR，2003）

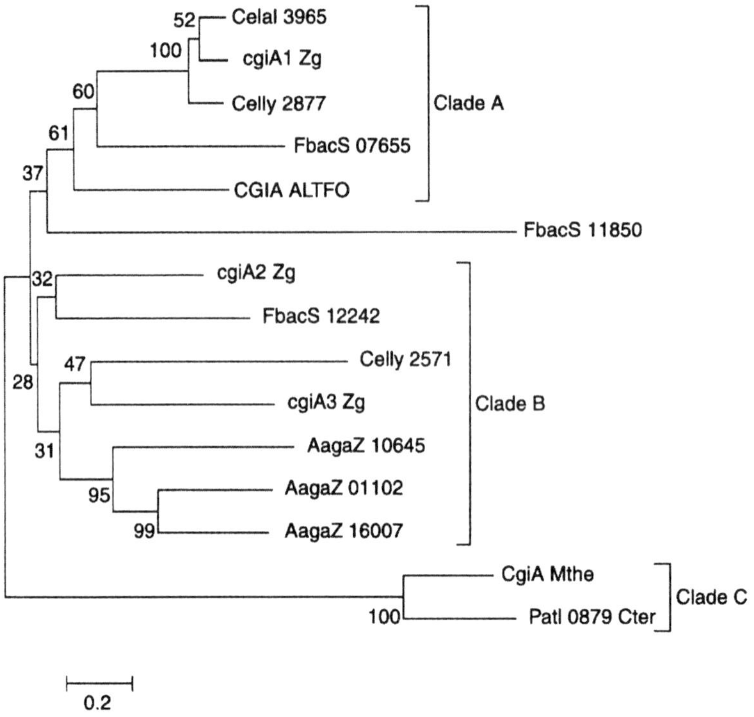

图 16.4　GH82 iotacarrageenase 的无根系统树

注：这个系统树依靠 PhyML 程序的最大可能性（ML）途径所派生出来（Guindon and Gascuel，2003），数字表明了在最大可能性分析中的引导值，这个片段标签和图 16.3 中相同.

Arg^{303}、Glu^{310}、Tyr^{341}、Arg^{353}、Lys^{394}和 Gly^{423}在分支 A 艾欧塔酶中是高度保守的，证实了这些残基在 GH82 家族中参与底物艾欧塔-卡拉胶结合中起重要作用图 16.3 和图 16.4。

拉姆达-卡拉胶酶

拉姆达酶活性最早是在 40 多年前，在海洋细菌 *P. carrageenovora*（Weigl and Yaphe，1966））中发现的。开始认为这种酶活性是由细胞外的一种酶复合物组成，包含 3 种水解酶（Johnston and McCandless，1973），但是后来纯化出一种细胞外 98 kDa 的单体蛋白质，表现出水解拉姆达家族卡拉胶的活性（Greer and Yaphe，1984a；Potin，1992）。我们最终成功克隆了 *P. carrageenovora* 中拉姆达酶的基因（*cglA*），表明该酶属于一个新的糖苷水解酶家族，与 GH16 家族卡帕酶和 GH82 家族艾欧塔酶（Guibet et al.，2007）是毫无关系的。这一基因产物比野生型的细胞外拉姆达酶要大一些（105 kDa），能与其他两个独立的模块形成弱结合的复合物，将二者连接在一起，其中一个是 N-末端的能够折叠成β-螺旋桨状的区域，一个 C-末端功能不明确的区域。CgIA 的作用方式是内切作用模式，机制是改变异头物构型的 inverting 机制。酶催化反应的主要终产物拉姆达-新角叉四糖和拉姆达-新角叉六糖（Guibet et al.，2007）。第二个拉姆达酶的基因也是从一种未鉴定的深海生物

Psedoalteromonas 菌株 CL19 中克隆的。这个蛋白与 *P. carrageenovora* 中发现的拉姆达酶是高度相似的（98%的序列相似性），而且表现出相同的作用模式，产生相同的终产物（Ohta and Hatada，2006）。由于这两种蛋白几乎是相同的，因此拉姆达酶还没有形成一个 CAZY 数据库（Bernard Henrissat，personal communication）中的家族编号。然而，这种现状即将发生变化。确实，我们课题组最近在几种细菌的基因组中鉴定出新的拉姆达酶：在海洋细菌 *Lentisphaera raneosa* HTCC2155（LNTAR-16468，Thrash et al.，2010）中，刺尾鱼属肠道共生菌 *Epulopiscium* sp.（Epulo_08588）中，和人肠道细菌 *Bacteroides ovatus* CL02T12C04（HMPREF1069-02051，Gevers et al.，2012）菌株中发现一个基因，在海洋产黄菌属 S85 菌株中发现 3 个基因（FbascS-08875，FbacS-08885 and FbacS-012272；（Oh et. al.，2011））。相应的基因产物与 *P. carrageenovora* 中的拉姆达酶在全长上是同源的，序列相似性从 25%（LNTAR_16468）至 47%（FbacS_08885）。拉姆达酶家族的催化性残基可能是属于在这 8 个序列中是高度保守的 10 个 Asp 和 3 个 Glu 中的一些残基（Asp-89，-93，-246，-250，-342，-471，-506，-631，-752，-873，-927；Glu-452，-533，-598，Cg1A 中的编号）。但是，预测 Cg1A 中哪个结构域在这个新的 GH 家族中是催化性区域仍是非常困难的。

16.4 其他海洋细菌多糖降解酶

在这一部分，我们将对其他特异性降解海洋多糖的酶进行综述。

16.4.1 褐藻胶裂解酶

褐藻胶是最重要的工业生产海洋多聚糖之一，从褐藻中分离。褐藻胶的工业应用与其保水能力、凝胶作用、黏稠性和稳定功能有关。另一方面，在正在开发的生物技术领域的应用，既基于褐藻胶分子本身的特殊的生物学功能，又基于其在多价阳离子（如 Ca^{2+}）存在下所具有的独特的温和的温度-非依赖性的固体/胶体（sol/gel）相转化。作为一个不分枝的线性共聚物家族，褐藻胶包含由 1，4 链接的 α-D-甘露糖醛酸（M）和 α-L-古罗糖醛酸（G）残基，在成分和序列上是高度变化的。该多糖可被看做是模块聚合物，由均一的 M 和 G 聚合区域组成，分别被称为 M 区和 G 区，其中散布着交替结构的区域（MG-区）。因此，褐藻胶降解酶能够表现出对这些多聚物结构特征的特异性（有趣的是，迄今为止还没有发现褐藻胶水解酶）这种多聚糖可被褐藻胶裂解酶降解，也称作褐藻胶酶或者褐藻胶裂解酶，能够通过 β-消除机制催化褐藻胶的降解，该催化机制还没有被完全阐明。

大量的褐藻胶裂解酶已经从海洋藻类或者无脊椎动物中被分离、进行性质研究（在第 13 章中也有描述），在革兰氏阴性土壤细菌中也有发现，该酶针对的是细菌外表面多糖中的褐藻胶荚膜，而不是海洋褐藻的细胞壁成分（Wong et al.，2000）。不管怎样，一些海洋细菌褐藻胶裂解酶已经被发现，尽管这些例子仅仅反映了众多未发现的酶中的极少部分。关于褐藻胶裂解酶分离和性质研究的详细的综述可以参考 Wong 等（2000）。最早的海洋细菌褐藻胶裂解酶是于 19 世纪 70 年代末期在 *Alginovibrio aquatilis*（Davidson et al.，1976；Stevens and Levin，1977）中发现的，随后，20 世纪 80 年代，又从其他海洋细菌（Doubet and Quatrano，1984）和马尾藻相关的海藻细菌（Romeo and Preston，1986）中分离出。因为褐藻胶能够作为大多数细菌的碳源和能源，许多研究最初都是以褐藻胶作为生长培养基中的唯一的碳源，诱导褐藻胶的产生。然而，在一些海洋细菌中，发现褐藻胶裂解酶活性是组成型的（Boyen et al.，1990）。由于筛选模式的原因，通常认为每种细菌生物中只有一种褐藻胶裂解酶。但是随着基因组数据库（Fernandez-Gomez et al.，2012；Kim et al.，2012b；Shi et al.，2012；Wakabayashi et al.，2012；Yu et al.，2012）数量的增加，越来越清楚地表明许多能降解和利用褐藻胶细菌会编码褐藻胶降解系统，至少是产生几种酶共同来彻底地降解和代谢这一碳源（Preston et al.，1985；Sawabe et al.，1997）。在这种情形下，我们团队最近已经分清和定性研究了第一个海洋细菌褐藻胶操纵子（Thomas et al.，2012）。

与糖苷水解酶（Cantarel et al.，2009）的分类相似，裂解酶被归为多糖裂解酶家族（PL），迄今为止，已在 7 个不同的 PL 家族中发现了褐藻胶裂解酶（http：//www.cazy.org）。含有褐藻胶裂解酶的家族有 PL5、PL6、PL7、PL14、PL15、PL17 和 PL18，隶属多样性家族的海洋褐藻胶裂解酶在底物特异性、分子大小以及生物化学性质等方面也是不同的（Wong et al.，2000）。大多数酶是典型的细胞外或者是胞质酶。虽然大部分的酶表现出内切 poly（M）裂解活性，至少一种具有外切 poly（M）裂解酶活性的酶已被报道（Doubet and Quatrano，1984）。少数已被报道的裂解酶表现出内切 poly（G）裂解酶活性，至少有一种酶具有外切 poly（G）裂解活性。菌株 SFFB 080483 Alg-G 是一株与 *Sargassum fluitans* 相关的海洋细菌，能够产生 G-特异性裂解酶，具有外切活性，能够降解 poly（G）区主要产生三聚体（Brown and Preston，1991）。

7 个包含褐藻胶裂解酶的 PL 家族相对应于 4 种不同的结构形式，即 PL5 中的（α/α）4 结构（1HV6），PL6 中的β-螺旋结构（1DBG），PL7、PL14 和 PL18 中的β-三明治结构或 2 卷饼折叠（2CWS），PL15 中的多结构域（α/α）6-β-三明治结构（3AFL）。这种结构多样性也可以解释不同的褐藻胶裂解酶在大小和特性上巨大的多样性。然而，迄今为止仍然没有海洋细菌褐藻胶裂解酶的 3D 晶体结构被报道，上述列举的例子均是同源蛋白酶的结构。

16.4.2 岩藻聚糖酶

岩藻聚糖是硫酸酯多糖，含有 α-L-岩藻糖残基，是褐藻细胞壁的组成成分，在海洋无脊椎动物如海参、海胆中也有发现。棘皮类动物岩藻聚糖是线性的、相对单一的岩藻糖硫酸酯的多聚物（Pomin and Mourao，2008），而褐藻中的岩藻聚糖含有连续的高分支结构的多糖，既有糖醛酸含量高，硫酸酯含量低的多聚物，分子中含有高比例的 D-木糖，D-半乳糖和 D-甘露糖（xylo-fuco-glucuronan and eylo-fuco-glucans），也有高含量单岩藻糖硫酸酯分子（Usov and Bilan，2009；Michel et al.，2010；Popper et al.，2011））。*Fucales* 中的硫酸化岩藻糖分子中含有较长片段的 L-岩藻糖-2，3-二硫酸和 L-岩藻糖-2-硫酸二糖重复单位，交替性的由 α-1，3 和 α-1，4 糖苷键（Chevolot et al.，2001；Colin et al.，2006）连接。*Laminariales*（Nishino et al.，1991）和 *Ectocarpales*（Ponce et al.，2003）中的硫酸化岩藻聚糖主要是靠 3-连接的 α-L-岩藻糖残基组成，硫酸化位点在 C-4。

岩藻聚糖裂解活性最早是在已知能够降解其他藻类多糖的海洋细菌中被发现的，如 *Pseudoalteromonas atlantica* 具有琼胶降解活性和 *P. carrageenovora* 具有卡拉胶降解活性（Yaphe and Morgan，1959）。此后，相继分离到岩藻聚糖裂解细菌，特别是从褐藻和海参中，还有隶属下列两个门的细菌：Grammaproteobacteria（Bakunina et al.，2000，2002；Furukawa et al.，1992；Sakai et al.，2004）和 *Bacteroidetes*（Bakunina et al.，2000；Barbeyron et al.，2008b；Chang et al.，2010；Kim et al.，2008；Ohshiro et al.，2010；Sakai et al.，2002；Urvantseva et al.，2006）。在 *Verrucomicrobia* ‘*Fucophilus fucoidanolyticus*’（Sakai et al.，2003b）和 *Luteolibacter algae*（Ohshiro et al.，2012）中也已经发现了两种岩藻聚糖裂解菌。

仅有几种野生型岩藻聚糖酶已经被纯化，它们生物化学性质的多样性表明岩藻聚糖酶一词包含了不同种类的多糖降解酶。*Vibrio* sp.，N-5 中有 3 种岩藻聚糖酶，以外切的作用方式作用于褐藻 *Kjellmaniella crassifolia* 中的岩藻聚糖，释放出短的含有硫酸化 L-岩藻糖的寡糖，但是断裂的糖苷键的本质还没有被阐明（Furukawa et al.，1992）。来自 *Pseudoalteromonas citrea* 中的三株菌能够降解褐藻岩藻聚糖（Bakunina et al.，2002），*P. citrea* KMM 3296 中的岩藻聚糖酶对硫酸化 α-1，3 和 α-1，3 -α-1，4 岩藻糖是有活性的，能够以内切的方式断裂上述两种情况中的 α-1，3 糖苷键（Kusaykin et al.，2006）。Sakai 及其同事介绍了来自 3 种不同海洋细菌中的岩藻聚糖酶。‘*F. fucoidanolyticus*’中的酶是内切水解酶，能够断裂 *Cladosiphon okamuranus* 岩藻聚糖分子中的 α-1，3 糖苷键，释放出具有下述重复单位的寡糖混合物：L-岩藻糖-α-1，3-L-岩藻糖 4S-α-1，3-L-岩藻糖 4S-α-1，3-（D-葡萄糖醛酸-α-1，2）-L-岩藻糖（Sakai et al.，2003a）。*Alteromononaceae* 菌株 SN-1009 产生一种内切岩藻聚糖酶，能够降解 *K. crassifolia* 和其他昆布目岩藻聚糖（Sakai et al.，2004），但是其释放出的寡糖的结构在后来的报道中没有进行描述。来自第三种细菌 *Fucobacter marina* 中的“岩藻聚糖酶”不是糖苷键水解酶，而是多糖裂解酶，能够断裂由 β-1，2 键所连接的 D-葡萄糖醛酸和 D-甘露糖组成的重复单位之间的 α-1，4 键。D-甘露糖残基通常与硫酸化 L-岩藻糖通过 α-1，3 键（Sakai et al.，2003a，c）形成分支。来自 *K. crassifolia* 的这种新多糖被命名为硫酸化岩藻糖葡萄糖醛酸甘露糖（SFGM），形式上被划分为岩藻聚糖，但是其分子结构与 *Fucales* 和 *Laminariales* 中的硫酸化均一岩藻聚糖是完全不同的。

最后，我们课题组从一家水处理工厂分离了一株岩藻聚糖裂解菌（菌株 SW5），该厂主要回收利用从朗代诺地区一种植物中提取褐藻酸的工业废水（Brittany，France）（Descamps et al.，2006）。该菌株只能在海水中生长，被定义为一种新的 *Mariniflexile* 属中的产黄菌类，命名为 *M. fucanivorans*（Barbeyron et al.，2008b）。该海洋细菌分泌一种内切岩藻聚糖酶，能够降解褐藻 *Pelvetia caniculata*（Descamps et al.，2006）中的岩藻聚糖，该野生型酶已经被纯化，其内部 3 个肽段已经通过 Edman 降解法进行测序。利用从这些肽段推导出的核苷酸探针，我们能够对其相应的基因（fcnA）进行克隆和测序，这是第一个岩藻聚糖酶的基因（Colin et al.，2006）。FcnA 蛋白（1007 残基）包含有一个 N-端催化模块（~400 残基），有 3 个重复性结构域，经预测采用 IG 折叠，还有一段 80 个残基长度的 C-末端结构域，功能未知。一个包括 N-末端模块和 IG 折叠模块的重组蛋白在 *E. coli* 得到高表达，经纯化，表明该蛋白保留了与野生型酶相同的活性。该岩藻聚糖酶能够释放出四糖和六糖作为终产物，能够断裂重复单位 L-岩藻糖-2，3-二硫酸化-α-1，3-L-岩藻糖-2-硫酸之间的 α-1，4 糖苷键。N-末端的催化模块与专利报道的菌株 SN-1009 中的两个岩藻聚糖酶基因表现出约 25% 的序列相似性，这 3 个蛋白共同被定义为一个新的糖苷水解酶家族，GH107 家族（Colin et al.，2006）。我们近期在海洋细菌 *Shewanella violacea* DSS12 的基因组中鉴定出一个新的 GH107 家族的岩藻聚糖酶。FcnA 的催化区与汶蛋白（SVI_0379）有 28%的序列同源性。

16.4.3 石莼聚糖裂解酶

石莼聚糖是 *Ulva* 属绿藻细胞壁的主要成分（Popper et al.，2011）。这种复杂的硫酸化多糖，占干重的 38%~54%，主要由硫酸化 L-鼠李糖、D-葡萄糖醛酸及其 C5-差向异构体 L-艾杜糖醛酸（在动物的糖胺聚糖中也存在）和小部分的 D-木糖（Brading et al.，1954；Lahaye and Robic，2007；McKinnell and Percival，1962）组成。分子中 3 个主要的重复单位是 α-L-鼠李糖-3S-1，4-β-D-葡萄糖醛酸（石莼醛酸 A），α-L-鼠李糖-3S-1，4-α-D-艾杜糖醛酸（石莼醛酸 B）和 α-L-鼠李糖-3S-1，4-β-D-木糖（Lahaye and Robic，2007）。第一个被报道的石莼聚糖裂解微生物是一株革兰氏阴性海洋细菌，是在圣布里厄湾的一个绿潮点分离得到。这一细菌没有被进一步深入研究，但是一个部分纯化的酶能够断裂 L-鼠李糖-3-硫酸（Rha3S）和 D-葡萄糖醛酸（GlcA）之间的 β-(1，4)糖苷键，释放出的寡糖在非还原端带有一个不饱和醛酸。该酶属于多糖裂解酶，被定义为石莼聚糖裂解酶（Lahaye et al.，1997）。另一个石莼聚糖裂解酶是由 *Ochrobactrum* 类产生的，在 2010 年申请专利（Elboutachfaiti et al.，2010）。该酶与 Lahaye（1997）及其同事报道的石莼聚糖裂解酶具有相似的生化性能。近期，我们实验室从软体动物 *Aplysia punctata* 的粪便排泄物中分离出石莼聚糖裂解菌，该软体动物主要喂食 *Ulva* sp. 该海洋细菌经鉴定为 *Persicivirga* 属的一个新种 *P. ulvanivorans*（Barbeyron et al.，2011）。奇怪的是，该产黄细菌不能利用任何的碳水化合物作为碳源，问题就在于石莼聚糖裂解活性在该微生物体内发挥什么生物学作用？从 *P. ulvanivorans*（Nyvall Collen et al.，2011）的细胞外培养基中纯化出两个石莼聚糖裂解酶，分子量是 30 kDa 和 46 kDa。经质谱对肽链测序后发现这两个蛋白质有一个同源的催化模式。利用简并寡核苷酸引物克隆出第一个石莼聚糖裂解酶的基因（1 536 bp）。这个基因编码 56 kDa 的蛋白产物，大部分肽链序列与已经纯化出的 46 kDa 蛋白质的序列相一致。蛋白质的 N-末端有一个信号肽，后面是推断出的催化区（约 34 kDa)，接着是一个低复杂度的功能区域与 C-末端未知的模式分离开。其中推断出的催化区与公共序列数据库中的任何蛋白质都没有明显的相似性。针对 N-末端区的重组蛋白质在 *E. coli* 中得到较弱的表达，但是表现出明显的石莼聚糖裂解活性。最后，对两种野生型石莼聚糖裂解酶释放出的降解产物的分析表明裂解酶能够断裂 L-鼠李糖 3-硫酸和 D-葡萄糖醛酸或者 L-艾杜糖醛酸之间的 1，4 糖苷键（Nyvall Collen et al.，2011)。

16.5 海洋多糖酶在现代生物科技中独特/选择性的应用和海洋微生物的整体多糖降解系统

起初，降解海藻细胞壁多糖的酶被分离纯化主要用于海藻原生质体的生产，旨在用于对藻类细胞发育的生物学研究（Boyen et al.，1990；Davidson et al.，1976；Quatrano and Stevens，1976)。海洋细菌中藻类细胞壁降解酶（Ralet et al.，2009）就好比植物细胞壁降解酶，其另一个重要科学应用是用于研究有意义的多糖的精细结构，将酶的降解与寡糖分析检测的方法如色谱法，NMR 和质谱等相结合运用（Coorrec et al.，2011；Duckworth and Turvey，1969；Greer and Yaphe，1984b；Guibet et al.，2007；Jouanneau et al.，2010；Lahaye et al.，1986，1989)。出于这种目的，明确了解所应用酶的底物特性和控制酶作用的降解模式变得尤其重要。最近，在基因科技领域，通过利用 β-琼脂酶，已经开发出更温和、高效的从纯化的琼胶中提取 DNA 的方法（Gold，1992；Sugano et al.，1993b)。

显而易见，对能够特异降解藻类多糖的酶的研究兴趣日益增长是因为越来越多的最新研究报道了一些藻类衍生而来的寡糖的生理学功能（for review，see for example Courtois，2009；Ghosh et al.，2009；Pomin and Mourao，2008)。的确，藻类多糖/寡糖的硫酸化性质与人体糖胺聚糖（GAGs）中的硫酸化是非常相似的，导致了寡糖对人类的健康具有非常重要的生物学功能，如抗凝（Pereira et al.，1999)、免疫激活（Bhattacharyyaet al.，2010)、抗氧化（Hatada et al.，2006）或者抗炎（Berteau and Mulloy，2003）功能等。为继续探索这些功能，用于医疗目的的开发，能够以明确的、可重复的方式生产功能性寡糖是至关重要的。利用酶技术生产寡糖也因此成为科学研究越来越关注的领域（Zhang and Kim，2010)。

有关海洋细菌酶的研究的确在日益发展，最初的研究是筛选可以用于生物资源降解/再生旨在生产藻类生物燃料的海洋微生物开始的。在这种背景下，近期报道了一个非常有意义的生物工程项目，该项目旨在建立一株能够利用褐藻生产生物乙醇的 *E. coli* 菌株

(Wargacki et al., 2012)。在这项研究中，将 *Sphingomonas* sp. A1 和 *V. splendidus* 12B01 中的与褐藻酸-相关的基因共同引入 *E. coli* 中，以完全降解利用褐藻酸用于进一步的发酵。还有一个研究旨在利用海藻生产生物乙醇，报道了通过利用琼胶裂解酶的“鸡尾酒”组合，实现琼脂糖的彻底单糖化（Kim et al., 2012a)。已经建立起大基因组策略，来寻找‘the missing links’（酶)，以完善海洋细菌代谢途径，或者了解海洋微生物群体之间与沿海生物物质的彻底循环中的相互作用，（Kennedy et al., 2008, 2010）越来越多的新型海洋藻类多糖降解菌被分离和研究（Sakatoku, 2012; Shi et al., 2012; Wakabayashi et al., 2012; Yu et al., 2012)。总之，这些进一步的研究工作将有助于填补针对海洋细菌多糖酶尚未发现的多样性、分布和生化特征的知识空缺。

16.6 结论

尽管在本章中总结了发现海洋细菌中能够作用于藻类多糖的酶的漫长发现历程，显而易见的是在不久的将来会有更多的酶被发现。在对一些全球性问题的共同关注的驱动下，例如，开发可再生生物资源作为替代能源，或者其他的高附加值产品，或者研究全球性海洋碳循环及其对气候的影响，了解海洋微生物及其与环境的相互作用变得刻不容缓。微生物占到海洋初级生产力的98%左右，在海洋食物链以及碳循环和能量循环中发挥重要作用，微生物产生的多糖降解酶在上述环节中发挥关键作用。多聚糖降解酶在此过程中扮演重要角色。

参考文献

Abt, B., Lu, M., Misra, M., Han, C., Nolan, M., et al., (2011) Complete genome sequence of *Cellulophaga algicola* type strain (IC166). *Stand. Gen. Set.* 4, 72-80.

Aizawa, M. (2007) ‘Seaweed bioethanol production in Japan——the Ocean Stmrae Project’. In *Oceans* 2007 (eds K. Asaoka, M. Atsumi and T. Sakou), IEEE Conference Publications, Vancouver, BC, Canada. Published by Woodhead Publishing Limited, 2013.

AlloucK, J., Helbert, W., Henrissat, B. and Czjzek, M. (2004) Parallel substrate binding sites in a beta-agarase suggest a novel mode of action on double, helical agarose. *Structure*. 12, 623-632.

Anderson, N. S. and Rees, D. A. (1965) Porphyran——a polysaccharide with a masked repeating structure. *J. Chem. Soc.* 5880-5887.

Arnaud-Haond, S., Arrieta, J. M. and Duarte, C. M. (2011) Marine biodiversity and gene patents. *Science* 331, 1521-1522.

Azam, F. and Malfatti, F. (2007) Microbial structuring of marine ecosystems. *Nat. Rev. Microbiol.* 5, 782-791.

Bakunina, I., Nedashkovskaia, O. I., Alekseeva, S. A., Ivanova, E. P., Romanenko, L. A., et al., (2002) Degradation of fucoidan by the marine proteobacterium *Pseudoalteromonas citrea*. *Microbiol.* 71,

49-55.

Bakunina, I. , Shevchenko, L. S. , Nedashkovskaia, O. I. , Shevchenko, N. M. , Alekseeva, S. A. , et al. , (2000) Screening of marine bacteria for fucoidanases. *Microbiol.* 69, 370-376.

Barbeyron, T. , Carpentier, F. , L'Haridon, S. , Schuler, M. , Michel, G. and Amann, R. (2008a) Description of *Maribacter forsetii* sp. nov. , a marine Flavobacteriaceae isolated from North Seawater, and emended description of the genus *Maribacter. Intern. J. Syst. Evol. Microbiol.* 58, 790-797.

Barbeyron, T. , Gerard, A. , Potin, P. , Henrissat, B. and Kloareg, B. (1998) The kappa-carrageenase of the marine bacterium *Cytophaga drobachiensis.* Structural and phylogenetic relationships within family-16 glycoside hydrolases. *Mol. Biol. Evol.* 15, 528-537.

Barbeyron, T. , Henrissat, B. and Kloareg, B. (1994) The gene encoding the kappa-carrageenase of *Alteromonas carrageetiovora* is related to beta-1, 3-1, 4-glucanases. *Gene.* 139, 105-109.

Barbeyron, T. , L'Haridon, S. , Michel, G. and Czjzek, M. (2008b) *Mariniflexile fucanivorans sp. nov.* , a marine member of the Flavobacteriaceae that degrades sulphated fucans from brown algae. *Intern. J. Syst. Evol. Microbiol.* 58, 2107-2113.

Barbeyron, T. , Lerat, Y. , Sassi, J. F. , LePanse, S. , Helbert, W. and Collen, P. N. (2011) *Persicivirga ulvanivorans* sp. nov. , a marine member of the family *Flavobacteriaceae* that degrades ulvan from green algae. *Intern. J. Syst. Evol. Microbiol.* 61, 1899-1905.

Barbeyron, T. , Michel, G. , Potin, P. , Henrissat, B. and Kloareg, B. (2000) Iota-Carrageenases constitute a novel family of glycoside hydrolases, unrelated to that of kappa - carrageenases. *J. Biol. Chem.* 275, 35499-35505.

Baueir, M. , Kube, M. , Teeling, H. , Richter, M. , Lombardot, T. , et al. , (2006) Whole genome analysis of the marine Bacteroidetes *Gramella forsetii* reveals adaptations to degradation of polymeric organic matter. *Environ. Microbiol.* 8, 2201-2213.

Belas, R. (1989) Sequence analysis of the *agrA* gene encoding beta-agarase from *Pseudomonas atlantica. J. Bacteriol.* 171, 602-605.

Beltagy, E. A. , Youssef, A. S. , El-Shenawy, M. A. and El-Assar, S. A. (2012) Purification of kappa (k) - carrageenase from locally isolated *Cellulosimicrobium cellulans. Afr. J. Biotechnol.* 11, 11438-11446.

Benner, R. , Pakulski, J. D. , McCarthy, M. , Hedges, J. I. and Hatcher, P. G. (1992)

Bulk chemical characteristics of dissolved organic matter in the ocean. *Science* 255, 1561-1564.

Berteau, O. and Mulloy, B. 2003) Sulfa ted fucans, fresh perspectives: structures, functions, and biological properties of sulfated fucans and an overview of enzymes active toward this class of polysaccharide. *Glycobiol.* 13, 29-40.

Bhattacharyya, S. , Liu, H. , Zhang, Z. , Jam, M. , Dudeja, P. K. , et al. , (2010) Carrageenan-induced innate immune response is modified by enzymes that hydrolyze distinct galactosidic bonds. *J. Nutr. Biochem.* 21, 906-913.

Boyen, C. , Bertheau, Y. , Barbeyron, T. and Kloareg, B. (1990) Preparation of guluronate lyase from *Pseudomonas alginovora* for protoplast isolation in *Laminaria. Em. Microb. Technol.* 12, 885-890.

Brading, J. W. E. , Georg-Plant, M. M. T. and Hardy, D. M. (1954) The polysaccharide from the alga *Ulva lactuca.* Purification, hydrolysis, and methylation of the polysaccharide. *J. Chem. Soc.* 319-324.

Brown, B. J. and Preston, J. E, 3rd. (1991) L-guluronan-specific alginate lyase from a marine bacterium associated with *Sargassum*. *Carbobydr. Res.* 211, 91-102.

Campo, V. L., Kawano, D. F., Silva, D. B. J. and Carvalho, I. (2009) Carrageenans: biological properties, chemical modifications and structural analysis——a review. *Carbobydr. Pol.* 77, 167-180.

Cantarel, B. L., Coutinho, P. M., Rancurel, C., Bernard, T., Lombard, V. and Henrissat, B. (2009) The Carbohydrate-Active EnZymes database (CAZy): an expert resource for Glycogenomics. *Nuc. A. Res.* 37, D233-D238.

Chang, Y., Xue, C., Tang, Q., Li, D., Wu, X. and Wang, J. (2010) Isolation and characterization of a sea cucumber fucoidan-utilizing marine bacterium. *Lett. Appl. Microbiol.* 50, 301-307.

Chevolot, L., Mulloy, B., Ratiskol, J., Foucault, A. and Colliec-Jouault, S. (2001) A disaccharide repeat unit is the major structure in fucoidans from two species of brown algae. *Carbobydr. Res.* 330, 529-535.

Chi, W. J., Chang, Y. K. and Hong, S. K. (2012) Agar degradation by microorganisms and agar-degrading enzymes. *Appl. Microbiol. Biotechnol.* 94, 917-930.

Colin, S., Deniaud, E., Jam, M., Descamps, V., Chevolot, Y., et al., (2006) Cloning and biochemical characterization of the fucanase FcnA: definition of a novel glycoside hydrolase family specific for sulfated fucans. *Glycobiol.* 16, 1021-1032.

Correc, G., Hehemann, J. H., Czjzek, M. and Helbert, W. (2011) Structural analysis of the degradation products of porphyran digested by *Zobellia galactanivorans* beta-porphyranase A. *Carbobydr. Pol.* 83, 277-283.

Courtois, J. (2009) Oligosaccharides from land plants and algae: production and applications in therapeutics and biotechnology. *Curr. Opitt. Microbiol.* 12, 261-273.

Craigie, J. (1990) 'Cell walls'. In *Biology of the Red Algae*, Cole, K. and Sheath, R. (eds), Cambridge University Press, Cambridge, 221-257.

Dabin, J. (2008) Etude structurale et fonctionnelle des polysaccharidases de *Rhodopirellula baltica* (Structural and functional study of polysaccharidases from *Rhodopirellula baltica*). Phd thesis, University Pierre and Marie Curie -Paris 6.

Davidson, I. W., Sutherland, I. W. and Lawson, C J. (1976) Purification and properties of an alginate lyase from a marine bacterium. *Biockem.* 159, 707-713.

Day, D. F. and Yaphe, W. (1975) Enzymatic hydrolysis of agar: purification and characterization of neoagarobiose hydrolase and p-nitrophenyl alpha-galactoside hydrolase. *Canad. J. Microbiol.* 21, 1512-1518.

Descamps, V., Colin, S., Lahaye, M., Jam, M., Richard, C, et al., (2006) Isolation and culture of a marine bacterium degrading the sulfated fucans from marine brown algae. *Mar. Biotechnol.* 8, 27-39.

Doubet, R. S. and Quatrano, R. S. (1984) Properties of alginate lyases from marine bacteria. *Appl. Environ. Microbiol.* 47, 699-703.

Duckworth, M. and Turvey, J. R. (1969) The action of a bacterial agarase on agarose, porphyran and alkali-treated porphyran. *Biochem. J.* 113, 687-692.

Ekborg, N. A., Taylor, L. E., Longmire, A. G., Henrissat, B., Weiner; R. M. and Hutcheson, S. W. (2006) Genomic and proteomic analyses of the agarolytic system expressed by *Saccharophagus degradans* 2-40. *Appl. Environ. Microbiol.* 72, 3396-3405.

Elboutachfaiti, R., Pheulpin, P., Courtois, B. and Courtois-Sambourg, J. (2010) Method for enzyme cleavage of polysaccharides derived from green algae. US patent 2010/0261894.

Fernandez-Gomez, B. , Fernandez-Guerra, A. , Casamayor, E. O. , Gonzalez, J. M. , Pedros-Alio, C. and Acinas, S. G. (2012) Patterns and architecture of genomic islands in marine bacteria. *BMC Genom.* 13, 347.

Flament, D. , Barbeyron, T. , Jam, M. , Potin, P. , Czjzek, M. , et al. , (2007) Alpha-agarases define a new family of glycoside hydrolases, distinct from beta-agarase families. *Appl. Environ. Microbiol.* 73, 4691-4694.

Fu, X. T. and Kim, S. M. (2010) Agarase: review of major sources, categories, purification method, enzyme characteristics and applications. *Mar. Drugs.* 8, 200-218.

Fu, X. T. , Pan, C. H. , Lin, H. and Kim, S. M. (2009) Gene cloning, expression, and characterization of a beta-agarase, agaB34, from *Agarivorans albus YKW*-34. *J. Microbiol. Biotechnol.* 19, 257-264.

Furukawa, S. -I. , Fujikawa, T. , Koga, D. and Ida, A. (1992) Purification and some properties of exo-type fucoidanases from *Vibrio* sp. N-5. *Biosci. Biotechnol. Biochem.* 56, 1829-1834.

Gao, K. and McKinley, K. R. (1994) Use of macroalgae for marine biomass production and CO_2 remediation: a review. *J. Appl. Phycol.* 6, 45-60.

Gevers, D. , Knight, R. , Petrosino, J. F. , Huang, K. , McGuire, A. L. , et al. , (2012) i The human microbiome project: a community resource for the healthy human microbiome. *PLoS Biol.* 10, el001377.

Ghosh, T. , Chattopadhyay, K. , Marschall, M. , Karmakar, P. , Mandal, P. and Ray, B. (2009) Focus on antivirally active sulfated polysaccharides: from structure-activity analysis to clinical evaluation. *Glycobiol.* 19, 2-15.

Gilbert, H. J. , Stalbrand, H. and Brumer, H. (2008) How the walls come crumbling down: recent structural biochemistry of plant polysaccharide degradation. *Curr Opin Plant Biol.* 11, 338-348.

Glöckner, F. O. and Joint, I. (2010) Marine genomics in Europe: current status and perspectives. Microb. Biotechnol. 3, 523-530.

Glöckner, F. O. , Kube, M. , Bauer, M. , Teeling, H. , Lombardot, T. , et al. (2003) Complete genome sequence of the marine planctomycete *Pirellula* sp. strain 1. Proc. Natl. Acad. Sci. USA 100, 8298-8303.

Gloster, T. M. , Turkenburg, J. P. , Potts, J. R. , Henrissat, B. and Davies, G. J. (2008) Divergence of catalytic mechanism within a glycosidase family provides insight into evolution of carbohydrate metabolism by human gut flora. *Chetn. Biol.* 15, 1058-1067.

Gold, B. (1992) Use of a novel agarose gel-digesting enzyme for easy and rapid purification of PCR-amplified DNA for sequencing. *Biotechn.* 13, 132-134.

Gouet, P. , Robert, X. and Courcelle E. (2003) ESPrint/ENDscript: Extracting and rendering sequence and 3D information from atomic structures of proteins. *Nucleic Acids Res.* 31, 3320-3323.

Gran H. H. (1902) Studien Uber mearesbakterien. Ⅱ. tJber die hydroiyse des agars-agars durch ein neues enzym, die gelase. *Bergens Museums Arb* 2, 1-16.

Greer, C. W. and Yaphe, W. (1984a) Characterization of hybrid (beta-kappa-gamma) carrageenan from *Eucheuma gelatinae* J. Agardh (*Rhodophyta*, *Solieriaceae*) using carrageenases, Infrared and, ^{13}C-nuclear magnetic resonance spectroscopy. *Bot. Mar.* 27, 473-478.

Greer, C. W. and Yaphe, W. (1984b) Hybrid (iota-nu-kappa) carrageenan from *Eucheuma nudum* (*Rhodophyta*, *Solieriaceae*), identified using iota-and kappa-carrageenases and ^{13}C-nuclear magnetic resonance spectroscopy. *Bot. Mar.* 27, 479-484.

Groisillier, A., Herve, C., Jeudy, A., Rebuffet, E., Pluchon, P. F., et al., (2010) MARINE-EXPRESS: taking advantage of high throughput cloning and expression strategies for the post-genomic analysis of marine organisms. *Microb. Cell Fact.* 9, 45.

Groleau, D. and Yaphe, W. (1997) Enzymatic hydrolysis of agar: purification and characterization of beta-neoagarotetraose hydrolase from *Pseudomonas atlantica*. *Canad. J. Microbiol.* 23, 672-679.

Guibet, M., Colin, S., Barbeyron, T., Genicot, S., Kloareg, B., et al., (2007) Degradation of lambda-carrageenan by *Pseudoalteromonas carrageenovora* lambda-carrageenase: a new family of glycoside hydrolases unrelated to kappa-and iota-carrageenases. *Biochem. J.* 404, 105-114.

Guindon, S. and Gascuel, O. (2003) A simple, fast, and accurate algorithm to estimate large phylogenies by maximum likelihood. *Syst. Biol.* 52, 696-704.

Ha, S. C., Lee, S., Lee, J., Kim, H. T., Ko, H. J., et al., (2011) Crystal structure of a key enzyme in the agarolytic pathway, alpha-neoagarobiose hydrolase from *Saccharophagus degradans* 2-40. *Biochem. Biophys. Res. Commun.* 412, 238-244.

Harmsen, H. J. M., Prieur, D. and Jeanthon, C. (1997) Distribution of microorganisms in deep sea hydrothermal vent chimneys investigated by whole-cell hybridization and enrichment culture of thermophilic subpopulations. *Appl. Environ. Microbiol.* 63, 2876-2883.

Hatada, Y., Mizuno, M., Li, Z. and Ohta, Y. (2011) Hyper-production and characterization of the iota-carrageenase useful for iota-carrageenan oligosaccharide production from a deep-sea bacterium, *Microbulbifer thermotolerans* JAMB-A94T, and insight into the unusual catalytic mechanism. *Mar. Biotechnol.* 13, 411-422.

Hatada, Y., Ohta, Y. and Horikoshi, K. (2006) Hyperproduction and application of alpha-agarase to enzymatic enhancement of antioxidant activity of porphyran. *J. Agric. Food. Chem.* 54, 9895-9900.

Hehemann, J. H., Correc, G., Barbeyron, T., Helbert, W., Czjzek, M. and Michel, G. (2010) Transfer of carbohydrate-active enzymes from marine bacteria to Japanese gut microbiota. *Nature* 464, 908-912.

Hehemann, J. H., Correc, G., Thomas, F., Bernard, T., Barbeyron, T., et al., (2012a) Biochemical and structural characterization of the complex agarolytic enzyme system from the marine bacterium *Zobellia galactanivorans*. *J. Biol. Chem.* 287, 30571-30584.

Hehemann, J. H., Smyth, L., Yadav, A., Vocadlo, D. J. and Boraston, A. B. (2012b) Analysis of keystone enzyme in Agar hydrolysis provides insight into the degradation (of a polysaccharide from) red seaweeds. *J. Biol. Chem.* 287, 13985-13995.

Henshaw, J., Horne-Bitschy, A., van Bueren, A. L., Money, V. A., Bolam, D. N., et al., (2006) Family 6 carbohydrate binding modules in beta-agarases display exquisite selectivity for the non-reducing termini of agarose chains. *Biol. Chem.* 281, 17099-17107.

Jam, M., Flament, D., Allouch, J., Potin, P., Thion, L., et al., (2005) The endo-beta-agarases AgaA and AgaB from the marine bacterium *Zobellia galactanivorans*: two paralogue enzymes with different molecular organizations and catalytic behaviours. *Biochem. J.* 385, 703-713.

Jensen, A. (1993) Present and future needs for algae and algal products. *Hydrobiol.* 260/261, 15-23.

Johnston, K. and McCandless, E. (1973) Enzymatic hydrolysis of potassium chloride soluble fraction of carrageenan: properties of λ-carrageenases from *Pseudomonas carrageenovora*. *Canad. J. Microbiol.* 19, 779-788.

Jouanneau, D., Boulenguer, P., Mazoyer, J. and Helbert, W. (2010) Enzymatic degradation of hybrid iota-/

nu-carrageenan by *Alteromonas fortis* iota-carrageenase. *Carbohydr. Res.* 345, 934-940.

Kennedy, J., Flemer, B., Jackson, S. A., Lejon, D. P., Morrissey, J. P., et al., (2010) Marine metagenomics: new tools for the study and exploitation of marine microbial metabolism. *Mar. Drugs.* 8, 608-628.

Kennedy, J., Marchesi, J. R. and Dobson, A. D. (2008) Marine metagenomics: strategies for the discovery of novel enzymes with biotechnological applications from marine environments. *Microb. Cell Fact.* 7, 27.

Khambhaty, Y., Mody, K. and Jha, B. (2007) Purification and characterization of x-carrageenase from a novel λ-*Proteobacterium*, *Pseudomonas elongata* (MTCC 5261) syn. *Microbulbifer elongatus comb. nov. Biotechnol. Biopr. Eng.* 12, 668-675.

Kim, H. T., Lee, S., Kim, K. H. and Choi, I. G. (2012a) The complete enzymatic saccharification of agarose and its application to simultaneous saccharification and fermentation of agarose for ethanol production. *Bioresour. Technol.* 107, 301-306.

Kim, H. T., Lee, S., Lee, D., Kim, H. S., Bang, W. G., et al., (2010) Overexpression and molecular characterization of Aga 50D from *Saccharophagus degradans* 2-40: an exo-type beta-agarase producing neoagarobiose. *Appl. Microbiol. Biotechnol.* 86, 227-234.

Kim, S. M., Cho, S. J. and Lee, S. B. (2012b) Genome sequence of the unclassified marine Gammaproteobacterium BDW918. *J. Bacterial.* 194, 3753-3754.

Kim, W.-J., Kim, S.-M., Lee, Y.-H., Kim, H. G., Kim, H.-K., Moon, S. H., et al., (2008) Isolation and characterization of marine bacterial strain degrading fucoidan from Korean *Uttdaria pinnatifida* sporophylls. *J. Microbiol. Biotechnol.* 18 (4), 616-623.

Knutsen, S., Myslabodski, D., Larsen, B. and Usov, A. (1994) A modified system of nomenclature for red algal galactans. *Bot. Mar.* 371, 63-169.

Kobayashi, T., Uchimura, K., Osamu, K., Deguchi, S. and Horikoshi, K. (2012) Genetic and biochemical characterization of the *Pseudoalteromonas tetraodonis* alkaline k-carrageenase. *Biosci. Biotechnol. Biochem.* 76, 506-511.

Kusaykin, M., Chizhov, A. O., Grachev, A. A., Alekseeva, S. A., Bakunina, I. Y., et al., (2006) A comparative study of specificity of fucoidanases from marine microorganisms and invertebrates. *J. Appl. Phycol.* 18, 369-373.

Lahaye, M., Brunei, M. and Bonnin, E. (1997) Fine chemical structure analysis of oligosaccharides produced by an ulvan-lyase degradation of the water-soluble cell-wall polysaccharides from *Ulva* sp., (*Ulvales*, *Chlorophyta*). *Carbohydr. Res.* 304, 325-333.

Lahaye, M. and Robic, A. (2007) Structure and functional properties of ulvan, a polysaccharide from green seaweeds. *Biomacromol.* 8, 1765-1774.

Lahaye, M. and Rochas, C. (1991) Chemical structure and physico-chemical properties of agar. *Hydrobiol.* 221, 137-148.

Lahaye, M., Rochas, C. and Yaphe, W. (1986) A new procedure for determining the heterogeneity of Agar polymers in the cell-walls of *Gracilaria* spp. (*Gracilariaceae*, *Rhodophyta*). *Canad. J. Bot.* 64, 579-585.

Lahaye, M., Yaphe, W., Phan Viet, M. T. and Rochas, C. (1989) ^{13}C-N. M. R. spectroscopic investigation of methylated and chargerd agarose oligosaccharides and polysaccharides. *Carbohydr. Res.* 190, 249-265.

Lemoine, M., Nyvall Collen, P. and Helbert, W. (2009) Physical state of kappa-carrageenan modulates the mode of action of kappa-carrageenase from *Pseudoalteromonas carrageettovora*. *Biochem. J.* 419, 545-553.

Lin, B., Lu, G., Li, S., Hu, Z. and Chen, H. (2012) Draft genome sequence of the novel agarolytic bacterium *Aquimarina agarilytica ZC1*. *J. Bacteriol.* 194, 2769.

Liu, G.-L., Zhe, Y. L. and Chi, Z. M. (2011) Purification and characterization of k-carrageenase from the marine bacterium *Pseudoalteromonas porphyrae* for hydrolysis of k-carrageenan. *Proc. Biochem.* 46, 265-271.

Ma, C., Lu, X., Shi, C., Li, J., Gu, Y., et al., (2007) Molecular cloning and characterization of a novel beta-agarase, AgaB, from marine *Pseudoalteromonas sp. CY24. /. Biol. Chem.* 282, 3747-3754.

Ma, Y.-X., Dong, S.-L., Jiang, X.-L., Li, J. and Mou, H.-J. (2010) Purification and characterization of k-carrageenase from marine bacterium mutant strain *Pseudoalteromonas* sp. AJ5-13 and its degraded products. *J. Food Biochem.* 34, 661-678.

Mavromatis, K., Abt, B., Brambilla, E., Lapidus, A., Copeland, A., et al., (2010) Complete genome sequence of *Coraliomargarita akajimensis* type strain (04OKA010-24). *Stand. Gen. Set.* 2, 290-299.

McKinnell, J. P. and Percival, E. (1962) Structural investigations on the water-soluble polysaccharide of the green seaweed *Enteromorpha compressa. J Chem. Soc.* 3141-3148.

Michel, G., Barbeyron, T., Flament, D., Vernet, T., Kloareg, B. and Dideberg, 0. (1999) Expression, purification, crystallization and preliminary x-ray analysis of the kappa-carrageenase from *Pseudoalteromonas carrageenovora. Acta Crystallogr. Section D, Biol. Crystallogr.* 55, 918-920.

Michel, G., Chantalat, L., Duee, E., Barbeyron, T., Henrissat, B., et al., (2001a) The kappa-carrageenase of *P. carrageenovora* features a tunnel-shaped active site: a novel insight in the evolution of Clan-B glycoside hydrolases. *Structure* 9, 513-525.

Michel, G., Chantalat, L., Fanchon, E., Henrissat, B., Kloareg, B. and Dideberg, O. (2001b) The iota-carrageenase of *Alteromonas fortis*. A beta-helix fold-containing enzyme for the degradation of a highly polyanionic polysaccharide. *J. Biol. Chem.* 276, 40202-40209.

Michel, G., Flament, D., Barbeyron, T., Vernet, T., Kloareg, B. and Dideberg, 0, (2000) Expression, purification, crystallization and preliminary X-ray analysis of the iota-carrageenase from *Alteromonas fortis. Acta Crystallogr. Section D, Biol. Crystallogr.* 56, 766-768.

Michel, G., Helbert, W., Kahn, R., Dideberg, O. and Kloareg, B. (2003) The structural bases of the processive degradation of iota-carrageenan, a main cell wall polysaccharide of red algae. *J. Mol. Biol.* 334, 421-433.

Michel, G., Nyval-Collen, P., Barbeyron, T., Czjzek, M. and Helbert, W. (2006) Bioconversion of red seaweed galactans: a focus on bacterial agarases and carrageenases. *Appl. Microbiol. Biotechnol.* 71, 23-33.

Michel, G., Tonon, T., Scornet, D., Cock, J. M. and Kloareg, B. (2010) The cell wall polysaccharide metabolism of the brown alga *Ectocarpus siliculosus*. Insights into the evolution of extracellular matrix polysaccharides in Eukaryotes. *New Phytol.* 188, 82-97.

Nishino, T., Nagumo, T., Kiyohara, H. and Yamada, H. (1991) Structural characterization of a new anticoagulant fucan sulfate from the brown seaweed *Ecklonia kurome. Carbohydr. Res.* 211, 77-90.

Nyvall Collen, P., Sassi, J. F., Rogniaux, H., Marfaing, H. and Helbert, W. (2011) Ulvan lyases isolated from the *Flavobacteria Persicivirga ulvanivorans* are the first members of a new polysaccharide lyase family. *J. Biol. Chem.* 286, 42063-42071.

Ogawa, H. , Amagai, Y. , Koike, I. , Kaiser, K. and Benner, R. (2001) Production of refractory dissolved organic matter by bacteria. *Science* 292, 917-920.

Oh, C. , Kwoii, Y. K. , Heo, S. J. , De Zoysa, M. , Affan, A. , et al. , (2011) Complete genome sequence of strain s 85, a novel member of the family *Flavobacteriaceae*. *J. Bacteriol.* 193, 6107.

Ohshiro, T. , Harada, N. , Kobayashi, Y. , Miki, Y. and Kawamoto, H. (2012) Microbial fucoidan degradation by *Luteolibacter algae* HI 8 with deacetylation. *Biosci. Biotechnol. Biochem.* 76, 620-623.

Ohshiro, T. , Ohmoto, Y. , Ono, Y. , Ohkita, R. , Miki, Y. , et al. , (2010) Isolation and charaterization of a novel fucoidan-degrading microorganism. *Biosci. Biotechnol. Biochem.* 74, 1729-1732.

Ohta, Y. and Hatada, Y. (2006) A novel enzyme, lambda-carrageenase, isolated from a deep-sea bacterium. *J. Biochem.* 140, 475-481.

Ohta, Y. , Hatada, Y. , Ito, S. and Horikoshi, K. (2005) High-level expression of a neoagarobiose-producting beta-agarase gene from *Agarivorans* sp. JAMB-A11 in *Bacillus subtilis* and enzmic properties of the recombinant emzine. *Biotechbnol. Appl. Biochem.* 41, 183-191.

Pati, A. , Abt, B. , TesKima, H. , Nolan, M. , Lapidus, A. , ct al. , (2011) Complete Genome sequence of *Gellulophaga lytica* type strain (LIM-21) . *Stand. Gen. Sci.* 4, 221-232.

Pereira, M. S. , Mulloy, B. and Mourao, P. A. (1999) Structure and anticoagulant activity of sulfated fucans. Comparison between the regular, repetitive, and linear fucans from echinoderms with the more heterogeneous and branched polymers from brown algae. *J. Biol. Chem.* 274, 7656-7667.

Pomin, V. H. and Mourao, R A. (2008) Structure, biology, evolution, and medical importance of sulfated fucans and galactans. *Glycobiol.* 18, 1016-1027.

Ponce, N. M. , Pujol, C. A. , Damonte, E. B. , Flores, M. L. and Stortz, C A. (2003) Fucoidans from the brown seaweed *Adenocystis utricularis*: extraction methods, antiviral activity and structural studies. *Carbohydr. Res.* 338, 153-165.

Popper, Z. A. , Michel, G. , Herve, C. , Domozych, D. S. , Willats, W. G. , et al. , (2011) Evolution and diversity of plant cell walls: from algae to flowering 1 plants. *Annu. Rev. Plant Biol.* 62, 567-590.

Potin, P. (1992) Recherche, production, purification et caractérisation de galactane - hydrolases pour la préparation d'oligosaccharides des parois d'algues rouges (Identification, production, purification and characterisation of galactan-hydrolases to prepare digosaccharides from red algal cell walls) . Phd thesis, Université de Bretagne Occidentale (France) .

Potin, P. , Richard, C. , Barbeyron, T. , Henrissat, B. , Gey, C. , et al. , (1995) Processing and hydrolytic mechanism of the *cgkA*-encoded kappa-carrageenase of *Alteromonas carrageenovora*. *Eur. J. Biochem.* 228, 971-975.

Preston, J. F. r. , Romeo, T. , Bromley, J. C. , Robinson, R. W. and Aldrich, H. C. (1985) Alginate lyase-secreting bacteria associated with the algal genus *Sargassum*. *Dev. Ind. Microbiol.* 26, 727-740.

Quatrano, R. S. and Stevens, P. T. (1976) Cell wall assembly in fucus zygotes: I. Characterization of the polysaccharide components. *Plant Physiol.* 58, 224-231.

Ralet, M. C. , Lerouge, P and Quemener, B. (2009) Mass spectrometry for pectin structure analysis. *Carbohydr. Res.* 344, 1798-1807.

Rebuffet, E. , Barbeyron, T. , Jeudy, A. , Jam, M. , Czjzek, M. and Michel, G. (2010) Identification of catalytic residues and mechanistic analysis of family GH82 iota-carrageenases. *Biochemistry*. 49, 7590-7599.

Rebuffet, E., Groisillier, A., Thompson, A., Jeudy, A., Barbeyron, T., et al., (2011) Discovery and structural characterization of a novel glycosidase family of marine origin. *Environ. Microbiol.* 13, 1253-1270.

Rees, D. (1969) Structure, conformation, and mechanism in the formation of polysaccharide gels and networks. *Adv. Carbohydr. Chem. Biochem.* 24, 267-332.

Romeo, T. and Preston, J. F. r. (1986) Purification and structural properties of an extracellular (1-4) -D-mannuronanspecific alginate lyase from a marine bacterium. *Biochemistry*. 25, 8391-8396.

Sakai, T., Ishizuka, K., Shimanaka, K., Ikai, K. and Kato, I. (2003a) Structures of oligosaccharides derived from *Cladosiphon okamuranus* fucoidan by digestion with marine bacterial enzymes. *Mar. Biotechnol.* 5, 536-544.

Sakai, T., Kawai, T. and Kato, I. (2004) Isolation and characterization of a fucoidan-degrading marine bacterial strain and its fucoidanase. *Mar. Biotechnol.* 6, 335-346.

Sakai, T., Kimura, H. and Kato, I. (2002) A marine strain of flavobacteriaceae utilizes brown seaweed fucoidan. *Mar. Biotechnol.* 4, 399-405.

Sakai, T., Kimura, H. and Kato, I. (2003b) Purification of sulfated fucoglucuronomannan lyase from bacterial strain of *Fucobacter marina* and study of appropriate conditions for its enzyme digestion. *Mar. Biotechnol.* 5, 380-387.

Sakai, T., Kimura, H., Kojima, K., Shimanaka, K., Ikai, K. and Kato, I. (2003c) Marine bacterial sulfated fucoglucuronomannan (SFGM) lyase digests brown algal SFGM into trisaccharides. *Mar. Biotechnol.* 5, 70-78.

Sakatoku, A., Wakabayashi, M., Tanaka, Y., Tanaka, D. and Nakamura, S. (2012) Isolation of a novel *Saccharophagus* species (Myt-1) capable of degrading a variety of seaweeds and polysaccharides. *Microbiol.* 1, 2-12.

Sawabe, T., Ohtsuka, M. and Ezura, Y. (1997) Novel alginate lyases from marine bacterium *Alteromonas* sp. strain H-4. *Carbohydr. Res.* 304, 69-76.

Seok, J. H., Kim, H. S., Hatada, Y., Nam, S. W. and Kim, Y. H. (2012) Construction of an expression system for the secretory production of recombinant alpha-agarase in yeast. *Biotechnol. Lett.* 34, 1041-1049.

Shi, X., Yu, M., Yan, S., Dong, S. and Zhang, X. H. (2012) Genome sequence of the thermostable-agarase-producing marine bacterium *Catenovulum agarivorans* YM01T, which reveals the presence of a series of agarase-encoding genes. *J. Bacteriol.* 194, 5484.

Shin, M. H., Lee, D. Y., Wohlgemuth, G., Choi, I. -G., Fiehn, O. and Kim, K. H. Global metabolite profiling of agarose degradation by *Saccharophagus degradans* 2-40. *New Biotechnol.* 27, 156-168.

Stach, J. E. M., Maldonado, L. A., Ward, A. C., Goodfellow, M. and Bull, A. T. (2003) New primers for the class Actinobacteria: application to marine and terrestrial environments. *Environ. Microbiol.* 5, 828-841.

Stevens, R. A. and Levin, R. E. (1977) Purification and characteristics of an alginase from *Alginovibrio aquatilis*. *Appl. Environ. Microbiol.* 33, 1156-1161.

Sugano, Y., Matsumoto, T., Kodama, H. and Noma, M. (1993a) Cloning and sequencing of agaA, a unique agarase 0107 gene from a marine bacterium, *Vibrio* sp. strain JT0107. *Appl. Environ. Microbiol.* 59, 3750-3756.

Sugano, Y. , Terada, I. , Arita, M. , Noma, M. and Matsumoto, T. (1993b) Purification and characterization of a new agarase from a marine bacterium, *Vibrio* sp. strain JT0107. *Appl. Environ. Microbiol.* 59, 1549–1554.

Sun, F. X. , Ma, Y. X. , Wang, Y. and Liu, Q. A. (2010) Purification and characterization of novel kappa-carrageenase from marine *Tamlana* sp. HC4. *Chin. J. Oceanol. Limnol.* 28, 1139–1145.

Teeling, H. , Fuchs, B. M. , Becher, D. , Klockow, C. , Gardebrecht, A. , et al. , (2012) Substrate-controlled succession of marine bacterioplankton populations induced by a phytoplankton bloom. *Science* 336, 608–611.

Temuujin, U. , Chi, W. J. , Chang, Y. K. and Hong, S. K. (2012) Identification and biochemical characterization of Sco3487 from *Streptomyces coelicolor* A3(2), an exo-and endo-type beta-agarase-producing neoagarobiose. *J. Bacterial.* 194, 142–149.

Thomas, F. , Barbeyron, T. , Tonon, T. , Genicot, S. , Czjzek, M. and Michel, G. (2012) Characterization of the first alginolytic operons in a marine bacterium: from their emergence in marine Flavobacteriia to their independent transfers to marine Proteobacteria and human gut Bacteroides. *Environ. Microbiol.* 14, 2379–2394.

Thrash, J. C. , Cho, J. C. , Vergin, K. L. , Morris, R. M. and Gjiovannoni, S. J. (2010) Genome sequence of *Lentisphaera araneosa* HTCC2155T, the type species of the order *Lentisphaerales* in the phylum *Lentisphaerae*. *J. Bacteriol.* 192, 2938–2939.

Trincone, A. (2011) Marine biocatalysts: enzymatic features and applications. *Mar. Drugs.* 9, 478–499.

Urvantseva, A. M. , Bakunina, I. , Nedashkovskaia, O. I. , Kim, S. B. and Zviagintseva, T. N. (2006) Distribution of intracellular fucoidan hydrolases among marine bacteria of the family *Flavobacteriaceae*. *Appl. Biochem. Microbiol.* 42, 552–559.

Usov, A. and Bilan, M. I. (2009) Fucoidans-sulfated polysaccharides of brown algae. *Rus. Chem. Rev.* 78, 785–799.

Van de Velde, F. (2008) Structure and function of hybrid carrageenans. *Food Hydrocoil.* 22, 727–734.

Van de Velde, F. , Knutsen, S. , Usov, A. , Rollema, H. and Cerezo, A. (2002) 1H and 13C high resolution NMR spectroscopy of carrageenans: application in research and industry. *Tr. Food Technol.* 13, 73–92.

Wakabayashi, M. , Sakatoku, A. , Noda, F. , Noda, M. , Tanaka, D. and Nakamura, S. (2012) Isolation and characterization of *Microbulbifer* species 6532A degrading seaweed thalli to single cell detritus particles. *Biodegr.* 23, 93–105.

Wargacki, A. J. , Leonard, E. , Win, M. N. , Regitsky, D. D. , Santos, C. N. , et al. , (2012) An engineered microbial platform for direct biofuel production from brown macroalgae. *Science* 335, 308–313.

Wecker, P. , Klockow, C. , Schuler, M. , Dabin, J. , Michel, G. and Glockner, F. O. (2010) Life cycle analysis of the model organism *Rhodopirellula baltica* SH IT by transcriptome studies. *Microb. Biotechnol.* 3, 583–594.

Weigl, J. and Yaphe, W. (1996) The enzymic hydrolysis of carrageenan by *Pseudomonas carrageettovora*: purification of a kappa-carrageenase. *Canad. J. Microbiol.* 12, 939–947.

Weiner, R. M. , Taylor, L. E. , 2nd, Henrissat, B. , Hauser, L. , Land, M. , et al. , (2008) Complete genome sequence of the complex carbohydrate-degrading marine bacterium, *Saccharophagus degradans* strain 2–40 T. *PLoS Genet.* 4, el000087.

Wong, T. Y. , Preston, L. A. and Schiller, N. L. (2000) Alginate lyase: review of major sources and enzyme characteristics, structure-function analysis, biological roles, and applications. *Atinu. Rev. Microbiol.* 54,

289–340.

Yaphe W. and Morgan, K. (1959) Enzymic hydrolysis of fucoidin by *Pseudomonas atlantica* and *Pseudomonas carrageettovora*. *Nature* 183, 761–762.

Yu, M., Tang, K., Shi, X. and Zhang, X. H. (2012) Genome sequence of *Pseudoalteromonas flavipulchra* JG1, a marine antagonistic bacterium with abundant antimicrobial metabolites. *J. Bacterial*. 194, 3735.

Zhang, C. and Kim, S. K. (2010) Research and application of marine microbial enzymes: status and prospects. *Mar. Drugs*. 8, 1920–1934.

Zhou, M.-H., Ma, J.-S., Li, j., Ye, H.-R., Huang, K.-X. and Zhao, X.-W. (2008) A K'-Carrageenase from a newly isolated *Pseudoalteromonas*–Uke bactenum, WZUC10. Biottchnol. *Bioproc. Eng*. 13, 545–551.

第 17 章　高温高盐环境下微生物中海藻糖和糖甘油酸酯相容质的生物合成

Joana Costa, Nuno Empadinhas, Susana Alarico, Ana Nobre, Luciana Albuquerque, Milton S. da Costa, University of Coimbra, Portugal

DOI：10.1533/9781908818355.3.465

摘要：在原核生物中发现了一些适应极端环境的策略，极端环境下的一些生理生化指标值都高于或低于多数生物体内的标准值。高盐环境代表着对微生物生存的一种挑战，因为高盐环境中水活度低，无机盐浓度高，这些对细胞新陈代谢是有毒害的。微生物通过从培养基中摄取，或者从头合成等方式积累相容质来对渗透压作出反应。这些渗透作用的活性分子调节细胞的渗透势以维持细胞分裂所需的正膨压。相容质是一种小分子量的有机化合物，它能够被大量的积累而不会负面影响细胞新陈代谢。相容质有很高的多样性，但仅可以划分成几种主要的化学类别，一些广泛分布在自然界中，其他的主要存在于特定的生物群体。本章节主要讨论糖衍生物相容质的多样性和分布，即海藻糖和糖甘油酸酯渗透溶质，同时讨论了生存在高温高盐环境下的微生物体内相容质合成相关的基因和通路等方面的知识。近年来的一些关于海洋细菌通过积累 GG 和 MGG 应对渗透压的发现，推动着完全表征这一生物催化过程的研究，以达到产业化在应用的目标。

关键词：相容性溶质，海藻糖，葡萄糖甘油酸酯，甘露糖甘油酸酯

17.1　引言

我们星球有着大量的严酷环境，这些环境从人本中心论观点上被认为是“极端”，包括温度、酸碱值、渗透值、自由水或者压力各方面都有涉及（Kristjánsson and Hreggvidsson，1995）。所有这些物理因素一直延续，在这些极限条件下生存的有机体被 Macelroy（1974）命名为“极端生物”，极端环境的喜好者（希腊语中的 philo）。极端条件可以包括温度、辐射、静水压力、干旱、盐度、酸碱值、自由基、氧还电势、金属和毒气。然而，这些特殊的群落环境也被众多生物成功开拓成为栖息地。在 21 世纪初期，我们能够在很大范围的环境中检测到这 3 种生命域中的极端生物：古菌、细菌和真核生物

(Javaux, 2006)。因为好奇心激发着科学家们对新环境的探索，地球上的无生命空间已逐渐消失（Rothschild and Mancinelli, 2001）。

已经从多种多样的热环境中分离出嗜热菌和超嗜热菌。这些生物体属于细菌或者古生菌，并且古生菌比细菌包括了更多种类的生活在高温环境中的生物体。嗜热菌这一术语经常被用来定义最适生活温度在65~80℃的生物体，但超嗜热菌是最适生活温度超过80℃的生物体（Blöchl et al., 1995）。

大部分嗜热生物已经从陆地上的地热或者人工高温环境中分离出来，但也有一些被从海洋超高温环境中分离出，其中最有名的是海洋红嗜热盐菌 *Rhodothermus marinus* 和嗜热栖热菌 *Thermus thermophilus*（Alfredsson et al., 1998; da Costa et al., 2001; Alarico et al., 2005）。从陆地温泉喷发出来的水一般钠离子含量低，分离菌是淡水生物，在含有 NaCl 超过 1.0%（*w/v*）的培养基中很难生存。而一些从陆地淡水温泉中分离出的耐盐的生物，在不添加 NaCl 的培养基中生长最佳，但像 *Thermus thermophilus* 的菌株可以在 NaCl 浓度为 4.0%~6.0%的溶液中生长（Alarico et al., 2005）。

另一方面，超嗜热生物中除去下列几个典型的例子，如冰岛热棒菌 *Pyrobaculum islandicum*、*Thermococcus zilligii* 和硫化叶菌目中的物种，都来自浅滩或者深海海洋地热区域（Blöchl et al., 1995）。而深海环境中的水通常是咸水，但是盐度从低变化至海水浓度。按照推断源自这些环境中的生物，通常是稍耐盐的，需要盐分维持生长。最适宜的生长的 NaCl 浓度为 0.5%~2%。极少数的生物，如闪烁古生球菌 *Archaeoglobus fulgidus* 的 VC-16 突变体，在 NaCl 浓度高于 6.0%~8.0%的条件下可以生长（Gonçalves et al., 2003）。

中度嗜盐菌被定义为在 NaCl 浓量为 3%~15%的培养基中生长率较高的生物，而在 NaCl 含量超过 15%的培养基中生长率较高的为极端嗜盐菌（Ventosa et al., 1998）。能同时在高温条件下生存的中度或极端嗜盐菌还没被发现。我们还不能分离出来这样的生物可能是因为地球上很少存在或则根本不存在相应的生长环境。有些陆地温泉的含盐量比海水高，但是水温却通常比较低。红海的深海热盐水的温度高达 60℃，更高的温度仍只是估计而没有被实际测到（Hartmann et al., 1998）。后者的环境条件在将来可能成为真正嗜盐菌和极端嗜热菌的来源。到现在已知的最嗜热和嗜盐菌，包括从盐湖沉积物里分离出来的最适生长温度为 60℃、生长所需的盐浓度范围为 4%~20%的 *Halothermothrix orenii*（Cayol et al., 1994），以及从太阳能盐场里分离出来的最适生长温度为 65℃、生长需要的盐浓度为 2%~15%的 *Thermohalobacter berrensis*（Cayol et al., 2000）。

不论环境中的物理化学条件怎样，生命都离不开液体水，因此，水的限制代表了一种极端环境（Rothschild and Mancinelli, 2001; Javaux, 2006）。极端嗜热菌和嗜盐菌像所有的生活在水相环境中的生物一样，会面临由于溶解的盐和糖的水平的浮动而导致水活力的改变。为了适应环境中较低的水活度和由此导致的细胞质中水的降低，微生物必须积累细胞内的离子和有机溶质以重建细胞膨压和细胞体积，与此同时，保持酶的活性（Brown, 1990）。渗透压适应一词就被用来形容由于有机体对外界水压的感知而引起的一系列活动，包括胞内水的调节和对新环境的适应（da Coata et al.，1998；Empadinhas and da Costa，2008）。不同生物可以耐受的水压的水平变化很大，但是都能够在固有的限制范围内适应这些变化。由此，维持跨膜渗透压的平衡是细胞受到水压力挑战时的主要问题。

原核生物维持渗透压的平衡主要有两种方式，也反映了两种进化机制。一种机制依赖于从环境中选择性的 K^+内流，有时候甚至 K^+水平极高，被称为“细胞质盐”类型的渗透压适应（Galinski，1995；da Costa et al.，1998；Roeβler and Müller，2001）。这种渗透压调节方式出现在 *Halobacteriaceae* 家族中的极端嗜盐古生菌中，*Haloanaerobiales* 目的极端嗜盐细菌中，以及从太阳能盐场的结晶池中分离出来的极端嗜盐杆菌 *Salinibacter ruber* 中（Anton et al.，2002；Oren et al.，2002）。这些生物体内可以积累大量的无机盐离子，而且在进化过程中，结构上进行了很大的调整来应对所产生的高离子强度。由于这一原因，不仅他们体内的高分子适应了高盐的环境，他们还依赖于这种环境（Galinski，1995；da Costa et al.，1998；Empadinhas and da Costa，2008）。这种方式是非常严格的，并且这些生物对环境的适应性范围很窄。这种类型的盐渗透压适应方式只出现在两种不同的物种中，表明这种方式是独立发展的，很难想象它能作为一种古老的特征而被保留在少数分散的物种中，或者是一种通过大规模的横向基因转移而出现的特征（Santos and da Costa，2002）。从能量的角度来说，盐类在细胞浆中的累积代表了一种在相对恒定盐度环境下的一种适应渗透压的合适策略。

第二种已知的抗衡外部水活度降低及其导致的生物体内部膨胀压降低的途径是积累小分子量有机分子，被称为相容质（Brown，1976，1990；Ventosa et al.，1998）。这些化合物的获得是通过从环境中摄取，也可以从头合成，微生物优先选择前者，因为这样能耗低（Galinski，1995；da Costa et al.，1998；Oren，1999）。这些特殊的小分子量化合物可以被积累到很高的浓度，以保护蛋白质和其他细胞成分不受渗透压引起的脱水的危害，因此，在20世纪70年代布朗创造了“相容质”这一专业术语（Brown，1976，1990；Ventosa et al.，1998）。按照这种理论，微生物细胞内的分子不需要经过特殊的修饰就可以快速地适应环境渗透压的改变（da Costa et al.，1998）。一些低嗜盐或者中度嗜盐甲烷微生物和一些极端嗜热菌同时拥有这两种渗透适应方式，即是积累高浓度 K^+和中性、负电性的有机溶质（da Costa et al.，1998；Martin et al.，1999）。

大部分的微生物（古生菌、细菌、酵母、丝状体、真菌和藻类）专一的通过积累相容质调节渗透压，这表明这一途径是很成功的一种方式（Santos and da Costa，2002）。对相容质的研究表明，一些多样化的分子已经远远超出其在渗透适应领域的功能，还发挥着更广泛的生物学功能（da Costa et al.，1998；Elbein et al.，2003）。实际上，相容质可以被认为是压力保护剂，因为部分或者所有的相容质可以保护细胞和细胞内的成分不受冰冻、干燥、高温和氧自由基的影响（da Costa et al.，1998；Welsh，2000；Argüelles，2000；Santos and da Costa，2001；Benaroudj et al.，2001）。

从化学本质来看，相容性溶质是非常多样化的分子，可以分为几类，如多元醇及其衍生物，糖及其衍生物，氨基酸及氨基酸衍生物，甜菜碱和四氢嘧啶。一些是在微生物中广泛存在的，如海藻糖、甘氨酸甜菜碱和α-谷氨酸盐，其余的只存在于少数生物中。例如多元醇，广泛地分布在真菌和藻类中，在细菌中分布极少，在古生菌中没有发现。四氢嘧啶和羟基四氢嘧啶只在细菌中被发现。许多在嗜热菌中被发现的相容质，在嗜温性细菌中从来没有被检测到。相反也是如此，如海藻糖、甘氨酸甜菜碱、四氢嘧啶和脯氨酸在嗜温性细菌中很常见。反之亦然，有机溶质如二肌醇-磷酸、双甘油磷酸盐、甘露糖基甘油酸及其中性衍生物在极端嗜盐菌中和古生菌中经常被发现（da Costa et al.，1998；Santos and da Costa，2002）。

这一章将会分析在高温、高盐环境下生长的生物体体内积累的相容质的生物合成。也就是指特殊类型的带负电荷的含糖基甘油磷酸盐的有机相容质的合成，包括甘露糖基甘油酸（MG）以及其少见的衍生物甘露糖基甘油酯（MGA）和葡糖基甘油酸（GG），还有它们自发衍生物甘露糖基葡糖基甘油酸（MGG）和双葡糖基甘油酸（GGG）。本章还将讨论海藻糖的合成，因为海藻糖是一种分布广泛的相容质，但在嗜温生物体中罕有发现。

17.2 海藻糖普遍存在于各类生物中并在嗜热菌中富集

海藻糖是一种非还原性的葡萄糖二糖，在自然界广泛存在，并且是一部分生物体中的基本组成分子。特别的结构和独特的物理性质使这种二糖能成为合适的细胞和生物大分子的保护剂，保护它们免受包括热、氧化、脱水、低温等环境压力的危害。海藻糖是一种碳源和能源物质，同时也是一种信号分子（Singer and Linquist，1998；Argüelles，2000；Elbein et al.，2003）。

不同的功能或许与不同的有机体相关，但也有可能观察到海藻糖在一种生物体中发挥多种功能。在酵母中，海藻糖可以作为储备性化合物（Hounsa et al.，1998），而且还参与对不同类型的非生物逆境的适应性响应（De Virgilio et al.，1994；Hottiger et al.，1994；Hounsa et al.，1998；Argüelles，2000）。在昆虫中，海藻糖是血液淋巴（80%~90%）和胸肌中最丰富的糖，在飞行过程中，胸肌中的海藻糖就会被消耗（Becker et al.，1996；Richards et al.，2002）。在原核生物中，这种二糖常常被作为相溶质以抵消渗透压，同时也可以作为一种碳源（Argüelles，2000）。例如，生活在 NaCl 浓度约为 3%（*w/v*）的环境中的海洋生物体内，海藻糖和蔗糖共同成为主要渗透因子，如蓝藻细菌和不产氧的光能营养属中的一些物种，包括 *Thiocapsa*、*Thiocystis*、*Amoebobacter*、*Chromatium* 和 *Chlorobium*（Mackay and Norton，1984；Reed et al.，1984；Welsh and Herbert，1993）。从海洋热泉中分离出来的嗜热和耐盐菌株 *Thermus thermophilus*，以及极端抗辐射嗜热细菌 *Rubrobacter xylanophilus* 体内积累高水平的海藻糖和甘露糖基甘油酸。但是 *Thermus thermophilus* 菌株只在盐胁迫的情况下才会积累这些溶质，而 *R. xylanophilus* 积累这些溶质作为机体组分（Alarico et al.，2005；Empadinhas et al.，2007）。其他一些嗜热细菌，如嗜盐菌 *Rhodothermus marinus* 和一些超嗜热古菌 *Pyrococcus horikoshii* 和 *Thermococcus litoralis*，也为了响应渗透压力而积累海藻糖，但海藻糖不是主要的相容质（Lamosa et al.，1998；Empadinhas et al.，2001）。

17.2.1 海藻糖生物合成途径的发现

海藻糖的合成已经发现有5种不同的路径，分别是TPS/TPP途径，TreS途径，TreY/Tre/Z途径，TreT途径和TreP途径（Elbein et al.，2003；Qu et al.，2004，Avonce et al.，2006）。

（1）最先被发现且在自然界中分布最广泛的是两步反应途径TPS/TPP，首先在T6P合成酶（TPS）的作用下，将来自NDP-葡萄糖供体中的葡萄糖转化至葡萄糖-6-磷酸上，得到海藻糖-6-磷酸（T6P）。然后在第二个酶T6P磷酸酶（TPP）的作用下，将T6P脱磷酸形成海藻糖（Giæver et al.，1988；Silva et al.，2005）。

（2）TreS途径是由海藻糖合成酶（TreS）催化的分子内重排，由麦芽糖合成海藻糖（Tsusaki et al.，1996）。

（3）TreY/TreZ途径主要包含两种酶，麦芽糖寡糖基海藻糖合成酶（TreY）和麦芽糖寡糖基海藻糖海藻糖水解酶（TreZ）。TreY催化麦芽糖低聚糖或者糖原糖链的非还原末端的次末端葡萄糖和末端葡萄糖之间的糖苷键的重排，形成a-1，1连键，TerZ将末端的海藻糖切断（Maruta et al.，1995）。

（4）第四种被发现的，不很普遍的海藻糖合成途径是TreT途径，海藻糖糖基转移合成酶（TreT）催化NDP-葡萄糖和葡萄糖合成海藻糖（Qu et al.，2004）。

（5）TreP是被报道的第五种途径，在海藻糖磷酸化酶（TreP）的催化下利用葡萄糖和葡萄糖-1-磷酸合成海藻糖（Wannet et al.，1998；Avonce et al.，2006；Schwarz et al.，2001）（图17.1）。

TreS、TreT和TreP酶同样可以催化逆反应，可能主要参与体内海藻糖的降解，而不是海藻糖的合成（Wolf et al.，2003；Qu et al.，2004；Avonce，2006；Cardoso et al.，2007；Schwarz et al.，2007；Nobre et al.，2008）。

大部分微生物只有一条海藻糖合成途径，即TPS/TPP途径，另外一些如*Thermus thermophilus*和*Propionibacterium freudenreichii*有两条合成途径，在*Mycobacterium* sp. 和*Corynebacterium glutamicum*中有3条合成途径，甚至在*Rubrobacter xylanophilus*中有4条合成途径（De Smet et al.，2000；Wolf et al.，2003；Silva et al.，2003；Cardoso et al.，2007；Nobre et al.，2008）。在*T. thermophilus*中，TPS/TPP途径参与渗透压调节，而TreS的功能依然不清楚。另一方面，*Propionibacterium freudenreichii*依赖TPS/TPP途径合成海藻糖溶质以应对不同的环境压力，而TreS参与海藻糖的降解（Cardoso et al.，2007）。TPS/TPP和TreY/TreZ途径均被发现在分枝杆菌中为分支菌酸的合成提供海藻糖发挥作用（De Smet et al.，2000）。在*C. glutamicum*中，提出TreY/TreZ途径参与渗透压调节，TreS途径参与海藻糖的代谢，而TPS/TPP途径的作用还不确定，可能存在一种未知的调节机制（Wolf et al.，2003）。TreP和TreT途径在微生物中是最不常见的。TreP途径在真菌和极少数的细菌中被

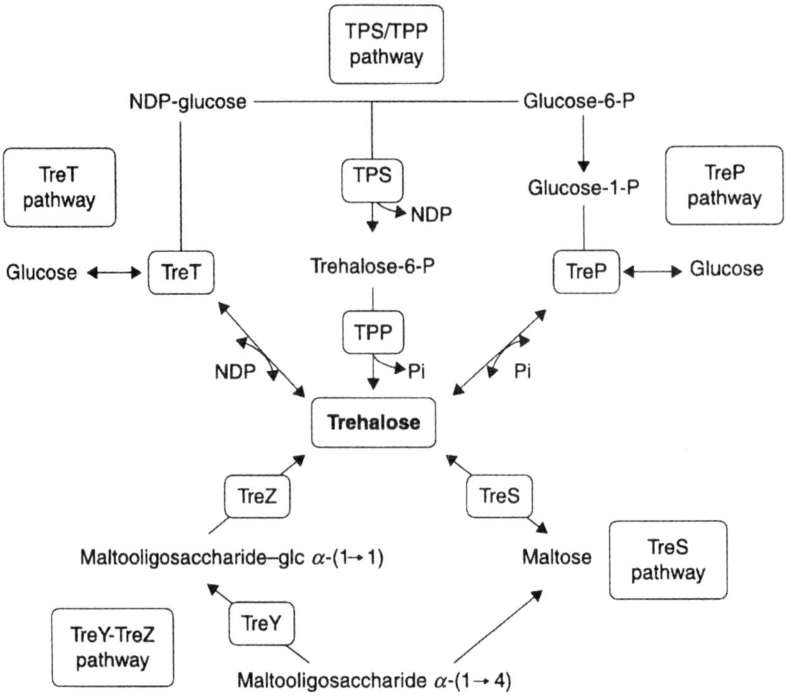

图 17.1 海藻糖合成通路

TPS：海藻糖-6-磷酸合酶；TPP：海藻糖-6-磷酸磷酸酶；TreS：海藻糖合酶；TreY：麦芽糖糖基海藻糖合酶；TreZ：麦芽糖糖基海藻糖海藻水解酶；TreT：海藻糖糖基转移合成酶；TreP：海藻糖磷酸化酶.

资料来源：Empadinhas et al.，2008；Nobre，2011.

发现（Avonce et al.，2006），而 TreT 途径只在超嗜热古生菌 *Thermococcus litoralis*、*Pyrococcus horikoshii*、*Thermoproteus tenax* 中被报道过，近期在嗜热菌 *Rubrobacter xylanophilus* 以及超嗜热菌 *Thermotoga maritima* 中也有报道（Qu et al.，2004；Ryu et al.，2005；Kouril et al.，2008；Nobre et al.，2008；Ryu et al.，2011）。除了 *T. tenax* 中的 TreT 途径，就像 *P. freudenreichii* 和 *C. glutamicum* 的 TreSs 途径，所有 4 种已被表征的 TreTs 途径也参与海藻糖的降解（Qu et al.，2004；Ryu et al.，2005；Kouril et al.，2008；Nobre et al.，2008；Ryu et al.，2011）。TreT 催化的海藻糖的水解是该酶的一个有趣的替代功能，因为它可以分别降低细胞中海藻糖的水平而增加葡萄糖的水平（Nobre et al.，2008）。

据报道，4 种海藻糖生物合成途径中酶的基因在嗜热、耐盐和超抗辐射/抗干旱的细菌 *R. xylanophilus* 中存在，TPS/TPP 途径和 TreT 途径在测定条件下很活跃：当向透析分离的细胞提取物中分别加入 NDP-葡萄糖和葡萄糖-6-磷酸或者葡萄糖后，可以检测到有海藻糖产生；当向细胞提取物中加入 GDP-葡萄糖、葡萄糖-6-磷酸（TPS）、海藻糖-6-磷酸（TPP）和 ADP-葡萄糖以及葡萄糖后也可以检测到海藻糖的产生，但是没检测到 TreS 和 TreY/TreZ 途径：向透析的细胞提取物中加入麦芽糖或者麦芽三糖、麦芽七糖、淀粉、糖原和直链淀粉，没有检测到海藻糖的产生（Nobre et al.，2008）。这种海藻糖合成途径的多样性是非常独特的，表明海藻糖在 *R. xylanophilus* 细菌中发挥着必须功能。

17.2.2 *Thermus thermophilus* 中海藻糖的生物合成和渗透调节

T. thermophilus 的 RQ-1 菌株基因文库的构建让我们首次鉴定出 *T. thermophilus* 中参与海藻糖合成的两种不同路径的基因：*otsA* 基因和 *otsB* 基因，它们分别编码 TPS 和 TPP 途径，以及编码 TreS 途径的 *treS* 基因。这 3 个基因以基因簇的方式组合，*otsB* 基因位于 *otsA* 基因的下游，*TreS* 基因在 *otsB* 基因下游。*T. thermophilus* 的 RQ-1 菌株通过 TPS/TPP 途径进行海藻糖合成，已经被证实是发挥作用的，但从未检测到 TreS 的活性（Silva et al.，2003）。

Nunes 等（1995）首次报道了在 *T. thermophilus* 菌株中为了响应 NaCl 浓度的升高而积累海藻糖和 MG，但是不同菌株的耐盐能力并不相同（表 17.1）。海藻糖生物合成基因和甘露糖基甘油酸合成基因在这些菌株中的分布不同，在其他 *Thermus* spp. 菌系中完全缺失，甘露糖基甘油酸是 *T. thermophilus* 菌株积累的第二大类溶质。体内有全部海藻糖和 *MG* 基因的菌株是最耐盐的，能在 NaCl 含量为 5%~6%的培养基中生长，缺少部分基因的菌株不能在 NaCl 含量高于 2%的培养基中生长（Alarico et al.，2005）。

表 17.1 70℃基本培养基上对于 *T. thermophilus* 菌株生长的最大 NaCl 浓度，兼容溶质积累与海藻糖的分布，甘露糖基甘油酸酯的生物基因合成

Thermus thermophilus 菌株	生长的 NaCl/%	相溶质	海藻糖/甘露糖基相关基因
HB8	2	MG	*otsB*，*treS/mpgs*，*mpgp*
HB27	2	MG	*mpgs*，*mpgp*
AT-62	2	MG	*otsB*，*treS/mpgs*，*mpgp*
GK24	2	MG	*otsB*，*treS/mpgs*，*mpgp*
RQ-1	5	MG and TRE	*otsA*，*otsB*，*treS/mpgs*，*mpgp*
B	5	MG and TRE	*otsA*，*otsB*，*treS/mpgs*，*mpgp*
Fiji3 A1	5	MG and TRE	*otsA*，*otsB*，*treS/mpgs*，*mpgp*
PRQ-14	6	MG and TRE	*otsA*，*otsB*，*treS/mpgs*，*mpgp*
T-2	5	MG and TRE	*otsA*，*otsB*，*treS/mpgs*，*mpgp*
CC-16	1	TRE	*otsA*，*otsB*，*treS*

海藻糖作为 *T. thermophilus* 菌株中一种决定性相容质，已由 *otsA* 干扰的 RQ-1 突变体所证实。后来的生理学研究证实，海藻糖参与了这些菌株对高水平 NaCl 的渗透压调节（Silva et al.，2003；Alarico et al.，2007）。

17.2.3 *Thermus thermophilus* HB27 中独特的海藻糖专一性 α-葡萄糖苷酶的作用

海藻糖作为碳源和能量来源，其消化吸收需要典型海藻糖酶（EC 3.2.1.28）的活性以水解二糖中的 α-1，1 键，海藻糖磷酸化酶（EC 2.4.1.64 和 EC 2.4.1.231）、少数情况下需要广泛专一性 α-葡萄糖苷酶（EC 3.2.1.20）。这些酶还有可能在调节细胞内海藻糖的水平中发挥重要作用（Inoue et al.，2002；Jorge et al.，2007）。

Thermus thermophilus HB27 在海藻糖生物合成上是一个自然突变菌种，这种二糖是从培养基中摄取的，但是并不是作为相容质，这与该种属中的大部分菌株不同。在这种生物体中，在高盐胁迫下 MG 是唯一的渗压剂，这表明海藻糖仅仅作为碳源被利用（Silva et al.，2005；Alarico et al.，2005）。但是，在 *T. thermophilus* HB27 基因组中还没有检测到海藻糖基因。在其他可能的候选基因中，α-葡萄糖苷酶或者糖苷水解酶可以表达并能优先水解海藻糖（Alarico et al.，2008）。有趣的是，GK24 和 HB8 菌株中的高度同源的酶不能水解这种二糖（Nashiru et al.，2001；Alarico et al.，2008）。定点突变技术已经证明，在 HB27 菌株中，α-葡萄糖苷酶中的 8 个保守氨基酸—LGEHNLPP 对该酶与海藻糖的特异性是至关重要的。这些突变对该酶和海藻糖的亲和力影响最大，因为该酶和其他底物之间的相互作用没有受到显著影响，进一步强调了其影响程度，但不排除这 8 个氨基酸对海藻糖的特异性。利用干扰 α-葡萄糖苷酶的 HB27 突变体的生理学研究证实了体外实验结果，强调了该酶在海藻糖消化中的作用，以及对异麦芽糖、蔗糖和异麦芽酮糖的消化作用，因为当这些二糖被作为单一碳源时，突变体的生长率会受到显著影响（Alarico et al.，2008）。

17.3 3 个生命领域分散群体中甘露糖基甘油酸酯的积累

17.3.1 红藻中首先发现甘露糖基甘油酸酯

第一次报道在海洋红藻 *Polysiphonia fastigiata*（*Rhodophyceae*）中存在一种低分子量的，由甘露糖和甘油酸组成的有机溶质，称为海人草素，后来被命名为甘露糖基甘油酸酯（MG），这篇报道发表于 1939 年。后来，Bouveng（1955）和他的同事确立了 MG 的结构。尽管 MG 最初在 Ceramiales 目的成员中被发现，而且被认为是该目的一个分类学指标，最近在 Gelidiales 目和 Gigartinales 目的其他红藻中检测到该溶质，因此 MG 在分类学中的重要性被推翻（Karsten et al.，2007）。由于，MG 的浓度并不总是能反映盐度的增加，因此其生理学功能有待进一步证实。

17.3.2 在低耐盐嗜热细菌中甘露糖基甘油酸酯的积累

甘露糖基甘油酸酯只在少数嗜热细菌如 *Termus thermophilus*、*Rhodothermus marinus* 和 *Rubrobacter xylanophilus* 中被发现，这几个细菌分属于不相关的细菌系谱，表明横向基因转移的现象（Nunes et al.，1995；Alarico et al.，2005；Empadinhas et al.，2007）。这其中只有 *R. marinus* 是唯一的低嗜盐菌，它的生长需要 NaCl，而 *T. thermophilus* 和 *R. xylanophilus* 都是低耐盐细菌，但是表现出完全不同类型的 MG 积累方式。*T. thermophilus* 仅仅在生长培养基中存在 NaCl 时才积累 MG，而 *R. xylanophilus* 中持续累积 MG。我们最近证实相关细菌 *R. radiotolerans* 的体内也能积累 MG（结果未发表）。这是一个史无前例的发现，因为该细菌的最佳生长温度是 40~50℃，这是所有已知能积累 MG 的生物中最低的最适温度。

17.3.3 甘露糖基甘油酸酯对 *Thermus thermophilus* 的渗透调节是非常必要的

Thermus 属的大部分种都是从陆地热液温泉排出的活水中分离出来，在含有 NaCl 的培养基中不能生存（da Costa et al.，2001；Alarico et al.，2005）。另一方面，许多 *Thermus thermophilus* 的菌株是从海洋热泉中分离出来的，尽管他们的生长不需要 NaCl，但他们能在 NaCl 含量为 3%~6%的培养基中生长（Alarico et al.，2005）。研究发现，他们都能够以盐浓度依赖性方式积累海藻糖、MG，或者二者皆有。一些能合成 MG 而不能合成海藻糖的菌株不能在盐浓度高于 2% NaCl 的培养基中生长，而那些具有能合成海藻糖和 MG 功能基因的菌株能在 NaCl 浓度高达 6%的培养基中生存。通过构建特异的海藻糖阴性或者 MG 阴性 *T. thermophilus* 突变体，明确表明 MG 和海藻糖对这些细菌的最佳渗透调节有协同作用（Silva et al.，2005；Alarico et al.，2007）。

栖热菌科也包含了 *Meiothermus*、*Vulcanithermus*、*Marinuthermus* 和 *Oceanithermus* 属的一些物种（da Costa et al.，2001；Nobre and da Costa，2001）。所有的 *Meiothermus* 物种是从深海热液喷口分离得到的，都是稍耐盐的。尽管在这些物种中至今没有报道 MG 的存在，基因组测序显示在 *Marinithermus hydrothermalis* 存在 *MG* 基因，提示着这些生物体可能合成和积累该溶质（Copeland et al.，2012）。

17.3.4 *Rhodothermus marinus* 积累甘露糖基甘油酸酯用于渗透和热调节

Rhodothermus marinus 是从海洋热温泉中分离出来的低耐盐嗜热细菌，是迄今为止唯一已知的 *Rhodothermus* 属的物种，Crenotrichaceae 科仅有的嗜热菌，Bacteroidetes 门中唯一的能够积累 MG 的成员（Nunes et al.，1995；Silva et al.，1999；Alfredsson et al.，1988）。但是，最近发现的物种 *R. profundi* 中也能积累该溶质（Marteinsson et al.，2010）。

在 *R. marinus* 中，MG 的浓度随着培养基中盐浓度的增加而增加，并有低水平的海藻糖和谷氨酸盐（Silva et al.，1999）。但是，在盐离子浓度接近最大耐受浓度时，MG 被一种称为甘露糖基甘油酰胺的中性衍生物所取代。后来发现当 *R. marinus* 在高于最适生长温度的环境中生长时，MG 的水平会增长，预测 MG 可能在高温下保护细胞内组分中发挥作用（Borges et al.，2004）。事实上，研究表明 MG 能在体外保护酶不受热变性，表明该溶质在生理环境下有助于蛋白的稳定性（Ramos et al.，1997；Borges et al.，2002；Faria et al.，2004）。

17.3.5 甘露糖基甘油酰胺是甘露糖基甘油酸酯少见的中性衍生物

在超过最适盐浓度的溶液中，特别是接近 *R. marinus* 最大耐受浓度时，负电性的 MG 会被中性的甘露糖甘油酰胺（MGA）所代替，可能 MG 的甘油酸部分增加了酰胺基团（Silva et al.，1999）。这种相容质还没有在其他任何生物体中被发现，对于盐胁迫如何调节 MG 转变成 MGA 和酰胺转移酶的鉴定还不清楚。但是，与 *R. marinus* 有亲缘关系很近的被命名为 *R. obamensis* 的菌株只能合成和积累 MG，不能合成或积累 MGA，尽管事实上两种菌株的盐耐受力都是相近的（Silva et al.，2000）。最近发现的 *R. profundi*，可能使这一困惑昭然若揭（Marteinsson et al.，2010）。

17.3.6 *Rubrobacter xylanophilus* 组成性地积累甘露糖甘油酸

Rubrobacter xylanophilus 是极端抗 γ-辐射细菌，最适生长温度是 60℃，是目前已知的最嗜热的放线菌属（Carreto et al.，1996；Ferreira et al.，1999）。这个细菌还能耐受极端脱水作用，*Rubrobacter* 菌株的 DNA 已经从沙漠和退化的生物遗迹中分离出来（Laiz et al.，2009）。此外，*R. xylanophilus* 是低耐盐的，它能在 NaCl 浓度为 6%～7%的环境中生存。当该生物生长在不同压力条件下，如低于或高于最适生长温度，盐胁迫，以及氧化胁迫、氮限制（未发表结果）等条件下时，体内相容质库极高且多样化，而对 MG 和海藻酸的水平相比于在最适生长条件下测定的水平影响不大（Empadinhas et al.，2007）。这些大量稳定的相容质库维持较高的细胞内部膨胀压，以此来抵消厚的肽聚糖层可能产生的机械作用，理论上，这有利于细菌的多极端耐受本质（Doyle and Marquis，1994）。

17.3.7 在极端嗜热古生菌中甘露糖基甘油酸酯的积累

在热球菌目中的 *Pyrococcus*、*Thermococcus* 和 *Palaeococcus* 菌株中都发现了甘露糖基甘油酸酯（Lamosa et al.，1998；Empadinhas et al.，2001；Neves et al.，2005）。*Pyrococcus furiosus* 是首个被报道的极端嗜热古菌，它能积累 MG 来响应盐胁迫，同时被发现积累高浓度的 DIP 作为在高于最适温度的环境中生存的一种保护措施，这也证实了 *P. woesei* 利用该溶质适应超热条件的意义（Scholz et al.，1992；Martins and Santos，1995）。因为细菌内 K^+

浓度也随着培养基盐度的升高而显著升高，K^+被认为是 MG 的抗衡离子。

17.3.8 *Rhodothermus marinus* 有两条合成 MG 的替代途径

在嗜热菌 *Rhodothermus marinus* 中研究了 MG 的生物合成，纯化出了催化 GDP-甘露糖和 D-甘油酸盐合成一步 MG 的糖基转移酶，命名为甘露糖基甘油酸酯合成酶（MGS）（Martins et al.，1999）（图 17.2）。MGS 部分氨基酸序列被用于在基因文库中鉴定其蛋白质序列。MGS 是一个少见的糖基转移酶，为其建立了新的 GT78 家族。后来，*R. marinus* 的 MGS 的三维结构揭示了一个独特的催化机制，处于 inverting 和 retaining 机制的边界（Flint et al.，2005）。在大量已存在的原核生物基因组中还没有检测到 MGS 同源蛋白质，但是 *Physcomitrella patens* 和 *Selaginella moellendorffii* 两个品系中含有功能性的 MGS 同源蛋白质。这一极罕见的酶在体外高效合成 MG 和 GG 类似物上有显著的和不可预测的特点，Nobre 和他们的同事（已提交结果）还在 *S. moellendorffii* 的基因组中发现了编码糖苷水解酶的基因，称为甘露糖基甘油酸酯水解酶（MgH）。这一新的真核生物酶有水解糖基甘油酸酯衍生物-MG 和 GG 的能力（Nobre，2011）。

将 *R. marinus* 提取物与 GDP-甘露糖和 3-磷酸甘油酸盐（而不是 D 甘油酸）共同孵育，也能通过一个磷酸化中间产物-甘露糖基-3-磷酸甘油酸盐合成 MG（Martins et al.，1999；Empadinhas et al.，2001）。这个通过形成磷酸化甘油中间物的两步合成过程是海藻糖、蔗糖和葡萄糖甘油合成中一种常见的过程（Pan et al.，1996；Curatti et al.，1998；Hagemann et al.，2001）。在 *R. marinus* 中存在两种合成 MG 的替代途径的现象表明，该溶质在生物体的生理上发挥至关重要的作用，而在 MG 合成的调控上有着高的灵活性（图 17.2）。实际上，这两条途径在 *R. marinus* 适应渗透压力和热胁迫上是分别调控的（Borges et al.，2004）。

17.3.9 极端嗜热古菌通过两步途径合成 MG

选用 *Pyrococcus horikoshii* 研究古菌中 MG 的生物合成、基因和相关的酶（Empadinhas et al.，2001）。*R. marinus* 能够利用 D-甘油酸盐或者 3-磷酸甘油酸盐作为受体底物合成 MG，与之不同的是，*P. horikoshii* 只能用 3-磷酸甘油酸盐来合成 MG。在天然酶的纯化过程中，检测到一个磷酸化化合物，经 NMR 鉴定为甘露糖-3-磷酸甘油酸酯（MPG），与之相关的酶命名为甘露糖-3-磷酸甘油酸酯合成酶（MpgS）。纯化的 MpgS 使我们进一步在基因组中鉴定出相关基因（*mpgS*），该基因组存在 4 个类操纵子基因结构，推断 2 个基因参与甘露糖的新陈代谢，1 个磷酸酶基因紧邻 *mpgS*，在其下游。这些基因在功能上分别被鉴定为 mpgS（EC 2.4.1.217）和甘露糖-3-磷酸甘油酸磷酸化酶（MpgP，EC 3.1.3.70）。mpgS 是一种新的糖基转移酶，为其设立新的 GT55 家族。分别来自 *P. horikoshii* 和 *T. thermophilus* 的 MpgS 和 MpgP 的三维结构最近已经被解析出来（Kawamura et al.，2008；

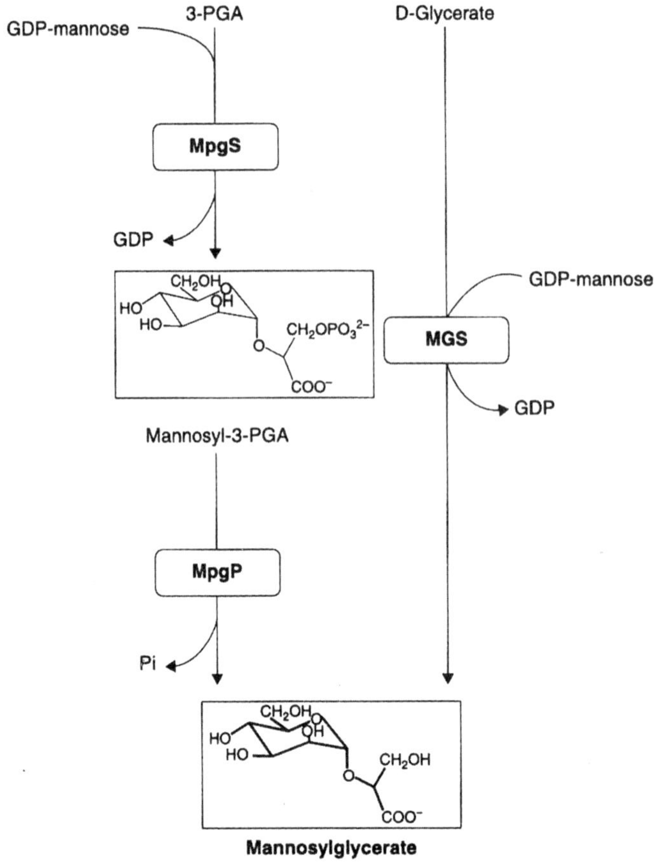

图 17.2 *Rhodothermus marinus* 中 MG 的合成的替代通路

MpgS：甘露糖基-3-磷酸甘油酸合酶；MpgP：甘露糖基-3-磷酸甘油酸磷酸酶；MG5：甘露糖基甘油酸合酶；3-PGA：D-3-磷酸甘油酸.

资料来源：Martins et al.（1999）和 Borges et al.（2004）.

Gonçalves et al.，2011）。

基因组测序数量的增加，使得我们在其他极端嗜热古生菌中检测到 *mpgS* 和 *mpgP* 的同系物，如 *Archaeoglobus profundus*，*Aeropyrum pernix* 和 *Staphylothermus* spp. 这些古菌大都能积累 MG（Santos et al.，2007）。根据 MpgP 的氨基酸序列进行 BLAST，在大量微生物的基因组中都检测到同系物，很多微生物都缺失相应的 *MpgS* 基因。因为 MpgPs 是 HAD 分解酶超家族的成员（Burroughs et al.，2006），该家族成员具有不同的底物特异性，这表明他们的同系物可能在宿主中有不同的活性。实际上，许多 MpgP 同系物的底物特异性仍然是未知的，据发现其他一些酶能够催化相应的相容质甘露糖基甘油酸酯前体化合物的去磷酸化。

17.3.10 嗜温细菌和古菌含有 *MG* 基因并可能积累该溶质

在嗜温细菌和古菌的基因组中检测到 MpgS 同系物，其中一些缺失 *mpgP* 基因。从寒冷的环境中，如海底或者森林土壤，分离出来的古菌的宏基因组中有 *mpgS* 基因（Quaiser et al.，2002；Hallam et al.，2004）。尽管它们的催化功能还不是很清楚，但是很有可能体内 MG 是在低温下合成的。

在"*Dehalococcoides ethenogenes*"基因组中发现 *mpgS* 同系物和 *mpgP* 基因融合在一起，该嗜温细菌能够净化多氯三苯污染物（Maymo-Gatell et al.，1997）。这一双功能基因被命为 *mgsD* 且在大肠杆菌中被表达，得到的重组酶从功能上被鉴定为 MG 合成酶。除此之外，在 *Saccharomyces cerevisiae* 中表达的 *mgsD* 能在体内合成和积累 MG，这表明在天然生物中 *mgsD* 有相似的功能（Empadinhas et al.，2004）。其他 *Dehalococcoides* 菌株的基因组和亲缘关系相近的嗜温细菌 *Dehalogenimonas lykanthroporepellens* 的基因组研究表明，这些是迄今为止发现的含有合成 *MG* 基因的绿弯菌门中的仅有成员（Moe et al.，2009）。有趣的是，该门类中的一些嗜温成员，如 *Thermomicrobium roseum*、*Chloroflexus aurantiacus*、*Roseiflexus castenholzii* 和 *Spherobacter thermophilus* 缺失合成 *MG* 的基因，且从未见报道能够积累 *MG*。在"*D. ethenogenes*"中存在的功能性基因反驳了最初的高温生物与 *MG* 合成特异性相关的假设。

17.3.11 *Rubrobacter xylanophilus* 中的 MpgS 可能是 MpgSs 和 GpgSs 的共同祖先

Rubrobacter xylanophilus 的基因组序列中没能找到任何已知的 MG 合成所需的基因（Martins et al.，1999；Empadinhas et al.，2001）。但是，从 GDP-甘露糖和 3-磷酸甘油酸可以合成 MG，这些底物也是参与 MpgS 和 MpgP 合成途径的底物。甘露糖-3-磷酸甘油酸合成酶（MpgS）与已知的 MpgSs 相比较有类似的特征，但有高度不同的序列（Empadinhas et al.，2001，2003；Borges et al.，2004）。值得一提的是，MpgS 也能合成葡萄糖基-3-磷酸甘油酸（GPG），它是 GG 的前体，催化速度较高，尽管在 *P. xylanophilus* 中从未检测到葡萄糖基甘油酸盐（GG）溶质。在生物体内，该前体通常是由 GpgS 催化合成的，进一步合成 GG。*R. xylanophilus* 的 MpgS 仅仅在放线菌基因组中发现过同系物，这些蛋白属于糖基转移酶家族 GT81，设立该家族以包括葡萄糖基-3-磷酸甘油酸合成酶（GpgS）。为了解该酶的特异性和体外底物杂泛性，将 MpgS 结晶并解析其三维结构（Sá-Moura et al.，2008；Empadinhas et al.，2011）。酶的结构和一个工程化三突变的 MpgS 突变体解释了它的核苷酸和糖供体的特异性，表明一个突环结构参与葡萄糖或甘露糖的识别。对该双底物 MpgS 的表征代表着在 MG 和 GG 生物合成领域里程碑式的发现，因为该酶可能代表着所有已知的 MpgSs 和 GpgSs 进化中的分支。

17.4 在高盐低氮环境下葡萄糖基甘油酸广泛存在于细菌和古菌中

有机溶质葡萄糖基甘油酸（GG）是一个 MG 的结构类似物，它广泛存在于古菌和细菌门（Empadinhas and Costa，2010）。它最早被鉴定为 *Mycobacterium phlei* 多聚糖的组分或者 *Nocardia otitidiscaviarum*（原名为 *Nocardia caviae*）中糖脂的组分，并且是海洋蓝细菌 *Synechococcus* sp. PCC 7002（原名为 *Agmenellum quadruplicatum*）中的一个游离分子（Saier and Ballou，1968；Kollman et al.，1979；Pommier and Michel，1981）。

GG 积累和氮的可利用性之间的关系最初是通过对轻度耐盐蓝细菌 *Synechococcus* sp. PCC 7002 的研究提供的，蓝细菌生长在低氮环境条件下（Kollman et al.，1979）。该蓝细菌中主要积累谷氨酸盐、葡萄糖基甘油和蔗糖。细胞内谷氨酸盐浓度在最初生长阶段显著增加，当细胞进入静止期后检测不到谷氨酸盐，被 GG 所代替（Kollman et al.，1979）。在该生物体内，很显然去氮培养基有利于 GG 的积累。

在 *Dicheya chrysanthemi* 3937 菌株（原名为 *Erwinia chrysanthemi* 或者 *Pectobacterium chrysanthemi*）中进一步观察到这种现象，该菌株导致大范围植物软腐病，大量研究表明对渗透压的调节是这些细菌在泥土中生存及其致病行为的重要因素（Gouesbet et al.，1995）。海藻糖是肠细菌中合成的唯一相容质，包括 *D. chrysanthemi* 菌株 ECC 和 SR 237，因此，在其他 *D. chrysanthemi* 菌株中也可能以海藻糖作为相容质（Prior et al.，1994）。尽管在分析 *D. chrysanthemi* 菌株 3937 的内源性渗透调节物时并没有检测到海藻糖，在肠细菌中却检测到两种不常见的渗透调节物-谷氨酸盐和 GG，该渗透调节物的比例随着培养基组成和生长阶段不同而有所变化。当该细菌在的盐胁迫条件下时，最适相容质是谷氨酰胺和谷氨酸盐，但是也能检测到低水平的 GG，这 3 种渗透调节物都随着培养基中盐度的增加而增加（Goude et al.，2004）。但是，当含盐培养基中氮的浓度降到最低水平时，谷氨酰胺和谷氨酸盐的浓度会突然降低至接近为零，并且被浓度急剧升高的 GG 取代，GG 成为唯一的相容质（Goude et al.，2004）。这些结果明确表明当缺少稳定氮源以供富含氮元素的氨基酸相容质的合成时，*D. chrysanthemi* 会转变方式，合成和积累不含氮的相容性溶质-糖衍生物 GG。这些发现提供了明确的证据，证明盐胁迫和氮缺乏之间存在密切的关系。

在轻度嗜盐蓝细菌 *Prochlorococcus marinus* 和 *Synechococcus* sp. PCC7002 品系（Klähn et al.，2010）中也观察到相类似的结果。海洋蓝细菌典型的相容质是中性葡萄糖基甘油，但是在上面提到的生物体中却不存在该相容质。Klähn 和他的同事（2010）研究发现 GG 的含量随着培养基盐度的增加而增加，特别是在氮限制的条件下。在这些生物体中，在盐胁迫和缺氮的情况下，GG 还可能取代谷氨酸盐作为 K^+的抗衡离子。

在极端嗜盐产甲烷古菌 *Methanohalophilus portucalensis* FDF-1 菌株中，GG 是次要的相容质，该菌株组成性积累氨基酸-α-谷氨酸盐和 β-谷氨酰胺，以及氨基酸衍生物-甘氨酸甜菜碱和Ne-乙酰β-赖氨酸。在盐胁迫条件下，当GG水平不能应对盐度的波动时，甘氨酸甜菜碱和Ne-乙酰β-赖氨酸的水平会适当的增加（Robertson et al.，1992）。尽管研究表明，氮的消化可能是渗透剂分布的关键因素，检测了不同的培养基构成以研究该生物体内的相容质库（Robinson and Roberts，1997），并没有得到有关该生物体内GG的合成和功能关系的证据。

在盐胁迫和氮限制双重条件下，从变形菌门到海洋蓝细菌，从盐杆菌到甲烷微菌等。积累GG作为一种应急性相容性溶质，现在看来似乎是一种散布的生存策略，由于GG的负净电性，极有可能GG不仅是作为一种相容质，也可能在氮限制条件下代替谷氨酸盐，以抗衡无机离子，如K^+和Na^+。尽管GG作为真正相容质的作用几十年来仍然没弄清楚，现在，GG被认为是一些生物体在面对特殊营养限制条件时的一种相容质，而在其他条件时，GG是一些重要大分子的前体。

与起初的猜测不同，随着可利用基因组数目的增加证明了GG的生物合成可能是一种散布的现象，因为编码GG合成的基因广泛分布在细菌的大部分分支，预期在研究GG在原核生物适应不同环境挑战中的生物学作用中有新的发现。另一方面，尽管该基因在所有已知*Halobacteriaceae*的基因组中都存在，它在古菌基因组中是很少见的，仅限于少数产甲烷菌。

17.4.1 海洋细菌 *Persephonella marina* 中GG合成的两条替代途径

从最初发现，GG的分布仅限于在嗜温细菌和古菌中（Saier and Ballou，1968；Kollman et al.，1979；Kamisango et al.，1987；Robertson et al.，1992），直到后来惊奇地在嗜热细菌*Persephonella marina*中鉴定到GG，它属于高度分支的Aquificales目的成员，从深海超高温喷射口排气口分离得到，在70～73℃，含有2.5% NaCl条件下生长最佳（Götz et al.，2002；Santos et al.，2007）。该分支的其他成员均不积累GG。Santos和他的同事（2007）发现，当*P. marina*生长在盐胁迫但是从营养上约束微需氧氧化条件时，GG是真正的相容质，这阻碍了对*P. marina*渗透调节动力学的更深入研究。

在*P. marina*中发现了两条不同的GG合成途径（Costa et al.，2007；Fernandes et al.，2007）。第一条被发现的是在自然界中分布最广泛的途径是GpgS-GpgP两步反应途径，之前在低嗜盐嗜冷产甲烷古菌*Methanococcoides burtonii*中介绍过该途径（Costa et al.，2006）并且在生物基因组鉴定到该基因。该两步反应途径包括葡萄糖-3-磷酸甘油酸合成酶（GpgS）催化将NDP-G和D-3-磷酸甘油酸（3-PGA）转变成葡萄糖基-3-磷酸甘油酸（GPG），然后在葡萄糖基-3-磷酸甘油酸磷酸酶（GpgP）催化下转变成GG（图17.3）。形成一种磷酸化中间代谢物是一种合成其他相容质的普遍方式，如海藻糖、蔗糖、葡萄糖基甘油和甘露糖基甘油酸酯（Empadinhas and da Costa，2008）。最近，在*Aquificales*，*Hydrogenivirga* sp.的另一个成员的基因组中发现了GpgS同系物，但是没有发现GpgP同系物。

第二种途径仅在该细菌中被发现，包括葡萄糖基甘油酸合成酶（GGS）催化的直接

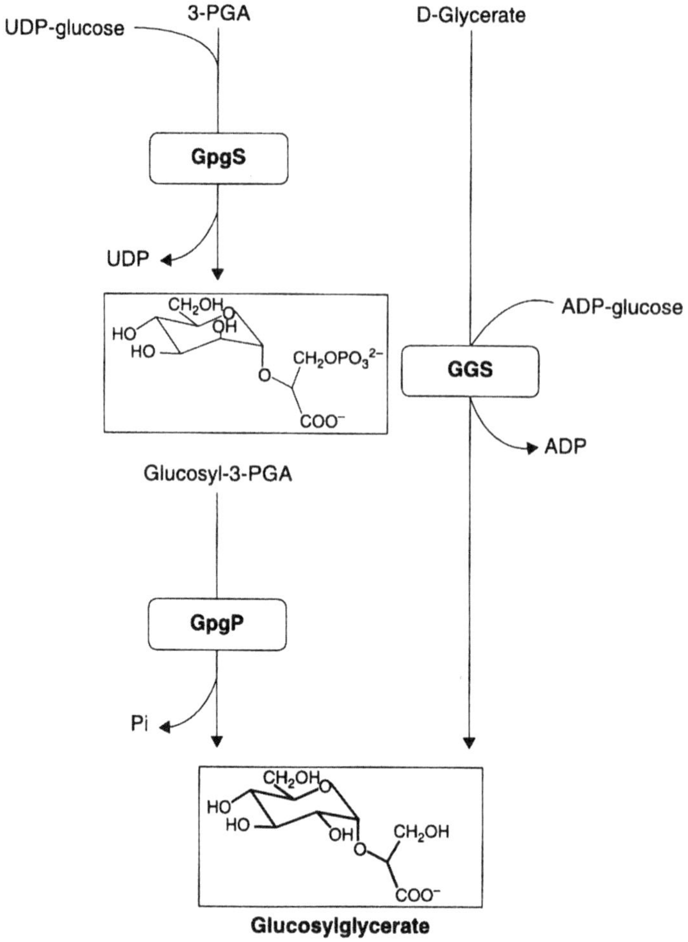

图 17.3 *Persephonella marina* 中 GG 的合成的替代通路

GpgS：葡糖基-3-磷酸甘油酸合酶；GpgP：葡糖基-3-磷酸甘油酸磷酸酶；GGS：葡糖基甘油酸合酶；3-PGA：D-3-磷酸甘油酸

资料来源：Costa et al.（2006）和 Fernandes et al.（2007）.

将 ADP-葡萄糖和 D-甘油酸转化为 GG 的一步合成途径（Fernandes et al.，2007）（图 17.3）。在 Thermotogales 目一些成员，如 *Thermotoga* spp.，*Petrotoga mobilis* 和 *Kosmotoga olearia*，和变形菌的基因组中也发现了葡萄糖基甘油酸合成酶同系物。在一些极端嗜热古生菌，如 *Thermococcus* 和 *Pyrococcus* 的基因组中也发现几个 GGS 的同系物。但是 GG 在这些生物体中的功能还不是很清楚。

P. marina 中存在两个可相互替代的 GG 合成途径意味着其在机体生理机能中起到重要作用，并且可能通过不同刺激进行 GG 库的动态调节。一种上游携带 GG 基因的操纵子类似结构中包含有磷酸吸收基因，因此推测磷酸的存在和 GG 的合成可能存在功能上相关的

假设（Costa et al.，2007）。

17.4.2 在 *Persephonella marina* 中葡萄糖基葡萄糖甘油酸（GGG）的积累

在研究嗜温细菌 *P. marina* 中的相容质库的过程中，Santos 及其同事发现该细菌积累 GG 和葡萄糖基-（1，6）-葡萄糖甘油酸 GGG 以响应盐胁迫（Santos et al.，2007）。但是，有关导致 GG 和 GGG 累积的条件等具体的细节还不清楚。之前发现在分支杆菌中存在痕量的相应化合物，这两种化合物被认为是甲基葡萄糖脂多糖生物合成的前体（Kamisango et al.，1987）。而且，在一些极端耐盐古生菌体内的糖脂的极性头部也发现了 GGG 分子中的 α-呋喃葡萄糖-（1-6）α-呋喃葡萄糖（Koga and Morii 2005）。理论上，GGG 也可能是 *P. marina* 中的大分子物质的前体，但是需要进一步的研究以证实该分子在进化中获得了其他的功能。

17.4.3 MGG 在 *Petrotoga mobilis* 中的积累和生物合成

在嗜热、厌氧细菌 *Petrotoga miotherma* 中发现了一种独特的从 GG 衍生而来的相容质，被鉴定为甘露糖基-（1，2）-葡萄糖甘油酸（MGG），该细菌属于 Thermotogales 目，是从热油水中提分离到的，最适生长温度为 55~60℃，培养基中 NaCl 浓度范围为 0.5%~10%（Jorge et al.，2007）。MGG 是主要的相容质，还有少量的谷氨酸和脯氨酸，仅在生物体对低于最适盐度时的调节发挥作用，因为 MGG 的浓度在高于最适盐浓度和温度生长条件下时有所下降，而此时甘氨酸和脯氨酸成为发挥主要作用的相容质（Jorge et al.，2007）。而且，该相容质并不参与 *P. miotherma* 对氧化胁迫的适应。最近，在 *P. mobilis* 中也发现了 MGG，但是与 *P. miotherma* 不同的是，MGG 的水平在高于最适盐度和最适生长温度条件下时是上升的（Fernandes et al.，2010）。

奇怪的是 MGG 合成的方式与 *Rhodothermus marinus* 中的合成方式并不类似，在 *Rhodothermus marinus* 中，先前存在的相容质 MG 被转化成不同的形式（MGA）。实际上，在 *P. miotherma* 或者 *P. mobilis* 中在任何条件下都没有检测到 GG。*P. mobilis* 基因序列中含有一个可能与糖基转移酶基因相关的 *GpgS* 基因，从功能上被鉴定为甘露糖基葡萄糖-3-磷酸甘油酸（MGPG）合成酶。命为 MggA（图 17.4）。该酶能利用 GpgS、葡萄糖基-3-磷酸甘油酸（GPG）和 GDP-甘露糖来产生甘露糖基葡萄糖-3-磷酸甘油酸（MGPG），这是磷酸化的 MGG 的前体。显著不同的是，在 *Spirochaeta thermophila*、*Spirochaeta smaragdinae* 和 *Coraliomargarita akajimesis* 的基因组中仅检测到 3 个 MggA 同系物，而且与 *P. mobilis* 中极罕见的 MggA 的氨基酸序列相似性不到 30%。尽管，在 *P. mobilis* 提取物中检测到 MGPG 的去磷酸产物 MGG，并部分纯化和表征了天然磷酸酶，但是还没有鉴定出该磷酸酶的基因（Fernandes et al.，2010）。

值得一提的是，最初只在 *Persephonella marina* 中发现的，编码 GGS 的基因，在 *Petrotoga mobilis* 中也被发现，而且部分纯化的天然磷酸酶和在大肠杆菌中表达的重组形式都表现出 GGS 活性，它们参与了 GG 合成的第一步。因为在 *P. mobilis* 中，在任何条件下都没有检测到 GG，GG 也不是 MggA 的底物，我们猜想存在一种非磷酸化途径以合成 MGG（图 17.4）。

我们推断鉴定到相应的基因，其编码的产物与 MggA 有一些序列相似，但是该基因不能重组表达 *P. mobilis* 中的功能酶，我们把这个酶称为甘露糖基葡萄糖甘油酸合成酶 MggS。相反的，来自 *Thermotoga maritima* 的同源基因（37%氨基酸相似性）被成功表达，同时也证实了利用合成 GDP-甘露糖和 GG 合成 MGG（Fernandes et al., 2010）。该酶在已知基因组中有几个同系物，但大都属于 *Thermotogales*。由于该 MggS 酶还能以很低的速率水解 MggS，生成 GG 和甘露糖，MggS 可能参与体内 MGG 的代谢。但是，*T. maritima* 不能积累 GG 或者 MGG，唯一可能的解释是这些溶质可能是其他大分子的前体。通过揭示 *P. mobilis* 中两条合成 MG 替代途径，还观察到 *R. marinus* 中 MG 的合成以及 *P. marina* 中的 GG 合成具有灵活性，这些都为解决 *P. miotherma* 和 *P. mobilis* 中 MGG 的模糊的功能以及生物体对极端环境的适应等问题提供了基本研究手段。

17.5 催化糖基-甘油酸酯相容质合成的海洋酶类在生物技术和生物医学领域的潜在应用

一些相容质除了有渗透调节能力，还能保护蛋白质和生物膜等的变性（Luzardo et al., 2000; Borges et al., 2002; Hincha and Hagemann, 2004），这也解释了它们参与细胞被暴露于冷冻、脱水和高温环境时的适应过程（Eleutherio et al., 1993; Ko et al., 1994; Welsh and Herbert, 1999）。由于在高温条件下可以为细胞内大分子提供极好的保护作用，MG 被赋予该保护功能，并且被发现在体外 MG 具有很好的稳定作用，保护酶不受热变性（Ramos et al., 1997; Borges et al., 2002; Faria et al., 2003, 2004）。同时，MG 也能阻止蛋白质聚集，而且据预测，其应用是非常之大的，从生物工程领域至生物医学领域以及由于蛋白质错误折叠而导致的疾病（Faria et al., 2003; Cruz et al., 2006; Lentzen and Schwarz, 2006）。事实上，Ryu 及其同事的近期研究表明 MG 可以作为体外可溶性 β-淀粉样肽聚集的抑制剂，因此，有望在治疗与聚集相关的疾病中有所应用（Ryu et al., 2008）。

最近的研究还表明 GG 具有稳定作用，GG 可以保护乳酸脱氢酶不受热失活，其效果比海藻糖和葡萄糖基甘油酸还要好（Sawangwan et al., 2010）。与此同时，MG 和 GG 以及它们的衍生物都在被研究，有望成为具有广泛应用的物质，从化妆品到生物工程领域，作为保湿成分或者提高微芯片的质量（Lentzen and Schwarz, 2006; Mascellani et al., 2007）。但是，能够带来满意效果的相容质的浓度太高，甚至限制了其在体内应用的尝试。因此需要我们继续发现新的分子，在较低的浓度下也是有效的、天然的或者从天然分子的基础上

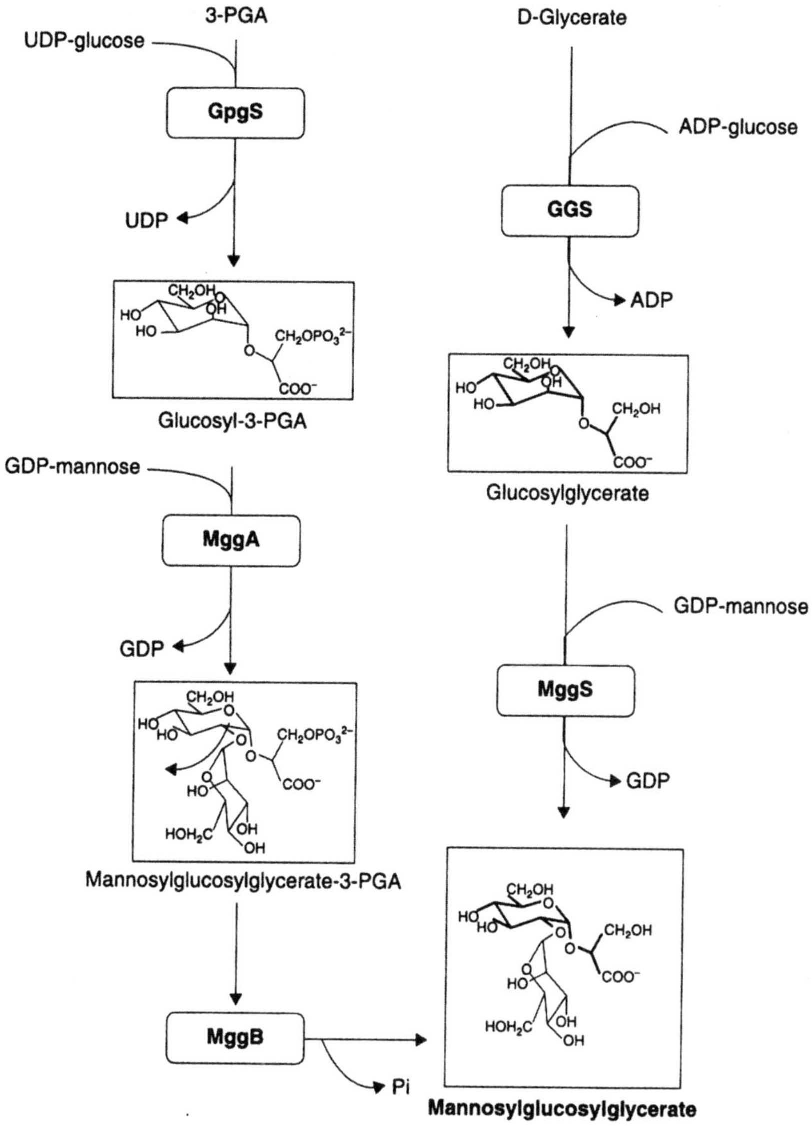

图 17.4　MGG 的合成替代通路

GpgS：葡糖基-3-磷酸甘油酸合酶；GGS：葡糖基甘油酸合酶；MggA：甘露糖基核糖基-3-磷酸甘油酸合酶；MggB：甘露糖基核糖基-3-磷酸甘油酸磷酸酶；MggS：甘露糖基蔗糖甘油酸合酶；3-PGA：D-3-磷酸甘油酸

资料来源：Fernandes et al.（2010）.

化学修饰而来的（Faria et al.，2008）。由于能够产生这些独特的相容质的天然生物通常都很难生长，而且产量不能满足大规模研究的需要，目前正在研究提高其产量的方法（Sauer and Galinski，1998；Empadinhas et al.，2004；Lentzen and Schwarz，2006；Egorova et al.，2007；Pospís et al.，2007；Faria et al.，2008；Sawangwan et al.，2010）。

最近研究发现，来自扁平菌门的一些海洋细菌能够积累 GG 和 MGG 以应对渗透压力，这是一组不常见但是广泛分布的细菌，被发现与微生物生态和进化、海洋地理学和污水处理（Arrigo，2005；Strous et al.，2006）等研究领域越来越相关，这促使我们清楚地阐明该生物催化过程，以提高 GG 和 MGG 的产量。

同时，在近几年的研究中，通过利用蓝藻菌中具有不同耐盐性的模式菌株，已经阐明了相容质积累的生物化学和分子机理。为将来生产生物能源，蓝藻菌和藻类的大规模培养将优先在盐水中开展（Hagemann，2011）。详细了解这些生物体对恶性环境条件的自然性适应，将有助于设计高级的用于生物技术目的的蓝藻菌生产菌株。

17.6 结论

本章针对含有海藻糖和糖基甘油酸的相容质的多样性、功能及合成代谢途径进行了综述。对海藻糖、MG、GG 和 MGG 等相容质的不同或替代合成途径进行鉴定，证实了其在生物体生理上的关键作用。例如，海藻糖被认为是广泛的能对抗渗透压，热氧化和脱水胁迫等压力的保护分子，而且迄今为止，已经鉴定出至少 5 种不同海藻酸生产系统，部分生物体包含两种、3 种甚至 4 种途径（De Smet et al.，2000；Elbein et al.，2003；Silva et al.，2005；Nobre et al.，2008）。

10 年前，认为含有糖基-甘油酸的相容质-MG，GG 和及其衍生物在自然界是很罕见的。除了两种嗜热细菌和超嗜热古菌，*Thermococcales*，只在少数红藻中鉴定到 MG。GG 更为罕见，只在两种嗜盐的微生物和蓝细菌中作为微不足道的溶质被检测到，它在两种放线菌中作为多糖和糖脂的组分，不同于能够累积 MG 的超/嗜热微生物，在分支上临近“生物树”根部，那些产 GG 微生物都有嗜温的生命现象。例外的是，报道称嗜热细菌 *Persephonella marina* 能够积聚 GG。MG 的积累通常是对盐胁迫的反应，在一些生物中当温度高于最适生长温度时也会积累 MG，GG 似乎是当特定营养物贫乏时，在其他生物适应盐胁迫中发挥关键作用。此外，一些生物体利用 GG 作为合成糖脂和多糖等大分子的前体，有些在宿主中具有重要功能。

然而，特别是由于对合成 MG 和 GG 的关键基因的鉴定和基因组数量的升级，发现了越来越多的能够合成这些溶质的新生物。在盐胁迫和氮缺失协同条件下，积累 GG 作为“突发事件”下的相容质，现在看来似乎是一个从肠道菌群到海洋蓝细菌中散布的求生策略。研究基因和相容质的合成途径，这些化合物生物合成的调控，和这些自然制造者之间的系统发生关系，将继续扩展我们了解原核生物在适应压力上的进化。

致谢

我们要向很多共同作者和我们合作者表达真挚的感谢，感谢他们使我们更好地了解了相容质生物学。本工作由 Fundação para a Ciência e a Tecnologia, FCT, Portugal, COMPETE and QREN, Project PTDC/BIA-MIC/105247/2008 提供支持，一并感谢。

参考文献

Alarico, S., Empadinhas, N., Simões, C., Silva, Z., Henne, A., et al., (2005) Distribution of genes for the synthesis of trehalose and mannosylglycerate in *Thermus* spp. and direct correlation with halotolcrance. *Appl. Environ, Microbiol.* 71, 2460-2466.

Alarico, S., Empadinhas, N., Mingote, A., Simões, C, Santos, M. S. and da Costa, M. S. (2007) Mannosylglycerate is essential for osmotic adjustment in *Thermus thermophilus* strains HB27 and RQ-1. *Extremophiles* 11, 833-840.

Alarico, S., da Costa, M. S. and Empadinhas, N. (2008) Molecular and physiological role of the trehalose-hydrolysing alpha-glucosidase from *Thermus thermophilus* HB27. *J. Bacteriol.* 190, 2298-2305.

Alfredsson, G. A., Kristjánsson, J. K., Hjörleifsdottir, S. and Stetter, K. O. (1988) *Rhodothermus marinus*, gen. nov., sp. nov., a thermophilic, halophiiic bacterium from submarine hot springs in Iceland. *J. Gen. Microbiol.* 134, 299-306.

Anton, J., Oren, A., Benlloch, S., Rodriguez-Valera, F., Amann, R. and Rosselló-Móra, R. (2002) *Salinibacter ruber* gen. nov., sp. nov., a novel, extremely halophiiic member of the Bacteria from saltern crystallizer ponds. *Lnt. J. Syst. Evol. Microbiol.* 52, 485-491.

Argüelles, J. C. (2000) Physiological roles of trehalose in bacteria and yeasts: a comparative analysis. *Arch. Microbiol.* 174, 217-224.

Arrigo, K. R. (2005) Marine microorganisms and global nutrient cycles. *Nature* 437, 349-355.

Avonce, N., Mendonza-Vargas, A., Morett, E. and Iturriaga, G. (2006) Insights on the evolution of trehalose biosynthesis. *BMC Evol. Biol.* 6, 109.

Becker, A., Schlöder, P., Steele, J. E. and Wegener, G. (1996) The regulation of trehalose metabolism in insects. *Experientia* 52, 433-439.

Benaroudj, N., Lee, D. H. and Goldberg, A. L. (2001) Trehalose accumulation during cellular stress protects cells and cellular proteins from damage by oxygen radicals. *J. Biol. Chem.* 276, 24261-24267.

Blöchl, E., Burggraf, S., Fiala, G., Lauerer, G., Huber, G., et al., (1995) Isolation, taxonomy and phylogeny of hyperthermophilic microorganisms. *World J. Microbiol. Biotechnol.* 11, 9-16.

Borges, N., Ramos, A., Raven, N. D., Sharp, R. J. and Santos, H. (2002) Comparative study of the thermostabilizing properties of mannosylglycerate and other compatible solutes on model enzymes. *Extremophiles* 6, 209-216.

Borges, N., Marugg, J. D., Empadinhas, N., da Costa, M. S. and Santos, H. (2004) Specialized roles of the two pathways for the synthesis of mannosylglycerate in osmoadaptation and thermoadaptation of *Rhodothermus marinus*. *J. Biol. Chem.* 279, 9892-9898.

Bouveng, H., Lindberg, B. and Wickberg, B. (1955) Low-molecular carbohydrates in algae. *Acta Chem. Scattd.* 9, 807-809.

Brown, A. D. (1976) Microbial water stress. *Bacteriol. Rev.* 40, 803-846.

Brown, A. D. (1990) Microbial Water Stress Physiology: Principles and Perspectives. John Wiley & Sons, Chichester.

Burroughs, A. M., Allen, K. N., Dunaway-Mariano, D. and Aravind, L. (2006) Evolutionary genomics of the HAD superfamily: understanding the structural adaptations and catalytic diversity in a superfamily of phosphoesterases and allied enzymes. *J. Mol. Biol.* 361, 1003-1034.

Cardoso, E. S., Castro, R. F., Borges, N. and Santos, H. (2007) Biochemical and genetic characterization of the pathways for trehalose metabolism in *Propionibacterium freudenreichii*, and their role in stress response. *Microbiology* 153, 270-280.

Carreto, L., Moore, E., Nobre, M. F., Wait, R., Riley, P. W., et al., (1996) *Rubrobacter xylanophilus* sp. nov., a new thermophilic species isolated from a thermally polluent effluent. *Lnt. J. Syst. Bacteriol.* 46, 460-465.

Cayol, J. -L., Ollivier, B., Patel, B. K. C., Prensier, G., Guezernnec, J. and Garcia, J. -L. (1994) Isolation and characterization of *Halothermothrix orettii gen.* nov., sp. nov., a halophilic, thermophilic, fermentative strictly anaerobic bacterium. *Lnt. J. Syst. Bacteriol.* 44, 534-540.

Cayol, J. -L., Ducerf, S., Patel, B. K. C., Prensier, G., Garcia, J. -L., et al., (2000) *Thermohalobacter berrensis* gen. nov., sp. nov., a thermophilic, strictly halophilic bacterium from a solar saltern. *Lnt. J. Syst. Evol. Microbiol.* 50, 559-567.

Copeland, A., Gu, W., Yasawong, M., Lapidus, A., Lucas, S., et al., (2012) Complete genome sequence of the aerobic, heterotroph *Marinithermus hydrothermalis* type strain (T1 (T)) from a deep-sea hydrothermal vent chimney. *Stand Genomic Sci.* 6, 21-30.

Costa, J., Empadinhas, N., Gonsalves, L., Lamosa, P., Santos, H. and da Costa, M. S. (2006) Characterization of the biosynthetic pathway of glucosylglycerate in the archaeon *Methanococcoides burtonii. J. Bacteriol.* 188, 1022-1030.

Costa, J., Empadinhas, N. and da Costa, M. S. (2007) Glucosylglycerate biosynthesis in the deepest lineage of the Bacteria: characterization of the thermophilic proteins GpgS and GpgP from *Persephonella marina. J. Bacteriol.* 189, 1648-1654.

Curatti, L., Folco, E., Desplats, P., Abratti, G., Limones, V., et al., (1998) Sucrose-phosphate synthase from *Synechocystis* sp. strain PCC6803: identification of the *spsA* gene and characterization of the enzyme expressed in *Escherichia coli. J. Bacteriol.* 180, 6776-6779.

Cruz, P. E., Silva, A. C., Roldao, A., Carmo, M., Carrondo, M. J. and Alves, P. M. (2006) Screening of novel excipients for improving the stability of retroviral and adenoviral vectors. *Biotechnol. Prog.* 22, 568-576.

da Costa, M. S., Santos, H. and Galinski, E. A. (1998) An overview of the role and diversity of compatible solutes in Bacteria and Archaea. *Adv. Biochem. Eng. Biotechnol.* 61, 117-153.

da Costa, M. S., Nobre, M. F. and Rainey, F. A. (2001) The genus *Thermus*. In *Bergey's Manual of Systematic Bacteriology*. Boone, D. R. and Castenholtz, R. W. (eds), 2nd edn, New York: Springer, vol. 1, 404-414.

De Smet, K. A. L., Weston, A., Brown, I. N., Young, D. B. and Robertson, B. D. (2000) Three pathways for trehalose biosynthesis in mycobacteria. *Microbiol.* 146, 199-208.

De Virgilio, C., Hottiger, T., Dominguez, J., Boiler, T. and Wiemken, A. (1994) The role of trehalose synthesis for the acquisition of thermotolerance in yeast. Ⅰ. Genetic evidence that trehalose is a thermoprotectant.

Eur. J. Biochem 219, 179-186.

Doyle, R. J. and Marquis, R. E. (1994) Elastic, flexible peptidoglycan and bacterial cell wall properties. *Trends Microbiol*, 2, 57-60.

Egorova, K., Grudieva, T., Morinez, C., Kube, J., Santos, H., et al., (2007) High yield of mannosylglycerate production by upshock fermentation and bacterial milking of trehalose-deficient mutant *Thermus thermophilus* RQ-1. *Appl, Microbiol. Biotechnol.* 75, 1039-1045.

Elbein, A. D., Pan, Y. T., Pastuszak, I. and Carroll, D. (2003) New insights on trehalose: a multifunctional molecule. *Glycobiology* 13, 17R-27R.

Eleutherio, E. C., Araujo, P. S. and Panek, A. D. (1993) Protective role of trehalose during heat stress in *Saccharomyces cerevisiae. Cryobiology* 30, 591-596.

Empadinhas, N., Marugg, J. D., Borges, N., Santos, H. and da Costa, M. S. (2001) Pathway for the synthesis of mannosylglycerate in the hyperthermophilic archaeon *Pyrococcus horikoshii.* Biochemical and genetic characterization of key enzymes. *J. Biol. Chem.* 276, 43580-43588.

Empadinhas, N., Albuquerque, L., Henne, A., Santos, H. and da Costa, M. S. (2003) The bacterium *Thermus thermophilus*, like hyperthermophilic archaea, uses a two-step pathway for the synthesis of mannosylglycerate. *Appl. Environ. Microbiol.* 69, 3272-3279.

Empadinhas, N., Albuquerque, L., Costa, J., Zinder, S. H., Santos, M. A. S., et al., (2004) A gene from the mesophilic bacterium *Dehalococcoides ethenogenes* encodes a novel mannosylglycerate synthase. *J. Bacteriol.* 186, 4075-4084.

Empadinhas, N., Mendes, V., Simoes, C., Santos, M. S., Mingote, A., et al., (2007) Organic solutes in *Rubrobacter xylanophilus*: the first example of *di-myoinositol*-phosphate in a thermophile. *Extremophiles* 11, 667-673.

Empadinhas, N. and da Costa, M. S. (2008) Osmoadaptation mechanisms in prokaryotes: distribution of compatible solutes. *Int. Microbiol.* 11, 151-161.

Empadinhas N. and da Costa, M. S. (2010) Diversity, biological roles and biosynthetic pathways for sugar-glycerate containing compatible solutes in bacteria and archaea. *Environ. Microbiol.* 13, 2056-2077.

Empadinhas, N., Pereira, P. J. B., Albuquerque, L., Costa, J., Sá-Moura, B., et al., (2011) Functional and structural characterization of a novel mannosyl-3-phosphoglycerate synthase from *Rubrobacter xylanophilus* reveals its dual substrate specificity. *Mol. Microbiol.* 79, 76-93.

Faria, T. Q., Knapp, S., Ladenstein, R., Maçanita, A. L. and Santos, H. (2003) Protein stabilisation by compatible solutes: effect of mannosylglycerate on unfolding thermodynamics and activity of ribonuclease A. *Chembiochem.* 4, 734-741.

Faria, T. Q., Lima, J. C., Bastos, M., Maçanita, A. L. and Santos, H. (2004) Protein stabilization by osmolytes from hyperthermophiles: effect of mannosylglycerate on the thermal unfolding of recombinant nuclease a from *Staphylococcus aureus* studied by picosecond time-resolved fluorescence and calorimetry. *J. Biol. Chem.* 279, 48 680-48 691.

Faria, T. Q., Mingote, A., Siopa, F., Ventura, R., Maycock, C. and Santos, H. (2008) Design of new enzyme stabilizers inspired by glycosides of hyperthermophilic microorganisms. *Carbohydr. Res.* 343, 3025-3033.

Fernandes, C., Empadinhas, N. and da Costa, M. S. (2007) Single-step pathway for synthesis of glucosylglycerate in *Persephonella marina*. *J. Bacteriol.* 189, 4014-4019.

Fernandes, C., Mendes, V., Costa, J., Empadinhas, N., Jorge, C., et al., (2010) Two alternative pathways for the synthesis of the rare compatible solute mannosylglucosylglycerate in *Petrotoga mobilis*. *J. Bacteriol.* 192, 1624-1633.

Ferreira, A. M., Wait, R., Nobre, M. F. and da Costa, M. S. (1999) Characterization of glycolipids from *Meiothermus* spp. *Microbiol.* 145, 1191-1199.

Flint, J., Taylor, E., Yang, M., Bolam, D. N., Tailford, L. E., et al., (2005) Structural dissection and high-throughput screening of mannosylglycerate synthase. *Nat. Struct. Mol. Biol.* 12, 608-614.

Galinski, E. A. (1995) Osmoadaptation in bacteria. *Adv. Microb. Physiol.* 37, 272-328.

Giaever, H. M., Styrvold, O. B., Kaasen, I. and Strøm, A. R. (1988) Biochemical and genetic characterization of osmoregulatory trehalose synthesis in *Escherichia coli*. *J. Bacteriol.* 170, 2841-2849.

Gonçalves, L. G., Huber, R., da Costa, M. S. and Santos, H. (2003) A variant of the hyperthermophile *Archaeoglobus fulgidus* adapted to grow at high salinity. *FEMS Microbiol. Lett.* 218, 239-244.

Gonçalves, S., Esteves, A. M., Santos, H., Borges, N. and Matias, P. M. (2011) Three-dimensional structure of mannosyl-3-phosphoglycerate phosphatase from *Thermus thermophilus* HB27: a new member of the haloalcanoic acid dehalogenase superfamily. *Biochemistry* 50, 9551-9567.

Götz, D., Banta, A., Beveridge, T. J., Rushdi, A. I., Simoneit, B. R. and Reysenbach, A. L. (2002) *Persephonella marina* gen. nov., sp. nov. and *Persephonella guaymasensis* sp. nov., two novel, thermophilic, hydrogen-oxidizing microaerophiles from deep-sea hydrothermal vents. *Int. J. Syst. Evol. Microbiol.* 52, 1349-1359.

Goude, R., Renaud, S., Bonnassie, S., Bernard, T. and Blanco, C. (2004) Glutamine, glutamate, and α-glucosylglycerate are the major osmotic solutes accumulated by *Erwinia chrysanthemi* strain 3937. *Appl. Environ. Microbiol.* 70, 6535-6541.

Gouesbet, G., Jebbar, M., Bonnassie, S., Hugouvieux-Cotte-Pattate, N., Himdi-Kabbab, S. and Blanco, C. (1995) *Erwinia crysanthemi* at high osmolarity: influence of osmoprotectants on growth and pectate lyase production. *Microbiology* 141, 1407-1412.

Hagemann, M., Effmert, U., Kerstan, T., Schoor, A. and Erdmann, N. (2001) Biochemical characterization of glucosylglycerol-phosphate synthase of *Synechocystis* sp. strain PCC 6803: comparison of crude, purified, and recombinant enzymes. *Curr. Microbiol.* 43, 278-283.

Hagemann, M. (2011) Molecular biology of cyanobacterial salt acclimation. *FEMS. Microbiol. Rev.* 35, 87-123.

Hallam, S. J., Putnam, N., Preston, C. M., Detter, J. C., Rokhsar, D., et al., (2004) Reverse methanogenesis: testing the hypothesis with environmental genomics. *Science* 305, 1457-1462.

Hartmann, M., Scohlten, J. C., Stoffers, P. and Wehner, F. (1998) Hydrographic structure of brine-filled deeps in the Red Sea——new results from the Shaban, Kebrit, Atlantis Ⅱ and Discovery Deep. *Mar. Geol.* 144, 311-330.

Hincha, D. K. and Hagemann, M. (2004) Stabilization of model membranes during drying by compatible solutes involved in the stress tolerance of plants and microorganisms. *Biochem. J.* 383, 277-283.

Hottiger, T., De Virgilio, C., Hall, N. M., Boiler, T. and Wiemken, A. (1994) The role of trehalose synthesis for the adquisition of thermotolerance in yeast. Ⅱ. Physiologycal concentrations of trehalose increase the thermal stability of proteins *in vitro*. *Eur. J. Biochem.* 219, 187-193.

Hounsa, C. G., Brandt, E. V., Thevelein, J., Hohmann, S. and Prior, B. A. (1998) Role of trehalose in survival of *Saccharomyces cerevisiae* under osmotic stress. *Microbiol.* 144, 671-680.

Inoue, Y., Yasutake, N., Oshima, Y., Yamamoto, Y., Tomita, T., et al., (2002) Cloning of the maltose phosphorylase gene from *Bacillus* sp. strain RK-1 and efficient production of the cloned gene and the trehalose phosphorylase gene from *Bacillus stearothermophilus* SK-1 in *Bacillus subtilis*. *Biosci. Biotechnol. Biochem.* 66, 2594-2599.

Javaux, E. J. (2006) Extreme life on Earth-past, present and possibly beyond. *Res. Microbiol.* 157, 37-48.

Jorge, C. D., Lamosa, P. and Santos, H. (2007) Alpha-D-mannopyranosyl-(1→2)-alpha-D-glucopyranosyl-(1→2)-glycerate in the thermophilic bacterium *Petrotoga miotherma*-structure, cellular content and function. *FEBS J.* 274, 3120-3127.

Kamisango, K., Deell, A. and Ballou, C. E. (1987) Biosynthesis of the mycobacterial O-methylglucose lipopolysaccharide. *J. Biol. Chem.* 262, 4580-4586.

Karsten, U., Gors, S., Eggert, A. and West, J. A. (2007) Trehalose, digeneaside, and floridoside in the *Florideophyceae* (*Rhodophyta*) —a reevaluation of its chemotaxonomic value. *Phycologia* 46, 143-150.

Kawamura, T., Watanabe, N. and Tanaka, I. (2008) Structure of mannosyl-3-phosphoglycerate phosphatase from *Pyrococcus horikoshii*. *Acta Crystallogr. D Biol. Crystallogr.* 64, 1267-1276.

Klähn, S., Steglich, C., Hess, W. R. and Hagemann, M. (2010) Glucosylglycerate: a secondary compatible solute common to marine cyanobacteria from nitrogen-poor environments. *Environ. Microbiol.* 12, 83-94.

Ko, R., Smith, L. T. and Smith, G. M. (1994) Glycine betaine confers enhanced osmotolerance and cryotolerance on *Listeria monocytogenes*. *J. Bacteriol.* 176, 426-431.

Koga, Y. and Morii, H. (2005) Recent advances in structural research on ether lipids from archaea including comparative and physiological aspects. *Biosci. Biotechnol. Biochem.* 69, 2019-2034.

Kollman, V. H., Hanners, J. L., London, R. E., Adame, E. G. and Walker, T. E. (1979) Photosynthetic preparation and characterization of ^{13}C-labeled carbohydrates in *Agmenellum quadruplicatum*. *Carbohydr. Res.* 73, 193-202.

Kouril, T., Zaparty, M., Marrero, J., Brinkmann, H. and Siebers, B. (2008) A novel trehalose synthesizing pathway in the hyperthermophilic Crenarchaeon *Thermoproteus tenax*: the unidirectional TreT pathway. *Arch. Microbiol.* 190, 355-369.

Kristjúnsson, J. K. and Hreggvidsson, G. O. (1995) Ecology and habitats of extremophiles. *World J. Microbiol. Biotechnol.* 11, 17-25.

Laiz, L., Miller, A. Z., Jurado, V., Akatova, E., Sanchez-Moral, S., et al., (2009) Isolation of five *Rubrobacter* strains from biodeteriorated monuments. *Naturwissenschaften* 96, 71-79.

Lamosa, P., Martins, L. O., da Costa, M. S. and Santos, H. (1998) Effects of temperature, salinity, and medium composition on compatible solute accumulation by *Thermococcus* spp. *Appl. Environ. Microbiol.* 64, 3591-3598.

Lentzen, G. and Schwarz, T. (2006) Extremolytes: natural compounds from extremophiles for versatile applications. *Appl. Microbiol. Biotechnol.* 72, 623-634.

Luzardo, M. C., Amalfa, F., Nunez, A. M., Diaz, S., Biondi De Lopez, A. C. and Disalvo, E. A. (2000) Effect of trehalose and sucrose on the hydration and dipole potential of lipid bilayers. *Biophys. J.* 78, 2452–2458.

Macelroy, R. D. (1974) Some comments on the evolution of extremophiles. *Biosystems* 6, 74–75.

Mackay, M. A. and Norton, R. S. (1984) Organic osmoregulatory solutes in cyanobacteria. *J. Gen. Microbiol.* 130, 2177–2191.

Marteinsson, V. T., Bjornsdottir, S. H., Bienvenu, N. and Kristjannsson, J. L. (2010) *Rhodothermus profundi* sp. nov., a new thermophilic bacterium isolated from a deep sea hydrothermal vent in the Pacific Ocean. *Lnt. J. Syst. Evol. Bacteriol.* 60, 2729–2734.

Martin, D. D., Ciulla, R. A. and Roberts, M. F. (1999) Osmoadaptation in archaea. *Appl. Environ. Microbiol.* 65, 1815–1825.

Martins, L. O., Empadinhas, N., Marugg, J. D., Miguel, C., Ferreira, C., et al., (1999) Biosynthesis of mannosylglycerate in the thermophilic bacterium *Rhodothermus marinus*. Biochemical and genetic characterization of a mannosylglycerate synthase. *J. Biol. Chem.* 274, 35407–35414.

Martins, L. O. and Santos, FI. (1995) Accumulation of mannosylglycerate and di–myo–inositol–phosphate by *Pyrococcus furiosus* in response to salinity and temperature. *Appl. Environ. Microbiol.* 61, 3299–3303.

Maruta, K., Nakada, T., Kubota, M., Chaen, F I., Sugimoto, T., et al., (1995) Formation of trehalose from maltooligosaccharides by a novel enzymatic system. *Biosci. Biotechnol. Biochem.* 59, 1829–1834.

Mascellani, N., Liu, X., Rossi, S., Marchesini, J., Valentini, D., et al., (2007) Compatible solutes from hyperthermophiles improve the quality of DNA microarrays. *BMC Biotechnol.* 7, 82.

Maymo–Gatell, X., Chien, Y., Gossett, J. M. and Zinder, S. H. (1997) Isolation of a bacterium that reductively dechlorinates tetrachloroethene to ethene. *Science* 276, 1568–1571.

Moe, W. M., Yan, J., Nobre, M. F., da Costa, M. S. and Rainey, F. A. (2009) *Dehalogenimonas lykanthroporepellens* gen. nov., sp. nov., a reductively dehalogenating bacterium isolated from chlorinated solvent–contaminated groundwater. *Lnt. J. Syst. Evol. Microbiol.* 59, 2692–2697.

Nashiru, O., Koh, S., Lee, S. –Y. and Lee, D. –S. (2001) Novel α–glucosidase from extreme thermophile *Thermus caldophilus* GK24. *J. Biochem. Mol. Biol.* 34, 347–354.

Neves, C., da Costa, M. S. and Santos, H. (2005) Compatible solutes of the hyper thermophile *Palaeococcus ferrophilus*: osmoadaptation and thermoadaptation in the order *Thermococcales*. *Appl. Environ. Microbiol.* 71, 8091–8098.

Nobre, A. (2011) Pathways for the synthesis and hydrolysis of some compatible solutes in the bacterium *Rubrobacter xylanophilus* and in the plant *Selaginella moellendorffii* – characterization of some recombinant enzymes. Ph. D. Thesis University of Coimbra, Portugal.

Nobre, A., Alarico, S., Fernandes, C., Empadinhas, N. and da Costa, M. S. (2008) A unique combination of genetic systems for the synthesis of trehalose in *Rubrobacter xylanophilus*: properties of a rare actinobacterial TreT. *J. Bacteriol.* 190, 7939–7946.

Nobre, M. F. and da Costa, M. S. (2001) 'The genus *Meiothermus*'. In Bergey's *Manual of Systematic Bacteriology*, Boone, D. R. and Castenholtz, R. W. (eds), 2nd ed., Springer, New York, vol. 1, pp. 414–420.

Nunes, O. C., Manaia, C. M., da Costa, M. S. and Santos, H. (1995) Compatible solutes in the thermophilic bacteria *Rhodothermus marinus* and '*Thermus thermophilus*' . *Appl. Environ. Microbiol.* 61, 2351-2357.

Oren, A. (1999) Bioenergetic aspects of halophilism. *Microbiol. Mol. Biol. Rev.* 63, 334-348.

Oren, A., Heldal, M., Norland, S. and Galinski, E. A. (2002) Intracellular ion and organic solute concentrations of the extremely halophilic bacterium *Salinibacter ruber. Extremophiles* 6, 491-498.

Pan, Y. T., Drake, R. R. and Elbein, A. D. (1996) Trehalose-P synthase of mycobacteria: its substrate specificity is affected by polyanions. *Glycobiology* 6, 453-461.

Pommier, M. T. and Michel, G. (1981) Structure of 2′, 3′-di-O-acyl-α-D-glucopyranosyl- (1 →2) -D-glyceric acid, a new glycolipid from *Nocardia caviae. Eur. J. Biochem.* 118, 329-333.

Pospísl, S., Halada, P., Petrfcek, M. and Sedmera, P. (2007) Glucosylglycerate is an osmotic solute and an extracellular metabolite produced by *Streptomyces caelestis. Folia Microbiol.* (*Praha*) 52, 451-456.

Prior, B. A., Hewitt, E., Brandt, E. V., Clarke, A. and Mildenhall, J. P. (1994) Growth, pectate lyase production and solute accumulation by *Erwinia chrysanthemi* under osmotic stress: effect of osmoprotectants. *J. Appl. Bacteriol.* 77, 433-439.

Qu, Q., Lee, S. J. and Boos, W. (2004) TreT, a novel trehalose glycosyltransferring synthase of the hyperthermophilic archaeon *Thermococcus litoralis. J. Biol. Chem.* 279, 47890-47897.

Quaiser, A., Ochsenreiter, T., Klenk, H. P., Kletzin, A., Treusch, A. H., et al, (2002) First insight into the genome of an uncultivated crenarchaeote from soil. *Environ. Microbiol.* 4, 603-611.

Ramos, A., Raven, N. D. H., Sharp, R. J., Bartolucci, S., Rossi, M., et al., (1997) Stabilization of enzymes against thermal stress and freeze - drying by mannosylglycerate. *Appl. Environ. Microbiol.* 63, 4020-4025.

Reed, R. H., Chudek, J. A., Foster, R. and Stewart, W. D. P. (1984) Osmotic adjustment in cyanobacteria from hyper saline environments. *Arch. Microbiol.* 138, 333-337.

Richards, A. B., Krakowka, S., Dexter, L. B., Schmid, H., Wolterbeek, A. P. M., et al., (2002) Trehalose: a review of properties, histroy of use and human tolerance, and results of multiple safety studies. *Food Chem. Toxicol.* 40, 871-898.

Robertson, D. E., Lai, M., Gunsalus, R. P. and Roberts, M. F. (1992) Composition, variation, and dynamics of major osmotic solutes in *Methanohalophilus strain* FDF1. *Appl. Environ. Microbiol.* 58, 2438-2443.

Robinson, P. M. and Roberts, M. R (1997) Effects of osmolytes precursors on the distribution of compatible solutes in *Methartohalophilus portucalensis. Appl Environ. Microbiol.* 63, 4032-4038.

Roeβler, M. and Müller, V. (2001) Osmoadaptation in bacteria and archaea: common principles and differences. *Environ. Microbiol.* 3, 743-754.

Rothschild, L. J. and Mancinelli, R. L. (2001) Life in extreme environments. *Nature* 409, 1092-1101.

Ryu, S. I., Park, C. S., Cha, J., Woo, E. J. and Lee, S. B. (2005) A novel trehalose-synthesizing glycosyltransferase from *Pyrococcus horikoshii*: molecular cloning and characterization. *Biochem. Biophys. Res. Commun.* 329, 429-436.

Ryu, J., Kanapathipillai, M., Lentzen, G. and Park, C. B. (2008) Inhibition of b-amyloid peptide aggregation and neurotoxicity by a-D-mannosylglycerate, a natural extremolyte. *Peptides* 29, 578-584.

Ryu, S. I., Kim, J. E., Kim, E. J., Chung, S. K. and Lee, S. B. (2011) Catalytic reversibility of *Pyrococcus horikoshii* trehalose synthase: efficient synthesis of several nucleoside diphosphate glucoses with enzyme *recycling. Prog. Biochem.* 46, 128–134.

Saier, M. H. Jr and Ballou, C. E. (1968) The 6–Omethylglucose–containing lipopolysaccharide of *Mycobacterium phlei*. Identification of d–glyceric acid and 3–Omethyl–D–glucose in the polysaccharide. *J. Biol. Chem.* 243, 992–1005.

Sauer, T. and Galinski, E. A. (1998) Bacterial milking: a novel bioprocess for production of compatible solutes. *Biotechnol. Bioeng.* 57, 306–313.

Sά-Moura, B., Albuquerque, L., Empadinhas, E., da Costa, M. S., Pereira, P. J. B. and Macedo–Ribeiro, S. (2008) Crystallization and preliminary crystallographic analysis of mannosyl–3–phosphoglycerate synthase from *Rubrobacter xylanophilus. Acta Crystallographica F*64, 760–763.

Santos, H. and da Costa, M. S. (2001) Organic solutes from thermophiles and hyperthermophiles. *Methods. Enzymol.* 334, 302–315.

Santos H. and da Costa, M. S. (2002) Compatible solutes of organisms that live in hot saline environments. *Environ. Microbiol.* 4, 501–509.

Santos, H., Lamosa, P., Borges, N., Faria, T. Q. and Neves, C. (2007) 'The physiological role, biosynthesis and mode of action of compatible solutes from (hyper) thermophiles'. In *Physiology and Biochemistry of Extremophiles*, Gerday, C. and Glandorff, N. (eds), Washington, DC, USA: ASM Press, pp. 86–103.

Sawangwan, T., Goedl, C. and Nidetzky, B. (2010) Glucosylglyceroland glucosylglycerate as enzyme stabilizers. *Biotechnol. J.* 5, 187–191.

Scholz, S., Sonnenbichler, J., Schäfer, W and Hensel, R. (1992) Di–wyo–inositol–1, 1′–phosphate: a new inositol phosphate isolated from *Pyrococcus woesei. FEBS Lett.* 306, 239–242.

Schwarz, A., Goedl, C., Minani, A. and Nidetzky, B. (2007) Trehalose phosphorylase from *Pleurotus ostreatus*: characterization and stabilization by covalent modification, and application for the synthesis of alpha, alpha–trehalose. *J. Biotechnol.* 129, 140–150.

Silva. Z., Borges, N., Martins, L. O., Wait, R., da Costa, M. S. and Santos, H. I (1999) Combined effect of the growth temperature and salinity of the medium on the accumulation of compatible solutes by *Rhodothermus marittus and Rhodothermus obamensis. Extremophiles* 3, 163–172.

Silva, Z., Horta, C., da Costa, M. S., Chung, A. P. and Rainey, F. A. (2000) Polyphasic evidence for the reclassification of *Rhodothermus obamensis* Sako et al 1996 as a member of the species *Rhodothermus marittus* Alfredsson et al., 1988. *Int. J. Syst. Evol. Microbiol.* 50, 1457–1461.

Silva, Z., Alarico, S., Nobre, A., Horlacher, R., Marugg, J., et al., (2003) Osmotic adaptation of *Thermus thermophilus* RQ–1: lesson from a mutant deficient in synthesis of trehalose. *J. Bacteriol.* 185, 5943–5952.

Silva, Z., Sampaio, M. M., Henne, A., Bohm, A., Gutzat, R., et al., (2005) The high–affinity maltose/trehalose ABC transporter in the extremely thermophilic bacterium *Thermus thermophilus* HB27 also recognizes sucrose and palatinose. *J. Bacteriol.* 187, 1210–1218.

Singer, M. A. and Lindquist, S. (1998) Thermotolerance in *Saccharomyces cerevisiae*: the Yin and Yang of trehalose. *Trends. Biotechnol.* 16, 460–468.

Strous, M., Pelletier, E., Mangenot, S., Rattei, T., Lehner, A., et al., (2006) Deciphering the evolution and metabolism of an anammox bacterium from a community genome. *Nature* 440, 790–794.

Tsusaki, K., Nishimoto, T., Nakada, T., Kubota, M., Chaen, H., et al., (1996) Cloning and sequencing of trehalose synthase gene from *Pimelobacter* sp. R48. *Biochim. Biophys. Acta* 1290, 1–3.

Ventosa, A., Nieto, J. J. and Oren, A. (1998) Biology of aerobic moderately halophilic bacteria. *Microbiol. Mol. Biol. Rev.* 62, 504–544.

Wannet, W. J. B., Op den Camp, H. J. M., Wisselink, H. W., van der Drift, C., Van Griensven, L. J. L. D. and Vogels, G. D. (1998) Purification and characterization of trehalose phosphorylase from the commercial mushroom *Agaricus bisporus*. *Biochim. Biophys. Acta* 1425, 177–188.

Welsh, D. T. and Herbert, R. A. (1993) Identification of organic solutes accumulated by purple and green sulphur bacteria during osmotic stress using natural abundance 13C nuclear magnetic resonance spectroscopy. *FEMS Microbiol. Ecol.* 13, 145–149.

Welsh, D. T. (2000) Ecological significance of compatible solute accumulation by micro-organisms: from single cells to global climate. *FEMS Microbiol. Rev.* 24, 263–290.

Welsh, D. T. and Herbert, R. A. (1999) Osmotically induced intracellular trehalose, but not glycine betaine accumulation promotes desiccation tolerance in *Escherichia coli*. *FEMS Microbiol. Lett.* 174, 57–63.

Wolf, A., Kramer, R. and Morbach, S. (2003) Three pathways for trehalose metabolism in *Corynebacterium glutamicum* ATCC13032 and their significance in response to osmotic stress. *Mol. Microbiol.* 49, 1119–1134.